Ecology of North America

*Dedicated to our wives, Sandy and Elizabeth,
whose love, forbearance, and creativeness
inspire and sustain us.*

Ecology of North America

Brian R. Chapman

Senior Research Scientist
Texas Research Institute for Environmental Studies
Sam Houston State University
Huntsville, Texas

Eric G. Bolen

Professor Emeritus
Department of Biology and Marine Biology
University of North Carolina Wilmington
Wilmington, North Carolina

SECOND EDITION

WILEY Blackwell

Library of Congress Cataloging-in-Publication data

Chapman, Brian R., author.
  Ecology of North America / Brian R. Chapman and Eric G. Bolen. – Second edition.
    pages cm
  Includes bibliographical references and index.
  ISBN 978-1-118-97154-3 (cloth)
1. Ecology–North America. I. Bolen, Eric G., author. II. Title.
  QH102.B65 2015
  577.097–dc23
                                        2015006625
A catalogue record for this book is available from the British Library.

Cover image: The spectacular Teton Range in Wyoming offers a centerpiece for the unspoiled places
remaining in North America. Lacking foothills, these mountains rise majestically 2200 m (7000 ft) above
Jackson Hole that together form Grand Teton National Park—a sanctuary protecting more than 100 species
of vascular plants, 300 species of birds, and 60 species of mammals. Some 100 species of grasses and
wildflowers enrich the sagebrush flats on the valley floor. Photograph reproduced by permission of Sandra
S. Chapman.

Set in 8.5/10.5pt Meridien by SPi Global, Pondicherry, India
Printed and bound in Singapore by Markono Print Media Pte Ltd

1 2015

Contents

Foreword

As a professor, scientist, and student of wildlife for over four decades, I cannot think of a more important body of work to understand than ecology. This field of science is the basis for all natural history professions including anything that involves flora and fauna in its native habitat. It would be hard to impossible to understand, conserve, or manage coastal areas, deserts, forests and woodlands, grasslands, or tundra (essentially the landscapes that contribute to habitats for wildlife) and the plants and animals that occupy them without a solid understanding of ecology. Brian Chapman and Eric Bolen have brought these systems and others to life in this new edition of *Ecology of North America*.

Eric Bolen published the first edition of the book in 1998. This new edition is not just a makeover of Bolen's earlier work. Chapters were updated where needed and new information was added, including a new chapter on coastal environments to round out the landscapes examined. Another exciting addition is the inclusion of information boxes that call out special individuals and events in ecology that will be of interest to readers. Individuals highlighted include the "Fathers" of animal ecology, ecosystem science, grassland ecology, wildlife management, and prominent movers and shakers in the field such as Samuel Hearne, Rachel Carson, E. Lucy Braun, and John Muir. There are 18 information boxes scattered throughout the book with intriguing titles that are engaging, interesting to read, and informative, including: "The elk that saved a forest", "Spirit bear, the other white bear", and "Agave, margaritas, and bats". Not only are they interesting, but these selected topics are written in such a way as to make the reader to want to know more.

This book is not the typical ecology text that only gives a cursory treatment to the important landforms of North America. It is unique in that it briefly describes the basics of ecology in the first chapter, and then delves into the ecology of the major landscapes of our continent in detail. College students and layman alike can easily understand the descriptions. The writing is clear and interesting, and would be of value to anyone interested in the places they live or new places they will be visiting. For example, I have taken many visitors with me during my studies in the deserts of North America, and often hear statements like "Wow, there is nothing here – no wonder they call it a desert". It is only after visitors spend time looking at the vast amount of life and signs of life that their tunes change to one of wonder. The authors

believe that people will have a much better appreciation for the landscapes they occupy – from the fragile cryptogamic soils in deserts to freezing frogs in tundra – by reading this book prior to visiting new areas (biomes). This new volume is chock-full of interesting information including how alligators create habitat with "'gator holes", introducing the reader to forests in the ocean, the importance of "scuzz" in enriching forests on land, and how *Lagerstätten* fossils and packrat middens can help humans understand their past. The writing brings together the interactions of plants, animals, and habitats (as ecology should), such as the explanations of how brown bears, salmon, and forests interact, or uses birds and thousand-year-old pines or cicadas to introduce ecological terms to readers.

The advent of computers and models has greatly changed the way students are taught and how they learn in North American universities and colleges. As more flora and fauna are reduced from living, breathing organisms to pixels on a computer screen, there seems to be less attention devoted to actual field study. Some universities even have labs in the natural sciences online. I cannot imagine learning something as basic, yet complex, as ecology from a computer, and that is clearly not the intent here. Brian Chapman and Eric Bolen bring the fascinating landscapes humans share with flora and fauna to life and also explain how mankind influences these systems.

The authors are well-known ecologists with a long history together; both have dedicated their lives to understanding ecology around the United States and passing on that knowledge to others. Because of their long-term association, the writing is seamless as though written by a single author. The book will certainly give budding ecologists and natural resource professionals a sense and better understanding of "place". The authors had a goal of following Leopold's example of emphasizing the importance of landscapes to the biologic community and to the human spirit. *Comme il faut*!

Paul R. Krausman
Certified Wildlife Biologist®
Fellow and Past-president of The Wildlife Society
Emeritus Professor of Wildlife Ecology,
University of Arizona
Boone and Crockett Professor of Wildlife Conservation,
University of Montana

Preface

About this book

Ecology of North America stems from our belief that many college students – and non-students – lack much awareness about the natural world in which they live. Thanks to the effects of mass media, some may indeed deplore the destruction of far-off tropical rain forests or the plight of whales; however, too few are acquainted with the all-but-vanished inland sea of tallgrass prairie, the 'gator holes of the Everglades, or the enigmatic population cycles of lynx (and other) animals in the spruce-fir forests of North America here at home. Our conviction rests, collectively, on more than 60 years of college teaching and countless public presentations. Moreover, until the first edition appeared no single text designed for classroom use had attempted to survey the ecological diversity of North America's vast landscape. The first edition of *Ecology of North America*, drafted while one of us (EGB) taught a course of the same name, was therefore conceived and eventually born as a tool fashioned for undergraduate instruction.

Shortly after the first edition appeared preparation began for a second edition, but various circumstances precluded its completion (not least EGB's retirement). Hence, if a new edition was to materialize, a coauthor was needed to revitalize the work. BRC then entered the scene when, after careers at other universities, he joined the faculty at UNCW as provost and professor of biology. Notably, his teaching and research experiences paralleled the basic themes expressed in *Ecology of North America*, and a marriage of interests thereafter followed. Our association in fact began decades ago when BRC was completing his doctoral work at Texas Tech University, where EGB was then serving on the faculty and as a member of his advisory committee. This edition is therefore the result of a long-standing partnership founded on both a deeply shared interest in plant and animal ecology and a collegial and personal friendship.

As in the previous edition, the geographical coverage begins in the far north (the frigid tundra), continues southward into boreal and deciduous forests swings westward into the interior grasslands and deserts, and continues to the western mountains before traveling onward to chaparral and temperate rain forests. Sites along North America's long and varied coastline conclude our regional tour. We present a selection of unique areas (e.g., the Grand Canyon), expanded in this edition to include the fossil history offered by the La Brea tar pits and Burgess Shale. We also visit far-flung locations such as the Polar Ice Cap, Mississippi River, and Florida Keys, as well as stopping to investigate pitcher plant bogs and other habitats of interest on the way. As appropriate to a new edition, we have supplemented and updated the previous text with a variety of subjects ranging from cicada cycles and the moth-eating habits of bears, to the mysterious disease now devastating bat populations and the havoc wrought to the Everglades by introduced pythons.

Some entirely new features highlight the second edition including a chapter dedicated to coastal environments, among them seagrass communities, tide pools, and barrier islands, and ecological portraits of Chesapeake Bay and the Laguna Madre. The new chapter also describes phenomena such as the vital relationship between horseshoe crabs and migrant shorebirds. "Infoboxes" likewise appear for the first time; these present stand-alone descriptions, including biographies, as disparate as spirit bears, Lucy Braun, and the conservation successes of gray whales and Maine's puffin colonies. To our delight, color photos now complement the gallery of black and white images.

Each chapter ends with "Readings and references," some of which present basic descriptions of community structure. These appeared early in the development of ecology, yet still provide an essential foundation for many readers. Others sources which have been published more recently represent new discoveries and refinements of previous concepts, such as: revelations about the hunting success of spirit bears; evidence of inbreeding depression in Isle Royale's isolated wolf population; the biotic community in the canopy of towering redwoods; and expanding knowledge of the beneficial impacts of nutrients derived from salmon carcasses. We also list works of greater scope for those seeking deeper insight into the subjects at hand.

Our focus

This edition, updated with both new and expanded coverage, broadly targets an audience of both undergraduate students and the general public. In doing so, we continue writing in a "user-friendly" format that appends Latin taxa and literature sources instead of embedding these within the text in the style of scientific journals. Understandably, some professional biologists may fault our format, but we aim to hold the attention of readers who may place higher priorities on the primary message of the book. For the same reason, we shied from marching lockstep through a catalog of species, soil types, and weather regimes for every unit; these are provided only when they seem important. Otherwise, our steady focus is on key or at least interesting plants and animals and their interactions.

Finally, we were guided by the thoughts of Aldo Leopold who, in *Sand County Almanac* (1949), entwined three thoughts into a common theme: "land is a community" that forms the fabric of ecology; land should be cherished and respected as an "extension of ethics"; and "land yields a cultural harvest." Leopold wedded science, ethics, and aesthetics. In doing so, he established an understanding of nature that we, in our own small way, have tried to nourish.

Brian R. Chapman, Huntsville, Texas
Eric G. Bolen, Wilmington, North Carolina

Acknowledgments

Our task would surely have foundered without the generous advice, knowledge, and contributions of many colleagues. Merely to list their names seems far too superficial, but we are nonetheless indebted to all, including anyone we have inadvertently omitted.

Our continued thanks are extended to those who contributed to the first edition (and in some cases, this edition as well): C. Davidson Ankney, Michael J. Armbruster, Stephen F. Arno, Guy A. Baldassarre, Thomas C. Barr, Jr, Mark Blumenthal, Carlton M. Britton, Lincoln P. Brower, C. Alexander Brownlow III, Dana C. Bryan, George C. Carroll, Donald D. Clark, Robert T. Coupland, John A. Crawford, James E. Deacon, Leonard F. DeBano, Ronald I. Dorn, Harold E. Dregne, Christine S. Dutton, Ellen Faurot-Daniels, Maeton C. Freel, Peter M. Frenzen, Ron R. George, Paul L. Gersper, Paul N. Gray, Lance H. Gunderson, Craig A. Harper, Wendy Hodgson, John R. Holsinger, Jon E. Keely, Fritz L. Knopf, Paul R. Krausman, Melissa Kreighbaum, Douglas W. Larson, Gordon S. Lind, Erik Lindquist, Glenn Longley, Charles Lowe, Richard Mack, John Mangimeli, Paul S. Martin, Roger Martin, Joseph R. McAuliffe, Burruss McDaniel, Robert W. Mitchell, Robert. H. Mohlenbrock, Elizabeth M. Morgan, Fritz P. Mueller, Richard Murzin, Nalini M. Nadkarni, S. Kim Nelson, David J. Odz, Karl A. Perry, James Petterson, Linda Pin, Donald Pinkava, Thomas L. Poulson, Donald R. Progulske, William O. Pruitt, Jr, John Riley, William L. Robinson, William H. Romme, Donald H. Rusch, Susan Shultz, John G. Sidle, John C.F. Tedrow, Jack W. Thomas, Walter M. Tovell, Merlin D. Tuttle, Gary Valentine, Thomas R. Van Devender, B.J. Verts, Doug Waid, Frederic H. Wagner, John R. Watson, Ronald R. Weeden, Nathaniel R. Whitney, Edward O. Wilson, M. Eugene Wright, Paul J. Young, and Paul Zedler.

For their contributions to the second edition, we also thank Kathy L. Allen, Donna Anstey, Angie Babbit, Candice Bressler, Sanford Brown, Lisa C. Chapman, William R. Clark, Sherry Cosper, Tara Cuvelier, Pamela Dibble, John Dokken, Catherine Evans Debbie R. Folklets, W. Mark Ford, Eric Hallerman, Richard W. Halsey, Thomas Henry, Susan Jacobson, Rick Kesterman, Jack L. Kindinger, Roel Lopez, Jim Lovett, D. A. Mascarelli, Joyce Maschinski, Vincente Mata, Laurie McBurnie, Danny L. McDonald, Thomas C. Michot, Thomas E. Moore, William F. Moore, Ashley R. Morgan, Paige A. Najvar, T. J. Pernas, Luis Prado, Kristen Quarles, T.P. Quinn, W. Scott Richardson, Celia Rozen, Charles H. Smith, Mike Smith, Autumn J. Smith-Herron, Clyde Sorenson, Nancy P. Stewart, James Stubbendieck, Orly (Chip) Taylor, Monte L. Thies, Nancy Thompson, Jace Tunnell, David Welch, Josh Westerhold, Don White, Jr, Richard Whyte, and Kathy S. Williams.

Colleagues at the University of North Carolina at Wilmington generously provided advice and assistance as one or both of the editions of the book were being prepared. Special mention goes to W. Franklin Ainsley, Walter J. Biggs, Thomas F. Britt, Lawrence B. Cahoon, Troy M. Clites, Mike Durako, D. Wilson Freshwater, Jocelyn Gaudet, Gilbert S. Grant, Melissa Grey, Neil F. Hadley, William B. Harris, Paul E. Hosier, John R. Huntsman, Stephen T. Kinsey, David L. LaVere, David G. Lindquist, John J. Manock, R.D. McCall, Ian A. McLaren, William McLellan, Daniel W. Noland, D. Ann Pabst, David E. Padgett, Beth A. Pardini, James F. Parnell, Joseph R. Pawlik, Martin H. Posey, David J. Sieren, Amanda Southwood, Peter N. Thomas, Carmelo Tomas, W. David Webster, and Barbara L. Wilson. The staff at the Randall Library at UNCW expertly – and patiently – accommodated requests for often obscure bibliographic information and materials.

Associates at Sam Houston State University, the University of Houston-Downtown, and colleagues at several other universities thoughtfully provided essential information, assistance with photographic needs, and various kinds of advice and support during preparation of the second edition. Among these, special thanks go to Loren K. Ammerman, Jerry L. Cook, Quenton R. Dokken, Richard Eglsaer, Jaimie L. Herbert, David R. Hoffpauir, William I. Lutterschmidt, Lisa A. Shipley, Melissa S. Sisson, and John W. Tunnell, Jr. Sandra S. Chapman painstakingly proofread the manuscript, and we appreciate her keen eye and invaluable suggestions.

Persons and/or agencies providing photographs are acknowledged in the appropriate figure captions, as are the artists who prepared maps and line drawings; special thanks go to Tamara R. Sayre. Many of her illustrations appear again in the second edition as testimony to her talent and, certainly, to the importance of her work to the content of the book.

We benefitted from constructive comments provided by colleagues who reviewed the first edition as well as those who examined draft chapters prepared for the second edition, among them Leonard A. Brennan, Paul R. Krausmann, Bruce D. Leopold, Joyce Maschinski, Eric Ribbins, Fred E. Smeins, and an anonymous reviewer.

We are especially indebted to Paul R. Krausman who provided advice and constructive suggestions during the preparation of both editions, and who graciously wrote the Foreword for the second edition.

Rachael Ballard, Executive Commissioning Editor at John Wiley and Sons, unhesitatingly authorized this project, and we enjoyed working with her and Delia Sandford, Managing Editor. We also thank Fiona Seymour, Senior Project Editor, and Audrie Tan, Project Editor, who guided the manuscript through the start-to-finish maze of production. We owe a special debt of gratitude to our copyeditor, Elaine Rowan, whose thoughtful improvements made what we were attempting to say far more intelligible.

Clearly, many have contributed, but we alone remain responsible for any errors that escaped unnoticed.

Finally, our wives, Sandy and Elizabeth, endured more neglect than we care to admit while this project inched toward completion. We marvel at our good fortune to find such wonderful, loving, and supportive partners. To them, we dedicate this book (albeit that it seems so little in return for so much).

CHAPTER 1
Introduction

In all things of nature there is something of the marvelous.
Aristotle

Think big for a moment. Imagine a transect running north to south, stretching across the midpoint of pristine North America in the year 1500. We will use the 100th meridian, which traces the right-hand edge of the Texas Panhandle, as our line of travel. At its northern end – the North Pole – the transect begins in a desolate cap of ice and snow and then crosses a vast Arctic landscape of tundra before traversing a wide band of spruce and fir known as the Boreal Forest. Wolves howl in the distance; the scattered remains of their most recent kill – a sickly moose – lie still fresh atop the deep snow. South of these dark forests spread the plains, grasslands grazed by millions of bison and even larger numbers of prairie dogs. After leaving the plains, our excursion takes us across the narrow, clear rivers and low hills of the Edwards Plateau before dropping into a region studded with low, thorny trees. The transect ends, for our purposes, when it reaches northern Mexico and the yucca-spiked Chihuahuan Desert.

A similar transect running east to west across the waist of North America at first encounters the sandy beaches and dunes of the Atlantic seashore. Here, at the latitude of Virginia, the transect crosses what once was an immense forest of oak and hickory (Fig. 1.1). According to folklore, an ambitious squirrel could have journeyed from the Atlantic seaboard to the Mississippi River 300 years ago without the necessity of ground travel, false testimony to the idyllic notion that an unbroken canopy of deciduous forest once stretched across more than a third of North America.

Westward, with the Appalachians, Cumberland Plateau, and Mississippi River behind us, the forest gradually thins and our transect enters the grasslands. Our trek skirts the southern edge of the Prairie Pothole Region – North America's famed "duck factories" – and pushes onward toward the plains. As we cross our north–south transect in western Kansas, courting prairie chickens dance and boom in the background and

black-tailed jackrabbits crouch, ears lowered, to escape our notice. Still farther west rise the Rocky Mountains with their rather distinctive zones of vegetation, after which we enter into a desolate terrain of sagebrush in the Great Basin. In the distance loom the peaks of the Sierra Nevada where the largest of trees, the giant sequoia, almost defy description. In a quiet grove of these immense trees, we might allow our imagination to behold the ghost of John Muir (1838–1914), the hard-trekking Scot who championed wilderness preservation.

By the time we reach the rocky seashore of the Pacific Ocean, our transect will have dropped into the Central Valley of California and then risen upward to cross the Coastal Range that rims the western edge of the continent. Offshore, frisky sea otters float above submerged kelp forests. North of where we stand are the old-growth forests of Sitka spruce and western hemlock, heavily draped with epiphytes, replete with spotted owls and maples. On the Alaskan coast, vulnerable hoards of migrating salmon attract giant bears to the rushing streams each year. To the south are chaparral-covered hills, and beyond these are the beckoning Joshua trees in the Mojave Desert. Still farther south at a small site in the mountains of Mexico are firs cloaked each winter with millions of slumbering butterflies. Our telescoped journey across North America has been brief to be sure, but perhaps it is long enough to preview the contents of this book.

A brief overview of ecology

Ecology is the branch of biology that investigates the interrelationships between organisms and their environment. The original name *oekologie*, based on the Greek word *oikos* meaning "home", was coined in 1866 by German zoologist Ernst Haeckel (1834–1919). An ecological study of any species involves a detailed examination of an organism's life history and biological requirements, the physical

Ecology of North America, Second Edition. Brian R. Chapman and Eric G. Bolen.
© 2015 John Wiley & Sons, Ltd. Published 2015 by John Wiley & Sons, Ltd.

Figure 1.1 A vast expanse of eastern deciduous forest once covered much of North America east of the Mississippi River, but the romantic notion of an unbroken canopy 300 years ago is inaccurate. Indeed, Native Americans and natural processes, such as fire, periodically cleared many areas, some quite large, within the primordial forest. A mountaintop bald, visible atop the Smoky Mountain ridge (right), represents a natural opening described in more detail in Chapter 3. Photo courtesy of Brian R. Chapman.

environment in which the organism lives, and its interactions with the other species that occupy the same area. Physical features of the environment (e.g., temperature, soil type, and moisture) influence the distribution and abundance of organisms, but all living things process materials from the environment and transform energy as they grow and reproduce.

The ecosystem

In 1935, English botanist Arthur G. Tansley (1871–1955) proposed the term **ecosystem** to characterize the flow of energy and matter through a network of **food chains** collectively known as a **food web**. Energy passes through a series of **trophic levels** (i.e., feeding levels), the functional parts of an ecosystem. These rest on a foundation of non-living matter, the **abiotic level**, which consists of air, soil, and water. When fueled by sunlight, the abiotic level provides the fundamental components required for **photosynthesis** by species known as **primary producers**, which are represented by green plants such as algae, grasses, and trees and first in the series of trophic levels. **Primary consumers** – rabbits, deer, or other **herbivores** that consume the energy and matter bound in green plants – represent the next trophic level, followed by **secondary consumers**; the latter are predators such as foxes or hawks. **Tertiary consumers**, sometimes known as apex predators, are represented by animals such as polar bears or mountain lions. The final trophic level, **decomposers**, is populated by scavengers, bacteria, and fungi that return the tissues of dead plants and animals to their elemental form (Fig. 1.2). An ecosystem is therefore an area or volume in which energy and matter are exchanged between its living and non-living parts.

Figure 1.2 In forests decomposition is commonly thought to originate with logs, but it often begins when a dead tree remains upright as a snag. The bracket fungi on this snag possess enzymes that break down lignin and other complex chemicals found in wood, and the fungi absorb the nutrients for their growth and reproduction. When the snag eventually falls, other decomposers on the forest floor will return its elements to the soil. Photo courtesy of Brian R. Chapman.

Abiotic limits

Many environmental influences – such as moisture, temperature, the availability of nutrients, wind, and fire – limit the kinds and abundances of organisms that populate an ecosystem. In 1840, German organic chemist Justus von Liebig (1803–1873) was the first to recognize the role of abiotic limitations in nature. After studying the relationships between surface soils and agricultural plants, Leibig concluded: "The crops of a field diminish or increase in exact proportion to the diminution or increase of the mineral substances conveyed to it in nature." Liebig recognized that each plant requires not only certain minerals, but each must also be present in the proper quantity for the plant to flourish. If a required nutrient is absent, the plant will not survive. Moreover, if the essential food substance is present only in a minimal amount, the plant's growth will be correspondingly minimal. In later years, this concept became known as the **law of the minimum**.

Later studies indicated that the growth and survival of living organisms also may be limited by an overabundance of a required substance (e.g., terrestrial plants require moisture, but die when waterlogged for a prolonged period). Plants and animals are successful only when they live in an environmental range between too much and too little, that is, within their limits of tolerance. Victor E. Shelford (1877–1968; Infobox 1.1) incorporated the concept of maximum

Infobox 1.1 Victor E. Shelford (1877–1968), Father of animal ecology

The scientific career of Victor Shelford began in 1899 when he enrolled at West Virginia University where his uncle, the assistant state entomologist, influenced his lifelong interest in insects. In 1901, however, the president of the university moved to the University of Chicago where he offered a scholarship to Shelford who accepted and eventually earned a Ph.D. (in 1907) dealing with tiger beetles and dune vegetation on the shores of Lake Michigan. Likely for the first time, this study associated animals with the successional changes in a plant community, a concept pioneered by his academic mentor Henry C. Cowles (1869–1939). Shelford thereafter joined the faculty at Chicago but moved to the University of Illinois in 1914, where he spent the remainder of his academic career.

In 1913, Shelford published his first book, *Animal Communities in Temperate America*, a landmark of its day. He helped found the Ecological Society of America and, in 1916, became its first president.

Along with his university duties, Shelford served as the laboratory supervisor for the Illinois Natural History Survey during 1914–1929 and, during alternate summers between 1914 and 1930, as director of marine ecology at the Puget Sound Biological Station. His research interests included topics ranging from benthic communities in both freshwater and marine environments to lemming populations in Arctic tundra. His experiments covered the physiological and behavioral responses of animals to temperature and other variables in climate-controlled chambers. In the field, he used photoelectric cells to determine light penetration into seawater. His research, which often employed novel equipment, led to a manual entitled *Laboratory and Field Ecology* (1929). On the practical side, Shelford also studied termites and other insect pests, as well as the response of fishes to sewage treatment. He eventually reduced his lab work in favor of spending more time on field studies, especially those that concerned food chains, structure, and other analyses of natural communities.

In the 1930s, Shelford began collaborating with Frederic E. Clements (1874–1945), a prominent plant ecologist. Their relationship was cordial but strained at times as, much to Shelford's dismay, Clements harbored reservations about the role animals played in vegetational development. Nonetheless, in 1939 the two produced *Bio-Ecology*, a book that integrated plants and animals into the formation of communities.

Shelford's marathon field trips were legendary experiences for students; some included several weeks of camping while visiting prairie, forest, desert, or tundra. Indeed, he wanted his students to study first hand every major biome in North America. Reelfoot Lake in Tennessee, created by a cataclysm of earthquakes in the winter of 1811–1812, was a regular stop for many years; the site included opportunities to study oxbow lakes, cypress sloughs, and floodplain forests. Wherever they went, however, Shelford steadfastly donned shirt and tie in the field. Shelford's career represented a major link in a chain of renowned ecologists that began with Cowles and continued with Shelford, followed by his student S. Charles Kendeigh (1904–1986) and, in turn, the latter's student Eugene P. Odum (1913–2002), who is regarded as the founder of modern ecology.

Shelford retired from university life in 1946, but he hardly remained inactive. His passion for preserving natural areas with fully intact communities initiated the Ecologist's Union, which evolved into *The Nature Conservancy* whose mission is now global in scope. His association with Clements helped Shelford develop his appreciation for the concept of biomes and, after years of work, resulted in a monumental treatise, *Ecology of North America* (1963), the inspiration and namesake for the book at hand.

Soon after Shelford completed his graduate studies at Chicago, he was advised by a prominent scientist to "discontinue this field of ecology" and instead pursue biology with a traditional focus on individual organisms. Fortunately, Victor E. Shelford chose otherwise.

and minimum limits on environmental condition into the **law of tolerance** in 1913. Some organisms are capable of living within wide ranges of conditions of one or more environmental factors, whereas others have narrow limits of tolerance. Certain species are capable of acclimatizing to different environmental limits as seasons or conditions change.

To complicate things further, an organism may tolerate a wide range of conditions for one environmental factor, but be restricted by a narrow range for another. In the latter case, the distributional range of the species will be restricted by the condition for which it possesses the narrowest tolerance.

Climate and topography

The climatic conditions of any region are determined by many factors, including latitude, seasonal temperature ranges, the amount and periodicity of rainfall, and location on the continent (e.g., interior versus coastal). Geographical features, such as proximity to mountain ranges or bodies of water, can alter local conditions enough to influence the type of plant associations that can exist. Thus, vegetation maps of North America (Fig. 1.3) illustrate the influence of climate. The northern region of North America has two broad, east–west belts of natural vegetation: tundra and boreal coniferous forest. These follow a gradient correlated with temperature patterns, but the vegetative zones on the remainder of the continent are more closely associated with the availability of moisture. Precipitation decreases from east to west while evaporation, which is largely influenced by temperature, increases; this interaction results in a series of north and south vegetation bands. Along the coasts, regions of higher humidity support forests. The broadest zone of forest is on the eastern half of the continent where rainfall exceeds evaporation.

Mountains intercept winds and directly influence regional climates. As winds ascend a mountain slope, the air mass cools and gradually becomes saturated; cool air holds less moisture than warm air. The windward sides of mountains usually receive rain at mid- to higher elevations as the moisture is, in effect, wrung out of the air. On the leeward side, descending cool dry air absorbs moisture from the soil and vegetation creating **rain shadow**, an area where evaporation greatly exceeds precipitation. The Great Basin Desert on the eastern (leeward) side of the Sierra Nevada Range illustrates the impact of a rain shadow, whereas the western slopes of the mountains support lush forests.

Topography and vegetation create local environmental conditions that can differ substantially from the overall climate of a region. For example, the **microclimate** under a dense clump of bushes does not share the same temperature, moisture, and wind conditions as may be found in an open area just a short distance away. Similarly, the north-facing slope on a mountain does not receive the same solar radiation as a south-facing slope. Because of the difference in exposure, the evaporation rate on north-facing slopes may be 50% lower, the temperature is lower, and soil moisture is higher. Thus, two sides of a mountain, even at the same elevation, are often occupied by different compositions of plant and animal species.

Soils and soil profiles

Climatic features, especially temperature and precipitation, influence the formation of soils. Soil is a complex mixture of minerals, organic matter, water, and air, forming a substrate harboring bacteria, fungi, and other small organisms. Minerals are derived from the weathering of parent material, usually rock, by chemical and physical processes. The microorganisms (e.g., decaying organisms) contribute to soil formation and development by breaking down organic matter, fixing atmospheric nitrogen, and contributing to nutrient cycling.

The US Department of Agriculture (USDA) recognizes twelve textural classes of soil based on various compositions of three particle sizes of minerals – clay, silt, and sand – but soil also may contain larger mineral particles ranging in size from pebbles to large fragments of rock. Of the textural classes, loam (composed of varying percentages of clay, silt, and sand) is the most valuable because it supports many agricultural crops. The USDA also developed a soil classification system that places all the soil types in the world into twelve **soil orders**. These soil groups are defined using a set of characteristics that includes texture as well as physical and chemical properties. The name of each soil order ends in *–sol*, which is derived from the Latin word *solum* meaning "soil". For example, about 16% of the world's surface is covered by entisols which have moderate to low fertility; mollisols, which are the most productive agricultural soils, only occur on about 7% of the Earth's surface.

Anyone who has dug a posthole or planted a tree knows that soils are organized into distinct layers or **horizons**, descending from the ground surface down to the bedrock. These layers form a **soil profile**, and are the product of weathering and the actions of vegetation, temperature, rainfall, and microorganisms acting for millennia on parent material in a specific locality. The organic material – leaf litter and decaying plant and animal matter – on the ground surface forms the **O horizon**. The O horizon is generally thin or absent in deserts and grasslands, but may be quite deep in forests. The first layer of soil, the **A horizon**, is often called "topsoil" and is usually rich in **humus**, the dark-colored products of decomposed organic materials. The A horizon is absent in most deserts, but can be up to 1 m (3 feet) deep in the fertile wheat-growing prairies of Washington State. Progressively downward in the soil profile, the E, B, and C horizons represent zones where the vertical processes of leaching gradually reduce the

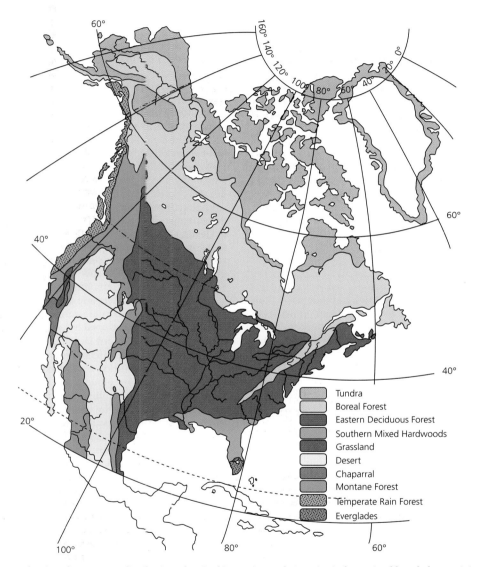

Figure 1.3 The predominately east–west distribution of major biomes in North America is determined largely by precipitation. In contrast, the north–south arrangement of major plant associations in the middle and southern regions of the continent result more from interactions of temperature, evaporation, and elevation. Illustration courtesy of Tamara R. Sayre and Brian R. Chapman, based on Brown et al. (2007).

organic content of each successive layer, thereby increasing the amount of mineral matter. Bedrock is reached at the R horizon, which typically represents parent material.

Biotic community

A **biotic community**, the living part of an ecosystem, is composed of many species, but the number of species and their relative abundance diminish "upward" through a food chain. More species of plants exist in the producer level than do herbivores functioning as primary consumers. Likewise, herbivores are more numerous than the carnivores in the upper trophic levels. These relationships result in a pyramid-shaped community, with numerous (both in species and abundance) green plants at its base, fewer herbivores, and fewer still **carnivores** at its apex (e.g., mountain lions). In Chapter 3, we will describe a relatively simple pyramid of vegetation, moose, and wolves on Isle Royale.

Community succession

Plant **succession** is the relatively predictable sequence of vegetational development within a geographic region. The composition of the local fauna also changes as the vegetation develops and, to express the full ecological impact, the process is best regarded as biotic succession (i.e., sequential changes in both plants and animals). In practice, ecologists usually refer to the concept simply as succession and use the term more in reference to plants than to animals.

Notions about succession are embedded in several 19th century sources, including discourses of Henry David Thoreau (1817–1862) who described the regrowth of forests near Concord, Massachusetts. The first ecologist (at least in North America) to describe succession was Henry C. Cowles (1869–1939), who based his observations on the changes in dune vegetation on the shores of Lake Michigan. Later, Frederic Clements (1874–1945) proposed a complex hierarchy to describe the development of vegetation over time, but only a few of his terms remain widely used today.

Succession is currently viewed as a process of multiple invasions. As certain species replace others because of their own particular **adaptations** (e.g., tolerance to shade in an early growth stage, or greater ability to disperse), the composition of the community changes. The process of succession is also strongly influenced by local environmental factors such as soil type and the availability of water or certain nutrients.

Plant succession: from pioneer to climax

To perceive succession visually, consider a cleared landscape such as an agricultural field in a region with a moderate climate. Left unplowed, the exposed soil is soon covered with so-called weeds – dandelions and asters are typical invaders in many areas – that represent **pioneer** vegetation. Such herbaceous immigrants have three general characteristics: (a) they thrive on disturbed soils where they tolerate harsh conditions (e.g., high soil temperatures, limited moisture and, in many cases, few nutrients); (b) they produce large numbers of seeds with adaptations for widespread dispersal and rapid germination but, when necessary, their seeds can persist in the soil for long periods; and (c) they are commonly, but not always, annuals. Some ecologists regard these as "opportunistic species", plants that quickly take advantage of opportunities to colonize (or re-colonize) sites where competitive species are absent. The pioneer community accordingly consists of relatively few species; recently abandoned fields, for example, are often covered by blankets of single species.

Eventually, the pioneers give way to perennial species, which include plants with established roots systems, storage organs, and defensive structures. These species are better competitors and they eliminate the pioneers. Broomsedge and goldenrod are representative species. Shrubs soon appear, to be replaced later by trees. In many locations, a pine forest develops first, followed by a forest of oaks or other deciduous hardwoods. Of course, the species in these sequences vary by region, soil type, and other factors, but the pattern is consistent.

The final stage in succession is known as the **climax**, a term ecologists use to characterize regional vegetation. Climax communities, such as the Oak–Hickory Association in the Eastern Deciduous Forest, are mature ecosystems consisting of characteristic plants as well as animals (e.g., wild turkeys and eastern gray squirrels). When compared to pioneer and other successional stages, climax communities are (a) neither as **hydric** (wet) or **xeric** (dry) as the earlier stages and instead represent **mesic** conditions; (b) more complex and better organized; they have complicated food webs and more interspecific relationships; (c) include more species, which tend to be relatively large and long-lived and have low reproductive rates; and (d) comparatively stable (e.g., resistant to invading species).

Primary and secondary succession

Old-field succession on abandoned farmlands is a real-world event often described to illustrate the sequence of community replacements. Fallow fields, which were widely available in the late 1930s, are useful sites to study succession because fields in various stages of re-growth are available (i.e., the date of abandonment, and hence the "age" of each field, and its vegetation can be determined from court records). One of the earliest studies of old-field succession on an abandoned farm in the Piedmont region of North Carolina revealed a sequence of herbaceous pioneers such as aster and ragweed, followed by the arrival of pine between years 5 and 15, and then full development of an oak–hickory forest 150 years after the fields were last cultivated.

Old-field succession illustrates **secondary succession**, the recovery of a previously vegetated but disturbed site. Secondary succession also occurs after fires remove the previous vegetation. These sites generally have well-developed soils that can support a variety of plants. The abandoned fields in North Carolina described in the previous paragraph were later cleared of forest vegetation and cultivated for many years thereafter. **Primary succession**, in contrast, occurs at sites devoid of previous vegetation, examples of which include dunes, atolls, volcanic cones, and land exposed by retreating glaciers. Such sites typically lack mature soils and have few nutrients. Primary succession, because it usually requires soil building, spans long time periods before the climax stage is reached.

Climax vegetation, once established, should not be viewed as static. Instead, it experiences a state of dynamic equilibrium, that is, it continually replaces itself. A climax community will maintain its composition so long as the climate remains stable and there are no major disturbances such as fire that return the community to an earlier stage of development. In some places, fire recurs often enough to inhibit the development of a climax community; such a setting is said to be a **subclimax** community.

Succession and species abundance

The number of species in a community increases steadily as succession progresses. Because of their harsh environment, pioneer communities typically contain only a few species but in large numbers. A barren field, for example, receives the full impact of solar radiation during the day creating surface temperatures ill-suited to, or even fatal for, many organisms, but at night the barren field cools rapidly due to a lack of both solar radiation and an insulating blanket of vegetation. The bare soil is also exposed to the full force of wind and rain, often creating unstable conditions (e.g., erosion). Consequently, only a limited number of species are suited to such environments and those few form the pioneer community. Indeed, pioneer communities are sometimes monotypic in their composition.

Physical conditions become less harsh after pioneer vegetation gains a foothold. The herbaceous blanket ameliorates extremes in temperature, reduces runoff, and discourages wind erosion, enabling the soil to hold more moisture. Other plants find suitable conditions and the number of species, including insects, rodents, and other animals, steadily increases during the middle stages of succession.

While it is tempting to conclude that climax communities contain a maximum number of species, the peak is reached slightly before the climax reaches its full development. This relationship results from heightened biological competition, that is, some less-competitive species (e.g., goldenrod) are eliminated when the dominant climax species (e.g., oaks) reach their fullest development. The relative degree of species reduction at climax is somewhat greater in forests and other mesic communities than in communities where the physical conditions are limiting (e.g., Arctic Tundra).

The biome concept

Climax vegetation with similar characteristics develops in broad areas throughout the world. Although the species may vary, the vegetation over a large region exhibits a common appearance and structure (i.e., **physiogonomy**) in response to analogous physical environments. Because of their similarities, climax vegetation throughout the world may be grouped into one of several ecological units, each known as a **biome.** Thus, the grasslands in North America (Great Plains), Eurasia (steppes), South America (pampas), and Africa (veld) are united in the Grassland Biome. Similarly, the northern evergreen forests that form a band across North America and Eurasia form the Boreal Forest Biome. Of the ten terrestrial biomes that are generally recognized worldwide, the only other biomes represented in North America include Desert, Temperate Deciduous Forest, Temperate Rain Forest, and Tundra. Despite the primary focus on their respective climax communities, however, each biome is best viewed as a mosaic that includes areas of early and mid-successional communities resulting from various types of disturbances. For example, stands of longleaf pine, although clearly not representative of broad-leafed deciduous trees, represent a significant subclimax community within the Eastern Deciduous Forest.

Animals should of course not be overlooked when considering biomes but, by tradition, biomes are named to reflect their dominant vegetation. In the past, "Spruce–Moose" was proposed to identify the Boreal Forest Biome, thereby including a major animal in the name, but this and similar designations never gained acceptance. Moreover, biomes are named for climax species, which is not the case for moose (as noted in Chapter 3).

Biodiversity

Biodiversity is shorthand for biological diversity and refers to the number of species that naturally occurs in a defined space or ecological unit, taking into account the relative population size of each species. **Species richness** is a similar term, although it does not account for the relative abundance of each species. The ever-growing human population and accompanying consumption of resources place increasing pressure on the world's fauna and flora. As a result, biodiversity is declining. Cities and other developed areas are obvious examples of locations where biodiversity has become severely limited. The destruction of tropical rain forests eliminates large numbers of species, but many other human activities (e.g., acid rain) also reduce biodiversity in places where concrete, bulldozers, or chainsaws are not in evidence.

The "species richness gradient"

The correlation between number of species and latitude is one of the more intriguing features of biogeography. Alfred Russell Wallace (1823–1913), who discovered **natural selection** independently of Charles Darwin (1809–1882), was among the first to highlight that "animal life is, on the whole, far more abundant and varied within the tropics than in any other part of the globe." This pattern, which applies equally to both plants and animals, begins with the incredible number of species in the tropical rain forests at the equator. North or south of the equator, species richness gradually declines toward the poles, where biotic communities consist of a rather small number of species. In the Northern Hemisphere, a steady reduction in the number of nesting birds is reflected at intervals along a gradient northward from the tropics: Columbia (1400 species); Panama (1100); Guatemala (470); New York (195); Newfoundland (81); and Greenland (56 species). A gradient of similar magnitude occurs in the biota of the Southern Hemisphere, reaching an extreme in Antarctica where terrestrial vertebrates and vascular plants are completely absent.

What might explain this pattern? Some short answers follow, but the debate continues. First, tropical climates are relatively stable year round and offer favorable conditions for life (i.e., warm and humid), whereas seasonal differences become steadily more apparent at higher latitudes. Tropical regions are also relatively free of dramatic disturbances. Conversely, disturbances such as drought and hard winters and, over the long term, glaciations occur with greater frequency in the middle and higher latitudes. Such disturbances hinder the course of diversification. While the influence of abiotic factors is lessened in a stable climate, biotic interactions – especially competition, herbivory, and predation – increase in importance. Under these conditions, tropical regions become a "diversity pump" where resources are partitioned, **speciation** increases, and **extinction** decreases. For example, as predation increases, the number of individuals in a prey population decreases, which enables a greater number of other species to coexist. At higher latitudes, however, the highly variable physical environment limits specialization and results in fewer species.

Time is the second factor to interact with climatic stability. Because diversity is the product of **evolution**, greater diversity results when and where evolution can proceed uninterrupted for long periods of time. Long periods of stability provide areas such as tropical rain forests and coral reefs with adequate time to produce rich biotas. Elsewhere, however, the regions today occupied by Tundra and Boreal Forest biomes only recently (from a geological perspective) emerged from a glacial blanket and have had much less time to evolve a mature biota.

Abundance and availability of food also contributes to species richness along a latitudinal gradient. In tropical regions, for example, plants of one kind or another bear fruit all year round, thereby providing an abundant and consistently available food source for **frugivorous** species. These in turn become the food base for various predators and scavengers. Conversely, fruiting is markedly seasonal in temperate regions, and there are fewer frugivores. Such a relationship is evident when the species diversity of bat fauna is compared by latitude. In temperate areas, bats feed on insects and must either migrate or hibernate when winter limits their food supply; in tropical areas, where both fruit and insects are abundant and available all year round, bats can specialize and more species have evolved. A correlated factor is the increase in primary productivity along a gradient from pole to equator; as solar radiation increases, so does plant production. Consequently, more consumers can exist which encourages competition and other biological interactions (see above). Interestingly, the latitudinal pattern for terrestrial species is reversed for marine mammals, but nonetheless remains based on food abundance. Species diversity of baleen whales and carnivorous pinnipeds (e.g., seals) is greater at higher latitudes where the abundance of their prey is directly or indirectly related to an immense base of planktonic foods.

The pattern in which food or other resources is distributed within a landscape, referred to as patchiness, also seems to be involved. Greater patchiness, typical in the tropics but much less so in the Boreal Forest or Arctic Tundra, may increase diversity by accelerating the development of subspecies and species among isolated populations. Habitat patches promote isolation, and isolation promotes speciation.

To summarize, terrestrial species diversity shows a global pattern of increase from the poles to the equator. In terms of the North American biota, a band stretching from Labrador to Alaska contains far fewer species than a band of the same width stretching from Virginia to California. Various factors contribute to this relationship including environmental stability, biological interactions, solar energy, and patchiness.

Biodiversity "hotspots"

Biodiversity is widely regarded as a global resource, yet extinction rates soar as the march of expanding human populations steadily degrades natural habitats. Some conservationists estimate a species is lost every day while others believe the rate is nearly one per hour. To help stem this destruction, some 25 areas around the world with important but endangered environments – notably those with large numbers of **endemic** species – have been designated as biodiversity hotspots. Each hotspot

Figure 1.4 Madrean pine–oak woodlands, which occur in deep canyons and remote sky-islands in the rugged mountains of northern Mexico and the southwestern United States, represent sites with an amazing diversity of flora and fauna. Usually surrounded by arid habitats, these isolated hotspots – this one in Hell's Canyon in the Davis Mountains, Texas – illustrate an insular type of distribution. Photo courtesy of Brian R. Chapman.

(a) contains vegetation that includes at least 0.5% (or alternately 1500) species of endemic vascular plants; and (b) has already experienced a significant loss of biodiversity.

Two regions in North America meet these criteria: the California Floristic Province, a broad strip lying between the Sierra Nevada Range and the Pacific Coast and extending from southern Oregon to northern Mexico, contains about 3500 species of flowering plants of which approximately 61% are endemic; and the Madrean Pine–Oak Woodland features a series of unique habitats in the rugged mountain ranges and deep canyons in southern Arizona, southwestern Texas, and northern Mexico (Fig. 1.4). Unfortunately, only a small percentage of the land within each of these hotspots is currently protected, but national and international conservation organizations, including The Nature Conservancy and Sierra Club, are actively working to protect additional habitats in both regions.

Patterns of distribution

Organisms are not randomly distributed but instead show more-or-less distinctive distributional patterns. Those with restricted distributions are said to be endemic to the area (e.g., pronghorns are endemic to western North America). Conversely, the patterns for some organisms are more complex and often difficult to explain. Discontinuous distributions, for example, are characterized by the presence of a single species in two or more widely separated areas. The range of wood ducks covers eastern North America, then skips the western plains and Rocky Mountains, but resumes along the northern Pacific coastline (California, Oregon, Washington, and southern British Columbia).

Continental patterns
The discontinuous distribution of some organisms involves even greater distances, which may even span parts of two or more continents. A few of the more common distributional patterns are briefly described below and, although they are traditionally applied to animals, they characterize plant distributions just as well.

Organisms whose distributions occur exclusively in North America are known as **Nearctic** species; prairie dogs and Gila monsters are good examples, as are pronghorns and wild turkeys. For comparison, **Palearctic** species occur in Eurasia. Organisms found only in the Northern hemisphere – Nearctic and Palearctic together – are said to have **Holarctic** distributions.

Holarctic distributions are most obvious in species associated with the boreal forest (e.g., lynx). Physical linkages between segments of boreal forest in North America and Eurasia are virtually absent, however, and thereby preclude movement of organisms between these regions. Lacking contact, these populations have varying degrees of reproductive isolation and progress towards speciation. Thus, Holarctic species are often distinguished by one or more subspecies (e.g., caribou and elk in the Nearctic and their Palearctic counterparts, reindeer and red deer). In other cases, however, closely related species may represent Holarctic distributions (e.g., two species of beaver, one in North America and another in Europe).

Late in the 18th century, Holarctic species contributed to a transoceanic debate between Thomas Jefferson (1743–1826) and French naturalist Georges-Louis Buffon (1707–1788). Buffon, a scientific giant of his century, steadfastly maintained that the biota of North America was a degenerate and smaller counterpart of life in Europe. Jefferson, a champion of scientific rigor and America's equality, responded by measuring specimens from both continents, which revealed no disparity in size. In order to avoid his data being challenged, the resourceful Jefferson sent Buffon a North American representative of a Holarctic species: the remains of a fully grown bull moose!

Typical **Neotropical** species are endemic to Central and South America (e.g., sloths and anteaters). Sometimes included in this group, however, are species that migrate to the tropical areas of the Western Hemisphere (= New World). For example, several species of migratory birds, among them tanagers and wood warblers, breed in North America and overwinter in the tropical climes of Central and South America. A few Neotropical species have expanded their distribution range by moving into North America (the armadillo is a prime example). Some groups of organisms (e.g., monkeys) occur in tropical zones around the world; these exhibit a pattern known as a pantropical distribution.

The distribution of some species forms a more-or-less circular pattern around the North Pole. These are known as **circumpolar** species, and their ranges lie along the northern zones of North America and Eurasia. The distribution of some circumpolar species also includes the polar icecap itself. Continental land masses are in close proximity in the far north, where pack ice enables the movements and genetic mingling of cold-adapted species across the expanse of Arctic environments. Arctic foxes and polar bears are circumpolar, as are walruses, beluga whales, and several other species of marine mammals. With the exception of marine organisms, species associated with the Arctic Tundra offer some of the best examples of circumpolar distributions. These species include mammals, especially lemmings, and resident birds such as willow ptarmigan and snowy owls. Circumpolar distributions also include the breeding ranges of migrant birds such as ruddy turnstones and lesser golden plovers.

Included among **cosmopolitan** species are organisms whose geographical distributions are essentially worldwide (excluding Antarctica and sometimes Australia in practice). Peregrine falcons and ospreys are each cosmopolitan species and the 11 species of barn owls exemplify a cosmopolitan family (in this case, including representatives in Australia). A grass known as common cane is an example of a cosmopolitan plant.

Geographical and ecological distribution

Plant and animal taxa (Infobox 1.2) have rather well-known geographical distributions that are often shown as shaded areas on maps but, within the confines of the mapped distributional range, a species may occupy only certain habitat types. Pronghorn, sometimes known as pronghorn antelope, is a species whose geographical distribution occurs only in western North America. Within

this large area, however, pronghorn are specifically associated with grasslands and desert communities; they are noticeably absent elsewhere in western North America (e.g., forest communities and mountainous terrain). Within the larger context of their geographical distribution, pronghorn have a smaller, habitat-specific ecological distribution. Wild turkeys also occur widely in North America, and one race – the eastern wild turkey – has an ecological distribution associated with eastern deciduous forests.

Some ecological concepts

Niches

In everyday use, a niche is a recess in a church wall where a religious icon rests securely in a place of its own. Such a site-specific relationship mirrors the concept of **niche** as an ecological term. In ecology, a niche represents the role played by an organism in its environment. Some herbivores occupy grazing niches in grasslands; other species fill grazing niches in forests, deserts, or aquatic environments (e.g., sea urchins graze on kelp). It is therefore useful to describe an organism's niche in terms of its habitat as well as its role in that habitat. Accordingly, niches are sometimes defined as the "job and address" of an organism.

Niches can be narrow and highly specialized, as in the case of a plant or animal that survives only in a limited range of conditions. Snail kites, described in Chapter 12, prey only on a single kind of food. Although snail kites may forage over a large area of open wetland, their niche as a wetland predator is actually quite narrow. A single tree may contain several feeding niches, as determined for five species of warblers in the Boreal Forest (Chapter 3). Organisms with narrow niches, that is, niche specialists, often have evolved physical features that "match" their niches (e.g., birds with uniquely shaped bills represent adaptations for obtaining specific foods). Conversely, niches may be wide as is the case for raccoons, opossums, and white-tailed deer, which occur in many types of environments and consume a variety of foods (i.e., they are generalists).

As described more fully in Chapter 9, some organisms can tolerate a broad set of conditions but competition from other organisms limits them to a narrower range. Indoor experiments with potted plants, for example, have demonstrated that some species of plants grow equally well in both saline and non-saline soils. In nature, however, the same plants grow only in saline soils because species of plants with greater competitive abilities but less tolerance for salt exclude them from other locations. In short, niches sometimes must be

Infobox 1.2 Taxonomy and the binomial system

Plants and animals are organized in taxonomic hierarchies. The largest group is a kingdom, of which Animalia and Plantae are foremost in our discussions, with Fungi mentioned less often. Following, in ever-smaller groups, are phylum (for animals) and division (for plants), class, order, family, genus (plural genera), and species. Subspecies may be designated in cases where a species exhibits two or more slight but consistent variations, typically in body size or coloration; such forms occupy separate areas within the overall distribution of the species (e.g., the diminutive Key deer in Florida). Each of these groups – kingdom to subspecies – is known as a taxon (plural taxa). Beginning with phylum, the taxa for white-tailed deer include:

Phylum: Chordata
Class: Mammalia
Order: Artiodactyla
Family: Cervidae
Genus: *Odocoileus*
Species: *Odocoileus virginianus*

For animals, the family taxon always ends in –idae; for plants, -aceae. Other taxa, however, are not readily identified by a consistent suffix. Genus and species names are always italicized. The names are Latin (or latinized, as when Smith appears as *smithii*) as it is a "dead language" and therefore remains unchanging.

This arrangement, known as the binomial system of nomenclature, replaced a confusing hodge-podge of names. Until the mid-18th century, species were described with unwieldy strings of Latin adjectives (which were often altered at will by naturalists of the day). The common wild briar rose, for example, once was designated *Rosa sylvestris alba cum rubore, folio glabro*. Adoption of the binomial system resulted from the work of the Swede, Carolus Linnaeus (1707–1778), published for plants as *Species Plantarum* (in 1753) and for animals as *Systema Naturae* (the 10th edition appeared in 1758). Thanks to Linnaeus, the briar rose became *Rosa canina*.

Soon after publication of his seminal works, the names proposed by Linnaeus gained wide acceptance and, although many were later revised as taxonomic science improved, many others remain unchanged to the present time. Some taxonomists are regarded as "splitters", those who make finer distinctions between taxa, typically "splitting" a genus into more species, whereas "lumpers" consolidate similar species, thereby reducing the number of species within a genus. In any event, Carolus Linnaeus once and for all established the binomial system as the standard format for nomenclature, and he thereby is honored as the Father of Taxonomy.

Linnaeus, trained in medicine, held a professorship at Uppsala, where he expanded the university's botanical gardens. His interest in botany inspired a generation of students, several of whom were commissioned to explore various areas of the world in search of new species. Pehr (Peter) Kalm (1716–1779) was among these and spent three years traveling in America's northeastern colonies. The genus *Kalmia*, created by Linnaeus, recognizes Kalm's contributions to the botany of North America; the taxon includes mountain laurel (*K. latifolia*), today the state flower of both Pennsylvania and Connecticut.

Plants and animals are identified using keys in which one in a series of paired choices, called couplets, is selected to match the distinctive features of the specimen at hand (e.g., leaves with spines versus leaves lacking spines). The progression of choices steadily eliminates other possibilities until only one remains.

Classification, a discipline closely allied with taxonomy, determines relationships among taxa (e.g., uniting closely related species in the same genus or, conversely, assigning a species to another genus). DNA techniques are strong tools used in making these decisions. In sum, classification deals with kinship and taxonomy with nomenclature.

considered in terms of the influences exerted by other species in the community.

On occasion, certain niches may be considered "unoccupied." This is most noticeable on oceanic islands (e.g., Hawaii), where certain types of organisms cannot colonize from mainland areas. Large grazing animals and ground-dwelling predators are among those often missing from the fauna of oceanic islands. Indeed, the only mammals in the original fauna of some islands are bats. Unfortunately, humans often disrupt the biota of ocean islands by introducing cattle, goats, or pigs. This usually leads to the rapid deterioration of the native vegetation, which evolved without incorporating adaptations for suddenly coping with new influences. Similarly, native animals, often birds, are eliminated or vastly reduced in number after predators gain access to oceanic islands under human influences (e.g., rats escaping from ships at dock). Introduced species that fill an unoccupied niche typically exploit the new situation and increase rapidly. Thus the adage "Nature abhors a vacuum."

Ecological equivalents

Not uncommonly, an organism may have a counterpart in another community or biome where it functions in a similar niche. Such pairs of organisms, although not closely related, are known as **ecological equivalents**. Forests, for example, represent the typical habitat for most kinds of woodpeckers, which excavate cavities in tree boles for their nests. The Sonoran Desert lacks trees, but two species of woodpeckers nest there by chiseling cavities in giant saguaro, a cactus that reaches tree-sized proportions (see Chapter 7). At least in terms of woodpecker habitat, these large cacti function as trees in the desert and thereby illustrate the concept of ecological equivalents.

As a further example, the world's major grassland communities each include grazing species. Many of these animals are insects (e.g., grasshoppers), but the grazers usually include at least one large-bodied herbivore. Grazing species of North America include bison, elk, and pronghorn; a much larger number of grazing species exist on Africa's grasslands, however (e.g., gazelles and several other kinds of antelope, as well as zebras). These animals, continent by continent, also represent ecological equivalents. Not all of these are closely related, yet they share some common characteristics typical of large grazing animals: hooves, complex stomachs, good eyesight, and usually horns or antlers. They also run rapidly and move in herds. In Australia, some species of kangaroo occupy grazing niches and, although they lack most physical features typical of large grazing animals, kangaroos are the ecological equivalents of bison and gazelles in grassland ecosystems.

Bergmann's rule

Ecologists discovered long ago that body size increases in the cooler portions and diminishes in the warmer parts of each species' distributional range. This relationship, known as **Bergmann's rule**, applies only to **endothermic** animals. In general, "cooler" and "warmer" conditions equate to the northern and southern latitudes of North America, respectively. Differences in body size are therefore best shown in species with distributions covering large areas of the continent (e.g., white-tailed deer, which are larger-bodied in Maine and Michigan than in Florida or Texas).

Bergmann's rule is usually explained in terms of the ratio of volume to surface area. Larger-bodied animals have less surface area per unit of volume and, when compared to smaller-bodied animals, they lose proportionately less body heat because less of their surface is exposed. In the cooler (= northern) regions of North America, larger-bodied animals gain an advantage over smaller-bodied animals. At lower latitudes, where the climate is warmer (e.g., southern North America), the advantage is reversed and smaller-bodied animals are better matched to ambient temperatures. Natural selection steadily favors animals whose body sizes are best suited to each climatic regime, north to south, resulting in a corresponding gradient of body sizes. However, exceptions do exist. The large body mass of the African elephant may be resistant to a significant increase in core body temperature in the warm climates within its range.

Recall that Bergmann's rule generally concerns endotherms, but some studies indicate that a few species of **ectotherms** may also conform to the rule. The reverse of Bergmann's rule applies to most reptiles and other ectothermic animals, but many large-bodied snakes and

Figure 1.5 Two closely related species of dark geese with white cheek patches evolved from a common ancestor in North America. The larger of these – the Canada goose – subsequently differentiated into several subspecies, three of which are shown in the bottom row. The smaller species – the cackling goose – evolved into four subspecies, of which one is shown at the top. The larger forms lose less heat relative to body size than smaller forms, and the limits of the winter distribution of the larger geese accordingly lie farther north in comparison to the smaller forms. The winter distribution of these related taxa reflects Bergmann's Rule. Photo courtesy of Victor E. Krantz and the National Museum of Natural History.

lizards that occur in tropical regions maintain a constant, relatively high body temperature because of their greater volume-to-surface ratio. This relationship, sometimes called **gigantothermy** or ectothermic heterothermy, allows leatherback sea turtles to retain the heat generated by muscular activity and exploit highly productive marine feeding habitats in the cold waters of high latitudes or great depths. Animals in Australia and elsewhere conform to Bergmann's rule, but some ecologists do not accept the traditional explanation of heat loss per unit of body size.

Nonetheless, an interesting example of Bergmann's rule occurs in Canada geese, which are represented by numerous subspecies in North America. The plumages of each subspecies show some distinctive features, but differences in body size are far more obvious (Fig. 1.5). Canada geese evolved as migratory birds and, after breeding, they fly south to spend the winter (the season when heat loss exerts its greatest influence on their survival) in warmer environments. Accordingly, body size emerges as a factor determining the northernmost limits of the winter distributions of each

subspecies. A large-bodied subspecies winters as far north as Minnesota, but a small subspecies spends winter no farther north than Oklahoma. When heat losses were calculated for these two subspecies, the results indicated that the larger geese lost about 40% less heat per hour per unit body weight than the smaller birds. In other words, the smaller geese would more rapidly expend their energy reserves if they wintered as far north as the larger birds. Because of this potential energetic cost, the smaller geese overwinter where temperatures are milder. In this instance, Bergmann's rule applies to the *winter* distribution of Canada geese rather than to their summertime breeding ranges.

Allen's rule
The extremities of endotherms are relatively shorter in colder regions when compared to those of the same or related species living in warmer areas. For example, the ears of hares and rabbits living in the Arctic and sub-Arctic areas are much shorter than those of jackrabbits living in the warm regions of North America (Fig. 1.6).

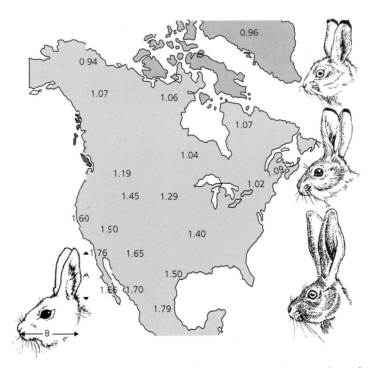

Figure 1.6 An example of Allen's Rule is shown in the ear lengths of rabbits and hares along a north–south temperature gradient across North America. The numbers are ratios determined by dividing ear length (A) by skull length (B), as shown lower left. Rabbits in warm regions have longer ears that readily dissipate excess body heat, whereas shorter ears help conserve body heat in cool areas. Illustrated by Tamara R. Sayre, based on Hesse (1928).

Thermoregulation is the underlying influence for this relationship. In warm areas, rabbits readily dissipate excess body heat from their large ears, whereas shorter ears help conserve body heat where the climate is colder. An anatomical explanation for **Allen's rule** is based on the relative amount of cartilage incorporated into the extremities. The growth rate of cartilage in developing endotherms is partially dependent upon temperature and blood flow. Young mammals raised in warmer regions of their range have more blood flowing to their warmer extremities during development and produce more cartilage, thus increasing the length of the body part. Bill sizes in 214 species of birds also follow the pattern predicted by Allen's rule.

Readings and references

Across North America

Barbour, M.G. and W.D. Billings (eds). 2000. *North American Terrestrial Vegetation*, second edition. Cambridge University Press, New York, NY. (A detailed review of major vegetational units, accompanied by extensive bibliographies for each.)

Chapman, J.A., B.C. Thompson, and G.A. Feldhamer (eds). 2003. *Wild Mammals of North America: Biology, Management, Conservation*, second edition. Johns Hopkins University Press Baltimore, MD.

De Bilj, H.J. 2005. *Atlas of North America*. Oxford University Press, USA, NY.

Dice, L.R. 1943. *The Biotic Provinces of North America*. University Michigan Press, Ann Arbor, MI.

Merriam, C.H. 1898. *Life Zones and Crop Zones of the United States*. Bulletin 10, Bureau of Biological Survey, US Department of Agriculture, Washington, DC. (An early attempt to classify ecological areas based on temperature data.)

Ricketts, T.H., E. Dinerstein, D.M. Olson, C. J. Loucks, et al. 1999. *Terrestrial Ecoregions of North America: A Conservation Assessment*. Island Press, Washington, DC.

Shelford, V.E. 1963. *The Ecology of North America*. University Illinois Press, Urbana. (A landmark work; Shelford is widely regarded as a founding father of animal ecology.)

Vankat, J.L. 1992. *The Natural Vegetation of North America: An Introduction*, reprint. Krieger, Malabar, FL.

A brief overview of ecology

Boul, S.W., R.J. Southard, R.J. Graham, and P.A. McDaniel. 2003. *Soil Genesis and Classification*, fifth edition. Iowa State Press-Blackwell, Ames, IA.

Brown, D.E., T.C. Brennan, and P.J. Unmack. 2007. A digitized biotic community map for plotting and comparing North American plant and animal distributions. Canotia 3: 1–12.

Hoffman, J., T. Tin, and G. Ochoa. 2005. *Climate: The Force that Shapes our World and the Future of Life on Earth*. Rodale Books, London.

Singer, M.J. and D.N. Munns. 2005. *Soils: An Introduction*, sixth edition. Prentice-Hall, NY.

Smith, T.M. and R.L. Smith. 2008. *Elements of Ecology*, seventh edition. Pearson Benjamin Cummings, San Francisco, CA.

Soil Survey Staff. 1999. *Soil Taxonomy: A Basic System of Soil Classification for Making and Interpreting Soil Surveys*, second edition. US Department of Agriculture, Soil Conservation Service, Washington, DC.

Stelia, D. and T.E. Pond. 1989. *The Geography of Soils, Formation, Distribution and Management*, second edition. Rowman & Littlefield, Savage, MD.

Townsend, C.R., M. Begon, and J.L. Harper. 2008. *Essentials of Ecology*, third edition. Blackwell Publishing, New York, NY.

Community succession

Billings, W.D. 1938. The structure and development of old field shortleaf pine stands and certain associated physical properties of the soil. Ecological Monographs 8: 441–499.

Clements, F.E. 1916. *Plant Succession*. Publication 242, Carnegie Institute, Washington, DC.

Clements, F.E. 1916. Nature and structure of the climax. Journal of Ecology 24: 252–284.

Connell, J.H. and R.O Slatyer. 1977. Mechanisms of succession in natural communities and their role in community stability and organization. American Naturalist 111: 1119–1144.

Cowles, H.C. 1899. The ecological relations of the vegetation on the sand dunes of Lake Michigan. Botanical Gazette 27: 95–117, 167–202, 281–308, 361–391. (The watershed study of plant succession; but for a reevaluation, see Olson, J.S. 1958. Rates of succession and soil changes on southern Lake Michigan sand dunes. Botanical Gazette 119: 125–170.)

Drury, W.H. and I.C.T. Nisbet. 1973. Succession. Journal of the Arnold Arboretum 54: 331–368.

Huston, M. and T. Smith. 1987. Plant succession: life history and competition. American Naturalist 130: 168–198.

MacDougall, A.S., S.D Wilson, and J.D. Bakker. 2008. Climatic variability alters the outcome of long-term community assembly. Journal of Ecology 96: 346–354.

Odum, E.P. 1960. Organic production and turnover in old field succession. Ecology 41: 34–49.

Sousa, W.P. 1984. The role of disturbance in natural communities. Annual Review of Ecology and Systematics 15: 353–391.

Spurr, S.H. 1952. Origin of the concept of forest succession. Ecology 33: 426–427.

Tansley, A.G. 1935. The use and abuse of vegetational concepts and terms. Ecology 16: 284–307.

Thoreau, H.D. 1860. The succession of forest trees. (This essay was delivered as an address to the Middlesex Agricultural Society in Concord in September 1860 and subsequently appeared in the society's *Transactions*. See pp. 72–92 in Henry David Thoreau, the natural history essays, with an introduction by R. Sattelmeyer, 1980, Pergrine Smith, Salt Lake City, UT.)

Biodiversity

Gaston, K.J., and J.I. Spicer. 2004. *Biodiversity: An Introduction*, second edition. Blackwell Publishing, Cambridge, MA.

Mittermeier, R.A., P.R. Gil, M. Hoffman, J. Pilgrim, T. Brooks, C.G. Mittermeier, J. Lamoreux, G.A.B. da Fonseca, and P.A. Seligman. 2005. *Hotspots Revisited: Earth's Most Biologically Richest and Most Endangered Terrestrial Ecoregions.* Conservation International, Arlington, VA.

Myers, N., R.A. Mittermeier, C.G. Mittermeier, G.A.B. da Fonseca, and J. Kent. 2000. Biodiversity hotspots for conservation priorities. Nature 403: 853–858.

Possingham, H. and K. Wilson. 2005. Turning up the heat on hotspots. Nature 436: 919–920.

Wilson, E.O. 1992. *The Diversity of Life.* W.W. Norton, New York, NY.

Wilson, E.O. (ed.) and F. M. Peter (assoc. ed). 1988. *Biodiversity.* National Academy Press, Washington, DC.

Patterns of distribution

Bedini, S.A. 1990. *Thomas Jefferson, Statesman of Science.* Macmillan, New York, NY. (Provides details concerning Jefferson's securing and shipping of a moose specimen to France.)

Currie, D.J. 1991. Energy and large-scale patterns of animal- and plant-species richness. American Naturalist 137: 27–49.

Currie, D.J. and V. Paquin. 1987. Large-scale biogeographical patterns of species richness in trees. Nature 329: 326–327.

Fischer, A.G. 1961. Latitudinal variations in organic diversity. American Scientist 49: 50–74.

France, R. 1992. The North American latitudinal gradient in species richness and geographical range of freshwater crayfish and amphipods. American Naturalist 139: 342–354.

Gaston, K.J. 2000. Global patterns in biodiversity. Nature 405: 220–227.

Hagmeier, E.M. and C. D. Stults. 1964. A numerical analysis of the distributional patterns of North American mammals. Systematic Zoology 13: 125–155.

MacArthur, R.H. 1963. *Geographical Ecology: Patterns in the Distribution of Species.* Harper and Row, New York, NY.

Macpherson, E. 2002. Large-scale species-richness gradients in the Atlantic Ocean. Proceedings of the Royal Society of London B 269: 1715–1720.

McCoy, E.D. and E.F. Conner. 1980. Latitudinal gradients in the species diversity of North American mammals. Evolution 34: 193–203.

Rabenold, K.N. 1979. Revised latitudinal gradients in avian communities of eastern deciduous forests. American Naturalist 114: 275–286.

Simpson, G.G. 1989. Species density of North American recent mammals. Systematic Zoology 13: 57–73.

Stevens, G.C. 1898. The latitudinal gradient in geographical range: how do so many species exist in the tropics. American Naturalist 133: 240–256.

Turner, J.R.G. 2004. Explaining the global biodiversity gradient: energy, area, history and natural selection. Basic and Applied Ecology 5: 435–448.

Some ecological concepts

Barnett, R.J. 1977. Bergmann's rule and variation in structures related to feeding in the gray squirrel. Evolution 31: 538–545.

Frair, W., R.G. Ackman, and N. Mrosovsky. 1972. Body temperature of *Dermochelys coriacea*: warm turtle from cold water. Science 177: 791–793.

Geist, V. 1986. Bergmann's rule is invalid. Canadian Journal of Zoology 65: 1035–1038.

Griffing, J. P. 1974. Body measurements of black-tailed jackrabbits of southeastern New Mexico with implications of Allen's Rule. Journal of Mammalogy 55: 674–678.

Hesse, R. 1928. *Die ohrmuscheln des elefanten als wärmeregulator.* Zeitschrift für wissenschaftliche Zoologie 132: 314–328. (Includes ear:skull ratios for rabbits and hares in North America.)

Hutchinson, G.E. 1957. Concluding remarks. Cold Spring Harbor Symposium on Quantitative Biology 22: 415–427. (Includes a notable discussion of niches.)

Hutchinson, G.E. 1959. Homage to Santa Rosalia, or why are there so many kinds of animals? American Naturalist 43: 145–159.

LeFebvre, E.A. and D.G. Raveling. 1967. Distribution of Canada geese in winter as related to heat loss at varying environmental temperatures. Journal of Wildlife Management 31: 538–546.

Mayr, E. 1956. Geographical character gradients and climatic adaptation. Evolution 10: 105–108.

Olalla-Tárraga, M.Á., M.Á. Rodríguez, and B. A. Hawkins. 2006. Broad-scale patterns of body size in squamate reptiles of Europe and North America. Journal of Biogeography 33: 781–793.

Paladino, F.V., M.P. O'Connor, and J.R. Spotila. 1990. Metabolism of leatherback turtles, gigantothermy, and thermoregulation of dinosaurs. Nature 344: 858–860.

Scholander, P.F. 1955. Evolution of climatic adaptation in homeotherms. Evolution 9: 15–26.

Serrat, M.A., D. King, and C.O. Lovejoy. 2008. Temperature regulates limb growth in homeotherms by directly modulating cartilage growth. Proceedings of the National Academy of Sciences 105: 19348–19353.

Stevenson, R.D. 1986. Allen's rule in North American rabbits (*Sylvilagus*) and hares (*Lepus*) is an exception, not a rule. Journal of Mammalogy 67: 312–316. (Advantages of improved locomotion may outweigh adaptation to cold.)

Symonds, M.R.E. and G.J. Tattersall. 2010. Geographic variation in bill size across bird species provides evidence for Allen's rule. American Naturalist 176: 188–197.

Yom-Tov, Y. and H. Nix. 1986. Climatological correlates for body size of five species of Australian mammals. Biological Journal of the Linnaean Society 29: 245–262.

Infobox 1.1. Victor E. Shelford (1877–1968), Father of animal ecology

Croker, R.A. 1991. *Pioneer Ecologist, the Life and Work of Victor Ernest Shelford 1877–1968.* Smithsonian Institution Press, Washington, DC.

Kendeigh, S.C. 1968. Victor Ernest Shelford, eminent ecologist, 1968. Bulletin of the Ecological Society of America 49: 97–100.

Infobox 1.2. Taxonomy and the binomial system

Mayr, E. and P.D. Ashlock. 1991. *Principles of Systematic Zoology*, second edition. McGraw-Hill College, New York, NY.

Schuh, R.T. and A.V.Z. Brower. 2009. *Biological Systematic: Principles and Applications*, second edition. Cornell University Press, Ithaca, NY.

Simpson, M.G. 2010. *Plant Systematics*, second edition. Academic Press, New York, NY.

Wilkins, J.S. 2009. *Species: A History of the Idea*. University of California Press, Berkeley, CA.

CHAPTER 2
Tundra

Picturing the Arctic in your imagination is easy: in summer, flat green Tundra without trees meeting a green ocean; in winter, flat white snow and ocean melding together as one.

Charles Wohlforth

"The land God gave to Cain"; French explorer Jacques Cartier (1491–1557) thus likened the bleakness of the tundra in Labrador to a biblical wasteland. When Martin Frobisher (1535?–1594), the intrepid voyager for Elizabeth I, arrived at Baffin Island, he was not too overwhelmed with the landscape either: "There is verie little plain grounde and no grasse.... There is no woode at all." The first impressions of Cartier and Frobisher match those of later visitors, and aptly reflect that tundra is derived from *tunturia*, the Finnish term for "treeless plain" (Fig. 2.1).

The Tundra Biome is one of two circumpolar biomes – the Boreal Forest Biome is the other – ringing the Northern Hemisphere. Arctic Tundra lies immediately below the polar cap of perpetual ice and snow. In North America, a broad wedge of tundra extends from the western shore of Hudson's Bay near Churchill, Manitoba, to the mouth of the Coppermine River in the Northwest Territories (Fig. 1.3) Other zones of tundra occur east of Hudson's Bay, across northern Alaska, and on numerous islands in the High Arctic (e.g., Ellesmere Island). A somewhat ambiguous border known as the **tree line** marks the southern edge of the Tundra Biome. The Boreal Forest Biome lies beyond.

The coldest of all the biomes, Tundra is a harsh landscape of few resources. The biota, adapted to the rigors of the far north, is not as species rich in comparison to most plant and animal communities developing elsewhere in North America. Plants face not only extreme cold, shallow and nutrient-deficient soils, but also severe aridity and a short (50–60 days) growing season. Annual precipitation, including water from melting snow, is about 15–25 cm (6–10 inches), similar to most deserts. Relatively few species of animals live all year round in the Tundra and, of these, the list of vertebrates adapted to the biting cold includes just a single ectothermic species. Food chains are short, and a snowy owl faced with a short supply of lemmings may have few choices for an alternate source of food.

Climatic and glacial influences

Arctic Tundra represents the youngest biome in North America. It developed its biological features following the relatively recent withdrawal of the Wisconsin glacial advance some 9000–11,000 years ago. At its maximum 11,000–15,000 years ago, the Wisconsin glacier buried much of North America beneath a thick sheath of ice. During this period of cooler climatic conditions, the area currently covered by Arctic Tundra was greatly reduced. Alpine vegetation also moved downward into lower elevations so that Alpine Tundra replaced forests on mountains at lower altitudes (e.g., southern Appalachians). In these circumstances, plants once associated with alpine and Arctic biotic communities undoubtedly mixed in the middle latitudes of North America. During this geological epoch (the Pleistocene), the tundra apparently persisted in a relatively narrow zone between the glacier's face and the remaining forest. Today, tundra vegetation covers a much larger area of North America than it did during the last glacial advance.

Soils and geological influences

The geologic strata exposed after the glacial retreat have not experienced the eons of weathering normally required for a well-developed soil structure. The soils in some areas of Tundra are classified as azonal because their stony or gravel soils intermixed with clays or silts and lack a profile distinguished by discernable horizons. Churning by frost, termed **solifluction**, and the lack of strong and long-term leaching inhibit the formation of horizons. In other locations, however, mature soils with distinctive profiles may develop. For example, Arctic Brown soil features a dark brown loam or sandy loam in the A horizon, grades with depth to a dark yellow-brown in the B horizon, and finally becomes a gray silt loam in the C horizon. The surface is acidic (sometimes reaching pH 3.5) and organic matter, which is high in the A horizon, decreases with depth. These well-drained soils are associated with mesic upland communities consisting of heath, dwarfed willows and birches, sedges, and

Ecology of North America, Second Edition. Brian R. Chapman and Eric G. Bolen.
© 2015 John Wiley & Sons, Ltd. Published 2015 by John Wiley & Sons, Ltd.

Figure 2.1 For early explorers, the vast tundra seemed no more than a bleak landscape lacking trees or other useful resources. Despite its comparative lack of biodiversity however, tundra provides nesting habitat each summer for countless thousands of migratory geese and shorebirds. Barberry leaves provide vivid color to the fall landscape. Photograph courtesy of Steve Hildebrand and the US Fish and Wildlife Service.

various lichens (see section on *Lichens and "reindeer moss"*). Nonetheless, the boundary between the Tundra and the forests lying to the south is not defined by well-marked changes in soil types.

Permafrost

The pre-eminent feature of tundra substrate is **permafrost**. The term was coined by S. W. Muller in 1943 as a convenient way to describe "permanently frozen ground" or, as most Arctic ecologists now prefer, "perennially frozen ground." His original definition is still acceptable:

> …a thickness of soil or other surficial deposit or even of bedrock, at a variable depth beneath the surface of the earth in which a temperature below freezing has existed continuously for a long time (from two to tens of thousands of years).

Permafrost reaches depths of about 1450 m (4657 feet) in Siberia, but the maximum known thickness in North America is nearer 1000 m (3280 feet). Although seemingly illogical at first glance, areas with thick permafrost do not coincide with the presence of glaciers during the Pleistocene Epoch. Permafrost instead developed from the exposure of unglaciated ground to long periods of extreme cold, a situation ongoing since the retreat of the last glacial advance. Near Nome, Alaska, for example, mine tailings developed permafrost to a depth of 18 m (60 feet) in a span of just 7 years.

Permafrost produces several ecological effects of importance to tundra vegetation. Because of its hardness, the frozen soil prevents the penetration of roots. Consequently, root systems are limited to the active layer, the upper few centimeters of soil that regularly thaw during the short period of warm weather. Aeration also is extremely poor in permafrost, and nutrients are locked in the ice as well. In fact, because most water exits as ice, the Tundra Biome is sometimes characterized as a physiological desert, an environment where water (although abundant) is not available in a state that will sustain most forms of life.

Plant cover is crucial to the integrity of permafrost in many areas. Where vegetation is removed or seriously disturbed, the frozen ground is exposed to solar radiation and quickly thaws then erodes. Erosion known as **thermokarst** is peculiar to permafrost. When permafrost thaws (e.g., under a heated building), ice in the soil profile melts and the ground slumps. Such processes may be natural as well as **anthropogenic** and shape an irregular landscape known as thermokarst terrain, which is usually characterized by the presence of thaw lakes.

For animals, permafrost limits burrowing behavior, making **fossorial** mammals uncommon in the Tundra Biome. Even where the active zone is otherwise deep enough for burrowing activities, the soggy soil may expose small rodents to prolonged wetting, often a fatal circumstance in a cold climate. Similarly, insects whose larval forms dig burrows (e.g., many kinds of beetles), are not well represented in tundra communities.

Patterned ground

Because permafrost forms an unyielding barrier just below the surface, water formed at the surface from rainfall or snowmelt cannot penetrate deeper into the soil profile. Moreover, large areas of Tundra have little relief – the

Figure 2.2 Patterned ground, although variable, highlights many areas of the tundra landscape. In some locations the polygons fill with water, creating a honeycomb of small lakes, whereas distinctive vegetation highlights the margins of the polygons in upland areas. Heath and lichens cover the interior of polygons with raised centers, and sedges rim the wetter perimeters of these formations. Rocks and stones edge the borders of others. As shown here, polygons rimmed with ice and up to 15 m (49 feet) in diameter typify patterned ground in the polar deserts of the High Arctic. Photograph courtesy of T. L. Pewe and the US Geological Survey.

aftermath of repeated glacial bulldozing – and there is little runoff of surface water resulting from changes in elevation. In summer, when temperatures remain above freezing, these features promote the accumulation of surface water in innumerable shallow ponds and lakes (Fig. 2.2) or in sheets of spongy bog. Thus, much of the tundra becomes a vast temporary wetland underlain by a bedrock of ice. Conversely, in places where there is significant topographical relief, runoff is rapid because the water cannot soak into the soil and, when the snow cover melts, Arctic streams and rivers experience heavy flooding.

A network of polygons etched into the soil surface represents the most widespread geological feature of the tundra landscape (Fig. 2.2). These multisided cells vary in size, but typically range from 2 to 5 m (7–11 feet) in diameter on sites covered with grasses and sedges and to 20 to 30 m (65–100 feet) in diameter on bare ground. They result from shrinkage, much in the same fashion as mud cracks on the bottom of a dried pool. In this case, the tundra surface contracts because of the extreme cold. The cracks that form fill with meltwater in the summer and, when the water freezes in winter, the wedges of ice at the perimeter expand the cracks even further. This process – summer thawing, winter freezing – is repeated year after year, producing slowly expanding and unique topographical features on the tundra floor. The cracks also provide **microhabitats**, small-scale sites with slightly different environmental conditions where some organisms may prosper. Additionally, small summer ponds often develop where the water-filled cracks join others. The result is an aquatic system of interlocking channels known as "beaded streams" (from the air, the ponds resemble a string of beads).

Frost heaving is another force associated with extreme cold, and this movement produces still other geometric patterns on drier areas across the tundra. Stones are forced upward from the rocky soil and, once at the surface, the larger of these slowly gravitate outward while the finer materials remain at the center in a somewhat mysterious process known as sorting. On level surfaces the result is a network of rock-rimmed polygons or circles, similar in appearance to those formed by ice wedges. On hillsides, the larger stones are sorted into more-or-less parallel rows. The overall result of these forces is a landscape pattern resembling a honeycomb.

On slopes, the tundra soil may slump during alternate periods of freezing and thawing, creating a staircase of solifluction terraces that generate a pattern geologists call "mass wasting." When thawed, the saturated soils simply flow downward and then later freeze in place, temporarily forming large lobes whose appearance is not unlike hardened lava. Solifluction creates highly unstable conditions for vegetation resulting in: (a) buried plants; (b) damaged roots and stems; (c) changed soil–water relationships; and (d) newly formed azonal soils. Solifluction also smooths and rounds hills in the tundra landscape.

Structures known as "pingos," based on the Inuit word for hill, sometimes rise from the tundra floor. These cone- or dome-shaped formations are produced when the unfrozen soil of a lake basin is exposed. As

Infobox 2.1 Samuel Hearne (1745–1792), Arctic explorer and naturalist

In 18th century England parents commonly enlisted their young sons as midshipmen, a term then denoting a servant for an officer in the Royal Navy. Thus, aged 11, Samuel Hearne began a career of adventure and exploration when he sailed for seven years in the English Channel and Mediterranean Sea, highlighted by action in the Seven Years' War. He left the navy in 1763, but nothing is known about his activities until three years later when he was employed by the Hudson's Bay Company whose commercial ventures were founded on North America's natural resources. Initially, he served as a mate on one of the company's trading ships followed by service aboard a whaling vessel. In 1768, the company utilized Hearne's navigational skills to survey Hudson Bay's coastline to improve the cod fishery.

With these experiences behind him, the Hudson's Bay Company charged Hearne to search for copper in the northwestern Arctic, a quest triggered by the nuggets Native People brought to the company's trading post located in a fort at Churchill, Manitoba. His first two attempts failed but Hearne then employed Chipewyan warriors to accompany his third expedition, including Chief Matonabbee as his guide. Eight of Matonabbee's wives functioned as "beasts of burden" to pull the sledges, cook, and otherwise serve the camp. Life was not easy on the trek and starvation was only thwarted when the expedition encountered migrating caribou. When the group reached the upper drainage of the Coppermine River in the "Barren Lands of the Little Sticks," to Hearne's horror the Chipewyans attacked an Inuit camp in what became known as the Massacre at Blood Falls. A few days later, Hearne reached the Arctic Ocean – the first European to do so from North America – where he discovered only one large lump of copper and little else of value. He returned by a different route, becoming the first European to visit Great Slave Lake. Hearne arrived back at Churchill on 30 June 1772 after trekking about 8000 km (5000 miles) and exploring more than 650,000 km² (250,000 square miles) of tundra wilderness. The trip was a commercial failure but his detailed journal was a treasure of its own.

Hearne observed much of nature during his service in Canada. Many notations concerned birds, including Arctic terns and willow ptarmigan, and he was the first to recognize the existence of two species of curlew, the Eskimo and Hudsonian. He described the Ross's goose years before the species was formally cataloged with a Latin name, and he recorded the frequent use of muskrat houses as a favored nesting site for Canada geese, just as they do today. Hearne was familiar with bird migrations and noted that trumpeter swans were the first waterfowl to arrive each spring, often as early as late March, when they sought ice-free water at rapids and below waterfalls.

Dissections also expanded his knowledge of natural history; his probing revealed that the trachea of a trumpeter swan loops into the breastbone before returning to the chest cavity to reach the lungs. In another case, Hearne indirectly stumbled on the unusual molt migration of giant Canada geese when he examined the undeveloped testes of birds appearing at Churchill late in the nesting season. Today biologists know that juveniles of this subspecies migrate northward shortly after fledging, traveling to molting areas some 1600 km (960 miles) northward from where they hatched, then return south again for the winter. Hearne's dissections therefore were of these juvenile geese far removed from their natal area.

Hearne returned to England in 1787 and later sold his journal to a publishing firm in London, but he died before the book appeared in 1795. Because of Hearne's keen eye, *A Journal from Prince of Wales's Fort in Hudson's Bay to the Northern Ocean* stands as a landmark not only for his time, but for ours as well.

permafrost develops downward from the surface, the force of expanding ice expels unfrozen water upward at the weakest point (i.e., the last remaining site of unfrozen soil in the lake basin). The process is somewhat similar to the way a frozen soft drink pushes upward through the neck of a bottle. When the water inside the uplift freezes, the dome-shaped pingo develops its characteristic core of ice. Pingos may reach a height of 45 m (150 feet), and many continue growing as their cores of ice further expand. In addition to presenting occasional relief to the generally flat topography of the tundra plain, pingos also are one of the few places on the tundra where caves might be located. After finding a denning cave in a pingo in northwestern Canada, tundra explorer Samuel Hearne (1745–1792; Infobox 2.1) became the first European to record the presence of barren-ground grizzly bears. Hearne also noted that the sod-covered summits of pingos – he called them "islands" because they protruded from wetland basins – provided excellent nesting habitats for migratory birds, safe from egg predators except for wolverines.

Eskers and tundra wildlife

Long, narrow ridges of gravel, sand, and rock are among the few landmarks arising from an otherwise featureless tundra plain. Eskers are geological formations that originated as stream beds that once twisted their way through icy tunnels *within* the retreating glaciers. The glaciers scraped up and stored gravel and other rocky debris when they were advancing. When the massive ice sheets began to melt, these materials were then released from their icy grip and were picked up and transported by streams of meltwater flowing within the ice. As the glaciers continued melting away, beds of rock and gravel settled in place in the streambeds, creating winding, elevated ridges sometimes more than 100 km (60 miles) in length and up to 100 m (330 feet) in height. Eskers resemble large abandoned railroad embankments because of their flat tops, steep sides, and flared bases.

Eskers often play key ecological roles in tundra communities where in summer they serve as elevated corridors across the soggy terrain. Good drainage is perhaps the most important ecological feature of eskers. In a land of permafrost, well-drained, unfrozen sites are uncommon and become locations where burrowing animals find soil conditions fit for digging their dens. Many eskers have become migratory routes followed by generations of caribou. Only the heartiest plants can survive on the exposed crests of eskers, where relentless winds readily dry the rocky soil and drive sand with abrasive force. Lichens, almost ubiquitous on rocky surfaces in the Arctic, are among the few plants that regularly colonize these harsh sites. On their protected sides, the ridges offer havens for blueberries, sedges, and other herbaceous plants, as well as sites where willows and birches can grow upright. Isolated spruces, although stunted, sometimes find refuge on the protected sides of tundra eskers lying north of the tree line.

Glacial refugia

Although extensive, Pleistocene glaciers did not completely cover the northern half of North America. In northwestern Alaska, a large area remained ice-free and in this **glacial refugium** many tundra organisms were isolated from the rest of the North American biota. The consequences of isolation are seen today in some segments of tundra communities. Grizzly bears, for example, seem to be expanding their range slowly eastward across Arctic Tundra from the large glacial refugium in the lower Yukon Valley of Alaska. A somewhat smaller refugium at the northern tip of Greenland – an area known as Pearyland – also remained ice-free during Wisconsin glaciation, as did Banks Island and some smaller islands in the High Arctic Archipelago. A rather narrow belt of Tundra also persisted far to the south at the leading edge of the ice sheet.

Features and adaptations

Despite the extremely cold winter temperatures, desert-like availability of moisture, shallow layer of active soil, and limited availability of nutrients, plants of many kinds are able to survive in the Arctic Tundra. Approximately 1700 species of plants have adapted to the rigors of the

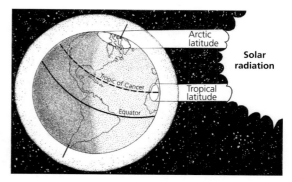

Figure 2.3 Schematic representation of how the Earth's tilt of 23.5° reduces the amount of solar radiation available for Arctic ecosystems. Sunlight, shown here by two incoming "beams" of the same intensity, falls on the Earth at different angles according to latitude. During the Arctic summer, solar radiation produces a large "footprint" but with relatively small amounts of energy per unit area. In contrast, more concentrated light falls all year round in tropical areas and therefore incorporates more energy per area into tropical ecosystems. The tilt also accounts for pronounced changes in day length between summer and winter in Arctic latitudes as the Earth moves around the Sun. Illustrated by Tamara R. Sayre.

cold climate. Many kinds of plants in biomes at more southerly latitudes cope with low temperatures during dormant seasons (e.g., usually during winter) but, unlike tundra vegetation, these plants require much higher ambient temperatures for the resumption of their growth. Two interacting factors account for the limited amount of solar radiation and, in turn, the low temperatures in the Arctic. The first of these concerns the Earth's rotation and the tilt of its axis. This combination results in days during winter north of the Arctic Circle (latitude 66° 30′ N) when the sun does not rise above the horizon. During this period, Arctic Tundra experiences a thermal deficit as it steadily loses heat to the atmosphere without being replenished by incoming solar energy. Second, the Sun's angle is so low during the continuous light of summer that the solar beams striking Earth are diminished of their energy (Fig. 2.3). For nearly half of the Arctic summer, the radiation resulting from the seasonally long days is devoted to thawing the soil surface and melting snow, after which most Arctic plants can resume growth. The result is a very short growing season for both plants and animals.

Plant adaptations to harsh conditions

Many plants are dwarfed by the rigors of the tundra environment. Species (e.g., birches) that would otherwise be much taller are no more than low bushes in the Tundra. Others produce stems that, instead of growing upright, creep parallel to the ground in a

(a)   (b)

Figure 2.4 (a) Woody plants often develop parallel to the surface in a growth form described as decumbent, a result of harsh physical conditions in tundra environments. The same species may grow upright where environmental conditions are more moderate. (b) Tundra plants often take advantage of protected areas on the leeward side of rocks and other kinds of shelter. Microhabitats such as these occur in communities elsewhere, but they are especially obvious in the rigorous and relatively exposed tundra landscapes. Note the branches hugging the protective surface of the boulder. Photographs courtesy of Elizabeth D. Bolen.

growth form known as **decumbent** (Fig. 2.4a). Wind, cold, and radiation gain easy access to the tissues of tundra plants as well as to soil and rock surfaces. In such circumstances, the effects of microclimate on plants and other tundra organisms in association with small changes in topography or shelter are heightened (Fig. 2.4b). For example, soil temperature, wind blast, and the moisture resulting from snowmelt vary greatly with topographic changes of just a few centimeters (e.g., a rock, hummock, or the rim of a polygon). All told, it is the physical environment rather than the biological environment that dominates tundra communities. Survival is based more on coping with aridity and cold temperatures than on competition with other organisms.

Low-growing herbaceous plants often develop in cushion-like shapes, a means of reducing their surface area and, in turn, limiting their exposure to the harsh climate. Examples include Arctic forget-me-not, moss-campion, and some species of saxifrage. Some plants counteract the Tundra's surprising aridity with leaves protected by water-saving cuticles.

Plant growth and reproduction

Perennial plants, rather than annual, characterize tundra vegetation. In the stark environment of North Greenland, for example, only one of nearly 100 plants is an annual. For many species in the Arctic flora, the growing season is too short for the yearly production of seeds, so many Tundra plants rely on asexual reproduction using rhizomes and other vegetative structures. Perennial plants also store energy in their underground tissues and, as another adaptation to the short growing season, can mobilize these stores for growth early the following spring (i.e., well before conditions are suitable for the resumption of photosynthesis). A surprising number of these perennial plants with well-developed root systems are herbaceous. Such plants survive using the simple strategy of dying back when the growing season ends, thereby avoiding exposure of their above-ground tissues to the ruthless winter. Woody plants, in contrast, remain exposed all year round and therefore must cope with winter's prolonged adversities. In some Arctic communities, 95% or more of the living plant tissues are underground, and the availability of stored energy so early in the year allows many plants to initiate their spring growth while still covered with snow.

Some tundra plants, including several grasses, have an additional adaptation: they develop pre-formed shoots and buds. These are initiated in the growing season the year prior to flowering, overwinter in a dormant state (usually protected by an insulating layer of dead leaves), and then quickly mature in the spring of the second year. As well as the obvious advantage of a quick response, this adaptation also allows plants to begin growth in years with favorable growing conditions and to complete their growth in the following years, even if conditions at that time are less than optimal. Tundra plants therefore respond rapidly when the long Arctic winter at last ends and the short growing season begins.

Although the summer is short, a few tundra plants do produce flowers and seeds. The seeds, however, may not germinate until much later when conditions favor the survival of seedlings (e.g., enough frost-free time to establish a root system). The seeds of tundra plants may have to retain their viability for considerable periods of time. Seeds of Arctic lupine recovered from permafrost in the Yukon Territory provide an indication of this persistence. The seeds were recovered from lemming burrows that had frozen suddenly some 10,000 years ago. A mining engineer who collected these materials kept them dry and at normal temperatures for another 12 years before they came to the attention of scientists at the National Museum of Canada. Impressed by the fresh-looking appearance of the seeds, the museum scientists placed several on wet filter paper, with the result that six germinated and produced normal plants. One of them bloomed 11 months later! The long-term viability of these and other seeds of tundra plants may explain some of the interesting patterns in the distribution of Arctic plants.

One species of buttercup enhances seed production by concentrating the weak Arctic sunshine toward the interior of its yellow flowers. The parabolic shape of the petals captures the rays, allowing the seeds to develop in a microhabitat several degrees warmer than the ambient temperature. The warmer environment also attracts insects, thereby ensuring pollination. In fact, the low temperatures in the Tundra Biome limits insect-induced pollination to only a few weeks each growing season, and this time may be even further limited to days with sunshine and temperatures warm enough for insects to be active (Infobox 2.2). On mountains high enough to support Alpine Tundra bees are common pollinators of flowering plants; in Arctic Tundra however, flies more commonly serve as pollinators.

Plants in temperate biomes often grow at rates that may produce large amounts of **biomass** (e.g., several thousand kilograms per hectare) each year, whereas willows growing north of the Arctic Circle produce only 30 kg ha^{-1} (62 lb/acre) in a year's time (i.e., only a few grams of new growth per plant per year). Such plants, despite their small size, may be exceptionally old. For example, Arctic willows reach more than 150 years of age and ground junipers often persist for 300 years, with an upper extreme of 544 years. The annual rings in the prostrate "trunks" of these "trees" are quite thin (0.1 mm), again reflecting the slow growth rate of most Arctic vegetation. Decumbent woody plants lack the same structural requirements of upright trees and can persevere in their modified form with only small amounts of new growth each year.

Decomposition and soil nutrients

Cold also retards the microbial activities necessary for **decomposition**, often resulting in a surface mat of plant litter sometimes characterized as "icebox mulch." Such slow decay means that there is a slow return of nutrients into the root zone of tundra vegetation. Indeed, the reduced presence and activities of bacterial and other microbial decomposers represent a weak link in tundra nutrient cycles. **Legumes** are uncommon in the tundra flora – a species of lupine being a notable exception – so tundra soils seldom benefit from **nitrogen fixation** in the mycorrhizae of leguminous plants. Minerals are also only available to plants in the active zone, further contributing to the overall low fertility of tundra soils.

Because soil nutrients are not abundantly available, sites on the tundra quickly respond to external sources of fertilization. A slowly decaying carcass of a single caribou or muskox may render an exceptionally green spot on the otherwise straw-colored tundra, and wastes concentrated at the entrances of dens often produce highly visible patches of green vegetation. Grizzly bears and Pacific salmon are considered linked **keystone species** in some Arctic **riparian** habitats because their activities move nutrients from the ocean to river margins. In some areas, up to 24% of riparian nitrogen budgets may be derived from half-eaten salmon carcasses and bear feces (see Chapter 10 for a fuller explanation). Such fertilization, because the cold greatly slows its incorporation into the soil, may show residual effects for some time after its sources are no longer at hand. For example, anthropologists sometimes locate ancient human habitations in the Arctic because these sites may still show lingering evidence of enrichment from food and other wastes.

Some animal adaptations

Various features and adaptations of animal – reproductive synchrony in migratory birds, for example – are covered in later sections, but a few others are mentioned here. Among these is the rapid growth of juveniles made possible in some species because the long period of daylight provides more time for feeding (i.e., a response triggered by **photoperiod**). Juvenile robins reared above the Arctic Circle, for example, leave their nests in just under an average of 9 days, whereas fledging usually requires 13 days elsewhere. The long day length allows adults to feed their young for more than 20 hours per day, thereby accelerating their growth rate.

For similar reasons, birds nesting in tundra generally lay more eggs per clutch than their counterparts do to the south. Yellow warblers, which have an extraordinarily large breeding range in North America, produce larger

Infobox 2.2 Of bumblebees and woollybears

Harsh Arctic winters exclude most invertebrates and several types of vertebrates from the Tundra Biome yet, surprisingly, 4 of about 250 species of bumblebees worldwide have adapted to short summers and extremely long, cold winters. Similarly, there exists a moth whose larva, the Arctic woollybear caterpillar, survives the rigors of not just one but several successive winters of subzero temperatures.

When air temperatures drop Arctic bumblebees generate body heat by "shivering," the tense contraction of their thorax muscles in a manner similar to isometric exercises. This tension, although barely visible, nonetheless creates heat as energy stores are metabolized. Arctic bumblebees are also covered with dense, fur-like protection that retards heat loss and helps maintain high abdominal temperatures. Some, in fact, maintain body temperatures of 20–30°C (69–86°F) *above* ambient and remain active when most other insects are incapable of movement. Arctic bumblebees are often active when air temperatures reach 10°C (50°F) or lower but, during sudden cold spells in spring, they may drop to the ground and remain dormant until the cold snap ends.

Arctic bumblebees usually nest on the ground (some are constructed within lemming burrows) at sites that remain dry even when the underlying strata are quite damp. A small entrance connects to a series of internal chambers constructed with a mud and wax mixture. Each chamber appears to have a specific purpose: a large chamber to house broods of eggs and larvae, and several smaller chambers to rear pupae or store honey or pollen reserves. Most Arctic bumblebee nests consist of a single queen and no more than 20–30 workers. After overwintering in their nests, queens leave in late May and begin foraging about two weeks before the workers emerge.

Two species of closely related and similar Arctic bumblebees, *Bombus polaris* and *B. hyperboreus*, have developed a curious relationship. The latter species evolved as a **kleptoparasite** (literally, "stealing parasite") that usurps the nest of the former in a manner somewhat analogous to the way brown-headed cowbirds lay their eggs in the nests of other birds. This happens when a female parasite bumblebee (*B. hyperboreus*) lays her eggs in the honey-rich nest cells of the pollen-collecting species (*B. polaris*), typically killing the parasitic queen in the process. This relationship, known ecologically as **Emery's rule**, occurs in other insects including about 29 mostly Eurasian species of so-called "cuckoo bumblebees" (like cowbirds, several kinds of cuckoos also parasitize the nests of other birds, hence the name).

The Arctic moth is endemic to Greenland and islands in Canada lying above the Arctic Circle (e.g., Ellsemere Island). The life cycle of typical moths follows an annual sequence in which eggs hatch (in spring) into larvae (caterpillars) which feed during the summer, then pupate (in fall and winter, inside cocoons), and concludes when the adults emerge and breed the following spring. In the Arctic moth, however, the life cycle extends to about seven years. After snowmelt, the newly hatched caterpillars – Arctic woollybears – spend only about three weeks feeding (primarily on the foliage of Arctic willows) until the end of June, then spin silken hibernacula (cocoons) in which they remain until the following spring when they emerge *still as caterpillars*. This process is repeated annually for six more years before adults finally emerge. Nearly half of the larvae share their hibernacula with another caterpillar. The hibernacula are anchored to the base of rocks, which absorb solar radiation and thereby may accelerate snowmelt and allow the early emergence of the woollybears in spring to begin foraging as soon as possible.

Interestingly, protection from subfreezing temperatures is not the primary function of the hibernacula. Instead, the hibernacula shield the caterpillars from two insect parasitoids (parasites that kill their hosts), in this case a type of wasp and type of fly, that together cause more than half of the mortality experienced by the larvae. The caterpillars, with timing honed by natural selection, begin spinning their hibernacula just before the parasitoids reach their peak of activity in July. Their development extends for several years because feeding is so curtailed by the threat of the parasitoids that several years are required for the caterpillars to accumulate enough energy to metamorphose into adults.

While providing adequate protection against parasitoids, the hibernacula is not sufficient to keep out the bitter cold of an Arctic winter. Instead, while inside their hibernacula, the caterpillars begin synthesizing cryoprotective compounds such as glycerol that act as antifreezes and enable the larvae to survive temperatures of –60°C (–76°F) or lower. Thus protected, winter mortality amounts to just 10–13% for a cohort during its seven-year development, far less than the 56% lost to parasitoids. All told, Arctic woollybears spend about 90% of their life frozen and just 5% feeding, with the balance spent unfrozen but inside their hibernacula in late summer.

clutches near Hudson Bay than anywhere else except Alaska. Moreover, the nests of yellow warblers in this area – a Tundra Forest **ecotone** – are larger, with a wall 54% thicker compared to those in southern Canada. The thicker insulation in these nests presumably provides better heat retention for protecting eggs and young from the harsh sub-Arctic climate.

Major vegetative communities

Plants in the heath family (Ericaceae) dominate much of the Tundra vegetation; Labrador tea is a representative species, as are several in the genus *Vaccinium* including bilberry and various species and relatives of blueberry. Laurel, bearberry, cassiope, and leatherleaf represent other heath vegetation. These low-growing shrubs, typically dwarfed by the rigors of the tundra environment, are also adapted to the generally acidic conditions prevailing in Arctic soils.

Communities of tundra vegetation, of course, may be divided and subdivided as much as those in any other biome. Nonetheless, the Arctic biota is somewhat less complex compared to more temperate locations; just three community types and one unique habitat association are briefly described below.

Shrub tundra

River terraces and other sites with well-drained soils, an appropriately deep active zone, and relatively abundant nutrients support stands of "tall shrub" tundra (i.e., about 2–5 m or 6–15 feet). The composition of this vegetation differs from site to site, but various species of birch, alder, and willow are pre-eminent. Regardless of the actual composition, this vegetation varies little in its thicket-like physiognomy. The understory is a diverse mixture of grasses and other herbaceous plants. Because of the food and cover resources they offer, sites with tall shrubs provide important habitat for several kinds of tundra wildlife. "Low shrub" tundra develops across other areas where heath mixes with small birches and willows to form ground cover about 50 cm (20 inches) high. Such vegetation typifies the seemingly endless "barren grounds" of the Tundra Biome.

Dwarf birch heath

Heath dominates this community, which includes both evergreen and deciduous-leaved species. In autumn, this community is a carpet richly hued with green, copper, and golden colors. Dwarfed birches are scattered throughout the heath, as are mosses, lichens, and dwarfed willows. Dwarf birch heath develops in relatively small patches and does not cover large tracts of landscape as other tundra communities do.

Cottongrass heath

One of the more common tundra communities, cottongrass heath, blankets large areas. This community is composed of heath generously mixed with cottongrass – actually a fluff-tipped sedge – numerous grasses, and other herbaceous vegetation. Because **graminoid** vegetation is so prominent in this community, some ecologists describe this as a "tussock heath." A complex of cottongrass and dwarfed heath covers the majority of Tundra in northwestern Alaska, where the flowers of such herbaceous plants as northern shooting star, buttercup Arctic lupine, and saxifrages add touches of color to the somewhat somber aspect of the vegetation. Mosses and lichens are well represented in the ground cover. Extensive meadows in which the vegetation is herbaceous and heath is rare or absent occupy wetter sites. Cottongrass tussocks, whose white tufts are a familiar summertime sight on the tundra plains, may rather uniformly vegetate large areas.

Fellfields

A sparse community of plants ekes out an existence atop the rocky, windswept ridges jutting from the tundra plain. Such exposed sites are known as fellfields, and their extreme aridity severely limits the development of vegetation. Only a few hardy plants, notably dryas, gain a foothold in the almost non-existent soils at these harsh locations. Dryas leaves are leathery, and the plants trail close to the ground. Rock surfaces are color-spotted with various crustose **lichens** and another lichen, *Alectoria*, may form black mats on the inhospitable terrain. Fellfields are fine examples of Arctic deserts and, as components of the landscape, they increase in frequency on a northward gradient (i.e., into the High Arctic). Eventually, in places such as Ellesmere and other islands in the Queen Elizabeth archipelago, nearly all plant cover disappears. The rocky surface, essentially a continuous fellfield, dominates the landscape. Lichens persist, however, and these, some mosses, and a few diminutive herbaceous species survive in protected microhabitats scattered across the severe environment of the Polar Desert.

Invertebrates and tundra ecology

The widespread occurrence of shallow lakes and wetlands during the summer creates ideal conditions for certain kinds of insects. Mosquitoes are particularly abundant and Arctic lore is replete with legends of caribou stampedes caused by clouds of those biting insects. Such stories may be somewhat exaggerated but, on a windless summer day, the mosquito hordes in search of a blood meal can certainly make life miserable for caribou and other tundra animals. Early in the century, the intrepid naturalist Ernest Thompson Seaton (1860–1946) developed a highly personalized means of determining the relative abundance of mosquitoes in the far north. Seaton simply counted the number of mosquitoes biting the back of his bare hand during a 5-second sampling period. Based on this index, mosquitoes increased northward from a beginning point of 5–10, then 15–25, and later 50–60 at Great Slave Lake, finally peaking at 100–125 in the "Arctic prairies," his name for Tundra. The mosquitoes at this point were so numerous that Seaton could complete his census "only by killing them and counting the corpses."

Mosquitoes pass the harsh winter in a desiccant-resistant egg stage, hatching as wriggling larvae only when warmer weather and moisture return the following spring. Mosquitoes, blackflies (vicious biting insects), and other flies (order Diptera) comprise about 50% of the insect taxa in the Tundra and beetles (order Coleoptera) only 10%. An even higher percentage

(95%) of dipterans in the Tundra's insect fauna exists at Barrow on the northern coast of Alaska. For comparison, dipterans form only about 20% of the insect taxa in temperate biomes, where beetles represent about 50% of the taxa.

Two other types of flies, both parasitic, are noteworthy in relation to the ecology of caribou: the larval stage of warble flies encapsulates just under the skin of caribou, and nose bots develop as larvae in the nasal passages of their hosts. Neither of these dipterans fly in swarms, but just a few adults of either group can panic caribou. Warble flies and nose bots follow the herds across the Tundra and some authorities believe that these flies, rather than mosquitoes, are the basis of stories about caribou stampedes. Warble flies are notable for another reason: Inuit and other native peoples regard their larvae as delicacies, and their children sought them as candy whenever caribou were skinned.

The dipteran fauna provides a crucial prey base in summer for the numerous species of migrant shorebirds nesting in tundra habitats. In fact, the breeding chronology of shorebirds corresponds with the seasonal peak in the tundra insect population – young birds hatch when an abundant food supply is readily available. This synchronization may also be beneficial to the insects, whose numbers "swamp" their avian predators. Because the various insect species emerge largely at the same time, the birds consume only a small fraction of the total insect population and the odds are lessened of any one individual falling victim to a hungry bird. It is therefore advantageous for insects of all kinds to emerge essentially at the same time, thereby producing the oft-described "clouds of insects" for which the Tundra is so infamous.

Butterflies (order Lepidoptera) are also a part of the tundra insect fauna. Their most striking feature is the long life of the larvae, which persist for two or more seasons before continuing their metamorphosis. The adults of many species are **melanistic**, presumably because the darker coloration improves the absorption of radiant heat (some experiments suggest that the relationship is not always strong). Melanism increases from south to north in various lepidopterans and also occurs in bumblebees and other kinds of tundra insects. The eyes of some moths that belong to night-flying taxa are modified to cope with the Tundra's long periods of summertime daylight.

Most kinds of plant-eating insects are rare in Arctic communities and the seasonally thawed soils above permafrost are not conducive to underground nest construction. Perhaps no more than four species of ants regularly endure tundra conditions. Dragonflies (order Odonata) are absent from Tundra, as are wood-boring insects. Overall, the tundra insect fauna is only 1–5% as rich in species when compared to a corresponding area at temperate latitudes.

Selected tundra mammals

The natural richness of organisms in a given region diminishes toward the poles. Although biodiversity is lower in the Arctic Tundra than elsewhere, each species represents a larger percentage of the total and some exert major influences on tundra ecology. The mammalian fauna of the Tundra consists of 48 species including shrews, rodents, and hares. Small herds of muskox and much larger herds of barren-ground caribou (known as reindeer in Eurasia) occur in parts of the Arctic. The major mammalian predators include polar and grizzly bears, wolverines, wolves and Arctic foxes. A few selected mammals are profiled in the following sections.

Lemmings

The rodent fauna of North America includes 12 species of lemmings. The ranges of most of these are northern in their distribution, but the ranges of a few lemmings extend southward to the middle latitudes. Three species are clearly associated with Arctic Tundra, and these lemmings are at the hub of tundra food webs. These hamster-sized rodents are staples in the diets of snowy owls, Arctic foxes, and other tundra predators.

The North American brown lemming, which retains its namesake color all year round, has an extensive range west of Hudson Bay, crossing the breadth of the Northwest and Yukon territories and all but the southern edge of Alaska. Collared lemmings, a complex of nine closely-related species, exchange their dark, summertime pelage for white in the winter. Of these, the northern collared lemming occupies the northern fringe of the continent west of Hudson Bay and some Arctic islands (including Greenland). Finally, the Ungava collared lemming exists in a similar habitat, but only east of Hudson Bay on the Ungava Peninsula.

That two similar species of collared lemmings have separate ranges in the far north – one on either side of Hudson Bay – offers an opportunity to examine the interworkings of **zoogeography** and speciation in two scenarios. In the first Hudson Bay provided a physical barrier, resulting in the postglacial isolation of a common stock. A single species of collared lemming sharing a common gene pool lived along the leading edge of the last continental glacier, where their genes mixed across a broad front. Later, when the glacier and the belt of Arctic Tundra at its southern edge retreated northward, the population of lemmings was split by Hudson Bay into eastern and western segments, and genes could no longer be exchanged. Thereafter, the two isolated populations gradually developed separate genetic features and full species status. Biologists apply the term **allopatric speciation** to situations of this type, that is, the formation of species at separate locations because a physical barrier isolates components of a once-shared gene pool.

A second scenario, supported by fossil evidence, may also explain the current distribution of the two species of collared lemmings, however. Fossil specimens associated with the Ungava collared lemming were discovered in Pennsylvania, which indicate that this species was already established when its population was pushed south ahead of the advancing glacier. When the glacier retreated, this species followed suit and moved northward. Meanwhile, northern collared lemmings persisted in the glacial refugium in Alaska when the Wisconsin glacier was at its maximum. When the glacier retreated, northern collared lemmings spread eastward from the refugium into their present range, displacing the western elements of the Ungava collared lemming in the process until Hudson Bay intervened. Irrespective of which scenario actually prevailed, we see how physical barriers – in this case, a continental glacier and a large body of water – influenced the past and present distributions of organisms.

Lemmings do not hibernate. They spend the winter scurrying for food beneath the relatively warm blanket of snow. However, lemmings occasionally venture to the surface, where they become vulnerable to snowy owls and other predators. Collared lemmings are particularly adept at tunneling through snow and, in winter, develop enlarged claws on the third and fourth digits on their front feet. These specialized claws, which are shed each spring, are used to comb snow and ice crusts from their fur.

In spring, the thaw exposes networks of well-worn trails in the cropped vegetation. The fluffy nests of lemmings provide a microhabitat for various kinds of arthropods (e.g., mites and insects known as springtails). Because of this association, these small arthropods are more numerous in Tundra where lemming nests are commonplace. When lemming population density is high, however, brown lemmings may strip enough vegetation to effect decreases in arthropod populations.

Rapid increases in the abundance of North America brown lemmings occur every 3–4 years. This "lemming cycle" in turn affects the populations of their major predators. Arctic foxes and snowy owls experience limited reproduction (i.e., reduced pup survival and smaller clutches, respectively) when lemming numbers are at their low ebbs. Explicit causes for the lemming cycle are difficult to determine, but food shortages and the consequences of frenzied dispersal and predation seem to be involved. In some years, female brown lemmings may raise five or more litters, and each litter may include up to 11 young. The female offspring in these litters are capable of breeding about 30 days later in some summers. Within a year, one pair of lemmings may produce thousands of descendants, thereby contributing to the sudden explosion of the overall population. Populations of collared lemmings generally fluctuate much less than those of brown lemmings.

The amplitude of these fluctuations varies considerably from peak to peak, at times registering 25- to 50-fold increases and 400- to 1000-fold in others. When lemming densities are near maximum levels, the summer population faces depleted food supplies and restless overcrowding. The hyperactive lemmings disperse, sometimes drowning when they attempt to cross large water barriers, but more often they succumb to starvation, predation, and, perhaps, disease. Populations are reduced to a low ebb – perhaps no more than one animal per 2.5 ha (6 acres) – before the population again begins to rebuild.

Among the many fables of Arctic lore are stories about mass suicides of lemmings. According to the legend, when the lemming populations peak every 3–4 years untold legions march into the sea, sacrificing themselves for the welfare of the remainder. The legend is myth of course, as such behavior cannot be transmitted genetically to the following generation (a necessity for such a trait to persist). In reality, lemmings disperse in response to the pressures of increased density, and some simply drown when encountering large bodies of water. Lemmings are able swimmers and can successfully cope with smaller ponds and streams.

Arctic ground squirrels

Moles and other strictly fossorial mammals are absent from the Tundra Biome, but a few species dig burrows where the permafrost lies deep enough to permit excavation. Among these is the Arctic ground squirrel, the northernmost and largest of the several species of ground squirrels in North America. These animals select sites where drainage keeps the spring snowmelt from flooding their burrows. Sandy areas along stream banks are attractive, as are eskers, but Arctic ground squirrels also den on rocky hillsides and other well-drained locations. They dig where the snow will accumulate, as the drifts provide insulation and keep the temperature from dropping below −12°C (10°F) in their underground chambers. Their extensive maze of tunnels may extend for 20 m (66 feet) or more and sometimes have several levels. Within the tunnel system, Arctic ground squirrels excavate a **hibernaculum**, a chamber that is lined with dry sedges and grasses that they "weave" into a thick, hollow ball-like nest. As might be expected, the squirrels respond to the long, cold tundra winter with an equally long period of hibernation. After plugging the entrance to the hibernaculum with soil, they snuggle into the thermally insulated nest and lie dormant for up to 9 months. Other species of ground squirrels living in more temperate climates have much shorter periods of hibernation, or do not hibernate at all.

Like most rodents, Arctic ground squirrels are important links in food chains. In Tundra, their predators include golden eagles and other raptors, Arctic foxes, wolves, and grizzly bears. The latter, as noted by Samuel Hearne, are forceful enough to disrupt the coarse soils in "long and deep furrows… like plowed ground" when searching for ground squirrels. Grizzly bears may excavate so much soil, especially on slopes, that they bury nearby vegetation creating bare areas on the tundra. Some hungry bears move boulders weighing up to 55 kg (120 pounds) in their persistent excavations for ground squirrels.

Arctic foxes

Few animals in North America have been more prized for their fur than the white fox of the far north. Beavers were certainly the stock-and-trade for the fur industry, but the prime pelts from Arctic foxes brought even greater rewards to doughty trappers who bent into the tundra wind.

Arctic foxes are a polymorphic species. While most adults have white winter pelage, a small percentage of others are known as "blue foxes" because of their darker, slate-colored fur. The blue-phase **morphs** represent a relatively rare combination of recessive genes. Because the blue fox occurs so infrequently, its pelt commanded a higher price on the fur market than its white counterpart, an economic situation not overlooked by trappers in Alaska. In the 1920s, as a means of producing a regular supply of highly valuable pelts, blue foxes were released on fox-free islands in the Aleutian chain. Inbreeding perpetuated the desired combination of recessive genes and produced populations of blue foxes. Unfortunately, however, Aleutian Canada geese nested on many of these same islands where there had been no mammalian predators to reckon with until the trappers introduced blue foxes. Nest success dropped in the goose colonies as a result of the heavy predation, and Aleutian Canada geese were eventually added to the list of endangered species. Fewer than 800 birds remained by 1975 but, with management, their numbers slowly increased. They were removed from the federal list of threatened and endangered species in 2001 after the population had rebounded to more than 37,000 geese.

In winter, Arctic foxes patrol coastal areas for food or venture well out onto pack ice far from land. In these settings, Arctic foxes are closely tied to the marine environment. Their winter diet consists largely of carrion washed up in the shoreline drift and leftovers from polar bear kills. In summer, however, Arctic foxes move inland where they can hunt for ground squirrels, lemmings, and bird eggs. On small coastal islands, eider ducks delay nesting until the surrounding ice melts, after which Arctic foxes cannot visit the islands and destroy the nests.

The interplay between the abundance of lemmings and Arctic fox populations is dynamic – both species increase and then ebb in what appears to be regular cycles. When and where lemming numbers peak litters of Arctic foxes include as many as 20 pups, but litter size drops to 10 or fewer in years when lemmings are scarce. Additionally, pups may experience significant first-year mortality in years when lemmings are in short supply, and fewer than 10% reach 2 years of age. Conversely, a larger percentage of fox pups survive when lemmings are plentiful. Interestingly, on the coastline of Greenland where Arctic foxes feed less extensively on lemmings, reproductive success as measured by litter size varies little from year-to-year.

Because permafrost must lie deep enough to permit excavation, and the unfrozen soils must be of a consistency suitable for tunneling (i.e., to avoid cave-ins), relatively few locations are suitable as fox dens. Such limitations require Arctic foxes to use the same dens generation after generation. In fact, the average "life" of fox dens in Canada's Northwest Territories extends for 330 years (similarly, radiocarbon dating of muskox bone from the den of an Arctic wolf indicated an age of 233 years).

Barren-ground caribou

Barren-ground caribou persist as the premier example of a migratory land mammal in North America. Vast herds of caribou once migrated each year between their wintering areas and their northern calving grounds, completing an age-old procession not ignored by wolf packs or spear-wielding humans. Today, the herds are much reduced from their former numbers, but many thousands of caribou still undertake their restless migrations (Fig. 2.5). Some herds follow ancient trails atop eskers, but networks of historic paths also cross open terrain. According to one estimate, the combined length of these trails in the Northwest Territories of Canada exceeds the total length of railroad tracks in the United States. The extent of these movements vary from herd to herd, but many caribou migrate a one-way distance of at least 400–500 km (250–310 miles), and some may travel twice as far. Some herds at times abandon their previous travel lanes and take other routes. Food shortages resulting from years of grazing and forest fires within their winter range may alter long-used pathways.

For the most part, barren-ground caribou overwinter within the northern edges of the Boreal Forest (Chapter 3), although some forego the shelter of the forest and wander about on open tundra at least during part of the winter. Caribou mate in late October and early November, about the time that they reach the southern edge of tundra and just before entering the forests for the winter. The cows begin to move northward to their calving grounds during March and April, reaching them in June and July. The bulls follow soon afterward. The return trip to the forest begins in August, this time with

Figure 2.5 Tundra serves as calving grounds for large herds of barren-ground caribou, including the Porcupine caribou herd shown here. Bands of female caribou in this herd depart their wintering grounds in Yukon Territory, followed several days later by the bulls. Much of the calving habitat is protected within the Arctic National Wildlife Refuge. Photograph courtesy of the US Fish and Wildlife Service.

the bulls leaving first. River crossings represent the greatest natural hazard for migrating caribou, and when the rivers are flood-swollen in spring several hundred drownings may occur. Strong swimmers, caribou are aided by their wide hooves which also serve in season as snowshoes and as effective tools for pawing through snow for ground-dwelling lichens.

One of the larger caribou migrations involves the "Porcupine herd," so-called because its route crosses the Porcupine River. In spring, this herd travels north from the Yukon Territory and into adjacent Alaska to reach calving grounds on tundra bordering the Beaufort Sea. Most of the herd's calving grounds lie within the Arctic National Wildlife Refuge, which was established to protect habitat vital to the Porcupine herd. Because of large oil reserves beneath the refuge however, political pressures for the development of these reserves are high and the future of the Porcupine herd may swing in the balance.

Selected tundra birds

Many North American migratory birds seek nesting habitat in the Tundra Biome. In fact, more than 100 species breed in the Tundra. Sandpipers and other kinds of shorebirds are well represented in this group. One particularly long-distance migrant that breeds in the Tundra is the lesser golden plover, a species wintering in South America. Among waterfowl, long-tailed ducks,

king and common eiders, lesser snow geese, white-fronted geese, and some smaller races of Canada goose nest in tundra as does the namesake species, Tundra swan. Only a few hearty birds, such as the rock ptarmigan and common raven, remain to endure the long winter period.

Several kinds of avian predators, including seabirds known as skuas and jaegers, occur in the Tundra Biome, where they prey in season on lemmings and on the eggs and young of geese and other birds. Other predators, more typical of groups occurring in other biomes, include hawks and owls. Two of these, the gyrfalcon and snowy owl, are noteworthy because they typically remain all year round in the Arctic; peregrine falcons, merlins, and rough-legged hawks leave the tundra after nesting however, and migrate far southward for the winter months.

Gyrfalcons

Partial migrants, gyrfalcons that nest at the extreme northern edge of the Arctic (i.e., north of about 70°N latitude) return southward for the winter, although they still remain in the Tundra. Others remain in their breeding areas as more-or-less year-round residents.

The gyrfalcon, largest of the world's falcons, is a circumpolar species. The males and females are similar in plumage, but like many other species of raptors, gyrfalcons exhibit a reversal of the usual pattern of **sexual dimorphism** in size: males are about 65% of the size of females in body weight. Gyrfalcons are a polymorphic species, with white, grey, and black morphs. This variation in plumage color once led to the designation of subspecies for each group (as it also did once for lesser snow geese), but the species is no longer subdivided on the basis of coloration.

The principle food of gyrfalcons is ptarmigans, and their breeding range overlaps remarkably with the distribution of rock ptarmigan. Ducks, shorebirds, ground squirrels, and rabbits are also regular items in gyrfalcon diets. Their deep and powerful wingbeats belie the exceptional swiftness of hunting gyrfalcons and they are also adept on the ground, at least for a falcon. They occasionally alight and run short distances after their prey.

Gyrfalcon nests are often no more than a shallow scrape on a cliff or other outcropping of rock, but many use stick nests originally constructed by ravens or other large birds. Some nests may be used generation after generation, which steadily builds deep accumulations of bones and debris at such sites. Because reproduction demands extra energy gyrfalcons may not breed every season, skipping years when prey is scarce.

Females may cache food during the breeding season, often storing the food behind vegetation within 100 m (110 yards) of their nest sites. The cached food is later retrieved and fed to the chicks. Like other raptors, gyrfalcons regurgitate ("cast") undigested material – feathers, fur, and bones – in well-formed pellets. These pellets, when carefully teased apart, offer biologists a means of determining the kinds of foods raptors obtain and the frequency with which prey are consumed. For example, lemming remains occurring in 15–20 gyrfalcon pellets represent a 75% **frequency of occurrence**, a standard statistic in food habits studies.

Old World monarchs highly prized gyrfalcons as hunters of large prey. Today, the status of gyrfalcons seems secure. Most live in remote regions and suffer few disturbances from humans. Gyrfalcons escaped the misfortunes of egg-shell thinning from pesticide contamination and the resulting population declines experienced by some other species of falcons.

Snowy owl

Another circumpolar raptor, the snowy owl – the world's only white owl – breeds throughout the full range of tundra environments, from the edges of polar waters southward to the beginnings of forest vegetation. In winter, snowy owls move nomadically southward. Some are observed regularly in the northern Great Plains (e.g., Montana and the Dakotas), but less so elsewhere in the northern half of the United States. Their unpredictable winter movements are apparently in response to the changing year-to-year abundance of lemmings, their primary prey, although the relationship may not be as synchronous as once believed. Snowy owls often appear in both southern Canada and the northern United States

about once every 4 years, which suggests a connection with the cycle of lemming abundance. Because of the long days in the Arctic summer and the equally long nights of northern winters, snowy owls hunt as much in daylight as in darkness. Their physical adaptations include the protection afforded by thick plumage on the legs and feet, and long facial plumage also obscures all but the very tip of the bill.

Nests are often situated on a prominence, where the wind keeps the site snow-free and where flooding is unlikely. Such locations also afford the birds a commanding view of the surrounding area. The nest itself is only a scrape, scratched out by the female on either turf or bare soil and lacking an insulating layer of down or vegetation.

The breeding biology of snowy owls, as well as their winter movements, seems closely tied to the abundance of their prey. In this regard, snowy owls demonstrate two effective adaptations. First, the clutch size of snowy owls varies from year to year: 3–5 eggs are laid when food is limited, but much larger clutches (7–11 eggs) are produced when lemmings or other prey are abundant. Nesting may not be initiated at all when food is extremely scarce. Second, and no matter whether the clutch is large or small, the eggs of snowy owls are laid at 2-day intervals, but incubation begins with the appearance of the first egg. As a result, the clutch hatches with the same asynchrony as it was produced and the young are of staggered ages. The older, larger birds therefore have a competitive advantage over their younger siblings when food is limited. The youngest birds typically starve, but this ruthlessness ensures that at least part of the brood survives; all might starve if the limited food supply was distributed equally. These adaptations, which also occur in other species of raptors, seem pronounced in the short food chains of the Arctic, where snowy owls depend so heavily on the "boom or bust" lemming population for food.

Ross goose

Harsh environments require special adaptations to ensure survival. Earlier, we outlined the synchronization and its apparent adaptive value between the peak emergence of insects and the hatching of shorebirds. The synchronized nesting behavior of the Ross goose is another example of adaptation to the short Arctic summer that provides only a brief window of opportunity for successfully rearing young.

Ross geese form pairs and copulate *before* reaching their tundra nesting grounds, so the birds spend no time in courtship after they arrive. They are physiologically ready to begin laying eggs almost immediately (Fig. 2.6). Strictly speaking, therefore, the Tundra is a nesting area for these geese and not a breeding area. Nest construction, completed in a single morning, begins just 3 days after the birds arrive, although capricious weather conditions may cause delays for a few

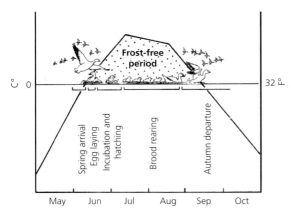

Figure 2.6 Close fit between the nesting activities of Ross geese and the frost-free period in the Perry River region. Northwest Territories, Canada. Egg laying and other phases in the nesting cycle take place during the period when the mean temperature is above freezing. Illustrated by Tamara R. Sayre, based on Ryder (1967).

more days. Clutches of 3–4 eggs are laid within 8–9 days, again with much synchronization throughout the colony. Incubation begins immediately and, about 22 days later, the goslings hatch.

All phases of nesting proceed almost simultaneously throughout the entire colony, thereby ensuring optimum use of the short period available for producing offspring successfully. The short period of favorable weather acts as a strong **selection pressure** affecting the reproductive activities of virtually all Arctic plants and animals, but such an adaptation to brief Arctic summer at times may backfire. Because of their synchronous state of development, for example, all nests in a goose colony are equally vulnerable to a midsummer storm. Should disaster strike, there is no time to start again because the reproductive systems of the adult birds regressed shortly after they arrived. In any case, goslings hatching from a second nest later in the summer could not mature and migrate before freezing weather returned. Consequently, populations of Ross geese and Atlantic brant frequently undergo "boom or bust," increasing when conditions during the "window" are favorable but failing when they are not.

Highlights

Absentees: amphibians and reptiles

Because they are ectotherms, amphibians and reptiles are essentially absent from the Tundra Biome; the climate is simply too cold for the existence of these groups. Except for a single species of frog, the Tundra lacks salamanders, toads, snakes, lizards, or turtles.

The distribution of the wood frog, which is distinctively marked with a dark patch behind the eye but otherwise variably colored, extends farther north than any other amphibian or reptile in North America. The northern edge of their extensive distribution includes tundra ponds, but some subspecies also range as far south as northern Georgia. In the northern populations the rear legs are proportionally shorter and the wood frogs are toad-like in appearance and hopping abilities.

In his 18th century chronicle of tundra exploration, Samuel Hearne noted that wood frogs survive the winter in a frozen state but, once thawed, cannot withstand freezing again in the same season. While seemingly impossible, animal physiologists later confirmed Hearne's keen observation: wood frogs do freeze in winter. When hibernating at subzero temperatures, up to two-thirds of the water in a wood frog's body may turn to ice for periods of several weeks. Normally, freezing kills or badly damages animal tissues when metabolic processes stop and ice crystals rupture the cell's organelles and membranes. To survive in the Tundra's cold environment wood frogs supply their cells with a cryoprotectant, a natural antifreeze of glucose. The production of cryoprotectant is triggered by the formation of extracellular ice. Liver glycogen synthesizes glucose (a sugar) and, as winter wears on, the reserves of glycogen in the hibernating frog are steadily diminished. Once thawed a frog cannot protect itself again – just as Hearne observed – until it rebuilds its reserves of glycogen while feeding during the summer. Because of this remarkable adaptation to cold, wood frogs are the only North American amphibian found north of the Arctic Circle.

Lichens and "reindeer moss"

Existing as a partnership between algae (or sometimes cyanobacteria) and fungi, lichens are flowerless plant-like organisms found in terrestrial environments ranging from deserts to rain forests, and are also prominent in the Tundra. The two types of organisms live together as a single unit, each providing essential materials to the other. The photosynthetic partner manufactures food as do other green plants and the fungi contributes water, carbon dioxide, minerals, and the structural surface on which the algae exists. This relationship is an example of **mutualism**, one type of a larger category of interactions known as **symbiosis**. Scientists classify these unusual organisms based on features of the fungal component, but the approximately 25,000 species of lichens are not easily identified to species.

Some lichens grow directly on the soil, but the surfaces of rocks and trees are habitats for many others. Lichens lack roots, stems, and leaves, but botanists use physical appearance to separate them into three groups. Foliose lichens resemble a small clump of leaves, and much of each leaf-like structure is attached to the underlying substrate. Fruticose lichens look like miniature shrubs, with

Figure 2.7 A small clump of reindeer moss, a fruticose lichen, grows among other tundra plants near Churchill, Manitoba. Note the crustose lichens appearing as spots on the rock in the background. Photograph courtesy of Elizabeth D. Bolen.

a cluster of upright stalks attached to their substrate by only a small part of their underside. Crustose lichens lie flat against their substrate, appearing not unlike roughened splashes of paint. Lichens that are exposed to high light intensity are often colorful, adding bright touches of red or yellow to the tundra vegetation, while others are gray-green, brown, or black. A bright, orange lichen grows at sites enriched by the droppings of perching birds.

Lichens are particularly adapted to cope with aridity and may lie dormant for long periods. When water becomes available – dew is a common source – it moves quickly from the fungus into the photosynthetic chemistry within the algal cells without having to pass through a longer, less direct system of soil, roots, and stems. Most lichens grow slowly, especially in harsh environments. The radius of some crustose lichens increases by no more than 0.3 mm (0.01 inches) per year, and some of the slower-growing species may be as much as 4000 years old.

Lichens are crucial in tundra ecology. They cover much of the soil surface, preventing the thawing and erosion of the underlying permafrost. Several species also are important links in tundra food chains. Among these are lichens known as "reindeer moss," which are

not true mosses (Fig. 2.7). Reindeer moss is the primary winter food of caribou, the North American counterpart of the Eurasian reindeer.

Snow goose "eat outs"

Cold temperatures and the short growing seasons at high latitudes profoundly influence soil conditions and vegetation. Most plants are perennials and the majority of these reproduce asexually. Seed-eating birds are uncommon, and the several kinds of migrant birds nesting in the Tundra Biome instead feed heavily on the summer flush of insects.

An interesting exception concerns the diet of lesser snow geese. Populations of the birds boomed in recent years as a result of increased yields of corn and other grain crops on farms where snow geese migrate and overwinter. The additional food supplies provide an "agricultural subsidy" that significantly increases survival rates at those times of the years when food would otherwise check population growth. Today, nearly 3 million snow geese breed on the western shore of Hudson Bay, where the carrying capacity remains much less than on farmlands. The current breeding population greatly exceeds its food supply.

After their spring arrival and before renewal of aboveground growth, the geese grub for the energy-rich roots and rhizomes of tundra plants, placing considerable pressure on the vegetation near their breeding grounds. A single goose may strip the roots and rhizomes from 1 m² (1.2 square yard) of meadow per hour of feeding time. The removal of this amount of plant material is significant because the rhizomes (rather than seeds) provide the means of vegetative regeneration each year. Large areas of peat are exposed in what biologists call "eat outs", sites where animals completely denude the vegetation. The peat surface at these locations soon dries, and mosses replace the grubbed-out tundra meadows. In some breeding areas foraging snow geese have completely eliminated coastal vegetation, which in turn has drastically altered the soils to the extent that desert-like areas can be detected with LandSat imagery. Unfortunately, the present pattern of habitat alteration in these areas of tundra is likely to continue for many years and, even if the snow goose population is significantly reduced, decades will pass before the disrupted soils and vegetation can be restored.

In contrast, another tundra community thrives under grazing pressure. Later in the breeding season, snow geese graze heavily on the new growth of sedges along the Hudson Bay shoreline where the aptly named "goose grass" forms turf communities in the high-tide zone. The result is a "grazing lawn" (i.e., closely cropped vegetation). At these locations, however, only above-ground production is removed. This actually stimulates further growth and maintains the structure of the wetland community. In fact, when experiments using wire frames known as exclosures prevent the geese from feeding in these wetlands, growth rates of the plants rapidly decline and the composition of the community changes to other species.

Alpine Tundra

Tundra communities also develop atop the taller mountains in North America well south of Arctic latitudes (e.g., peaks in the Rocky Mountains and the Sierra Nevada). With appropriate environment, Tundra also develops on mountains not usually regarded as alpine in nature (e.g., location in New Hampshire). The physiognomy of Alpine Tundra closely resembles that of Arctic Tundra, but there are physical differences between the two environments (Table 2.1).

Permafrost and patterned ground are absent from or rare in Alpine Tundra where water normally drains downslope, precluding formation of a permanent layer of subsurface ice. Solar radiation is greater in Alpine Tundra, and mountaintop soils are generally warmer than those at high latitudes. The fauna of Alpine Tundra includes fossorial species of which marmots, ground squirrels, pikas, and pocket gophers are representative groups. Similarly, the root zone available to alpine vegetation is not limited to a small area near the top of the soil profile. However, both Alpine and Arctic Tundra erode easily when their protective cover is damaged.

Alpine communities are not dominated by the heath family, and neither are willow or birch thickets prominent (although dwarfed species of both occur in alpine habitats). The alpine flora includes a variety of seed-producing plants with colorful flowers. Insect life is more diverse and includes more pollinators, although some species have reduced wings (i.e., strong winds at high elevation act via **natural selection** against long-winged insects). The flora on rock fields and other exposed sites includes several herbaceous species resembling cushions. Because of their low streamlined profile, cushion plants offer little resistance to the strong winds. This conserves moisture while still exposing large areas of leaf surface to sunlight. Because of their compact hemispheres of dense branches, temperatures within the cushions are several degrees warmer than those at the outer edges.

The scattered trees persisting at the timber line (i.e., the montane analog to the tree line in Arctic Tundra) on peaks with Alpine Tundra are relentlessly tortured by high winds. Trees – typically conifers – in these locations become twisted and stunted and generally survive only in the partial protection of a rocky outcrop, boulder, or other microhabitat. This ground-hugging vegetation shaped by the fierce, cold wind, is known as **krummholz**, or "crooked wood." With its unique appearance, krummholz indicates the beginning of Alpine Tundra and the terminus of those communities whose physiognomy typifies forest vegetation (Fig. 2.8). Strong winds also produce "flag trees" on exposed sites at high elevations, as well as some sites in the Arctic Tundra. Branches emerging on the exposed side of these trees quickly wither, leaving the windward side of their trunks barren, but those emerging from the opposite side extend straight out from the trunk like flags in a stiff wind.

Alpine Tundra in the Rocky Mountains as far south as northern New Mexico is home for white-tailed ptarmigan, whereas the ranges of two counterparts – rock and willow ptarmigans – lie solely in Arctic regions. All three species are similar in appearance and assume white plumage in the winter. They have similar food habits, with each species depending heavily on the buds of willows and other low-growing shrubs. Alpine Tundra, however, lacks large herds of grazing animals equivalent to barren-ground caribou, or even the relatively small herds of muskox, but in summer mountain goats scatter about rocky sites above the timber line in the northern Rockies.

Table 2.1 Comparison of physical features between Arctic and Alpine Tundra in North America. Adapted from various sources.

Feature	Arctic Tundra	Alpine Tundra
Photoperiod	Seasonal extremes greater than in other biomes	Not different from other biomes at same latitude
Surface drainage	Poor	Rapid
Permafrost	Typical	Rare or absent
Patterned ground	Common	Rare or absent
Precipitation, especially snowfall	Low	High
Winds	Moderate	High to extreme
Atmospheric oxygen	Normal	Low
Atmospheric pressure	Normal	Low[1]

[1]At high elevations, water will boil at temperatures well below 100°C (212°F).

Figure 2.8 Krummholz ("crooked wood") forms when buds and twigs projecting above the protective cover of snow are killed and the plant subsequently becomes twisted and gnarled as it grows laterally instead of vertically; those shown here are primarily dwarfed Engelmann spruce. Krummholz, which can form in other ways, prevails as a distinctive feature of timber line zones, especially in Alpine Tundra. Photograph courtesy of Brian R. Chapman.

Fragile Tundra

Despite the rigorous environments in which they develop, tundra communities are often described as fragile. Why? In the first place, there are few species in the food chains, so biological materials and energy are transferred rapidly from producers to consumers. With such a short "fuse," so to speak, elimination, reduction, or alteration of any link in the tundra's food web produces significant repercussions throughout the system. For example, when the nuclear plant at Chernobyl exploded in 1986, lichens in the Tundra of Lapland became "radiation sponges" for the fallout of cesium 137. The reindeer, which had no other source of food, fed on the radioactive lichens. Some Laplanders destroyed their reindeer, but others gambled with leukemia and other forms of cancer by consuming meat and milk from contaminated reindeer herds. Unfortunately, the half-life of cesium 137 is about 30 years, and the threat of radioactively contaminated food chains resulting from the Chernobyl disaster may extend for many decades. Lichens in the Alaskan tundra similarly accumulated strontium 90 and cesium 137 from nuclear testing in the 1940s and 1950s. The Inuit were thereby exposed to the effects of radiation, even though the bombs were exploded as far away as the South Pacific.

Second, tundra vegetation evolved almost entirely without the influence of humans. The present distribution of Arctic Tundra in North America developed at about the same time that humans ventured across the Beringian land bridge from Asia. Since early humans could not sustain an agricultural economy in the tundra ecosystem, they were hunter-gatherers. Many lived in association with the somewhat richer biotic resources of coastal areas, and ventured inland only seasonally (e.g., at the time of caribou migrations). Human populations remained at low densities under the constraints of such a regime. In modern times, however, industrial and recreational technology changed the scenario rather drastically. Technological advances allow the exploitation of mineral and biological resources once protected by distance and the harsh climate.

Impacts of human activity

Roads and pipelines today snake hundreds of miles across what was formerly a trackless barren ground, and hunters motor conveniently to remote gamelands. The technology that permits this was developed in temperate regions.

Roadways in temperate climates, for example, are cleared with bulldozers and then covered with surface materials. But in tundra, with its underlying foundation of permafrost, removal of vegetation quickly leads to serious and long-lasting erosion. Roadways had to be built by other means, a problem eventually resolved by burying the mat of tundra vegetation beneath a layer of gravel, thereby preserving the requisite layer of insulation.

Construction of the Trans-Alaskan pipeline raised concerns about many economic and environmental issues, including the migrations of caribou. Would the herds cross the new obstruction? For the most part, caribou migrations were not interrupted, although cows and calves often avoided the structure. However, where a roadway paralleled the pipeline, vehicular traffic interrupted caribou migrations.

The potential for oil spills along the 1287 km (800 miles) of the 122 cm (48 inch) diameter pipeline, at any of its 11 pump stations or along its feeder lines, concerned environmentalists prior to construction. By 2010, approximately 16 billion barrels of oil had been transported from Prudhoe Bay to Valdez, but not without leakage, however. Since the pipeline went into service in 1977, several notable incidents involving leakage and at least three large spills of an oil–saltwater mixture have occurred. The largest of these was in 2006 when 757,000 L (200,000 gallons) was spilled at Prudhoe Bay.

Contamination and pollution tend to linger on the tundra for considerable periods of time. Spilled oil, for example, decomposes very slowly in cold climates, as does the decay of refuse. The Arctic climate in such circumstances functions as an immense refrigerator, preserving wastes and pollution instead of perishable commodities. Oil development in the Arctic has had a somewhat unintended impact on nesting birds. The construction of facilities to obtain oil and house workers attracts opportunistic predators such as Arctic foxes, common ravens, and several species of gulls. These species benefitted from what amounts to "subsidized housing" and dietary supplements from human garbage to increase their populations near oil infrastructure. As a result, nest survival for several species of birds (e.g., Lapland longspur, red phalarope, and red-necked phalarope) declined when their nests were within 5 km (3 miles) of oil-related facilities in the Tundra.

Anthropogenic erosion also imperils tundra. Excessive disturbance of the protective plant cover exposes the underlying soil and, without this insulation, the upper layers of permafrost soon thaw and erode severely with spring runoff. As a simple illustration, the disturbance of a dog tethered for just 10 days caused subsidence of the immediate area to a depth of more than 30 cm (1 foot). Unfortunately, this example is only a microcosm within the greater expanse of Arctic Tundra. Vehicle trails on the tundra created during winter seismic explorations for oil persisted for two decades after disturbance, and vegetative recovery and vehicular tracks made prior to World War II have eroded so significantly that some are now lakes. The impacts of human foot trails can be just as severe in Alpine Tundra where the effect of slope intensifies erosion. The resulting damage continues for years or even decades afterward, leaving ugly scars on the landscape. Natural restoration of vegetation at such sites may require many thousands of years to replace eroded tundra soils.

Global warming

The long-term trend in global warming is well-documented, even though debate continues about the cause. Because of its fragility, the Tundra may well be the first biome to experience significant changes in response to climate change. Over the past 150 years, the mean surface temperature of Earth has increased by 0.4°C (0.8°F), but the average temperature in the Arctic has increased two to three times that amount. As a result of this environmental change, seasonal sea ice in the Arctic has declined by about 45,000 km^2 a^{-1} (17,400 square miles per year). The sea ice breaks up earlier and freezes later, and similar trends in tundra snow cover also occur. In some areas of the Tundra, plant flowering and insect appearance occur about 20 days ahead of schedules documented for generations. In certain parts of the Arctic, populations of Arctic foxes are declining as red foxes expand their range northward, likely in response to climatic shifts. Warming also results in the range expansion of shrubs and may eventually alter the location of the tree line separating the Tundra from the Boreal Forest.

Large amounts of carbon are stored in the undecomposed vegetation mats and soils of the Arctic Tundra. Many regions of the Tundra are now warmer than they have been in the past, resulting in a gradual lowering of the water table. Soil decomposition is occurring at a greater rate, allowing more carbon dioxide to be released into the atmosphere. Consequently, the Arctic is becoming a net source of carbon dioxide to the atmosphere rather than a carbon "sink," a role that it has played since the last glacial period. The release of greenhouse gases from tundra soils may further accelerate global climate change.

In summary, tundra is "fragile" for both ecological and anthropogenic reasons that often interact. These include: (a) short food chains (i.e., few species); (b) limited annual plant growth; (c) vulnerability to disturbances that produce thermokarst; (d) human development and activities that are ill-suited to regional conditions; and (e) alterations of climatic conditions as a consequence of global climate change.

Readings and references

The Tundra Biome
Hearne, S. 1795. A journey from Prince of Wales Fort, in Hudson Bay, to the northern ocean. Edited several times since its original publication, including R. Glover (ed.) 1958 Macmillan, Toronto. 301 pp. (A book based on Hearne's journals describing his three expeditions across the Tundra; includes fascinating descriptions of Chipewyan Indians as well as perceptive natural history.)

Ives, J.D. and R. G. Barry. (eds) 1974. *Arctic and Alpine Environments*. Metheun, London.

Pruitt, W.O., Jr. 1978. *Boreal Ecology*. Institute of Biology, Studies in Biology 9, Edward Arnold Ltd., London.

Quinn, J.A. 2008. *Arctic and Alpine Biomes (Greenwood Guide to Biomes of the World)*. Greenwood Publishing Group, Westport, CT.

Ricketts, T.H., E. Dinerstein, D.M. Olson, et al. 1999. *Terrestrial Ecoregions of North America: A Conservation Assessment*. Island Press, Washington, DC.

Soils and geological influences
Britton, M.E. (ed.) 1973. *Alaskan Arctic Tundra*. Arctic Institute of North America Technical Paper 25, Washington, DC.

Tedrow, J.C.F. 1977. *Soils of the Polar Landscape*. Rutgers University Press, New Brunswick, NJ.

Tedrow, J.C.F. and D.E. Hill. 1955. Arctic brown soil. Soil Science 80: 265–275.

Tedrow, J.C.F. and H. Harris. 1960. Tundra soil in relation to vegetation, permafrost and vegetation. Okios 11: 237–249.

Permafrost
Britton, M.E. (ed.) 1973. *Alaskan Arctic Tundra*. Arctic Institute of North America Technical Paper 25, Washington, DC.

Ives, J.D. and B.D. Fahey. 1971. Permafrost occurrence in the Front Range, Colorado Rocky Mountains, USA. Journal of Glaciology 10: 105–111.

Jenness, J.L. 1952. Problems of glaciation in western islands of Arctic Canada. Geological Society of America Bulletin 63: 939–951. (Suggests that post-Pleistocene climates are largely responsible for permafrost.)

McKay, J.R. 1972. The world of underground ice. Annals of the Association of American Geographers 62: 1–22.

Muller, S.W. 1943. Permafrost or permanently frozen ground and related engineering problems. US Geological Survey Special Report, Strategic Engineering Study 62, Office, Chief of Engineers, US Army, Washington, DC.

Pewe, T.L. 1957. Permafrost and its effect on life in the North. In: *Arctic Biology* (H.P. Hansen, ed.). Annual Biological Colloquium 18, Oregon State University, Corvallis, pp. 12–25.

Ray, L.L. 1951. Permafrost. Arctic 4: 196–203.

Taber, S. 1943. Perennial frozen ground in Alaska–its origin and history. Geological Society of America Bulletin 54: 1433–1548.

Patterned ground
Kerfoot, D.E. 1972. Thermal contraction crack in an Arctic Tundra environment. Arctic 25: 142–150.

Lachenbruch, A.H. 1962. Mechanics of thermal contraction cracks and ice-wedge polygons in permafrost. Geological Society of America, Special Paper 70: 1–69.

Porsild, A.E. 1938. Earth mounds in unglaciated northwestern America. Geographical Review 28: 46–58.

Eskers and tundra wildlife
Jakimchuk, R.D. and D.R. Caruthers. 1983. A preliminary study of the behavior of barren-ground caribou during their spring migration across Contwoyto Lake, NWT, Canada. Acta Zoologica Fennica 175: 117–119.

Krajick, K. 1996. An esker runs through it. Natural History 105(5): 28–36. (A popular account of these landforms and their ecology. See also Kay, J. and P. 1976. On Arctic eskers. Nature Canada 5(3): 33–37.)

Morrison, D. and G.H. Germain. 1995. *Inuit, Glimpses of an Arctic Past*. Canadian Museum of Civilization, Hull, Quebec.

Mueller, E.P. 1995. Tundra esker systems and denning by grizzly bears, wolves, foxes, and ground squirrels in the central Arctic, Northwest Territories. File Report 115, Department of Renewable resources, Yellowknife, NWT.

Sharp, H.S. 1977. The caribou-eater Chipewyan: bilaterality, strategies of caribou hunting, and the fur trade. Arctic Anthropology 14: 35–40.

Glacial refugia
Ehlers, J. and P.L. Gibbard. 2004. *Quaternary Glaciations: Extent and Chronology 2: Part II North America*. Elsevier, Amsterdam.

Features and adaptations
Plant adaptation to harsh conditions
Billings, W.D. 1973. Arctic and alpine vegetation: similarities, differences, and susceptibility to disturbance. BioScience 23: 697–704.

Billings, W.D. and H.A. Mooney. 1968. The ecology of Arctic and alpine plants. Biological Review of the Cambridge Philosophical Society 43: 481–529.

Bliss, L.C. 1962. Adaptations of Arctic and alpine plants to environmental conditions. Arctic 15: 117–144.

Britton, M.E. (ed.) 1973. *Alaskan Arctic Tundra*. Arctic Institute of North America Technical Paper 25, Washington, DC.

Hansen, H.P. (ed.) 1967. *Arctic Biology*. Annual Biological Colloquium 18, Oregon State University, Corvallis.

Holmen, K. 1957. The vascular plants of Peary Land, North Greenland. Meddelelser om Grønland 124: 1–149.

Jefferies, R.L. 1977. The vegetation of salt marshes at some coastal sites in Arctic North America. Journal of Ecology 65: 661–672.

Plant growth and reproduction
Dennis, J.G. and P.L. Johnson. 1970. Shoot and rhizome-root standing crops of Tundra vegetation at Barrow, Alaska. Arctic and Alpine Research 2: 253–266.

Hansen, H.P. (ed.) 1967. *Arctic Biology*. Annual Biological Colloquium 18, Oregon State University, Corvallis.

Hodgson, H.J. 1966. Floral initiation in Alaskan Graminae. Botanical Gazette 127: 64–70.

Mosquin, T., and J.E.H. Martin. 1967. Observations on the pollination biology of plants on Melville island, NWT, Canada. Canadian Field Naturalist 81: 201–205.

Porsild, A.E., C.R. Harrington, and G.A. Mulligan. 1967. *Lupinus arcticus* Wats. grown from seeds of Pleistocene age. Science 158: 113–114.

Wilson, J.W. 1957. Arctic plant growth. Advancements in Science 53: 383–388.

Decomposition and soil nutrients

Banks, T.P., II. 1953. Ecology of prehistoric Aleutian village sites. Ecology 43: 246–264.

Helfield, J.M. and R.J. Naiman. 2006. Keystone interactions: salmon and bear in riparian forests of Alaska. Ecosystems 9: 167–180.

Quinn, T.P., S.M. Carlson, S.M. Gende, and H.B. Rich, Jr. 2009. Transportation of Pacific salmon carcasses from streams to forests by bears. Canadian Journal of Zoology 87: 195–203.

Some animal adaptations

Briskie, J.V. 1995. Nesting biology of the yellow warbler at the northern limits of its range. Journal of Field Ornithology 66: 531–543.

Karplus, M. 1949. Bird activity in the continuous daylight of Arctic summer. Bulletin of the Ecological Society of America 30: 66.

Murie, O.J. 1959. *Fauna of the Aleutian Islands and Alaska Peninsula*. North American Fauna 61, US Fish and Wildlife Service, Washington, DC.

Remmert, H. 1980. *Arctic Animal Ecology*. Springer-Verlag, New York, NY.

Shelford, V.E. and A.C. Twomey. 1941. Tundra animal communities in the vicinity of Churchill, Manitoba. Ecology 22: 47–69.

West, G.C. and D.W. Norton. 1975. Metabolic adaptations of Tundra birds. In: *Physiological Adaptations to the Environment* (F.J. Vernberg, ed.). Intext Educational Publications, New York, pp. 301–329.

Major vegetative communities

Britton, M.E. 1967. Vegetation of the Arctic Tundra. In *Arctic Biology* (H.P. Hanson, ed.). Oregon State University Press, Corvallis, pp. 67–130.

Bryson, R.A., W.N. Irving, and J.A. Larsen. 1965. Radiocarbon and soil evidence of former forest in the southern Canadian Tundra. Science 147: 46–48.

Elliot-Fisk, D.L. 1983. The stability of the northern Canadian tree limit. Annals of the Association of American Geographers 73: 560–576.

Hanson, H.C. 1953. Vegetation types in northwestern Alaska and comparisons with communities in other Arctic regions. Ecology 34: 111–140.

Hustich, I. 1953. The boreal limits of conifers. Arctic 6: 149–162.

Lescop-Sinclair, K. and S. Payette. 1995. Recent advance of the Arctic treeline along the eastern coast of Hudson Bay. Journal of Ecology 83: 929–936.

Marr, J.W. 1948. Ecology of the forest-Tundra ecotone on the east coast of Hudson Bay. Ecological Monographs 18: 117–144.

Taylor, R.L. and R.A. Ludwig. (eds) 1965. The evolution of Canada's flora. University of Toronto Press, Toronto. (See pages 12–27 for vegetational development in northern North America.)

Invertebrates and tundra ecology

Downes, J.A. 1965. Adaptations of insects in the Arctic. Annual review of Entomology 10: 257–274.

Holmes, R.T. and F.A. Pitelka. 1968. Food overlap among coexisting sandpipers on northern Alaska Tundra. Systematic Zoology 17: 305–318.

Hurd, P.D., Jr and F.A. Pitelka. 1954. The role of insects in the economy of certain Arctic Alaskan birds. Alaskan Science Conference 3: 136–137.

MacLean, S.F., Jr and F.A. Pitelka. 1971. Seasonal patterns of abundance of Tundra arthropods near Barrow. Arctic 24: 17–40.

Seton, E.T. 1911. *The Arctic Prairies*. Charles Scribner's Sons, New York. (A somewhat romanticized account of tundra, including a back-of-the-hand census of mosquitoes by one of the best-known nature writers of his era.)

Selected tundra mammals

Lemmings

Chitty, D. 1996. *Do Lemmings Commit Suicide? Beautiful Hypothesis and Ugly Facts*. Oxford University Press, New York.

Guilday, J.E. 1963. Pleistocene zoogeography of the lemming, *Dicrostonyx*. Evolution 17: 194–197. (Proposes an explanation, based on fossil evidence, for the current distribution of collared lemmings on either side of Hudson Bay.)

Maher, W.J. 1970. The pomarine jaeger as a brown lemming predator in northern Alaska. Wilson Bulletin 82: 130–157.

Musser, G.G. and M. D. Carleton. 2005. Superfamily Muroidea. In: *Mammal Species of the World: A Taxonomic and Geographic Reference* (D.E. Wilson and D.M. Reeder, eds). Johns Hopkins University Press, Baltimore, MD, pp. 894–1531.

Pitelka, F.A. 1973. Cyclic pattern in lemming populations near Barrow, Alaska. In: *Alaskan Arctic Tundra* (M.E. Britton, ed.). Arctic Institute of North America Technical Paper No. 25, Washington, DC, pp. 199–215.

Weber, N.A. 1950. The role of lemmings at Point Barrow, Alaska. Science 111: 552–553.

Arctic ground squirrels

Bee, J.W. and E.R. Hall. 1956. Mammals of northern Alaska on the Arctic Slope. Miscellaneous Publications, University of Kansas Museum of Natural History 8: 1–309.

Butterworth, B.B. 1958. Molt patterns in the Barrow ground squirrel. Journal of Mammalogy 39: 92–97.

Carl, E.A. 1971. Population control in Arctic ground squirrels. Ecology 52: 395–413.

Iwen, F.A. 2003. Arctic ground squirrel: *Spermophilus parryii*. In: *The Smithsonian Book of North American Mammals* (D.E. Wilson and S. Ruff, eds). University of British Columbia Press, Vancouver, Canada, pp. 427–429.

MacClintock, D. 1970. Squirrels of North America. Van Nostrand Reinhold, New York, NY.

Arctic foxes

Braestrup, F.W. 1941. A study of the Arctic fox in Greenland. Meddelelser om Grønland 131: 1–101.

Dalerum, F. and A. Angerbjörn. 2000. Arctic fox (*Alopex lagopus*) diet in Karupelv Valley, East Greenland, during a summer with low lemming density. Arctic 53: 1–8.

Kapel, C.M.O. 1999. Diet of Arctic foxes (*Alopex lagopus*) in Greenland. Arctic 52: 289–293.

Larson, S. 1960. On the influence of the Arctic fox *Alopex lagopus* on the distribution of Arctic birds. Oikos 11: 276–305. (Suggests the distribution of purple sandpipers and other tundra-nesting birds result from fox predation.)

Macpherson, A.H. 1969. The dynamics of Canadian Arctic fox populations. Canadian Wildlife Service Report Series 8: 1–52.

Nielsen, S.M., V. Pedersen, and B.B. Klitgaard. 1994. Arctic fox (*Alopex lagopus*) dens in the Disko Bay area, West Greenland. Arctic 47: 327–333.

Smith, C.A.S., C.M.M. Smits, and B.G. Slough. 1992. Landform selection and soil modification associated with Arctic fox (*Alopex lagopus*) dens in Yukon Territory, Canada. Arctic and Alpine Research 24: 324–328.

Smits, C.M.M. and B.G. Slough. 1993. Abundance and summer occupancy of Arctic fox, *Alopex lagopus*, and red fox, *Vulpes vulpes*, dens in the northern Yukon Territory, 1984–1990. Canadian Field-Naturalist 107: 13–18.

Smits, C.M.M., C.A.S. Smith, and B.G. Slough. 1988. Physical characteristics of Arctic fox (*Alopex lagopus*) dens in northern Yukon Territory, Canada. Arctic 41: 12–16.

Barren-ground caribou

Ballard, W.B., M.A. Cronin, and H.A. Whitlaw. 2000. Caribou and oilfields. In: *The Natural History of an Arctic Oil Field–Development and the Biota* (J.C. Truett and S.R. Johnson, eds). Academic Press, New York, pp. 85–104.

Banfield, A.W.F. 1954. Preliminary Investigation of the Barren Ground Caribou. Wildlife Management Bulletin, Series 1 (10A), Canadian Wildlife Service, Ottawa.

Fancy, S.G. and K.R. Whitten. 1991. Selection of calving sites by Porcupine herd caribou. Canadian Journal of Zoology 69: 1736–1743.

Harper, F. 1955. The barren ground caribou of Keewatin. Miscellaneous Publications, University of Kansas Museum of Natural History 6: 1–163.

Kelsall, J.P. 1968. The migratory barren-ground caribou of Canada. Canadian Wildlife Service Monograph Series 3: 1–340.

Murphy, S.M., and J.A. Curatolo. 1987. Activity budgets and movement rates of caribou encountering pipelines, roads, and traffic in northern Alaska. Canadian Journal of Zoology 65: 2483–2490.

Smith, W.T. and R.D. Cameron. 1985. Reactions of large groups of caribou to a pipeline corridor on the Arctic coastal plain. Arctic 38: 53–57.

Speer, L. 1989. Oil development and the Arctic National Wildlife Refuge. Environment 31: 42–43.

Whitten, K.R. and R.D. Cameron. 1983. Movements of collared caribou, *Rangifer tarandus*, in relation to petroleum development on the Arctic Slope of Alaska. Canadian Field-Naturalist 97: 143–146.

Whitten, K.R., G.W. Garner, F.J. Mauer, and R.B. Harris. 1992. Productivity and early calf survival in the Porcupine Caribou Herd. Journal of Wildlife Management 56: 201–212.

Selected tundra birds

Gyrfalcons

Booms, T.L., T.J. Cade, and N.J. Clum. 2008. Gyrfalcon (*Falco rusticolus*). *The Birds of North America Online* (A. Poole, ed.). Cornell Laboratory of Ornithology, Ithaca, NY. Available at: http://bna.birds.cornell.edu/bna/species/114 (accessed 16 December 2014).

Snowy owl

Kerlinger, P. and M.R. Lein. 1988. Population ecology of snowy owls during winter in the Great Plains of North America. Condor 90: 866–874.

Kerlinger, P., M.R. Lein, and B.J. Sevick. 1985. Distribution and population fluctuations of wintering snowy owls (*Nyctea scandiaca*) in North America. Canadian Journal of Zoology 63: 1829–1834.

Parmalee, D. 1992. Snowy Owl. In: *The Birds of North America No. 10* (A. Poole, P. Stettenheim, and F. Gill, eds). Academy of Natural Sciences, Philadelphia, and American Ornithologists' Union, Washington, DC.

Pitelka, F.A., P.Q. Tomich, and G.W. Treichel. 1955. Ecological relations of jaegers and owls as lemming predators near Barrow, Alaska. Ecological Monographs 25: 85–117.

Watson, A. 1957. The behavior, breeding, and food ecology of the snowy owl, *Nyctea scandiaca*. Ibis 99: 419–462.

Ross goose

Ryder, J.P. 1967. The breeding biology of Ross' goose in the Perry River region, Northwest Territory. Canadian Wildlife Service Report Series, No. 8, Ottawa.

Highlights

Absentees: amphibians and reptiles

Constanzo, J.P. and R.E. Lee, Jr. 1993. Cryoprotectant production capacity of the freeze-tolerant wood frog, *Rana sylvatica*. Canadian Journal of Zoology 71: 71–75.

Remmert, H. 1980. *Arctic Animal Ecology*. Springer-Verlag, New York, NY.

Storey, K.B. and J.M. Storey. 1985. Freeze tolerant frogs: cryoprotectants and tissue metabolism during freeze-thaw cycles. Canadian Journal of Zoology 64: 49–56.

Lichens and "reindeer moss"

Brodo, I.M., S.D. Sharnoff, and S. Sharnoff. 2001. *Lichens of North America*. Yale University Press, New Haven, CN.

McCune, B. and L. Geiser. 2003. *Macrolichens of the Pacific Northwest*. Oregon State University Press, Corvallis, OR.

Snow goose "eat outs"

Abraham, K.F., R.L. Jefferies, and R.T. Alisaukas. 2005. The dynamics of landscape change and snow geese in mid-continent North America. Global Change Biology 11: 841–855.

Ankney, C.D. 1996. An embarrassment of riches: too many geese. Journal of Wildlife Management 60: 217–223.

Brazley, D.R. and R.L. Jefferies. 1986. Changes in the composition and standing crop of salt-marsh communities in response to the removal of a grazer. Journal of Ecology 74: 693–706.

Cargill, S.M. and R.L. Jefferies. 1984. The effects of grazing lesser snow geese on the vegetation of a subArctic salt-marsh. Journal of Applied Ecology 21: 669–686.

Handa, I.T., R. Harmsen, and R.L. Jefferies. 2002. Patterns of vegetation change and the recovery potential of degraded areas in a coastal marsh system of the Hudson Bay Lowland. Journal of Ecology 90: 86–99.

Jefferies, R.L., A. Jensen, and K.F. Abraham. 1979. Vegetational development and the effect of geese on vegetation at La Perouse Bay, Manitoba. Canadian Journal of Botany 57: 1439–1450.

Jefferies, R.L., R.F. Rockwell, and K.F. Abraham. 2004. Agricultural food subsidies, migratory connectivity and large-scale disturbance in Arctic coastal systems: a case study. Integrative and Comparative Biology 44: 130–139.

Kotanen, P. and R.L. Jefferies. 1997. Long-term destruction of sub-Arctic wetland vegetation by lesser snow geese. EcoScience 4: 179–182.

Krebs, R.H., P.M. Kotanen, and R.L. Jefferies. 1990. Destruction of wetland habitats by lesser snow geese: a keystone species on the west coast of Hudson Bay. Journal of Applied Ecology 27: 242–258.

Menu, S., G. Gauthier, and A. Reed. 2002. Changes in survival rates and population dynamics of greater snow geese over a 30-year period: implications for hunting regulations. Journal of Applied Ecology 39: 91–102.

Smith, T.J., III. 1983. Alteration of salt marsh community composition by grazing snow geese. Holarctic Ecology 6: 204–210.

Alpine Tundra

Bliss, L.C. 1963. Alpine plant communities of the Presidential Range, New Hampshire. Ecology 44: 678–697.

Mooney, H.A., G. St. Andre, and R.D. Wright. 1962. Alpine and subalpine vegetation patterns in the White Mts of California. American Midland Naturalist 68: 257–273.

Shabot, B.F. and W.D. Billings. 1972. Origins and ecology of the Sierran alpine flora and vegetation. Ecological Monographs 42: 163–199.

Zwinger, A.H. and B.E. Willard. 1972. *Land Above the Trees, A Guide to American Alpine Tundra*. Harper & Row, New York.

Fragile Tundra

Impacts of human activity

Anonymous. 2000. Arctic National Wildlife Refuge long-term monitoring of recovery of trails from winter seismic exploration. Arctic Research of the United States 14: 32–33.

Bliss, L.C. 1970. Oil and the ecology of the Arctic. In: *The Tundra Environment*. Transactions of the Royal Society of Canada, 4th Series, Vol. VII, Toronto, pp. 1–12.

Chernobyl Forum Expert Group 'Environment.' 2006. Environmental consequences of the Chenobyl accident and their remediation: twenty years of experience. Radiological Assessment Reports Series, International Atomic Energy Agency, Vienna, Austria.

Conservation of Arctic Flora and Fauna (CAFF). 2103. Arctic biodiversity assessment: report for policymakers. CAAF, Akureyri, Iceland.

Emers, M., J.C. Jorgenson, and M.K. Raynolds. 1995. Response of Arctic plant communities to winter vehicle disturbance. Canadian Journal of Botany 73: 905–919.

Gersper, P.I. and J.L. Challinor. 1975. Vehicle perturbation effects upon a Tundra soil-plant system: I. Effects on morphological and physical environmental properties of the soils. Soil Science Society of America Proceedings 39: 737–744.

Hanson, W.C. 1967. Cesium 137 in Alaskan lichens, caribou and Eskimos. Health Physics 13: 383–389.

Klein, D.R. 1979. The Alaska oil pipeline in retrospect. Transactions of the North American Wildlife and Natural Resources Conference 44: 235–246.

Lachenbruch, A.H. 1970. Some estimates of the thermal effects of a heated pipeline in permafrost. US Geological Survey Circular 632: 1–13.

Liebezeit, J.R., S.J. Kendall, S. Brown, C.B. Johnson, P. Martin, T.L. McDonald, D.C. Payer, C.L. Rea, B. Streever, A.M. Wildman, and S. Zack. 2009. Influence of human development and predators on nest survival of Tundra birds, Arctic Coastal Plain, Alaska. Ecological Applications 19: 1628–1644.

National Research Council. 2003. *Cumulative Environmental Effects of Oil and Gas Activities on Alaska's North Slope*. The National Academies Press, Washington, DC.

Sigafoos, R.S. 1951. Soil instability in Tundra vegetation. Ohio Journal of Science 51: 281–298.

Stephens, S. 1987. Lapp life after Chernobyl. Natural History 96(12): 33–40.

Willard, B.E., D.J. Cooper, and B.C. Forbes. 2007. Natural regeneration of alpine Tundra vegetation after human trampling: a 42-year data set from Rocky Mountain National Park, Colorado. Arctic and Alpine Research 39: 177–183.

Global warming

Christensen, T.R., T. Johansson, H.J. Åkerman, and M. Mastepanov. 2004. Thawing of sub-Arctic permafrost: effects on vegetation and methane emissions. Geophysical Research Letters 31(L04501): 1–4.

Diaz, H.F., and J.K. Eischeid. 2007. Disappearing "alpine Tundra" Köppen climatic type in the western United States. Geophysical Research Letters 34(L18707): 1–4.

Foster, J.L. 1989. The significance of the date of snow disappearance on the Arctic Tundra as a possible indicator of climate change. Arctic and Alpine Research 21: 60–70.

Jorgenson, J.C., J.M. der Hoef, and M.T. Jorgenson. 2010. Long-term recovery patterns of Arctic Tundra after winter seismic exploration. Ecological Applications 20: 205–221.

Kennedy, M., D. Mrofka, and C. von der Borch. 2008. Snowball Earth termination by destabilization of equatorial permafrost methane clathrate. Nature 453: 642–645.

Macpherson, A.H. 1964. A northward range extension of the red fox in eastern Canadian Arctic. Journal of Mammalogy 45: 138–140.

Oechel, W.C., S.J. Hastings, G. Vourlitis, et al. 1993. Recent change of Arctic Tundra ecosystems from a net carbon dioxide sink to a source. Nature 361: 520–523.

Pamperin, N.J., E.H. Follmann, and B. Petersen. 2006. Interspecific killing of an Arctic fox by a red fox at Prudhoe Bay, Alaska. Arctic 59: 361–364.

Post, E., M.C. Forchhammer, et al. 2009. Ecological dynamics across the Arctic associated with recent climate change. Science 325(5946): 1355–1358.

Skre, O., R. Baxter, R.M.M. Crawford, et al. 2002. How will the Tundra-tiaga interface respond to climate change? Ambio: A Journal of the Human Environment 31: 37–46.

Sturm, M., C. Racine, and K. Tape. 2001. Climate change: increasing shrub abundance in the Arctic. Nature 441: 546–547.

Suarez, F., D. Binkley, M.W. Kay, and R. Stottlemyer. 1999. Expansion of forest stands into Tundra in the Noatak National Preserve, northwest Alaska. Ecoscience 6: 465–470.

Tannerfeldt, M., B. Elmhagen, and A. Angerbjörn. 2002. Exclusion by interference competition? The relationship between red and Arctic foxes. Oecologica 132: 213–220.

Infobox 2.1. Samuel Hearne

Hanson, H.C. 1965. *The Giant Canada Goose*. Southern Illinois University Press, Carbondale.

Hearne, S. 1795. A journey from Prince of Wales Fort, in Hudson Bay, to the northern ocean. Edited several times since its original publication, including R. Glover (ed.), 1958 Macmillan, Toronto. (A book based on Hearne's journals describing his three expeditions across the Tundra.)

Houston, C.S. 2006. Once upon a time in American ornithology. Wilson Journal of Ornithology 118: 577–579. (An account of Hearne's observations of birds in 18th century Canada.)

McGoogan, K. 2004. *Ancient Mariner, the Amazing Adventures of Samuel Hearne, the Englishman who Walked to the Arctic Ocean*. Bantam Press, London, UK.

Infobox 2.2. Of bumblebees and woollybears

Bennett, V.A., R.E. Lee, Jr, J.S. Nauman, and O. Kukal. 2003. Selection of overwintering microhabitats used by the Arctic woollybear caterpillar. CryoLetters 24: 191–200.

Kukal, O. 1994. Winter mortality and the function of larval hibernacula during the 14-year life cycle of an Arctic moth, *Gynaephora groenlandica*. Canadian Journal of Zoology 73: 657–662. (Source for mortality estimates and the incidence of shared hibernacula.)

Kukal, O. and P.G. Kevan. 1987. The influence of parasitism on the life history of a High Arctic insect, *Gynaephora groenlandica* (Wocke) (Lepidoptera: Lymantriidae). Canadian Journal of Zoology 65: 156–0163. (Source of mortality rates for larvae.)

Kukal, O., J.G. Duman, and A.S. Serianni. 1989. Cold-induced mitochondrial degradation and cryoprotectant synthesis in freeze-tolerant Arctic caterpillars. Journal of Comparative Physiology B 158: 661–671.

Michener, C.D. 2007. *The Bees of the World*, second edition. Johns Hopkins University Press, Baltimore, MD.

Milliron, H.E. and D.R. Oliver. 1966. Bumblebees from northern Ellesmere Island, with observations on usurpation by *Megabombus hyperboreus* (Schönh.) (Hymenoptera: Apidae). Canadian Entomologist 98: 207–213.

Morewood, W.D. and R.A. Ring. 1998. Revision of the life history of the High Arctic moth *Gynaephora groenlandica* (Wocke) (Lepidoptera: Lymantriidae). Canadian Journal of Zoology 76: 1371–1381. (Notes that the life cycle – egg to adult – of the species is 7 and not 14 years as previously believed.)

CHAPTER 3
Boreal Forest

The vastness of the boreal forest region makes it one of the few remaining places on Earth where entire ecosystems function.

Peter Blancher and Jeffery Wells

Boreal Forest forms a nearly continuous belt across northern North America and covers much of Alaska, southern Canada, and New England. Fingers of Boreal Forest extend down western mountain ranges and the Appalachians. Boreal Forest is the largest biome in North America (Fig. 1.3) and, with its counterpart in Eurasia, forms the largest in the world. Despite its great size, the biome is remarkably uniform in its physiognomy, a result of a canopy dominated by relatively few species of spruce, firs, and other **conifers**. Thus, *northern coniferous forest* often is used to describe Boreal Forest. Still another term is *taiga*, although some ecologists apply this name to the forest tundra ecotone described later in this chapter instead of limiting its usage to the original Russian meaning: dense northern forests of spruce and fir.

The wide band of Boreal Forest occupies a formerly glaciated region that bears the tell-tale scars of glacial moraines, cold lakes, bogs, and rivers (Fig. 3.1). The region is characterized by long, severe winters in which temperatures may remain below freezing for up to six months. Short summers offer only 50–100 frost-free days, but temperatures may reach 32°C (90°F) for brief periods. Low evaporation rates create a humid climate despite limited rainfall (38–50 cm or 15–20 inches per year).

Unlike most types of forests Boreal Forest lacks a layered structure, a feature of its limited diversity and uniform physiognomy. Shrubs and other understory vegetation are poorly developed because little light penetrates the persistent canopy of needles, and the forest changes little in its appearance throughout the year. Virtually no species in the mature forest regenerates with sprouts, but sprouting is relatively commonplace in species associated with the early stages of succession. Tangles of windfalls after heavy storms may cover sizeable areas of Boreal Forest, opening areas to secondary succession.

Climatic boundaries and soils

Based on climate, the separation of Tundra from Boreal Forest seems closely associated with the summertime location of the Arctic air mass. In addition to temperature, the Arctic air mass features distinctive characteristics of moisture, turbidity, and structure; together, they form a recognizable meteorological entity. The seasonal locations of Arctic air masses and those in other areas can be charted, respective to each other, and thereby become useful for independent comparisons with the distribution of ecological units such as biomes. Accordingly, Boreal Forest begins to develop where the Arctic air mass, on average, reaches its southernmost extent between July and October. In winter the Arctic air mass moves farther south and, where it reaches its southernmost extent between November and February, the Boreal Forest ends and other types of vegetation prevail. The Boreal Forest therefore occupies a region of North America lying between the average summer and winter positions of the Arctic front.

Soils in cool, moist climates develop by **podzolization**, a process that leaches iron and aluminum from the A horizon and deposits these minerals in the underlying B horizon. The resulting soils, **spodosols**, are closely associated with Boreal Forest, but they also develop elsewhere. The root word *podzol* (Russian for "ash") refers to the gray color of these soils. Podzolization also depletes the surface horizon of clays and other fine particles, which migrate deeper into the soil profile and leave behind a sandy upper strata.

The cool climate of the Boreal Forest inhibits the activities of soil microorganisms, and a dense mat of needles persists on the forest floor. The ground litter produced by conifers lacks calcium and nitrogen, but contains lignin, wax, and resins, which resist decay. The tannins and organic acids in the ground layer further inhibit decomposition, and the upper soil profile is acidic. The

Ecology of North America, Second Edition. Brian R. Chapman and Eric G. Bolen.
© 2015 John Wiley & Sons, Ltd. Published 2015 by John Wiley & Sons, Ltd.

Figure 3.1 The terrain of the Boreal Forest is punctuated by rivers and bogs offering evidence of a past dominated by glaciers. Development of understory vegetation is limited by the short growing season and a dense canopy of black spruce, shown here, and other conifers. Photograph courtesy of Steve Hildebrand and the US Fish and Wildlife Service.

year-round shade of the evergreen canopy and the mat of ground litter impede evaporation, and the spodsol soils remain wet, a condition favoring the leaching process. Nutrient cycling is limited under these conditions.

Features and adaptations

Plant adaptations

Plants in Boreal Forest must survive long, frigid winters and short, dry summers. In addition, plants must withstand fire, as discussed later. Because few species are adapted to such rigorous conditions, the species richness of plants in the biome is low.

The needles of conifers, because of their thick cuticles and reduced surface area, are protected against freezing. This condition restricts the loss of moisture which is important, even in swampy areas where water may remain frozen for much of the year. The cone-like profile and flexible branches of spruce and fir also reduce the burden of heavy snowfall. Because they bear year-round foliage, conifers can quickly engage in photosynthesis any time ambient temperatures rise above freezing, an advantage that helps compensate for the otherwise short growing season in the cold climate of the Boreal Forest.

The shallow root systems of many trees represent an adaptation to the shallow soils in the Boreal Forest. For example, spruce trees can grow in soils just 51 cm (20 inches) or less in depth, but this ability also heightens their susceptibility to wind throw during strong storms.

Animal adaptations

Herbivores typically graze on succulent vegetation in the brief summer but switch to a diet of **browse** during the remainder of the year. The digestive systems of most

herbivores therefore must cope, at least seasonally, with diets high in fiber content. In winter, for example, structures known as caeca increase in length by almost 50% after spruce grouse begin feeding heavily on a fiber-rich diet of buds and needles. Caeca, a pair of slender blind tubes that branch off from the intestine, provide additional sites for bacterial decomposition of fibrous foods.

Cones produce most of the seeds in the Boreal Forest, and the bills of white-winged crossbills are remarkably adapted to extract conifer seeds. With their twisted bills, these birds wedge apart cone scales and extract the seeds with their tongues. An efficient crossbill can remove and consume as many as 3000 conifer seeds per day. White-winged crossbills travel in flocks as large as 10,000 birds and readily exploit local cone crops, but these nomadic birds move elsewhere when cone supplies are limited. White-winged crossbills breed opportunistically throughout the year, responding primarily to the availability of cones.

Snowshoe hares and lynx have oversized feet, which are an obvious benefit for travel during the long periods when deep snows cover the forest floor. Native Americans almost certainly used this adaptation as a model for snowshoes which, like the canoe, has not changed appreciably in design for centuries.

Frequent fires

Fire plays a major role in Boreal Forest and large areas often burn at a time, fueled by the vegetation's highly volatile resins and uniformity. After a fire, an understory of deciduous vegetation, typically birch, aspen, or willow, quickly blankets the burned area. Along the northern edge of Boreal Forest, fires also destroy the rich growth of lichens which are not fire-tolerant, including those on trees and the ground-dwelling species. The two fire-related responses – a sprouting understory and the destruction of lichens – initiate a third response, which affects caribou and moose. In winter, barren-ground caribou depend heavily on lichens for much of their diet, and large herds find shelter in mature Boreal Forest where lichens are abundant (Fig. 3.2). Fires can eliminate the dietary mainstay of caribou in a matter of hours and, because lichens are extremely slow-growing, decades pass before the carrying capacity for caribou returns to pre-burn levels. Meanwhile, after a burn moose thrive on the regrowth of birch and other succulent browse, and their population increases as secondary succession progresses through its early stages. Eventually however, the forest and its associated growth of lichens slowly replace the deciduous understory, and moose populations thereafter decline while caribou numbers rebound at the site.

This relationship demonstrates a fundamental ecological difference between two species of large animals: one adapted to climax vegetation (caribou) and the other to early successional growth (moose). Such ebbs and flows have been a natural part of Boreal Forest ecology for millennia, but human carelessness has increased the frequency of fires, as witnessed by tell-tale layers of charcoal in the soil profile.

Niches in the Boreal Forest

The niche concept explains how organisms avoid interspecific competition. By adapting a niche, one species utilizes resources (e.g., food) without constantly diverting its energies into a never-ending rivalry with another species occupying similar habitat. For example, some species of hawks prey on rodents during the day, whereas owls hunt rodents at night; this is an example of hunting niches based on activity patterns. Differences between spruce and ruffed grouse, species with overlapping distributions in the Boreal Forest Biome, illustrate niche segregation based on habitat preferences. Ruffed grouse select sites where deciduous trees are about as abundant as conifers, whereas spruce grouse clearly prefer sites with almost pure stands of coniferous trees. Ruffed grouse are associated with the early and middle stages of succession in Boreal Forest – typically sites burned 20–40 years previously – but a well-established climax community furnishes prime habitat for spruce grouse.

(a)

(b)

Figure 3.2 (a) Stands of black spruce at the northern edge of Boreal Forest provide winter food and shelter for caribou. Note the cushions of lichens (light areas) at ground level. (b) Heavy growths of lichens also cover the lower branches of spruce trees. Photographs courtesy of (a) Elizabeth D. Bolen and (b) Brian R. Chapman.

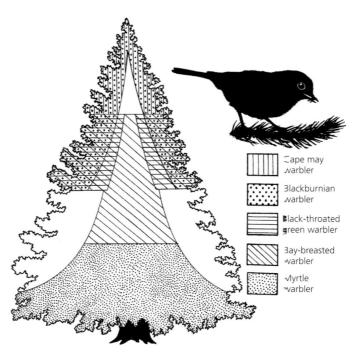

Figure 3.3 Five species of wood warblers feed at the same time on insects in the same spruce tree, which seems to violate the concept of niche segregation. However, close study revealed that each species has its own feeding niche in different parts of the tree and can therefore avoid competition from the other species. The marked areas show where each species spends at least half of its foraging time. Illustrated by Tamara R. Sayre, based on MacArthur (1958).

Niche segregation at a much finer scale was determined for five species of wood warblers feeding in spruce trees in northern New England. Wood warblers are small, insect-eating birds often feeding in the same trees at the same time. At first glance, these species appear to be competing with each other for food, thus defying the concept of niche segregation. However, careful observation revealed that each species spent at least half of its time foraging in separate parts of the trees, for example lower or upper, or inner or outer branches (Fig. 3.3). The five species exhibited some combination of these four alternatives, each thereby avoiding direct competition with other species feeding in the same tree. Each of the five species concurrently (but separately) utilized food resources available at a common site. Because of these subtle but real differences, several species of wood warblers coexist in the rather uniform habitat of a spruce forest.

Selected biotic communities

The continental belt of Boreal Forest is not rich in species despite extending east to west from Labrador to Alaska. Much of the eastern Boreal Forest is characterized by balsam fir and white and black spruce, whereas westward balsam fir drops out and paper (= white) birch, jack pine, and two species of aspen join the spruces. Larch (also known as tamarack) occurs throughout the Boreal Forest in North America, but gains in prominence toward the northern edge of the forest. At higher elevations in North America, Boreal Forest grades into communities in the Montane Forest where other species characterize the vegetation (Chapter 9). Still other species of dominant trees characterize the Appalachian Extension of Boreal Forest.

Tree line and forest tundra

Most maps of North America boldly include the designation tree line as the northern limit of the Boreal Forest Biome. The same line, of course, concurrently forms the southern limit of the adjacent biome to the north, the Tundra. Tree line is also the traditional line separating Arctic and sub-Arctic zones. Locating such a line on the ground in the real world of vegetational development is not without some difficulties, however.

The first of these concerns the question, "What is a tree?" Dwarfed trees (e.g., birch) are a common part of Tundra vegetation, but these are often ignored when tree lines are drawn on maps. Only the tall individuals are considered as trees, not the smaller individuals nestled close to the ground or growing in shrubby thickets. This, of course, begs the question, "How tall must a birch be before it represents a tree?" Some ecologists have proposed a vertical height of at least 2 m (6.5 feet) to define vegetation for this purpose. Still others suggest a "biological limit," which is the location where trees no longer form a continuous stand of forest vegetation. Yet another calls for a tree line based on the point where it is no longer profitable to harvest trees for commercial purposes; an economic limit therefore becomes an ecological boundary. Finally, there is the species limit, which is the line representing the last occurrence of the species, irrespective of an individual's height or shape at its final outpost. This of course means that there might be several tree lines, one for each tree species found in the forest. One ecologist, perhaps with tongue in cheek, observed that a map of Ontario could include at least 75 separate tree lines.

What might be concluded from all this? First, a tree line on a map cannot be interpreted without knowing what definition was used as a basis for the designation. Second, regardless of the definition, tree line is not a narrow zone, one side of which is vegetated with a plain of Tundra and the other with a curtain of forest (Fig. 3.4). Finally, ecological and other influences continually shape the location of the tree line. Various occurrences, especially fire but also shifting climatic changes, produce a dynamic zone in which the tree line ebbs and flows. Based on radiocarbon dating of soil charcoal, for example, the Boreal Forest extended as much as 280 km (174 miles) north of its present location in Canada between 1500 BC and AD 1000, suggesting a milder climate during part of this period, along with a cultural change in the distribution of native peoples. Conversely, fires in the forest tundra might quickly drive tree line southward for considerable distances, especially if the burns should happen during a period of global cooling. Boreal Forest is currently expanding northward in response to a 150-year-old trend of increasing temperatures in the northern regions of North America. Irrespective of what constitutes a tree line, ecologists recognize a zone of forest tundra lying between the closed canopy of the Boreal Forest and the open expanse of Tundra. Forest Tundra varies in width across the breadth of North America, being up to 250 km (155 miles) or more in parts of central Canada and somewhat narrower at the eastern and western ends. The zone, together with its Eurasian counterpart, is one of the world's largest **ecotones**. Patches of trees, commonly black spruce but also white spruce and larch, are scattered throughout the forest tundra landscape. Heath, lichens, cottongrass, and other species typical of tundra vegetation characterize the open "plains" between the scattered elements of forest.

Figure 3.4 *Tree line* is a gradual transition between heath and other tundra vegetation and the beginnings of forest communities. Spruces in these locations are not necessarily reduced in stature, as is evident in this ecotone in Alaska, but the forest becomes progressively thinner as it gives way to tundra vegetation. Photograph courtesy of Elizabeth T. Bolen.

Muskeg

Bog vegetation, popularly known as **muskeg**, occurs widely throughout the regions of Boreal Forest, but the ecological features of the communities are not always defined with great precision. The term (of Algonquian origin, meaning "trembling earth") is common stock in the lexicon of trappers, loggers, and others living and working in the north woods, where it is widely used to describe any sort of bog community. Muskegs, sometimes called mires, quagmires, or peat bogs, vary in size from just a few meters in diameter to those covering several hectares and are **ombrotrophic**, which means they are nourished entirely by rain. The substrate, typically spongy underfoot, is a layer of **peat** that provides the root zone for most of the vascular vegetation. Ecologists propose that the peat layer must be at least 40 cm (16 inches) thick for the site to qualify as muskeg.

Muskegs may be of several kinds, but a more-or-less representative type includes those with: (a) thin stands of black spruce; (b) an understory of heath; and (c) a cushioned ground cover of sphagnum moss over a layer of peat. Larch may join or replace black spruce in the overstory of many muskegs. The understory vegetation often includes Labrador tea, leatherleaf, and other species of heath as well as grasses and sedges; **forbs** are relatively uncommon. Stands of cattail may rim muskeg ponds, and yellow pond lilies and pondweeds often occur in deeper water.

Muskegs commonly develop as a result of **paludification**, the process of bog expansion caused by rising water tables. For example, paludification begins when conditions favor the establishment and expanding growth of sphagnum moss, which has an extraordinary capacity for retaining water. In response, the waterlogged substrate prevents aeration in the root zone, and the forest vegetation dies back. The blanket of sphagnum continues expanding and thickens, further increasing the waterlogged conditions and an ever-deepening layer of peat. Disturbances on the forest floor, such as the uprooting of trees, provide sites where paludification may initiate a muskeg, but other factors (e.g., beaver dams) leading to rising water tables or altered drainage patterns may produce similar results.

Sphagnum mosses represent a single genus of about 135 species, most of which are northern in distribution. About 50 species occur in North America. These plants are of ecological importance because they can alter their surroundings by: (a) retarding decomposition; (b) creating acidic conditions; and especially (c) holding large volumes of water. The water-holding capacity of sphagnum, depending on the species, ranges upward to almost 4000% of its dry weight. Sphagnum ground cover limits seed germination and, except for some ferns, muskeg often lacks an understory of vascular plants. Some Native American tribes once used these mosses as socks or as packing for diapers because of their immense absorptive powers.

Relatively few species of wildlife are closely dependent on the habitat offered in spruce muskegs, although northern bog lemmings are a notable exception. Moose visit muskegs and muskrats may build their lodges where open water and aquatic foods are sufficient. A large area with many muskegs in Wood Buffalo National Park in Canada serves as the nesting area for the remaining wild population of whooping cranes. Waterfowl also nest in lake-dotted muskeg areas, of which Old Crow Flats in Yukon Territory is noteworthy.

Coniferous swamps

Saturated ground is common throughout the Boreal Forest Biome, but the soils typically dry during at least part of the summer. Coniferous swamps develop where low-lying areas or large topographic basins maintain saturated soils for much of the growing season. These forested wetlands may be temporarily inundated by up to 30 cm (1 foot) of standing water and generally do not develop a mat of sphagnum moss. The dominant trees on such sites are northern white cedar and larch, but balsam fir or black spruce may be important components in some areas. The understory is often dominated by cinnamon fern, marsh fern, and several species of sedges.

Coniferous swamps were important to Native American tribes as source of many products used for food, medicine, and construction. Northern white cedar (Infobox 3.1) was especially important and the subject of sacred legends. Because the biological diversity of Boreal Forest swamps is high relative to surrounding upland areas, some ecologists consider these sites, especially those that have not been perturbed for hundreds of years, as "biodiversity hotspots." Although some coniferous swamps may develop more rapidly, a period of about 270 years may be required to establish the old-growth conditions that increase biological diversity.

Comparative ecology of lakes

Lakes and ponds in the Boreal Forest and certain locations elsewhere (e.g., alpine areas in otherwise temperature zones) typically differ in numerous ways from those in deciduous forests and grasslands (Table 3.1). Because of these differences, separate designations are assigned to each type although their characteristics may vary at some locations; **oligotrophic** (Greek for "poorly nourished") and **eutrophic** ("well nourished") are chosen to reflect the biological productivity of each type of lake in relation to the nutrients available in the surrounding landscape. Subjects such as these represent examples of limnology, a branch of ecology dealing with the chemical, physical, and biological properties of water in lakes, rivers, and other freshwater systems.

Oligotrophic lakes typically have low surface-to-volume ratios, and their waters are crystal clear, deep, and

Infobox 3.1 The tree of life: a cedar tonic

The crew of Jacques Cartier (1491–1557), the explorer of the St Lawrence River, experienced a severe attack of scurvy near present-day Montreal during the winter of 1535–1536. With his fleet icebound in the St Charles River, Cartier and his men spent four months in a primitive fort surviving on salted fish and game but without fresh fruit or vegetables. Of the 110 Frenchmen in his command, Cartier noted that by mid-February "not ten were well enough to help the others, a pitiful thing to see." Cartier's Iroquois hosts also suffered from the same deficiency of ascorbic acid, the source of vitamin C. About 50 of the Native Americans and 25 of the Frenchmen had died before the Iroquois survivors brewed a curative tea made from leaves of the northern white cedar.

The Iroquois chieftan's son Domagaya, who had accompanied Cartier to France the previous year, also suffered from the symptoms of scurvy but recovered rapidly after drinking the tea. Cartier persuaded Domagaya and the Iroquois chief to share the secret remedy and was provided the recipe for the tea infusion. Fearing that the tea contained poison some men refused to partake, but those who did recovered within days. Cartier described how his men, once the medicine had "proven true," fought over the foliage from a large cedar which provided enough tea for six days. The tea probably saved the lives of the remaining Frenchmen and saved the expedition from failure. Years later, Cartier wrote that the physicians and drugs in France could not have done so much in a year as that single tree had accomplished in six days, "for [the medicine] did so prevail that as many as used of it by the grace of God recovered their health."

Impressed with the curative powers of northern white cedar, Cartier subsequently carried seedlings on his return to Europe where the tree gained favor as an ornamental. Unfortunately, Cartier failed to mention the medicinal properties of northern white cedar in his original journal, and scurvy continued to plague many subsequent expeditions. Because of its apparent medicinal properties, the French eventually designated the species *l'arbre de vie* ('the tree of life'), a name persisting today as *arborvitae*.

Today, over 300 varieties of arbor vitae exist. The wood is commercially used for many building products and the oil from the tree is a constituent for soaps, insecticides, cleansers, and other applications. Tea mixtures rarely contain northern white cedar leaves because a neurotoxin contained within can be harmful if ingested for prolonged periods.

Table 3.1 Ecological differences between oligotrophic and eutrophic lakes. The comparisons are general in nature and represent idealized examples of each type of lake. Compiled from Carlson (1977), Dodds and Whiles (2010), and other sources.

Characteristic	Oligotrophic lake	Eutrophic lake
Water temperature	Cool or cold	Warm
Clarity	Clear	Murky to turbid with algae
Biological diversity	Low	High
Dissolved oxygen	High, especially at lower depths	Low, sometimes with dead zones and fish kills
Substrate	Inorganic, often stony	Organic, typically muddy
Primary productivity	Low, little algal growth	High, algal blooms are common
Nutrients	Low concentrations, especially N and K	High concentrations, especially N and K
Surface:volume ratio	Low, usually deep with steep sides	High, usually large surface area relative to depth
Littoral zone	Typically rocky, lacking emergent plants	Muddy, often marsh with many emergent plants

cold. They often appear blue to blue-green in sunlight. Although nitrogen may be abundant in oligotrophic lakes, phosphorus is limited and the overall nutrient content of the water is low. The terrestrial systems (e.g., spruce-fir forests) surrounding oligotrophic lakes provide few nutrients, which contributes to an aquatic system in which the primary producers yield little organic matter. Consequently, decomposer populations on lake bottoms are also reduced,

so oxygen concentrations stay relatively high near the bottom of these lakes. Bottom materials in oligotrophic lakes are largely inorganic and often include rocks and gravel; submerged as well as shoreline vegetation (e.g., cattails and reeds) is sparse or absent.

Oligotrophic lakes and ponds often harbor numerous species even though the population density of each species is usually low. The greatest diversity occurs at much greater depths in oligotrophic lakes than in most lake systems elsewhere; these typically include fishes requiring cold, well-oxygenated waters.

Eutrophic lakes lie at the other end of the spectrum. They occur where the surrounding environment (e.g., deciduous forest) is enriched with nutrients, hence the soils in the watersheds of eutrophic lakes are organic and fertile. High concentrations of nutrients, especially phosphorus and nitrogen, enable these lakes to support an abundance of algae and underwater vegetation. Eutrophic lakes have high surface-to-volume ratios and warm temperatures. Waters supporting aquatic vegetation may remain clear, but algal blooms – easily identified by green-appearing water – often decrease water clarity. Excessive or prolonged algal blooms also commonly cause hypoxia (i.e., lack of oxygen) at times, producing fish kills when the consumption of dissolved oxygen by benthic decomposers exceeds its production. A layer of organic matter often covers the muddy floors of eutrophic lakes, and the shoreline is commonly fringed with well-developed beds of marsh plants. Eutrophic lakes are populated by species that tolerate warmer waters and comparatively low oxygen.

During the course of their geological history, oligotrophic lakes may slowly transform into eutrophic lakes. This process – eutrophication – results when sediments slowly fill the lake basin, thereby reducing its volume and effecting a change in trophic status. Likewise, nutrients gradually accumulate and stimulate greater rates of primary productivity. These two events often act in concert, and the nature of a lake undergoing eutrophication may fall midway between the opposing characterizations summarized in Table 3.1.

Unfortunately, widespread and serious anthropological influences can greatly speed up eutrophication, too often to the detriment of the aquatic system. Influxes of phosphorus from external sources such as agricultural or urban runoff may transform even very large lakes (see *The Great Lakes*, Chapter 12). Currently, global changes in climate seem sure to impact oligotrophic lakes. Some oligotrophic lakes in Boreal Forest have already experienced blooms of a cyanobacterium, *Gloetricha echinulata*, which increase water turbidity and anoxia. These and related topics concerning global climate change are now a focus of limnologists investigating oligotrophic ecosystems.

Appalachian Extension

White and black spruce, along with balsam fir, serves as the primary indicator species for much of the typical Boreal Forest that stretches across North America. However, a somewhat similar forest pokes a long, and sometimes disjointed, finger southward and parallel to the continent's eastern coastline. This forest rides astride the upper elevations of the Appalachian chain, including the White, Adirondack, Blue Ridge, and Great Smoky mountains. Only in the Allegheny Mountains in Pennsylvania, where elevations are lower, is the ecological continuity interrupted. Otherwise, as noted by ecologist B.W. Wells (1884–1978), the settings offer "a glimpse of Canada in North Carolina."

Red spruce and Fraser fir are characteristic species of the Appalachian Extension of Boreal Forest. Red spruce occurs widely, north to south, in the Appalachian chain, but Fraser fir has strong southern affinities and reaches its northern limit on Mount Rogers in southwestern Virginia. Fraser fir favors higher elevations when compared with red spruce. Yellow birch replaces paper birch as the third species of importance in comparisons between spruce-fir forests of the northern and southern Appalachians. Eastern hemlock occurs throughout both the eastern Boreal Forest and Appalachian Extension. Forests of oaks and hickories, among other species, occupy lower elevations and represent Eastern Deciduous Forest (Chapter 4).

Moving north to south, the cool, moist conditions favoring spruce and fir become progressively more dependent on altitude and less on latitude. Spruce and fir forests develop at elevations of about 150 m (490 feet) in Maine, but in the Great Smoky Mountains of Tennessee and North Carolina these forests only find suitable conditions at elevations above 1524 m (5000 feet). A similar relationship can be shown for various kinds of vegetation in North America. As a rule, each 305 m (1000 feet) increase in altitude corresponds to a 322–400 km (200–250 mile) advance northward in latitude.

The Appalachian Extension permits a corresponding southward projection in the distributions of various animals otherwise associated with northern forests. Birds breeding in the spruce forests of Virginia (e.g., the blackburnian warbler) reflect a similar composition of species as found in Maine. The distribution of northern flying squirrels and snowshoe hares shows a similar pattern.

Mountain balds

Treeless sites known as "mountain balds" spot the mountainsides in some forested regions in the southern Appalachians (Fig. 3.5). However, not all of the open areas are surrounded by vegetation associated with the Appalachian Extension of the Boreal Forest. Some mountain balds also lie within the upper edges of oak, beech, and other deciduous vegetation.

Many mountain balds are covered in heath, in particular mountain laurel and various species of rhododendrons. Because of the latter, which includes azaleas, heath balds are ablaze with color in early June. Other mountain balds are either grassy or, more rarely, brushy

Figure 3.5 Treeless areas known as "balds" interrupt forest vegetation on some mountains in the Southern Appalachians (inset). Grasses cover some balds, as shown here, whereas many others feature shrub communities of rhododendron. Photograph courtesy of Craig A. Harper.

stands of alder or beaked hazelnut. Grassy balds are typically meadows of mountain oat grass.

Just how these interesting sites developed and thereafter persisted remains uncertain and controversial. Grassy balds and heath balds may differ significantly in their origins. Some ecologists have suggested origins associated with human activities (e.g., disturbances dating to prehistoric times), whereas others propose natural beginnings for these sites. A long history is suggested by the presence of many endemic species and northern relicts, and these areas may have originated from the grazing pressure of about 20 species of giant herbivores, including mammoths and mastodons, during the Pleistocene.

Highlights

The 10-year cycle

Population cycles, including those of lemmings (Chapter 2), have fascinated ecologists for many years. The 10-year cycle is of particular interest for several reasons, including its close association with the Boreal Forest Biome of North America and the adjacent aspen parkland ecotone described in Chapter 5.

Discovery of the 10-year cycle resulted from work published in 1924 by Charles S. Elton (1900–1991), which included an analysis of furs purchased by the Hudson's Bay Company. Conspicuous peaks in lynx numbers, as measured by the number of pelts brought to trading posts, occurred almost every 10 years (Fig. 3.6). Snowshoe hare population trends are closely synchronized with the cyclic pattern for lynx. The cycle seems based on the predator–prey relationship between the two species, especially because the hare population often peaks a year or two before lynx numbers reach their zenith. However, because the intervals between

peaks sometimes varied from 8 to 11 years (with a long-term average of about 9.6 years), as did population numbers at each peak, some ecologists believed cycles resulted simply from random events. This explanation was later dispelled by a mathematical study based on probability theory, however.

Understandably, the 10-year cycle stimulated a good deal of research and debate, and much remains unresolved. Nonetheless, the cycle seems to be initiated when the snowshoe hare population experiences food shortages during winter, begins to starve, and declines sharply in numbers hastened by poor reproduction. Lynx and other predators exert proportionally greater pressure on the declining hare population, which reaches its low point under the combined forces of predation and starvation. However, as the hares steadily decline, so do the numbers of lynxes, although they switch to ruffed grouse and other foods during this phase of the cycle. Eventually, aspen and other browse are no longer over-utilized, and the hare population rebuilds as its reproductive success increases and mortality decreases. The lynx population likewise recovers as its food supply becomes more plentiful. The cycle begins anew when the demands of the hare population again exceeds the food supply. In most cycles, the responses of the lynx population lag about 2 years behind the fluctuations in hare numbers.

The 10-year cycle therefore seems to be a result of a hare–vegetation interaction, followed by a hare–predator interaction. However, some larger issues remain puzzling, even when the explanation above is accepted unchallenged. These include: (a) Why is a similar cycle not obvious in northern Europe and Asia, where the species and ecological settings are essentially the same? (b) Why is the 10-year cycle restricted to the Boreal Forest and not evident in predator–prey relationships in

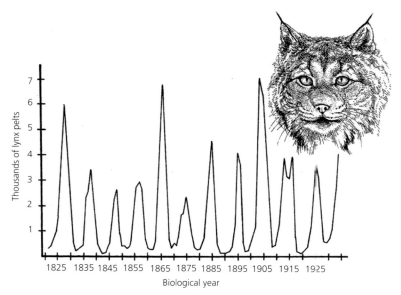

Figure 3.6 Long-term records of fur transactions by the Hudson's Bay Company clearly show oscillations of about 10 years in the numbers of lynx in the Boreal Forest of North America. Snowshoe hares and ruffed grouse display similar patterns. Some features of the 10-year cycle remain an enigma, but shortages of winter food for snowshoe hares apparently trigger the downward phase of the cycle, which thereafter continues because of predator–prey interactions. Illustrated by Tamara R. Sayre, based on Elton and Nicholson (1942).

grasslands, deserts, or other biomes? (c) Why is the cycle consistently 10 years and not 5 years, 12 years, or some other length?

Wolves and moose

The quaint designation of Spruce–Moose Biome for Boreal Forest brings to mind the exceptional investigations of moose and their principal predator on Isle Royale. This long-term research revealed a good deal about the mechanics of predation, one of the more controversial topics of ecology. In the process, much was learned about the behavior and ecology of wolves.

Isle Royale, a US national park since 1940, lies 24 km (15 miles) off the northwestern shores of Lake Superior. The island is a granite outcrop of 544 km² (210 square miles) forested primarily with spruce and fir, but also a number of other species. Moose are relative newcomers to Isle Royale. A few had immigrated to the island by 1912, and the fledgling populations thereafter soared to at least 1000 and perhaps 3000 animals by 1929. Such numbers lay well beyond the island's **carrying capacity** and, when food supplies were depleted, the moose population dwindled. A fire in 1936 resulted in a fresh supply of aspen and other browse and the rebuilding of the moose population.

Wolves are even newer arrivals to Isle Royale. The first solid evidence of their presence was noted in 1949 after an exceptionally cold winter likely produced an icy connection between island and mainland. Isle Royale emerged as a model laboratory for studying predator–prey relationships. The setting was unusual because it remained relatively free of complicating influences. In particular, neither moose nor wolves were subject to hunting or trapping, and there were few alternate prey choices for the wolves. Winter conditions also helped; snow allowed the animals to be tracked either on the ground or from a light plane (Fig. 3.7a), and the cold preserved dead moose for post-mortem examinations (Fig.3.7b). More important, the animals were confined to the island, that is, it was essentially a closed system allowing no additional immigration or emigration. In short, Isle Royale represented a giant test tube whose contents were ready for scientific scrutiny.

Four findings (of many) deserve mention here. The first concerns the hunting success of wolves, which could be determined from tracks in the snow as well as, on occasion, from direct observations. Some of the 160 moose for which records were obtained remained undiscovered by wolves simply on the basis of chance, whereas wolves located 82% of their potential prey (131 moose) by sight or smell. Of the latter, 77 (59%) were "tested" by the wolves, but most successfully resisted the challenge. In fact, only six (just 8%) of the tested moose were killed. The conclusion emerges that wolves are not

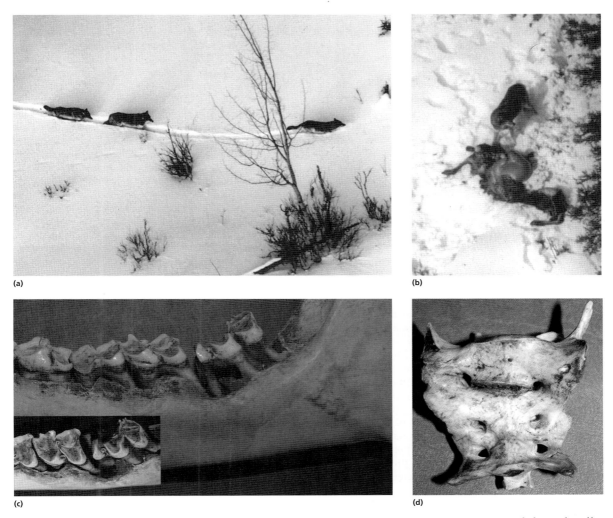

(a)

(b)

(c)

(d)

Figure 3.7 Wolf and moose populations on Isle Royale offer an excellent opportunity to study predation. (a) Snow helps track wolf movements, which sometimes lead to (b) kills of moose calves. (c) Examinations of the carcasses often reveal "lumpy jaw," a debilitating infection that commonly predisposes moose to predation. (d) The inbred population of gray wolves on the island was recently found to exhibit lumbosacral deformities, the result of genetic deterioration that limits movement and survival. Photographs courtesy of (a, b) Rolf O. Peterson, (c) Dale C. Lockwood, and (d) Jannikke Räikkönen.

particularly efficient predators, or at least were not a great threat to the moose population.

Second, the majority of the moose killed by wolves were either calves and therefore defenseless when separated from their cows, or infirm, many of which were older moose. Moose between 1 and 6 years of age were seldom victims. Nearly one-third of the adult moose killed by wolves suffered from "lumpy jaw," a necrotic and debilitating inflammation of the mandible (Fig. 3.7c). This and other impairments, including heavy lung infections of tapeworm cysts, often predisposed moose to successful wolf attacks. Wolf predation, for the most part,

acted as a culling agent, leaving a healthy moose herd intact. Predation in this context represents a rigorous form of natural selection.

Third, this more-or-less normal pattern of predation at times may be altered by chance events. To illustrate, a winter of unusually heavy snowfall produced an abrupt turn in events. Moose struggling in deep snow faced a greater disadvantage when defending themselves against wolves. As a result, even healthy animals were not always able to repel an attack. The dynamics of predation thus changed for a time, and the moose population was depressed to a greater degree than usual. This discovery highlights the importance of long-term research, without which the effect of heavy snowfall might never have emerged as part of the equation. Although years of exceptional snowfall (or fires, flooding, and similar phenomena) can be expected sooner or later, the dates of their occurrence cannot be predicted. Instead, these chance phenomena known as **stochastic** events are considered in terms of their mathematical probability. Most stochastic events concern unusual weather, but they also include outbreaks of disease (known as **epizootics**) and other biological phenomena.

Finally, the research at Isle Royale illustrates a case of dynamic equilibrium among the biological resources in a natural community. In the absence of wolves, moose numbers reached a maximum of about 3000 which resulted in overbrowsing, starvation, and depleted habitat. But with wolves, the moose population stabilized at about 600 animals and a healthy supply of browse. In such a situation, the prey population survives while producing a crop for the wolves to harvest.

How are wolf numbers limited? Wolves, like moose, certainly experience starvation, disease, and other debilitations, but social structure also regulates their populations. Within each pack, an alpha male reigns as the dominant individual, a status shared with his mate. Other wolves in the pack may form pair bonds, but only the alpha pair actually breed. On occasion, the second-ranking male (the beta male) may mate with the alpha female, but only with the forbearance of the dominant male. Reproduction therefore remains limited, well below the full potential of the pack, and the size of wolf packs remains fairly stable from year to year. The structure of wolf society therefore includes behavioral adaptations that thwart the obvious difficulty of a predator population exceeding its food supply.

Today the wolves of Isle Royale – and in turn the island's moose and forest ecosystem – face a challenge only recently revealed: the wolves suffer from **inbreeding depression**, which in this case includes seriously deformed vertebrae, along with indications of stillbirths and a type of blindness known as opaque eye. Fully 58% of the wolves on Isle Royale show some evidence of pronounced congenital malformation of the vertebral column (Fig. 3.7d), whereas only 1% of two wolf populations elsewhere exhibited similar deformities. This revelation indicated that genetic deterioration had remained undetected on Isle Royale for some time, despite long-term studies of the island's wolves and their role as predators. In dogs, these deformities lead to paralysis, back pain, and locomotor difficulties in the rear legs and tail.

By 2014, the wolf population had plummeted to just eight individuals, all related, whereas 24 represents its long-term average. As the wolves decline moose increase and soon damage the forest, especially balsam firs, which are replaced with stands of less palatable pine or spruce. In time, starvation triggers a crash in the numbers of moose and, with the fir gone, the remaining forest can no longer support enough moose to subsequently maintain a viable wolf population.

Among the potential responses – including doing nothing and simply seeing what happens – is to infuse new genes by releasing individuals from other wolf populations, as similarly practiced elsewhere for Florida panthers and other species with limited gene pools; this is an example of **conservation biology**. Such "genetic rescue" in fact happened when a single male crossed the ice from the mainland to Isle Royale in 1997, a chance event in one of the few years in recent decades when an ice bridge developed. The immigrant, a particularly fit male, took over one of the packs and sired 34 offspring before his death in 2006. In a single generation, the genes of this alpha male occurred in 56% of the population, an event characterized as a "genomic sweep" and evidence of what a planned introduction of new blood might accomplish. One expert believes the appearance of this single wolf saved the population for another 10–15 years. Without the regular formation of ice bridges between Isle Royale and the mainland, which seems unlikely in an era marked by rising temperatures, the odds of further immigration are however slim.

In April 2014, the National Park Service announced that it would not introduce wolves from the mainland to improve the genetic vitality of the island's wolf population. The decision will stand so long as the wolves on Isle Royale continue breeding (three pups were produced in 2013). However, the death of the last male or female in the remaining wolf population or signs of moose overbrowsing the island's vegetation may be sufficient evidence to reverse the current decision.

More than 55 years of continuous study at Isle Royale have revealed considerable knowledge about the relationships between predator and prey, among them that ecosystems lacking the membership of viable predator populations are neither whole nor wholesome. In the words of Thoreau, populations of prey species without predators resemble "a tribe of Indians that had lost all its warriors."

A wealth of salamanders

The cool, moist conditions favoring development of the Appalachian Extension of Boreal Forest, described earlier, also produced a wonderful collection of salamanders. In fact, the Great Smokies and other mountains in the southern Appalachians offer a natural laboratory in which to study the complexities of speciation. To set the stage, we must briefly consider two geological features of the region. First, the southern Appalachians are among the oldest mountains in North America. The mountains and their salamanders – which represent one of the more ancient taxa of vertebrates – have shared a long history with each other. Second, glaciers did not reach the southern Appalachians, which means that the salamander populations continued diversifying for millions of years after those to the north had been eliminated. Only after the glaciers retreated did a few species return to the northern mountains, while an unusually rich assembly of salamanders evolved in the southern Appalachians.

How did this happen? The answer lies in the isolation of the mountaintops. At the time of the Ice Age (i.e., the Pleistocene), southern climates were far cooler than today and habitats at lower elevations resembled those on the mountaintops. Cool, moist forests prevailed in much of the region, and animals that adapted to these conditions could move freely across the intervening valleys to adjacent mountains. More to the point, their gene pools continually mixed. However, the climate warmed when the ice sheets retreated, and suitable habitat for many salamanders remained only at higher elevations. The valleys became ecological barriers for travel between mountains, ending genetic exchange and fragmenting the salamander populations into mountaintop "islands."

Continued isolation produced a complex of salamanders on mountains only a short distance apart geographically, but widely separated ecologically. The woodland salamander complex provides a particularly interesting example of mountaintop speciation. Many species of salamanders in this complex dwell primarily in the spruce-fir zone where the females lay eggs in decaying logs instead of breeding in water, as is the case for many other kinds of salamanders. These salamanders also lack lungs and instead absorb oxygen exclusively through their moist skin. Of particular interest, their markings vary in color and pattern from one mountain range to another, and sometimes occur uniquely on a single peak. The body structure, behavior, and ecology vary little among the members of this complex, but each species is genetically distinct.

The several species of woodland salamanders in the southern Appalachians illustrate the mechanics of speciation, which result from the reproductive and subsequent genetic isolation of units within a larger group. In the initial stages of isolation, animals in each unit might still breed with the others if they could somehow still cross the valleys to find mates. In time, and with no further changes in climate and habitat, the isolation and continued inbreedings will produce mountaintop populations no longer genetically compatible with each other. Gradually, each isolated population will become a distinct species.

Red squirrels

The range of the red squirrel coincides with virtually all of the Boreal Forest, including the Appalachian Extension and many areas of Montane Forest (Chapter 9). Coniferous forests and red squirrels are also closely associated in large areas of Europe and Asia (i.e., a Holarctic distribution). Wherever they occur, red squirrels busily harvest and store cones from coniferous trees (Fig. 3.8).

Red squirrels feed opportunistically during the summer, and their varied diet includes mushrooms, berries, buds, lichens, seeds, sap, and insects. In winter however, red squirrels rely heavily on seeds from the cones of spruce, pine, and other coniferous trees. Cutting and gathering cones for their caches occupies as much as 80% of a red squirrel's daily activity between August and November. The caches – large piles of uncovered cones – often represent the accumulations of several generations of red squirrels. One such stockpile contained 15 bushels of cones, but the volume in most caches is somewhat smaller. A single red squirrel may cut and cache 12,000–16,000 cones each season and, from these, the seeds of almost 200 cones a day. Because the caches remain moist (the debris from previous feedings serves as mulch) the cones remain closed and the seeds may remain viable for several years. Foresters and horticulturalists have in fact exploited caches as a cheap source of large quantities of conifer seeds for planting. In some regions, the caches of red squirrels also provide grizzly bears with a ready source of energy-rich food (Chapter 9). Although uncommon, two consecutive years of poor cone crops may severely reduce local populations of red squirrels.

Red squirrels vigorously defend their caches from other squirrels by employing both vocal and visual displays followed, if needed, by spirited chases whenever other squirrels intrude. This is classic **territorial behavior**, which intensifies in red squirrels in later autumn and early winter when caches become their primary sources of food. Territorial behavior, that is, defense of an area, represents an effective way for red squirrels to protect their individual supplies of winter food. In contrast, gray and fox squirrels, which scatter and bury small hoards of large foods, such as a single acorn or hickory nut, have no central location to defend and lack well-developed territorial behavior.

Territorial behavior apparently regulates the densities of red squirrel populations. Each **territory** is occupied by a single red squirrel and defended against others, regardless of sex. Those individuals that cannot hold territories – and therefore lack caches – simply perish in winter, resulting in relatively stable year-to-year population densities. In contrast, the populations of several other animals in the Boreal Forest experience pronounced cyclic behavior, as noted earlier.

Figure 3.8 Red squirrels accumulate and vigorously defend stockpiles of cones known as middens, as shown here at the base of a fir. A single squirrel (inset) maintains these food reserves, but the middens often represent the work of several generations. Some middens may cover 20 m² (215 square feet) at a depth of 1 m (3 feet). Note the excavations where a squirrel searched for cones to its liking. Photographs courtesy of Sue Thomas (www.suethomasphoto.ca) and James F. Parnel .

Ecological challenges

All nations within the range of Boreal Forest rely on its ecological and economic values. These spurred geopolitical development beginning in the 16th century and intensified in the 20th century. During this period, human encroachment, logging, and mineral acquisition all impacted forest ecosystems, but here we briefly discuss the ecological effects of acid rain and the introduction of insect pests, using the spruce budworm and balsam woolly adelgid as examples (omitting others such as the jack pine budworm and the forest tent caterpillar). Global climate change will no doubt produce significant changes, but these are not yet fully clear and likewise deferred.

Acid rain

Unfortunately, the emissions from industrial centers – many in the midwestern "Rust Belt" of North America – have damaged large areas of Boreal Forest, including the Appalachian Extension. These emissions (primarily nitrogen oxides and sulfur dioxide) spew from coal-fired electric and manufacturing plants where they rise and mix with atmospheric moisture to form acidic compounds. When disbursed, what is widely known as acid rain often falls in areas far removed the point of origin.

The corrosive action of acid rain was first noted on limestone and marble structures in 17th century London, but the ecological effects of the phenomenon were not apparent until the 1960s. Acid rain affects forest vegetation, freshwater systems, and soils. Once reaching a pH of 5.0 or less, the acidity limits the hatching success of fish eggs and, at some locations, kills aquatic insects and trout and other fishes. Although the spodsol soils of Boreal Forest are normally acidic, the addition of still more acid eliminates soil microbes and increases leaching of essential nutrients and minerals. Forests, especially those at high altitudes in the Appalachian Extension, are often shrouded in clouds and fog heavily contaminated with acidic vapor. The acidic fog leaches calcium from the needles of red spruce and other conifers, concurrently lessening their tolerance to colds and increasing winter injury, circumstances often leading to their death. Trees weakened by acid rain are also prone to large-scale attacks from insects. For example, a chain reaction likely initiated by acid rain, followed by insect defoliation, produced a skeleton forest of dead and dying trees on Grandfather and other mountains in western North Carolina.

Beginning in the 1970s, government agencies began targeting the airborne pollutants producing acid rain.

Amendments (in 1990) to the Clean Air Act established a successful cap-and-trade program that limits emissions of sulfur dioxide and nitrogenous oxides. Atmospheric levels of these compounds currently fall below projected levels, and the costs to businesses and consumers proved significantly less than anticipated.

Spruce budworm and DDT

Another sort of chain reaction followed in the wake of insect plagues in Boreal Forest. Spruce budworms, true to their name, attack white and other spruces, but these insects actually favor balsam fir. The insects range from the Yukon Territory eastward to Newfoundland, coinciding with the distribution of both white spruce and balsam fir. The 20th century witnessed three outbreaks, each lasting 10 or more years. When outbreaks occur, the mass flights of moths resemble snow swirling around the tree crowns. Prevailing winds sometimes aid these flights, which may cover considerable distances; one extended from mainland Canada to Newfoundland.

Each female lays 100–200 eggs in groups of 10–20 per needle. After hatching, the larvae spend the winter hibernating under bark scales, enclosed in silken webs. When the caterpillars emerge the following spring, they eat older needles but switch to new foliage as it appears. The severed needles, held in place by silken webs, turn brown and appear fire-damaged. Outbreaks of spruce budworm occur irregularly and, because these last several years, the continual defoliation eventually kills the trees. Stand mortality reaches 75%, with the surviving trees thereafter experiencing years of curtailed growth.

In response, huge areas of Boreal Forest were sprayed with DDT, an insecticide widely used after WWII and made infamous by Rachel Carson (Infobox 3.2) in *Silent Spring*. DDT, a chlorinated hydrocarbon, persists in the environment for many years after its application. It sometimes kills directly but more often accumulates in fat deposits, where it is released when stressful events such as migration and winter storms demand extra energy.

Significantly, DDT accumulates into ever-greater concentrations upward through the food chain, a process known as **biomagnification.** Regrettably, repeated applications of DDT during plagues of spruce budworm in eastern Canada corresponded with marked reductions in woodcock populations. The number of young birds per adult female declined as the insecticide applications increased. DDT and its chemically related compounds, primarily heptachlor and dieldrin, also harmed other kinds of birds, from robins to bald eagles, throughout North America. These insecticides at times killed robins and other birds outright or, more insidiously, disrupted the formation of normal egg shells in brown pelicans and several species of raptors.

The title of *Silent Spring*, which suggested that DDT and other agricultural chemicals might cause cancer in humans as well as harm wildlife, refers a spring morning without the music of songbirds. This message initiated a public outcry and, in 1972, legal use ended for DDT and similar pesticides in the United States. The United Nations followed suit with enactment of the Stockholm Convention on Persistent Organic Pollutants, an international treaty that banned applications of persistent chemicals throughout most of the world. This pact, ratified in 2004, banned DDT for agricultural purposes but allowed its use to control insect-borne diseases such as malaria. The ban on DDT in the United States, coupled with the Endangered Species Act of 1973, helped reverse the threat of extinction of several birds, including bald eagles, peregrine falcons, and brown pelicans.

Balsam woolly adelgid

Native to silver fir forests in central Europe, the balsam woolly adelgid is an **exotic species** which was introduced to the balsam fir forests of eastern Canada and Maine around 1900. Once established in northeastern North America, populations of this insect spread westward and southward, soon becoming serious pests in the Fraser fir forests of the southern Appalachians. By the time balsam woolly adelgids were discovered at Mount Mitchell State Park, North Carolina, they had already killed thousands of fir trees on more than 2832 hectares (7000 acres) of spruce-fir forest.

Balsam woolly adelgids are **parthenogenic**, and an adult female can lay more than 200 eggs. After hatching, the first instar nymphs (known as "crawlers", the only stage when these insects move on their own) travel short distances from the branches where they hatched. At times, winds may blow the crawlers to other locations. The tiny, reddish-brown crawlers insert their tube-like mouthparts into the bark then enter a period of diapause, never again to move. Upon emerging from diapause, they cover their bodies with a waxy, wool-like secretion. Within this protective woolly mass, wingless adult females develop from the nymphs and lay their eggs. Balsam woolly adelgids are capable of producing two, and occasionally three, generations in a single summer.

Clusters of adelgids tend to congregate on either the main branches or the outer stems of tree crowns. The insects concentrate in bark fissures of trees larger than 4 cm (1.5 inches) in diameter at breast height, and large masses sometimes give the trees a whitewashed appearance. Trees respond to the insects' salivary secretions with an "allergic" reaction. In larger stems and branches, heartwood formation increases whereas sapwood constricts, accompanied by diminished water flow in the affected area. Concentrations of needles in the crown steadily thin and dieback occurs. The greatest mortality

Infobox 3.2 Rachel Carson (1907–1964), nature's advocate

Only a few individuals have influenced the course of history with the power of their writings but Rachel Carson was such a person. Born and raised on a farm in Springdale, Pennsylvania and the youngest of three children, she was introduced to the world of nature by her mother. Her passion for nature and her talent for writing quickly emerged when, at 10, she published her first magazine article.

While earning a BS degree in biology at Pennsylvania College for Women (now Chatham College), Carson held a summer fellowship at the US Marine Laboratory at Woods Hole, Massachusetts, an experience that triggered her special interest in oceanography and marine biology. After graduation, she received a scholarship enabling her to complete an MS degree from Johns Hopkins University in Baltimore.

Her first professional job was as script writer for the radio show *Romance Under the Waters*, produced by the US Bureau of Biological Survey (precursor of the US Fish and Wildlife Service). In 1936, Carson was the first woman to take and pass the civil service test, a milestone that qualified her to become a biologist with the Bureau of Fisheries. She rose steadily in rank, eventually serving as the chief editor for all publications issued by the US Fish and Wildlife Service. Even today, many of the descriptive brochures available at national wildlife refuges parks still contain paragraphs she wrote long ago. The success of her second book *The Sea Around Us*, which stayed on the NY Times' best-seller list for 81 weeks and appeared in 32 languages, persuaded Carson to resign her government position in favor of full-time writing.

During research for her three books about oceans, Carson became aware of the environmental consequences resulting from the indiscriminate use of pesticides. She realized that she had the economic freedom and independence to challenge these abuses and thus began documenting the ecological effects of DDT, one of the most widely applied pesticides at the time. Aided by two former employees of the Fish and Wildlife Service, Shirley Biggs and Clarence Cottam, Carson published her exposé *Silent Spring* in 1962. The book was based on science, but it was the powerful image of a spring unaccompanied by the melodies of songbirds that resonated with the public.

Silent Spring became a runaway bestseller but nonetheless stirred great controversy; as might be expected, it prompted an extensive campaign by the pesticide industry to discredit the book along with an attack on Carson's integrity and mental stability. Despite these attacks, President John F. Kennedy responded to the book's influence by directing federal agencies to test the pesticides mentioned in *Silent Spring* and Carson, along with other witnesses, testified before Congress about the dangers posed by pesticides. As a result, continued use of all organochlorine pesticides, including DDT, was eventually banned in the United States. *Silent Spring* also sparked creation of both the Environmental Protection Agency and a national environmental movement that peaked in the 1970s but continues today.

Carson and her works won many awards, including (among others) those from the National Audubon Society, Garden Club of America, and the New York Zoological Society. She was recognized by the National Wildlife Federation as the 1963 Conservationist of the Year.

After her death, she was further honored when the US Fish and Wildlife Service dedicated the Rachel Carson National Wildlife Refuge near her summer home on Maine's coastline. Her words, spoken in a television interview, provide everlasting insight into her conviction that "…man's endeavors to control nature by his powers to alter and destroy would inevitably evolve into a war against himself, a war he would lose unless he came to terms with nature."

occurred in stands of Fraser fir in the vicinity of Mt Mitchell, but adult firs on Mt Rogers, Virginia, proved resistant and were not affected.

Infestations of balsam woolly adelgids substantially altered communities in the southern Appalachians. In hard-hit areas, the insects virtually eliminated Fraser fir and reduced the canopy cover by more than half. Birds that normally foraged in the canopy and subcanopy were replaced by species better adapted to earlier stages of forest succession. The combined density for all breeding birds declined by half, and some species of birds were locally extirpated. Fortunately, avian populations in other forest types in the Great Smoky Mountains helped buffer the adelgid's indirect but harmful effects on the avian community.

The Boreal Forest Agreement

At the beginning of the current century Canada's Boreal Forest experienced extensive logging, primarily for pulp for paper products. Some estimates indicated the forest fell to the axe and saw a rate of more than 48 hectares (120 acres) per hour, 24 hours per day. Such widespread clear-cutting degraded habitat for many species of wildlife, but especially the wintering territory of barren-ground caribou.

Early in 2010, 21 major forest products companies and 9 leading environmental organizations in Canada agreed to a three-year moratorium that protected 72 million hectares (178 million acres) from further logging activities. During this period, the groups also agreed to identify and protect species at risk as well as the diversity of the Boreal Forest ecosystem. The plan focused on the preservation of caribou habitat on company-owned lands, with secondary objectives designed to develop sustainable, yet environmentally responsible, forest management. In sum, all parties to the agreement committed to a strong and competitive forest industry while protecting communities and sensitive species in the Boreal Forest.

Readings and references

The Boreal Forest Biome

Hoffman, R.S. 1958. The meaning of the word "taiga." Ecology 39: 540–541.

Larsen, J.A. 1980. *The Boreal Ecosystem*. Academic Press, New York, NY.

Quinn, J.A. 2008. *Arctic and Alpine Biomes*. Greenwood Press, Westport, CT.

Climatic boundaries and soils

Barry, R.G. 1967. Seasonal location of the Arctic front over North America. Geography Bulletin 9: 79–95.

Bryson, R.A. 1966. Air masses, streamlines, and the boreal forest. Geography Bulletin 8: 228–269.

Woodin, S.J. and M. Marquiss. 1997. *Ecology of Arctic Environments: 13th Special Symposium of the British Ecological Society*. Cambridge University Press, London.

Features and adaptations

Plant adaptations

Davis, R.B. 1966. Spruce-fir forests on the coast of Maine. Ecological Monographs 36: 79–94.

Larsen, J.A. 1980. *The Boreal Ecosystem*. Academic Press, New York, NY.

Animal adaptations

Benkman, C.W. 1987. Crossbill foraging behavior, bill structure, and patterns of food profitability. Wilson Bulletin 99: 351–368.

Benkman, C.W. 1989. Seed handling efficiency, bill structure, and the cost of specialization for crossbills. Auk 105: 715–719.

Pendergast, B.A. and D.A. Boag. 1973. Seasonal changes in the internal anatomy of spruce grouse in Alberta. Auk 90: 307–317.

Robinson, W.L. 1969. Habitat selection by spruce grouse in northern Michigan. Journal of Wildlife Management 33: 113–120.

Frequent fires

Scotter, G.W. 1964. Effects of forest fires on the winter range of barren-ground caribou in northern Saskatchewan. Canadian Wildlife Service Wildlife Management Bulletin Series 1(18): 1–111.

Scotter, G.W. 1970. Wildfires in relation to the habitat of barren-ground caribou in the tiaga of northern Canada. Proceedings of the Tall Timbers Fire Ecology Conference 10:85–105.

Wradle, D.A., O. Zackrisson, and M.C. Nilsson. 1998. The charcoal effect in boreal forests: mechanisms and ecological consequences. Oecologia 115: 419–426.

Niches in the Boreal Forest

Hardin, G. 1960. The competitive exclusion principle. Science 131(3409): 1292–1297.

Johnsgard, P.A. 1973. *The Grouse and Quails of North America*. University of Nebraska Press, Lincoln, NE.

MacArthur, R.H. 1958. Population ecology of some warblers of northeastern coniferous forests. Ecology 39: 599–619.

Selected biotic communities

Tree line and forest tundra

Bryson, R.A., W.N. Irving, and J.A. Larsen. 1965. Radiocarbon and soil evidence of former forest in the southern Canadian Tundra. Science 147: 46–48.

Elliot-Fisk, D.L. 1983. The stability of the northern Canadian tree limit. Annals of the Association of American Geographers 73: 560–576.

Lescop-Sinclair, K. and S. Payette. 1995. Recent advance of the Arctic treeline along the eastern coast of Hudson Bay. Journal of Ecology 83: 929–936.

Post, E., M.C. Forchhammer, M.S. Bret-Harte, T.V. Callaghan, et al. 2009. Ecological dynamics cross the Arctic associated with recent climate change. Science 325: 1355–1358.

Skre, O., R. Baxter, R.M.M. Crawford, T.V. Callaghan, and A. Fedorkov. 2002. How will the Tundra-tiaga interface respond to climate change? Ambio 31: 37–46.

Suarez, F., D. Brinkley, M.W. Kaye, and R. Stottlemeyer. 1999. Expansion of forest stands into Tundra in the Noatak National Preserve, northwest Alaska. Ecoscience 6: 465–470.

Muskeg

Crum, H. 1973. Mosses of the Great Lakes forest. Contributions of the University of Michigan Herbarium 10: 1–404.

Crum, H. and L.E. Anderson. 1981. *Mosses of Eastern North America*. Vol. 1. Columbia University Press, New York, NY. (Volume 1 includes *Spaghnum*.)

Jeglum, J.K. 1972. Boreal forest wetlands near Candle Lake, central Saskatchewan, I. Vegetation. Musk-Ox 11: 41–58.

Lewis, F.J. and E.S. Dowding. 1926. The vegetation and retrogressive changes of peat areas ("muskegs") in central Alberta. Journal of Ecology 14: 317–341.

Mark, A.F. 1958. The ecology of the southern Appalachian grass balds. Ecological Monographs 28: 293–336.

Neiland, B.J. 1960. The forest-bog complex in southeast Alaska. Vegetation 22: 1–64.

Noble, M.G., D.B. Lawrence, and G.P. Streveler. 1984. *Sphagnum* invasion beneath an evergreen forest canopy in southeastern Alaska. Bryologist 87: 119–127. (Cites data for the water-holding capacity of sphagnum mosses.)

Radforth, N.W. and C.O. Brawner. (eds) 1977. *Muskeg and Northern Environment in Canada*. University of Toronto Press, Toronto, Ontario, ONT.

Rigg, G.B. 1940. The development of sphagnum bogs in North America. Botanical Review 6: 666–693.

Segadas-Vianna, S. 1955. Ecological study of peat bogs of eastern North America: the *Chamaedaphne* community of Quebec and Ontario. Canadian Journal of Botany 33: 647–684.

Zach, L.W. 1950. A northern climax, forest or muskeg? Ecology 31: 304–306.

Coniferous swamps

Bunting, M.J., C.R. Morgan, M. Van Bakel, and B.G. Warner. 1998. Pre-European settlement conditions and human disturbance of a coniferous swamp in southern Ontario. Canadian Journal of Botany 76: 1770–1779.

Hörnberg, G., O. Zackrisson, U. Segerström, et al. 1998. Boreal swamp forests. BioScience 48: 795–801.

Northern coldwater lakes

Carey, C.C. and K. Rengefors. 2010. The cyanobacterium *Gloeotricha echinulata* stimulates the growth of other phytoplankton. Journal of Plankton Research 32: 1349–1354.

Carlson, R.E. 1977. A trophic state index for lakes. Limnology and Oceanography 22: 361–369.

Dodds, W.K. and M.R. Whiles. 2010. *Freshwater Ecology, Second Edition: Concepts and Environmental Applications of Limnology (Aquatic Ecology)*. Academic Press, New York, NY.

Fey, S.B., Z.A. Mayer, S.C. Davis, and K.L. Cottingham. 2010. Zooplankton grazing of *Gloeotricha echinulata* and associated life history consequences. Journal of Plankton Research 32: 1337–1347.

Appalachian Extension

Brown, D.M. 1941. The vegetation of Roan Mountain: phytosociological and successional study. Ecological Monographs 11: 61–97.

Crandall, D.L. 1958. Ground vegetation patterns of the spruce-fir area of the Great Smoky Mountains National Park. Ecological Monographs 28: 337–360.

Holt, P.C. (ed.) 1970. The distributional history of the southern Appalachians. Part II: Flora. Virginia Polytechnic Institute and State University Research Division Monograph 2: 1–414.

Oosting, H.J. and W.D. Billings. 1951. A comparison of virgin spruce-fir forest in the northern and southern Appalachian system. Ecology 32: 84–103.

Peattie, D.C. 1946. *A Natural History of Trees in Eastern and Central North America*. Houghton Mifflin, Boston, MA.

Stephenson, S.L. and H.S. Adams. 1949. The spruce-fir forest on the summit of Mount Rogers in southwestern Virginia. Bulletin of the Torrey Botanical Club 111: 69–75.

Stewart, R.E. and J.W. Aldrich. 1949. Breeding bird populations in the spruce region of the central Appalachians. Ecology 30: 75–82.

Mountain balds

Billings, W.D. and A.F. Mark. 1957. Factors involved in the persistence of montane treeless balds. Ecology 38: 140–142.

Brown, D.M. 1941. The vegetation of Roan Mountain: phytosociological and successional study. Ecological Monographs 11: 61–97.

Cain, S.A. 1930. An ecological study of the heath balds of the Great Smoky Mountains. Butler University Botanical Studies 1: 177–208.

Weigl, P.D. and T.W. Knowles. 1995. Megaherbivores and southern Appalachian grass balds. Growth and Change 26: 365–382.

Wells, B.W. 1961. The southern Appalachian grass bald problem. Castanea 26: 98–100.

Whittaker, R.H. 1963. Net production of heath balds and forest heaths in the Great Smoky Mountains. Ecology 44: 176–182.

Highlights

The 10-year cycle

Cole, L.C. 1954. Some features of random cycles. Journal of Wildlife Management 18: 2–24.

Elton, C. 1924. Periodic fluctuations in the numbers of animals: their causes and effects. British Journal of Experimental Biology 2: 119–163.

Elton, C. and M. Nicholson. 1942. The ten-year cycle in numbers of lynx in Canada. Journal of Animal Ecology 11: 215–244.

Hodges, K.E., C.J. Krebs, and A.R.E. Sinclair. 1999. Snowshoe hare demography during a cyclic population low. Journal of Animal Ecology 68: 581–594.

Keith, L.B. 1963. *Wildlife's Ten-Year Cycle*. University of Wisconsin Press, Madison, WI.

Keith, L.B. 1990. Dynamics of snowshoe hare populations. In: *Current Mammalogy* (H.H. Genoways, ed.). Plenum Press, New York. NY, pp. 119–195.

Keith, L.B., A.W. Todd, C.J. Brand, R.S. Adamcik, and D.H. Rusch. 1977. An analysis of predation during a cyclic fluctuation of snowshoe hares. Proceedings of the International Congress of Game Biology 13: 151–175.

Keith, L.B., J.R. Cary, O.J. Rongtsad, and M.C. Brittingham. 1984. Demography and ecology of a declining snowshoe hare population. Wildlife Monographs 90: 1–43.

Krebs, C.J. 1996. Population cycles revisited. Journal of Mammalogy 77: 8–24.

Krebs, C.J., S. Boutin, R. Boonstra, A.R.E. Sinclair, J.N.M. Smith, M.R.T. Dale, and R. Turkington. 1995. Impact of food and predation on the snowshoe hare cycle. Science 268: 112–115.

Wolves and moose

Adams, J.R., L.M. Vucetich, P.W. Hedrick, et al. 2011. Genomic sweep and potential genetic rescue during limited environmental conditions in an isolated wolf population. Proceedings of the Royal Society B 278: 3336–3344. (Genetic impact of a single immigrant whose fitness exceeded that of the resident population.)

Allen, D.L. 1979. *The Wolves of Minong*. Houghton Mifflin, Boston, MA.

Krefting, L.W. 1974. The ecology of the Isle Royale moose with special reference to the habitat. University of Minnesota Agricultural Experiment Station Forestry Series 15, Technical Bulletin 297: 1–75.

Marris, E. 2014. Iconic island study on its last legs. Nature 506: 140–141. (See also an editorial in the same issue, page 132.)

McLaren, B.E. and R.O. Peterson. 1994. Wolves, moose, and tree rings on Isle Royale. Science 266: 1555–1558.

Mech, L.D. 1966. The wolves of Isle Royale. National Park Service Fauna Series 7: 1–210.

Mlot, C. 2013. Are Isle Royale's wolves chasing extinction? Science 340: 919–921

Peterson, R.O. 1995. *The Wolves of Isle Royale, a Broken Balance*. Willow Creek Press, Minocqua, WI.

Post, E., R.O. Peterson, N.C. Stenseth, and B. McLaren. 1999. Ecosystem consequences of wolf behavioural responses to climate. Nature 401: 905–907.

Räikkönen, J., J.A. Vucetich, R.O. Peterson, and M.P. Nelson. 2009. Congenital bone deformities and the inbred wolves (*Canis lupus*) of Isle Royale. Biological Conservation 142: 1025–1031. (See for photos of the abnormal vertebrae.)

Thoreau, H.D. 1856. The Journal of Henry David Thoreau, available at http:// http://www.walden.org/Library/The_Writings_of_Henry_David_Thoreau:_The_Digital_Collection/Journal (accessed 8 January 2015). (See his journal entry for March 23 for the quote regarding predators as analogs of warriors. Thoreau's journals are widely available, including online.)

A wealth of salamanders

Beane, J.C., A.L. Braswell, J.C. Mitchell, W.R. Palmer, and J.R. Hairston, III. 2010. *Amphibians and Reptiles of the Carolinas and Virginia*, second ed. The University of North Carolina Press, Chapel Hill, NC.

Hairston, N.G. 1949. The local distribution and ecology of the plethodontid salamanders of the southern Appalachians. Ecological Monographs 19: 48–73.

Highton, R. 1971. Distributional interactions among eastern North American salamanders of the genus *Plethodon*. In: *The Distributional History of the Biota of the Southern Appalachians. Part III: vertebrates* (P.C. Holt, ed.). Virginia Polytechnic Institute and State University Research Division, Monograph 4: 139–188.

Organ, J.A. 1961. Studies of the local distribution, life history, and population dynamics of the salamander genus *Desmognathus* in Virginia. Ecological Monographs 31: 189–220.

Petranka, J.W., M. E. Eldridge, and K.E. Haley. 1993. Effects of timber harvesting on southern Appalachian salamanders. Conservation Biology 7: 363–370.

Red squirrels

Finley, R.B. 1969. Cone caches and middens of *Tamiasciurus* in the Rocky Mountain region. In: *Contributions in Mammalogy* (J.K. Jones, Jr, ed.). University of Kansas Museum of Natural History, Miscellaneous Publications 51: 233–273.

Kemp, G.A. and L.B. Keith. 1970. Dynamics and regulation of red squirrel (*Tamiasciurus hudsonicus*) populations. Ecology 51: 763–779.

Layne, J.N. 1954. The biology of the red squirrel, *Tamiasciurus hudsonicus loquax* (Bangs), in central New York. Ecological Monographs 24: 227–267.

Rusch, D.A. and W.G. Reeder. 1978. Population ecology of Alberta red squirrels. Ecology 59: 400–420. (Reviews territoriality and food supply, among other topics.)

Smith, C.C. 1970. The coevolution of pine squirrels (*Tamiasciurus*) and conifers. Ecological Monographs 40: 349–371.

Smith, M.C. 1968. Red squirrel responses to spruce cone failure in interior Alaska. Journal of Wildlife Management 32: 305–317. (Source of data on cone accumulations and consumption rates.)

Ecological challenges

Acid rain

Likens, G.E., and F.H. Borman. 1974. Acid rain a serious environmental problem. Science 184 (4142): 1176–1179.

Likens, G.E., F.H. Borman, and N.M. Johnson. 1972. Acid rain. Environment 14: 33–40.

Likens, G.E., C.T. Driscoll, D.C. Buso, et al. 2002. Long-term effects of acid rain: response and recovery of a forest ecosystem. Science 272 (5259): 244.

Spruce budworm and DDT

Anderson, R.F. 1960. *Forest and Shade Tree Entomology*. John Wiley & Sons, New York, NY.

Bouchard, M., D. Kneeshaw, and Y. Bergeron. 2006. Forest dynamics after successive spruce budworm outbreaks in mixedwood forests. Ecology 87: 2319–2329.

Boulanger, Y. and D. Arseneault. 2004. Spruce budworm outbreaks in eastern Quebec over the last 450 years. Canadian Journal of Forest Research 34: 1035–1043.

Burleigh, J.S., R.I. Alfredo, J.H. Borden, and S. Taylor. 2002. Historical and spatial characteristics of spruce budworm *Choristoneura fumiferana* Lepidoptera: Torticidae outbreaks in northeastern British Columbia. Forest Ecology and Management 168: 301–309.

Candau, J.N., R.A. Fleming, and A. Hopkin. 1998. Spatiotemporal patterns of large-scale defoliation caused by the spruce budworm in Ontario since 1941. Canadian Journal of Forest Research 28: 1733–1741.

Carson, R. 1962. *Silent Spring*. Houghton Mifflin, Boston, MA.

Graham, F. 1970. *Since Silent Spring*. Houghton Mifflin, Boston. MA.

Hickey, J.J. and D.W. Anderson. 1968. Chlorinated hydrocarbons and eggshell changes in raptorial and fish-eating birds. Science 16: 271–273.

Hickey, J.J. and L.B. Hunt. 1960. Initial songbird mortality following a Dutch elm disease control program. Journal of Wildlife Management 24: 259–265. (Cites the death of robins after their exposure to DDT.)

Irland, L.C., J.B. Diamond, J.L. Stone, J. Falk, and E. Baum. 1988. The spruce budworm outbreak in Maine in the 1970s—assessment and directions for the future. Maine Agricultural Experiment Station Bulletin 819: 1–119.

Royama, T., W.E. MacKinnon, E.G. Kettela, N.E. Carter, and L.K. Hartling. 2005. Analysis of spruce budworm outbreak cycles in New Brunswick, Canada, since 1952. Ecology 86: 1212–1224.

Sleep, D.J.H., M.C. Drever, and K.J. Szuba. 2008. Potential role of spruce budworm in range-wide decline of Canada warbler. Journal of Wildlife Management 73: 546–555.

US Department of State. 2010. Stockholm Convention on persistent organic pollutants. Available at: http://www.state.gov/e/oes/eqt/index.htm (accessed 8 January 2015).

Volney, W.J.A. and R.A. Fleming. 2000. Climate change and impacts of boreal forest insects. Agriculture, Ecosystems & Environment 82: 283–294.

Webb, F.E. 1959. Aerial chemical control of forest insects with reference to the Canadian situation. Canadian Fish Culturist 24: 1–14.

Williams, D.W. and A.M. Liebhold. 2000. Spatial synchrony of spruce budworm outbreaks in eastern North America. Ecology 81: 2753–2766.

Wright, B.S. 1960. Woodcock reproduction in DDT-sprayed areas of New Brunswick. Journal of Wildlife Management 24: 419–420.

Wright, B.S. 1965. Some effects of heptachlor and DDT on New Brunswick woodcock. Journal of Wildlife Management 29: 172–185.

Balsam wooly adelgid

Amman, G.D. 1962. Seasonal biology of the balsam woolly adelgid on Mt. Mitchell, North Carolina. Journal of Economic Entomology 55: 96–98.

Arthur, F.H. and F.P. Hain. 1984. Seasonal history of the balsam woolly adelgid (Homoptera: Adelgidae) in natural stands and plantations of Fraser fir. Journal of Economic Entomology 77: 1154–1158.

Rabenold, K.N., P.T. Fauth, B.W. Goodner, J.A. Sadowski, and P.G. Parker. 1998. Response of avian communities to disturbance by an exotic insect in spruce-fir forests of the southern Appalachians. Conservation Biology 12: 177–189.

Retnakaran, A., W.L. Tomkins, M.J. Primavera, and S.R. Palli. 1999. Feeding behavior of the first instar *Choristoneura fumiferana* and *Choristoneura pinus pinus* (Lepidoptera: Toryricidae). Canadian Entomologist 131: 79–84.

Smith, G.F. and N.S. Nicholas. 1998. Patterns of overstory composition in the fir and fir-spruce forests of the Great Smoky Mountains after balsam wooly adelgid infestation. American Midland Naturalist 139: 340–352.

Sprugel, D.G. 1976. Dynamic structure of wave-generated *Abies balsamea* forests in the northeastern United States. Journal of Ecology 64: 889–911.

The Boreal Forest Agreement

Forest Products Association of Canada. 2010. The Canadian Boreal Forest agreement. Available at: www.canadianborealforestagreement.com (accessed 18 December 2014).

Infobox 3.1. The tree of life: a cedar tonic

Cook, R. (ed.) 1993. *The Voyages of Jacques Cartier*. University of Toronto Press, Toronto, ONT.

Peattie, D.C. 1946. *A Natural History of Trees in Eastern and Central North America*. Houghton Mifflin, Boston, MA. (Includes a discussion of arbor vitae as a source of vitamin C.)

Trudel, M. 1966. Cartier, Jacques. In: *Dictionary of Canadian Biography*, Vol. 1. (G.W. Brown, M. Trudel, and A. Vachon, eds) University of Toronto Press, Toronto, pp. 154–172.

Trudel, M. 1973. *The Beginnings of New France, 1524–1663*. McClelland and Stewart, Ltd., Toronto, ONT.

USDA/NRCS Plant guide: northern white cedar, *Thuja occidentalis*. US Department of Agriculture. Available at: http://plants.usda.gov/plantguide/pdf/cs_thoc2.pdf (accessed 8 January 2015).

Infobox 3.2. Rachel Carson (1907–1964), nature's advocate

Brooks, P. 1972. *The House of Life, Rachel Carson at Work*. Houghton Mifflin, Boston, MA.

Carson, R. 1951. *The Sea Around Us*. Oxford University Press, New York, NY. (Revised in 1961.)

Carson, R. 1962. *Silent Spring*. Houghton Mifflin, Boston, MA.

Lear, L. 2009. *Rachel Carson: Witness for Nature*. Mariner Books, New York, NY.

Lytle, M.H. 2007. *The Gentle Subversive: Rachel Carson, Silent Spring and the Rise of the Environmental Movement*. Oxford University Press, New York, NY.

CHAPTER 4
Eastern Deciduous Forest

You see, the forests are the sanctuaries not only of wildlife, but also of the human spirit. And every tree is a compact between generations.

George H.W. Bush, 41st President of the US

The magnificent autumn coloration of Eastern Deciduous Forest was not lost on Henry David Thoreau (1817–1862), Concord's leading friend of woodchucks and nemesis of government. In his journal, the philosopher-naturalist took full note of the light shining through the forest, rendering to its leaves rich tinges "surpassing cathedral windows." These eastern woodlands also are known as the "summer green forest" because of the seasonal rhythm in their deciduous foliage (Fig. 4.1). The prominent trees are often called "hardwoods" and, while this is certainly true of the diverse collection of maples, oaks, hickories, and beech, it is not of others (e.g., basswood is easily whittled). In places, various species of pines represent distinctive units within the biome.

Eastern Deciduous Forest covers essentially all of the eastern United States except for the subtropical vegetation at the southern tip of Florida and an intrusion of grassland known as the prairie peninsula (Chapter 5). The irregular western edge of the forest runs from central Texas to Minnesota (Fig. 1.3). The Boreal Forest Biome forms its northern border.

This biome is listed as either Temperate Deciduous Forest or Temperate Broadleaf Deciduous Forest in most classifications of worldwide ecological units. In Asia and parts of Europe, most of these forests were cleared long ago and the remnants are species-poor. The only expression of this biome in the southern hemisphere occurs on the drier slopes of the southern Andes. In eastern North America, almost all of the forests are second growth yet still preserve the best example of biodiversity among the world's temperate deciduous forests. The richness of species in the Great Smoky Mountains underlies its designation as a World Biosphere Reserve.

Climatic boundaries and soils

When compared to polar regions or deserts, the temperate climates that characterize the Eastern Deciduous Forest Biome generally lack extremes of heat or cold. Summer temperatures reach highs ranging from 27 to 32°C (80–90°F), whereas the winter high temperatures usually lie between –1 and 15°C (30–55°F). The growing season varies from 140 to 200 days per year with 4–6 frost-free months. Precipitation is distributed fairly evenly throughout the year. Up to 152 cm (60 inches) falls in the southern Appalachians, but drops to less than 76 cm (30 inches) where the forest gives way to grassland. Snowfall varies with elevation and latitude, ranging from zero to about 75 cm (30 inches) per year.

Because the trees are deciduous, a thick layer of leaves accumulates on the forest floor every autumn. When the growing season begins in the spring, the ground litter decomposes rapidly and provides an abundance of mineral-rich humus. In northern regions once covered by glaciers, the humus content gives both A and B horizons a brown color. These extremely fertile soils (alfisols) are not as acidic as those under conifers. In regions unaffected by glaciers, soils developed during much longer periods of continuous weathering in a humid temperate climate. The soils of southeastern North America (ultisols) are typically acidic and deficient in major nutrients. An abundance of iron oxides gives these clay-rich soils a red to reddish-orange color. Ultisols were often further degraded in the past by poor agricultural practices.

Features and adaptations

Plants with broad leaves are ill-suited for enduring the rigors of winter. Maples, beech, and other broad-leaved plants drop their summer foliage and pass the winter free of tissues subject to frost damage, an adaptation that is the defining feature of Eastern Deciduous Forest. Leaf fall itself is a complex physiological process largely governed by photoperiod or shortening days in this case (see *Autumn leaves*, this chapter). Concurrently, a rapid increase of sugar content in the sap serves as a natural type of antifreeze for stems and other woody tissues, an

Ecology of North America, Second Edition. Brian R. Chapman and Eric G. Bolen.
© 2015 John Wiley & Sons, Ltd. Published 2015 by John Wiley & Sons, Ltd.

Figure 4.1 Bedecked in "summer green" foliage, an Eastern Deciduous Forest canopy filters light before it can reach the layers beneath. Shortened photoperiods in autumn signal a decrease in chlorophyll production and photosynthesis, and the coloration of hardwood leaves gradually shift to reds and yellows (inset). Eventually, the leaves dry, fall to the forest floor, and become part of the decomposer food chain. Reproduced by permission of Brian R. Chapman.

adaptation most evident in the enriched sap produced by sugar maples.

The variety and abundance of leaves on the forest floor offers niches for a diverse collection of consumers, especially invertebrates. The wealth of invertebrates, in turn, provides important links in the food webs of the Eastern Deciduous Forest, including several species of birds that forage in ground litter (e.g., brown thrashers and ovenbirds). Additionally, the many species of deciduous trees produce an equally diverse collection of fallen logs and **snags**, which in turn provide habitat for many forms of wildlife (e.g., centipedes, bees, woodpeckers, shrews, and bats). Snags, once a bane of intensive forest management, are today valued components of forests that are managed for more than timber production alone, notably those lands subject to the National Forest Management Act of 1976.

Eastern Deciduous Forest also produces large volumes of **mast**. Acorns, hickory nuts, and other woody fruits are major components in the food chain of forest animals (e.g., eastern gray squirrels and wild turkeys). In bottomland hardwood forests, the acorns of pin oak are a staple in the winter diets of wood ducks and mallards. Wildlife managers monitor mast production as a means of gauging the nutritional welfare of game populations.

Deciduous trees generally require greater amounts of nutrients when compared to conifers. Ash, walnut, basswood, elm, and tulip trees are among the species requiring high levels of nutrients. To meet their needs, deciduous trees develop extensive root systems with which they can exploit the rapid return of nutrients to the soil. The extensive network of roots is itself a source of nutrients, with dead and decaying roots sometimes adding more nutrients to the soil than ground litter.

Differences in root systems offer another contrast between deciduous and coniferous vegetation. For example, the amount of plant material below ground and the amount of fine roots in a forest of yellow poplars exceed those in a plantation of loblolly pines. Moreover, the root tips of deciduous trees are smaller which increases their ability to absorb nutrients (i.e., absorption per unit of surface area increases as the radius of the root decreases). Such adaptations enabling rapid nutrient cycling remain a significant feature of Eastern Deciduous Forest.

Finally, some species of deciduous trees tolerate long periods in saturated soils or submerged in flooded habitats. Bottomland hardwood forests develop in the immediate floodplains of rivers. After noting that flooded stands of pin oaks attracted large flocks of mallards and wood ducks, wildlife managers duplicated these conditions with "green tree reservoirs" by intentionally flooding bottomland vegetation in the fall and winter to provide ducks with a rich supply of acorns. The bottomlands are drained before the trees begin to regrow in spring, thereby ensuring that flooding does not impair timber production. Water oak, willow oak, and black oak are among the other species of value in green tree reservoirs.

The forest primeval

A widely held notion concerns the majesty of pristine environments in North America. According to folklore, an adventurous squirrel could once travel from the Atlantic coastline to the Mississippi River in the unbroken canopy of deciduous forest. This legend suggests that the Eastern Deciduous Forest was among the natural areas undisturbed by humans prior to European exploration and settlement.

This view is fashionable, even revered, but not entirely accurate. Native Americans had for millennia modified large areas of forest and other vegetation. The remarkable travels of John Lawson (1674–1711), William Bartram (1739–1823), and William Dunbar (1750–1810) included numerous visits to Native American "plantations," an obvious indication of a long history of land cleared for crop production. These clearings could scarcely be accomplished with the sorts of tools available prior to the arrival of the Europeans. Instead, the forests were burned – often twice a year – opening large areas for agriculture and other purposes. Native Americans set afire piles of fuel at the bases of stubborn trees, ensuring the outcome. Rows of corn, beans, and squash were planted around the lifeless snags, which were burned again until they eventually fell on their own accord.

Large areas were cleared, as reflected by the stores encountered by 17th century explorers and travelers. In what is today central New York, for example, Henry Hudson (c. 1560s–1611) saw a stockpile of Native-American-grown crops that was "enough to load three ships, besides what was growing in the fields." Some Europeans mentioned the size of the fields cleared by the Native Americans. Cornfields extended for 3.2 km (2 miles) on either side of an Onondaga village in New York, and a Seneca village stood in the center of a clearing 9.6 km (6 miles) in circumference. Land was also cleared for orchards; one planted by Cayugas included 1500 fruit trees. These and other reports are no doubt conservative as the early travelers likely visited only a relatively small area of the Eastern Deciduous Forest cleared by Native Americans. Moreover, forest vegetation re-grew when the fields were abandoned, and the original clearing of these sites probably went unnoticed.

The result of these intentional burnings was an Eastern Deciduous Forest Biome in which large areas were open and park-like. Such fires regularly reduced the accumulation of ground litter, which decreased the incidence of destructive wildfires. The limited supply of ground fuel also meant that the fires set by Native Americans burned at relatively low temperatures and could be extinguished before burning out of control. The successional sequence initiated by fire often produced thickets of berries, desired by humans and wildlife alike, and there is evidence to suggest that the Native Americans were well aware of the cause-and-effect nature of these responses. Native populations also depleted the supply of firewood from large areas near their villages. In response, they moved their encampments to new sites every 10–30 years, leaving behind an exhausted landscape. In any case, native cultures influenced the composition of large areas of forest.

The Eastern Deciduous Forest was rapidly exploited in the years following European settlement and expansion in North America; many species offered important commercial opportunities. The practice of marking the better trees with a blaze known as the "King's arrow," which reserved the tree exclusively for the English navy, reflected the importance of white pines as a source for ship masts. An entry dated 3 December 1666 in the famous diary of Samuel Pepys (1633–1701?) noted "… the very good news [that] four new England ships came home safe to Falmouth with masts for the King, which is a blessing…without which…we must have failed the next year." Oak wood, also scarce in the exhausted forests of Europe, was in great demand for barrels (the standard container for molasses, sugar, tobacco, spirits, and other staples for the period). The first sawmill in New England, and perhaps in North America, was built in today's York, Maine, in either 1623 or 1624 and, by 1682, Maine's forests supported 24 mills. The deciduous forests yielded other products, of which maple sugar and syrup remain among the best known.

The layered forest

The physical structure of every forest community forms distinctive vertical layers. These strata result from the density, distribution, and species of trees in the uppermost layer and the gradient of light that penetrates downward. When mature, eastern deciduous forests typically develop five layers which (from top to bottom) include: **canopy**; understory; shrub; herb; and the ground or litter layer.

Composed of dominant species, the canopy exerts the greatest influence on the composition of the underlying layers in the forest community. In addition to being the site where photosynthesis fixes the greatest amount of energy, the canopy's leaf density and cover area determine the amount of light reaching the lower layers. Well-developed understory, shrub layers, and herb layers occur when the canopy is fairly open and supplies of water and nutrients are adequate. In the eastern United States, the forest understory consists of young trees of the same species as those forming the canopy, along with short trees such as dogwood and hornbeam and tall shrubs such as witch hobble and yaupon. The composition of the shrub and herb layers varies from place to place throughout the forest. The shrub layer often includes members of the heath family, notably rhododendron, azaleas, and huckleberries, whereas ferns and spring-blooming perennial forbs commonly form the herb layer. In many areas, the flora of the ground layer features lichens, clubmosses, and true mosses.

Autumn leaves

Two characteristics figure prominently in the annual cycle in the Eastern Deciduous Forest: the spectacle of autumn colors and the deciduous behavior of the foliage (the latter event underlies the origin of "fall" as a term for the autumn months). These changes allow deciduous trees to survive desiccation during the cold months when the ground freezes and roots cannot readily absorb water. The broad surfaces of deciduous leaves, unlike conifer needles, are ill-suited for water conservation and, because of their thinness, deciduous leaves and the vital tissues therein are easily damaged by frost and prolonged freezing. When weighed against the chances of gaining additional growth during the winter, maples and other broad-leaved trees fare better by overwintering in a dormant state instead of risking irreparable damage.

The panoply of autumn color – a rich palette of golds, browns, oranges, and especially brilliant reds – stems from several kinds of pigments in leaf tissues. Chlorophyll, the major component of photosynthesis, provides the typical green coloration for leaves during the growing season. As long as conditions are favorable, photosynthesis continues and chlorophyll controls the coloration of the foliage. However, when the photoperiod shortens in autumn, chlorophyll production gradually ends until the chlorophyll progressively disappears. Carotenoid pigments – the sources of yellow, orange, and brown colors – then become visible in the leaves. Anthocyanins, another major group of pigments, provide the leaves of some species with a vivid showcase of reds and purples. Most anthocyanins are produced at the end of the growing season and their appearance coincides with the increased production of sugars as the trees ready for winter dormancy.

Autumn colors, particularly those produced by the anthocyanin group, are most vivid in years when the autumn weather includes a succession of warm, sunny days and cool (but not freezing) nights, a delightful period known as "Indian summer." Soil moisture and sunlight are other factors affecting leaf color and contribute to the year-to-year variations in autumn foliage. A moist growing season and autumn sunshine heightens the prospects for a brilliant display.

The foliage of the various groups and species of trees displays distinctive features. Oaks turn brown, russet, or red and acquire their colors late in the season. Dogwood is among several species with distinctive red foliage. Hickories become golden bronze, and yellow foliage is typical of ash, sycamore, and birch. Maples differ by species: red maple turns scarlet and sugar maple turns orange-red, but black maples become bright yellow. American beech leaves turn tan and parchment-like and remain attached long after the foliage of other trees has fallen. The leaves of elms shrivel and drop without acquiring more than drab brown coloration.

Leaf fall itself is triggered by the shortening photoperiod each autumn. The process begins with the formation of special cells – the abscission layer – at the base of the leaf's petiole and the slow closing of the veins connecting the leaves with the rest of the tree. Sugars, trapped in the leaves as the veins close, then promote the production of anthocyanin pigments. The leaf falls when enzymes weaken the cellular structure of the completed abscission layer. A layer of corky tissue concurrently forms a protective scar on the twig, thereby preventing the invasion of pathogenic agents. Beech and some oaks arrest leaf drop until early spring, but most deciduous species complete the process during a relatively short period in autumn.

Ground and leaf litter

The blanket of leaves falling each autumn forms a major link in the flow of nutrients within the Eastern Deciduous Forest. The biomass represented by ground litter, including branches, mast, and leaves, is quite large in most deciduous forests. About 6480 kg ha^{-1} (5767 lb/acre) of ground litter fell each year in a mature oak forest in New Jersey, and 5702 kg ha^{-1} (5080 lb/acre) of ground litter fell annually in a forest of sugar maples, beech, and other hardwoods in New Hampshire. Obviously, leaf and other litter may vary considerably between sites because of the forest's composition and year-to-year differences in production (e.g., mast). Oaks and beech are among the species typically generating large amounts of leaf litter, whereas birch, aspen, and willows produce lesser amounts. Periodic outbreaks of leaf-eating insects also influence the amount of foliage falling to the forest floor each autumn.

Most ground litter originates from the forest canopy. At the study area in New Hampshire, the overstory contributed 98% of the biomass; this material contained about 140 kg ha^{-1} (125 lb/acre) of nutrients, of which nitrogen, calcium, and potassium together represented about 81% of the total. **Micronutrients** such as zinc, iron, sodium, and copper represented less than 1% of the nutrients in the litter. These and other nutrients in the litter usually continue cycling in the forest ecosystem with little net loss.

Branches dying while still attached to the tree lose about 8% of their weight per year and, by the time they fall, have lost about 40% of their initial weight. Fungi begin to colonize dead branches in the canopy, but wood-boring insects invade soon after the branches fall. The rates of decomposition for branches and other kinds of ground litter are significantly influenced by the moisture and temperature regimes on the forest floor (e.g., microhabitats).

Fallen tree trunks are an important component in the litter accumulating on the forest floor. Logs, of course, decay more slowly than leaves, and forest ecologists have recognized five classes of decomposition ranging from "sound" (Class I) to "total decomposition" (Class V) based on: (a) the physical state of the wood; (b) evidence of new vegetation, especially the state of root development, growing on the log; and (c) the kind of fungi colonizing the log.

The rate at which logs decay varies with many factors, but an important element in decomposition is the extent of contact between the log and the underlying soil. Logs lying flush with the ground decompose more rapidly than those with only partial contact. Insects are important agents in the decomposition of logs. Boring insects quickly form galleries extending well into a log's interior, which hastens the penetration of moisture and other organisms. Insects also consume and digest the organic material, further breaking down the log's structure and increasing the surface area available to microbial attack. Finally, insects entering logs carry with them other fungi or decomposing agents, either on their extremities or in their feces.

All of the various agents of decomposition – fungi, insects, and microbes – promote the recycling of vital nutrients bound within a log back into the soil and forest ecosystem.

Logs falling into streams and wetlands similarly charge aquatic systems. Logs on the forest floor provide shelter and sources of food for small vertebrates such as salamanders, mice, and shrews, and those falling into streams likewise nurture minnows, bass, and other fishes. The tangle of woody debris, leaf litter, and understory vegetation also offers an advantage to organisms with color patterns – **cryptic coloration** – allowing concealment from prey and protection from predators (Fig. 4.2).

Mast

Mast refers to either the woody fruits (e.g., "hard mast," such as acorns and nuts) or edible reproductive parts (e.g., "soft mast," such as leaf buds, catkins, rose hips, true berries, and drupes) produced by woody species. Mast production reaches a pinnacle of both species diversity and quantity in the Eastern Deciduous Forest. The volume of hard mast represents a wealth of woody fruits from a variety of oaks, hickories, American beech, sweetgum, and hazelnut. The hardened berry-like fruits – drupes – of the various kinds of hollies, cherries, and wild plums represent soft mast.

The numbers of acorns falling to the forest floor may reach staggering proportions. Pin oaks, for example, produced up to 445,000 acorns/ha (180,000 acorns/acre) per year during a 14-year period. Mast production for a given species may vary greatly from one year to another. At a study site in a hardwood forest in Missouri, for example, black oaks produced an average of 136 kg ha^{-1} (55 lb/acre) one year, but only 15 kg ha^{-1} (6 lb/acre) the following year.

(a)

(b)

Figure 4.2 (a) The disruptive color patterns of some animals blend with forest floor litter providing camouflage from potential prey as well as predators. (b) Only when debris is removed can the cryptically colored timber rattlesnake, a lie-and-wait predator, be detected. Reproduced by permission of William I. Lutterschmidt.

While mast production certainly varies, distinctive "mast years" also occur; these are times when mast production is exceptionally abundant throughout the forest. The most common explanation for mast years is that seed-consuming animals, including various insects (e.g., nut weevils), might limit the number of acorns and other woody seeds available for germination if a steady year-to-year supply of mast was available (i.e., animal populations would stabilize, consistent with an equally stable food supply each year). However, when nut-bearing trees skip a few years between seasons of heavy mast production, populations of mast-dependent animals diminish in size. Then, when the trees suddenly produce a bumper crop of mast, the animals cannot fully exploit the resource and the odds favoring germination increase significantly. Year-to-year variations in mast production therefore seem to be a reproductive strategy resulting from the threat of animal depredations.

When available, mast forms a large percentage of the diet of wild turkeys, white-tailed deer, and other wildlife. Historical mention of "mass migrations" of eastern gray squirrels, the earliest appearing in the journals of Lewis and Clark, were often attributed to failures of the mast crop. However, when Lewis encountered squirrels swimming *en masse* across the Ohio River, he found no differences in the abundance of walnuts and hickory nuts on either side of the river. Whatever their cause, these movements now occur only rarely.

Passenger pigeons, now extinct, also fed heavily on mast. One hunter removed 30 acorns from the crop of a pigeon, and another bird contained 28 beechnuts. Alexander Wilson (1766–1813), who witnessed the migration of an estimated 2.2 billion passenger pigeons, calculated that such a flock might consume 614.7 million L (17.4 million bushels) of mast in a single day. John James Audubon (1785–1851) later halved Wilson's estimates, but the point remains clear: the Eastern Deciduous Forest once produced enormous amounts of mast.

Biotic associations

The Eastern Deciduous Forest is traditionally considered in smaller units known as associations, each characterized by two or three dominant species. Renowned botanist E. Lucy Braun (Infobox 4.1) recognized nine forest associations based on her assessment of the "original" (i.e., prior to European settlement) forest composition. Each of her forest regions was delineated primarily on the basis of physiognomy (i.e., overall similarities in composition, appearance, and structure) but some recent surveys, which are based on a variety of methods including satellite imagery, recognize as many as 35 associations. Still, Braun's associations remain one of the most widely cited classifications of the Eastern Deciduous Forest. In

the descriptions below, we follow a re-evaluation of Braun's work by James M. Dyer of Ohio University who described eight "regions" based on inventories of more than 100,000 forest plots. We use seven of Dyer's regions to describe the Eastern Deciduous Forest Biome, but retain Braun's term, "association," because of several similarities in the two classifications. Dyer's eighth region, the Subtropical Evergreen Region, is discussed in Chapter 11.

Northern hardwoods conifer forests

Along the northern border of the Eastern Deciduous Forest, two associations, Northern Hardwoods–Red Pine Association and Northern Hardwoods–Hemlock Association, form a broad transitional zone between the Boreal Forest and the communities of oaks and other deciduous trees in eastern North America. Ice covered much of the region during the Wisconsin glacial period so moraines, outwash plains, and distinctive watersheds (e.g., the Finger Lakes in New York) are among the glacial features of the landscape. The Northern Hardwoods–Hemlock Association extends from southeastern Canada westward to the Great Lakes where it intergrades with the Northern Hardwoods–Red Pine Association, which is sometimes designated the "Lake Forest."

Two conifers – eastern hemlock and red spruce – occur regularly in the eastern Northern Hardwoods–Hemlock Association, which also is characterized by sugar maple, red maple, American beech, and white ash (Fig. 4.3). The current composition of the Lake Forest – quaking aspen, red maple, balsam fir, and paper birch – reflects a history of disturbance. For example, extensive stands of paper birch typically indicate a history of fire at the site. American beech reaches its distributional limits at the western edge of the Lake Forest, and eastern hemlock becomes less common. Jack pine and aspen cover large areas where the original forest has been disturbed by extensive cutting and fire.

White pine is relatively shade tolerant, and the species is a normal component in the climax forest in these two associations. Although geographical distribution of red pine also coincides almost exactly with both northern hardwood associations, the conifer was selected to serve as the indicator species for the westernmost of the two units. White pine is especially valuable as timber, so this species was subject to heavy cutting immediately after European settlement. The harvest of white pine later continued westward into the Great Lakes Region, with the result that mature trees today are relatively uncommon in most areas of the two associations. With the prolonged absence of fire, however, hardwoods and hemlock eventually dominate sites within the two northern hardwood associations.

The ruthless exploitation of white pine and other species late in the 1800s produced serious ecological

Infobox 4.1 E. Lucy Braun (1889–1971), diva of forest ecology

Widely known for her ground-breaking work in plant ecology, E. (Emma) Lucy Braun was a life-long resident of Cincinnati, Ohio. She earned BA and MA degrees in geology, capped with a PhD in botany, at the University of Cincinnati where she thereafter joined the faculty. Braun specialized in the floristics of forest vegetation and traveled widely conducting surveys that resulted in her classic treatise, *Deciduous Forests of Eastern North America*, published in 1950. Overall, she produced more than 180 publications. In 1917, Braun founded the Wildflower Preservation Society of North America. She also discovered several new species, and her herbarium of nearly 12,000 plants was post-humously donated to the US National Museum in Washington, DC.

In recognition of her contributions, Braun became the first woman president (1933–1934) of the Ohio Academy of Science and, in 1950, the first woman to preside over the Ecological Society of America. Shortly before her death, she was inducted into the Ohio Conservation Hall of Fame, again being the first woman thus honored. On the occasion of its fiftieth anniversary, the Botanical Society of America presented Braun with its Certificate of Merit.

Braun's influential book traced the development of the eastern deciduous forest since the end of the Pleistocene (Ice Age). Most notably, she divided the forest into nine groups known as associations, among them a favorite that she named the Mixed Mesophytic Association. The designation of "mixed" was based on the predominance of 6–8 species of trees instead of the more typical situation where one or two species dominated the forest canopy (e.g., Oak–Hickory Association). "Mesophytic" describes plant life that thrives where there is neither too much, nor too little available water. A reviewer of her book noted "The interpretation of the deciduous forest in terms of its existing remnants, historical and fossil records, dynamic physiography and geological history is an outstanding accomplishment." Braun later published two other books, including *The Woody Plants of Ohio*.

Cove forests, the diverse vegetation developing on the deep soils that accumulate in "hollows" resulting from streams carving into mountainsides, were prized locations for her fieldwork. Some of the best examples of these sites occur in the Great Smoky Mountains where huge yellow poplar, hemlock, and other trees flourish amid an understory rich with wildflowers and other herbaceous plants. Braun marveled that even small plots in a cove forest often contained 75 species, whereas similar samples measured at other forest locations averaged only 30 species. She postulated the theory that cove forests served as refugia for forest vegetation during glacial periods, but her idea is not widely accepted.

Braun never married and instead lived with her older sister Annette, an entomologist who likewise had earned a PhD. The two spinsters held fast to a prim Victorian lifestyle and traveled together, at first in a horse-drawn buggy and later in a Model T, to field locations throughout the eastern United States. Seated side by side in their intrepid Ford, they lurched along primitive mountain roads in the Appalachians in search of interesting study areas; where roads were lacking, they hitched rides on logging trains. Despite the helpful train rides, Braun was distressed that logging was claiming some of the last tracts of virgin forest even as she was conducting her studies. E. Lucy Braun therefore dedicated the latter part of her career to preserving natural areas, including the few remaining prairie areas in Ohio, but forests nonetheless remained her enduring passion.

aftershocks. For example, the intensive logging left behind a thick accumulation of "slash" – branches and other trimmings – that soon fueled unchecked wildfires and reduced regeneration. White pine is also attacked by blister rust – a fungus that eventually kills the trees – and by weevils, which deform the trees into "cabbage pines" by killing the terminal shoots. The fungus survives on gooseberry and currant bushes; the latter serve as alternate hosts from which airborne spores disperse widely. Labor-intensive attempts to eliminate the alternate hosts and other control measures were unsuccessful. The development of genetic-resistant strains of white pine offers greater promise for ultimately curbing blister rust.

Beech–Maple–Basswood

Braun described two forest associations in the mid-western United States – the Beech–Maple Association and the Maple–Basswood Association – but Dyer united these because of their similarities. The Maple–Basswood

Association, the smallest of Braun's units, is restricted to an unglaciated area in Wisconsin where it forms the northwestern limit of the Eastern Deciduous Forest. Braun's Beech–Maple Association extends west of the Alleghenies across the northern two-thirds of Ohio into eastern Indiana and northward into the southern parts of Michigan and Wisconsin. It surrounds Lake Erie and forms a slender extension eastward along the southern shore of Lake Ontario. The southern border corresponds to the maximum glacial advance. Dyer extended the Beech–Maple–Basswood Association beyond the glacial boundary to include similar forests in northern Kentucky and central Tennessee.

Much of the area formerly occupied by this association has been converted to croplands. Although American beech, sugar maple, and American basswood are the forest components for which this association is named, they are no longer the dominant species. The dominate trees of the Beech–Maple–Basswood Association include

Figure 4.3 Although a conifer, eastern hemlock occurs throughout much of the Eastern Deciduous Forest. Note how the evergreen canopy in this stand of mature hemlock in Pennsylvania limits light penetration, which has curtailed development of understory vegetation. Reproduced by permission of USDA Forest Service.

American elm, white ash, black cherry, northern red oak, and white oak.

Mesophytic forest

This large region extends from southern Maine and the eastern half of Pennsylvania westward through most of West Virginia, southeastern Ohio, to the southern tip of Illinois. The southern border includes northern portions of Alabama, Mississippi, Georgia, and South Carolina and the eastern limits coincide with the western halves of North Carolina and Virginia. This association combines three of Braun's regions: Mixed Mesophytic,

Oak–Chestnut, and Western Mesophytic. Forests in this species-rich association develop on moist, but well-drained, sites. Many of the species in this region are oaks, but the mix of the 162 species includes sweet buckeye, American basswood, sugar maple, eastern hemlock, and red maple (Infobox 4.2). Red maple has increased markedly in abundance in the Mesophytic Forest and elsewhere in other eastern forest associations, where abandoned farmlands initiated secondary succession and fires have been suppressed.

Chestnut was once important in this and other hardwood associations, but an introduced pathogen essentially

Infobox 4.2 Red maple, a species on the move

A species once associated primarily with wet locations (e.g., swamps and other soggy locations) is steadily expanding its ecological niche into forests on well-drained sites. Red maple, aptly named for its brilliant autumn foliage, is transforming many areas within the Eastern Deciduous Forest, often at the expense of the otherwise dominant oaks and hickories.

The expansion results from the red maple's adaptable and aggressive abilities to respond to human influences, particularly fire suppression and disturbance. Today, fires that kill the thin-barked red oak seldom occur with the same widespread devastation as they once did, thanks to modern fire-fighting capabilities and a network of roadways that act as fire breaks. In contrast, the thicker-barked oaks and hickories are better able to withstand most fires. Historically, oaks and hickories often survived fires but red maple did not and was instead confined to wetter, fire-resistant sites within its large geographical range in eastern North America. Fire suppression now allows red maple to expand into habitats where it once could not survive.

Secondly, red maples spread easily (almost weed-like) into disturbed areas. The trees accordingly invade abandoned farmlands, roadsides, and other cleared areas. Other disturbances, notably plagues of gypsy moths, take a heavy toll of oaks and thereby open the forest to invasive species, whereas alkaloids in red maple leaves repel these noxious insects.

Red maples enjoy an important competitive advantage in comparison to other species of upland trees: early seed production. Each spring, even before they sprout leaves, red maples produce clusters of red (what else?) wind-dispersed seeds. Maple seeds, technically known as samaras, are flattened and equipped with a single, wing-like sail that "whirlybirds" when falling to the ground, but only those of the red maple have a reproductive head start. Red maples establish seedlings months before oak and hickory seeds have fallen to the forest floor. The seedlings are shade-tolerant and thereby continue growing under a canopy of other species. Additionally, even young trees (4–10 years old) produce seeds and soon yield crops of 12,000–91,000 seeds per year; one individual, 30 cm (12 inch) in diameter, produced nearly 1 million seeds. Wind effectively disperses the winged seeds, of which 85% or more germinate.

Red maples also thrive under a wide range of environmental conditions including soil properties (e.g., nutrient-poor as well as fertile soils), temperature (e.g., the frigid winters of southern Quebec to the summer heat of Florida), and elevations from sea level to 900 km (3000 feet). The species indeed may be favored with the broadest **ecological amplitude** in the Eastern Deciduous Forest. In essence, red maple is limited only by its intolerance of fire and the arid conditions associated with grassland vegetation.

The ascendancy of red maple is sure to trigger ripple effects in the forest community. For example, production of acorns and other mast will decline as red maples replace oaks and hickories, thereby reducing food resources available to squirrels, black bears, blue jays, and other wildlife. Additionally, the smooth bark of younger trees does not provide the same cracks and crevices for insects as the wrinkled bark of oaks and other species, again diminishing the food web.

In ecological terms, red maples are almost a perfect species: a super generalist. They possess features associated with early succession – rapid maturity, huge seed production, and invasion of disturbed sites – yet also those of climax vegetation (e.g., shade tolerance and long lived). Given these advantages, type maps for eastern forests may one day recognize an Oak–Hickory–*Maple* Association.

eliminated the species from North American forests (see *Destruction by exotic organisms*, this chapter). Prior to that disaster, a rich forest of chestnut supplemented by yellow poplar (= tuliptree), hickory, scarlet oak, and chestnut oak covered the mid- and lower slopes of the Appalachians from southern New England to Georgia. Red oak was among the major co-dominant species that found unusually favorable growing conditions after the chestnut declined. Red oak still dominates locations in southwestern Virginia.

In some particularly mesic sites in the southern Appalachians, communities known as "cove forests" or "cove hardwoods" develop in secluded places that extend back into mountain ranges; these locations are sometimes known locally as "hollows." The best examples of this vegetation occur in the Great Smoky Mountains

where yellow poplar, yellow birch, and American beech join other species typical of the Mesophytic Association (Fig. 4.4). Some cove forests also include large numbers of eastern hemlock, often with a shrub understory of rhododendron. A fine example of virgin cove forest is protected at Joyce Kilmer Memorial Forest in western North Carolina, but other undisturbed sites occur in the southern Appalachians of Georgia.

Oak–Hickory

The Oak–Hickory Association today is restricted to eastern Oklahoma and the interior highlands of Arkansas and Missouri. Most of the northern area (primarily in Illinois and southern Wisconsin) and the southern area in eastern Texas once forming most of this association have now been replaced by farmland.

Figure 4.4 Cove forests develop in the southern Appalachians at protected sites ("hollows") where local conditions foster better growth and greater diversity in both the over- and understory vegetation. As shown here, large trees such as the sugar maple (far left) and American basswood (right) dominate this cove forest in Great Smoky Mountains National Park. Yellow poplars likewise reach exceptional sizes in many cove forests. Eastern hemlocks are among the smaller trees in the background but also grow much larger at these sites. Reproduced by permission of USDA Forest Service.

Remnants of the association remain as small, isolated fragments. In this relatively dry, westernmost region of the Eastern Deciduous Forest, post oak, mockernut hickory, black oak, and white oak dominate the remaining stands.

Mississippi alluvial plain

The bottomland forests in the Mississippi River Valley south of the mouth of the Ohio River are distinct enough to warrant designation as a separate association. Braun included it as "bottomland forests" in her Southeastern Evergreen Forest Association. Nobel Prize-winning author William Faulkner (1897–1971) referred to the alluvial forest bordering the Mississippi and its tributaries as "the big woods." Before extensive logging and clearing, a broad strip of forest up to 130 km (80 miles) wide straddled the river. Periodic flooding deposited rich soils that nourished forests dominated by sweetgum, green ash, sugarberry, American elm, tupelo, bald cypress, and loblolly pine.

Today, the forest covers only about 32,000 km² (12,350 square miles), approximately 20% of the area it occupied 200 years ago. The remnants of the once-vast forest along the Mississippi River in Arkansas, Mississippi, and Louisiana are laced with bayous, oxbow lakes, and sandy ridges, all habitat for black bears, white-tailed deer, raccoons, waterfowl wild turkeys, and songbirds. Some hope that this dense forest may still harbor a relict population of the endangered ivory-billed woodpecker, but the last reliable record of the species occurred in the mid-1940s.

Southern Mixed Forest

Dyer combined Braun's Oak–Pine and Southeastern Evergreen associations into a single unit: Southern Mixed Forest. This widely distributed association covers most of the Piedmont region from Delaware to Georgia, then swings across the Gulf States and into eastern Texas after skipping across the Mississippi Alluvial Plain. Mockernut hickory, red hickory, shagbark hickory, and bitternut hickory are common species, widely accompanied by red, white, and black oaks. Where the forests thin into savannas as the woodlands yield to grasslands, blackjack oak and post oak in the southwest and bur oak in the northwest become important components.

Within the Southern Mixed Forest are communities of differing character, including those with evergreen magnolia, American beech, laurel oak, and flowering dogwood. In the cool, moist parts of its range, dogwood is attacked by dogwood anthracnose, a fatal disease. Caused by a fungus, the disease has spread across more than 1.6 million ha (4 million acres), thereby changing the composition and appearance of many southeastern woodlands. Wetlands within the southern coastal plain feature stands of bald cypress. At one location on the Black River of North Carolina scientists discovered bald cypress in excess of 1600 years of age, making these the oldest trees in eastern North America. Trees of several species in this region are often draped with Spanish moss, an **epiphyte** in the pineapple family and a well-known feature of coastal environments in the southern United States.

Large areas within the geographical boundaries of the Southern Mixed Association are covered with pine forests. Virtually all of these pine woodlands develop on the sandy coastal plains of the Atlantic and Gulf states. This situation illustrates a subclimax, a stage in vegetational development that differs markedly from the usual climax. The pine subclimax is maintained naturally by recurring fires, without which the fire-resistant pines would slowly give way to a hardwood climax. Because of the large-scale presence of pines, some authorities classify this vegetation separately (e.g., "Southern Evergreen Forest"). For similar reasons, Dyer recognizes an "Oak–Pine Section" on the southern Piedmont Plateau and in southeastern Texas.

Some associated communities

Longleaf pine forests

Forests of longleaf pine, which cover much of the area now designated as the Southern Mixed Forest Association, are of particular interest to ecologists and conservationists.

William Bartram (1739–1823) remarked more than two centuries ago in his *Travels* about the "vast plain…of the great longleaved pine, the earth covered with grass, interspersed with an infinite variety of herbaceous plants." The same view struck a chord with George Washington (1732–1799) during his "Southern Tour" of the United States. In his diary for 24 April 1791, the first president wrote of a "course [sic] grass…having sprung since the burning of the woods…." Washington had observed what today is known as wiregrass, the fire-dependent associate in the understory of longleaf pine forests.

Longleaf pines characteristically grow in park-like stands known as **savannas** (Fig. 4.5) and thrives on a variety of soils, notably those with high sand content. Once widespread, longleaf pine shaped the regional economy and culture, the result of its rich endowment of resins suitable for tar, pitch, turpentine, and other naval stores, products essential for the upkeep of wooden ships. The importance of naval stores continued into the early 20th century. When steel-hulled ships replaced wooden vessels, demands for naval stores diminished but not for the strong wood sometimes known as "southern yellow pine." In the late 1880s, the nation's expanding railroad system quickened the harvest and marketing of longleaf pine, which had been somewhat secure from wholesale cutting. Virtually all of the virgin stands of longleaf pine were gone by the 1930s.

However, overcutting was not the only difficulty; thousands of free-ranging wild pigs annually uprooted tens of thousands of seedlings for their rich starch content. One boar uprooted 800 seedlings in 10 hours. Foresters then (as now) preferred replacement stands of

Figure 4.5 Savannas of longleaf pine represent a prominent subclimax community within the Eastern Deciduous Forest, as shown here on the coastal plain of North Carolina. Frequent fires keep oaks from gaining a foothold and establishing a climax community in these areas. Wiregrass is typical of the understory vegetation. Reproduced by permission of Brian R. Chapman.

the fast-growing loblolly pine and established these instead of regenerating longleaf pine. Of the estimated 24–28 million ha (60–70 million acres) once present, longleaf pine communities today cover less than 1.4 million ha (3.5 million acres).

Longleaf pine requires fire for regeneration, without which it yields to oak forests in the successional march toward climax. Fire rids the forest floor of litter accumulations, exposing mineral soil as a seedbed for new growth and curtailing brown spot disease, a fungus harbored in dead needles that attacks pine seedlings. Most of all, fire eliminates hardwood trees that eventually shade the forest floor; longleaf pines are **shade intolerant** and cannot regenerate without adequate sunlight. Once past their first year, young pines can tolerate fires about once every 3 years. Because of the continued threat from brown spot, seedlings exposed to winter fires at intervals of 3 years produce twice as much growth as unburned seedlings.

Most growth during the first 5–7 years concentrates in the root systems, which nourish the production of new leaves whenever fires damage the existing foliage. Seedlings at this period are in the "grass" stage of development, during which their buds are protected from fire by a thick cluster of green needles. However, the seedlings rapidly shoot upward for the next 3 years, no longer resembling tufts of grass and becoming more tree-like in appearance. The young trees are susceptible to fire at this stage, but they become (and remain) extremely fire tolerant once they reach a height of 3 m (10 feet). Thereafter, the regular occurrence of fire eliminates the establishment of oaks and other fire-intolerant species from the understory. Fire therefore prevents the development of the oak climax and instead maintains a subclimax of longleaf pine.

Despite the dominance of a single species in the overstory, the ground cover in longleaf pine forests may be unusually rich; a square meter may contain 35 species, whereas an area of similar size elsewhere in North America likely contains half that number. Wiregrass is a particularly close associate of longleaf pine and, as might be expected, fire stimulates its growth and flowering. Among the other herbaceous plants are "specialties" such as insectivorous plants. Charles Darwin (1809–1882) characterized one of the latter, the Venus flytrap, as "one of the most unusual [plants] in the world." The species is endemic to a few counties in coastal North and South Carolina (Fig. 4.6). The habit of trapping and digesting spiders and insects represents adaptations for coping with the nutrient-poor soils where these plants occur. In particular, the arthropods supply Venus flytraps with a rich source of nitrogen, which is otherwise deficient in the sandy soils that typically support longleaf pine communities.

Longleaf pine forests are inexorably tied to the ecology of the endangered red-cockaded woodpecker, the only

Figure 4.6 Venus flytraps are endemic to a small coastal area in southeastern North Carolina and northeastern South Carolina, where they often occur in the understory of longleaf pine communities. These unique carnivorous plants obtain nitrogen from the small insects and spiders they capture and digest in their "traps" of specialized leaves. Reproduced by permission of Marjorie Boyd and the Plant Conservation Program, North Carolina Department of Agriculture and Consumer Services.

North American woodpecker that drills nest cavities in living trees (longleaf pine is the preferred species). A fungus (red heart rot) often weakens older longleaf pines and therefore expedites the excavation of holes in living wood. These birds – their namesake red markings are not obvious – chip away the bark surrounding their holes and then hammer shallow pits into the sapwood. The result is a disk of oozing pitch, an effective guard against several kinds of predators, especially rat snakes which are particularly adept climbers and predators of birds' nests. Patches of whitish pitch on the upper trunks of longleaf pines are telltale signs of an active cluster of red-cockaded woodpeckers.

For many years, fire suppression was an unyielding postulate of forest management and both natural and anthropogenic fires were vigorously suppressed. A highly successful campaign featuring a well-known bear wearing a ranger's hat molded public attitudes. However, when longleaf and other southern pines failed to regenerate under the prevailing fire policy and other species invaded the understory, a handful of forest ecologists proposed

(a)   (b)

Figure 4.7 (a) Dense stands of stunted pitch pine and tangles of blackjack oak develop dwarf forests in parts of the New Jersey Pine Barrens. (b) Pitch pines elsewhere in the Pine Barrens reach "normal" heights some three times greater than those in the dwarf forest. Fires maintain both types of communities but occur more often in the dwarf forest. The road cut reveals the coarse white sands forming the upper horizon in the soils throughout the Pine Barrens. Photographs courtesy of Joanna Burger.

the remedy – and heresy – of controlled burning. Today, now that the ecological circumstances are clear, conservationists manage longleaf pine communities, including gopher tortoises, gopher frogs, eastern indigo snakes, and clusters of red-cockaded woodpeckers, with prescribed burning. A somewhat similar situation occurs with the management of jack pine forests (see *Kirtland's warblers and fire*, this chapter).

New Jersey Pine Barrens

A sizeable area within a small and highly urbanized state represents a unique ecological site. This is the 5180 km² (2000 square miles) Pine Barrens representing more than 25% of New Jersey's area. The Pine Barrens are defined, in part, by their dry, sandy soils which are acidic and of low fertility. Although pines characterize the vegetation, the region falls under the greater umbrella of the Eastern Deciduous Forest. Pitch pine, the principal species, is joined by several oaks (e.g., black, scarlet, and chestnut oaks). Sheep laurel, blueberries, and other plants associated with acidic and/or sandy soils typify the shrubby understory. Despite the relatively harsh soil conditions, the flora of the Pine Barrens includes about 800 species. Of these, 14 species are northern plants at the southern limit of their range, but even more impressive are the 109 southern species that reach their northern limits in the Pine Barrens.

Among the ecological novelties in the Pine Barrens are two extensive areas of dwarf forest, the East Plains (2368 ha or 5920 acres) and West Plains (2467 ha or 6168 acres). The dwarf forest is sometimes known as the

"pygmy forest." The woody vegetation in the dwarf forest consists of two predominant species – pitch pine and blackjack oak – that seldom exceed 3–3.3 m (10–11 feet) in height (Fig. 4.7). Bear (= scrub) oak is also a frequent part of the diminutive overstory. The density of pitch pine is about three times greater in the dwarf forest than in other areas of the Pine Barrens. Besides the small stature of the trees, other characteristics of the dwarf forest include: (a) the absence of other pines and oaks (e.g., shortleaf pine, black oak, scarlet oak, and chestnut oak) otherwise present in the Pine Barrens; and (b) the scarcity of seedlings and, instead, the prevalence of basal sprouts. Carpets of pixie flowering-moss and shrubby thickets of broom crowberry appear in the understory throughout the Pine Barrens, but both of these plants occur with greater frequency in the dwarf forest.

Of the various rationalizations, only fire seems to explain the dwarfed nature of the forest vegetation. Pitch pine and blackjack oak can withstand fire frequencies of 10 years or less. The repeated damage produces a shrub-like growth, the result of vigorous sprouting from the root crowns of both species after a fire. The shoots that emerge from bases of the dominant trees quickly produce viable seeds (i.e., rapid maturity). Pitch pines growing in the dwarf forest also produce a high percentage of **serotinous** cones, which remain closed until heated. The seeds are released only after a fire and thereafter colonize favorable sites (i.e., unshaded areas). These features – brisk sprouting, rapid maturity, and serotinous cones – ensure the perpetuation of this vegetation in settings where fires recur with great frequency. Studies of fire frequency revealed that the Pine Barrens burns about once every 16 years, but fires sweep across the dwarf forest about twice as often. Pitch pine and the two oaks survive, but repeated fires keep these species from reaching their normal stature. Moreover, some evidence indicates that pitch pines within the dwarf forest represent genetically unique stock whose adaptations to fire perpetuate this distinctive forest community. If so, this vegetation will remain a dwarf forest even in the absence of repeated fires.

Peat bogs are also a part of the Pine Barrens. Various heath plants, including laurel, blueberry, and leatherleaf, are among the shrubs at these locations. Some bogs are also managed for the commercial production of cranberries, another species of heath vegetation. The ground cover includes sedges, pitcher plants, and a rich variety of herbaceous vegetation. Among the latter is curly grass fern, an unusually rare species occurring nowhere else in New Jersey. Red maples and black gum are common, but Atlantic white cedar is the centerpiece of the overstory. Atlantic white cedar typically forms almost pure stands of tall, straight trees, but many bogs were heavily logged years ago because of the demand for shingles and many other durable wood products; among these were decoys, once a cottage industry in southern New Jersey.

Because of their value, cedar logs that sank decades ago were later searched for and retrieved from Pine Barrens bogs. The wood in such logs was preserved in the acidic bog water, which is heavily stained by tannins and other organic acids. As rainwater soaks through the ground litter of the Pine Barrens, its increasing acidity leaches iron from the underlying sands. Eventually, the dissolved iron oxidizes in a bog or stream; when and where the oxides accumulate, ore is deposited. From this "bog iron" came the cannonballs for Washington's army and, for the following half-century, a steady flow of kettles, nails, and stoves for a growing nation.

Writer John McPhee summed up the Pine Barrens: "… from Bear Swamp Hill, where, in a moment's sweeping glance, a person can see hundreds of square miles of wilderness. The picture of New Jersey that most people hold in their minds is so different from this one, considered beside it, the Pine Barrens…become incongruous as they are beautiful."

Carolina bays

Several thousand clearly delineated depressions of 10 to more than 1000 ha (2500 acres) in size dot the Atlantic Coastal Plain of the southeastern United States. These landforms, known as Carolina bays, are remarkably consistent in their oval shape and common orientation along a northwest–southeast axis (Fig. 4.8). Carolina bays are found from southern New Jersey to northern Georgia, but about 80% occur in North and South Carolina.

The origin of Carolina bays remains something of a geological mystery. One of the more widely debated ideas describes how a meteor shower struck at an oblique angle, creating the oval shapes of the depressions along a common axis. Some depressions partially overlap others, just as might result if two meteors hit the same area one after another. Moreover, a lip of sand forms the southeast rim of many Carolina bays, again suggesting an impact of an object at an oblique angle. Despite the attractiveness of this idea, no Carolina bay has yet yielded the tell-tale evidence of nickel-iron fragments or shocked quartz. Current evidence suggests that Carolina bays resulted from the long-term effects of wind and water erosion

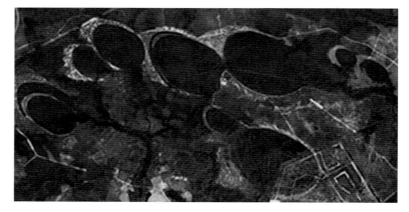

Figure 4.8 An aerial view of Carolina bays shows their predominant northwest–southeast alignment. Note their partial overlap, which is often much greater than shown here. These features suggest to some observers that an ancient meteor shower created the oval depressions, but their origin remains uncertain. The vegetation in Carolina bays shown here includes pines and extremely dense evergreen shrub forests. Reproduced by permission of Shane Freeman, Park Superintendent, Jones Lake State Park, North Carolina.

occurring when the Wisconsin glacier covered areas to the north, but geologists lack agreement.

Some Carolina bays are lakes, of which Lake Waccamaw in southeastern North Carolina is distinctive because of its endemic fauna (e.g., some fishes and a mollusk). Others are filled with peat and support various kinds of vegetation, including evergreen shrub forests. In the latter, broadleaved evergreens dominate the woody vegetation, particularly sweetbay, red bay, and loblolly bay. Titi, fetterbush, and a species of holly add to the thick vegetation, which is commonly entwined with greenbriar. Such vegetation becomes almost impenetrable for humans. Red maples and thin stands of pond pines sometimes jut from the understory.

Carolina bays apparently serve as mini-refuges for wildlife. Farmland or pine plantations surround many Carolina bays, representing monocultures in comparison with habitat within the bays. However, because of their thick vegetation, Carolina bays are relatively unstudied and their value as wildlife habitat remains more intuitive than factual. Large Carolina bays likely provide key habitat for black bears and gray foxes, whereas those of all sizes attract several species of songbirds. Amphibians reach exceptionally large numbers in some of the wetter Carolina bays. Some 100,000 southern leopard frogs frequented one Carolina bay in South Carolina with a perimeter of less than 450 m (490 yards) during the course of a single year, and as many as 11,000 mole salamanders completed their metamorphosis at another. At still another, a census included 500 ornate chorus frogs, 500 mole salamanders, and 5000 southern leopard frogs, all on a Carolina bay of less than 1 ha (2.5 acre) in size.

Because of their high water tables and thick vegetation, the Carolina bays apparently buffer the extremes of both summer and winter temperatures. As a result, the populations of some species may persist beyond the otherwise typical limits of their distributions. In other words, shade and evaporative cooling are available to northern species in the summer, whereas thick cover and heat retention ameliorate conditions for southern species in the winter.

Highlights

Acorns and blue jays

When Henry David Thoreau trudged, transit in hand, across the fields and woodlots near Concord, the some-time surveyor and full-time naturalist thoughtfully observed blue jays. His observations confirmed that blue jays are important disseminators of acorns and, for Eastern Deciduous Forest, serve as "useful agents in the economy of nature." In fact, blue jays carry quantities of acorns to caches up to 4 km (2.5 miles) from the oaks where the birds initially foraged.

Acorns contain organic compounds known as tannins, which apparently defend the seeds against herbivores. Tannins reduce the protein digestion and limit enzymatic activity in the gut of some animals and actually damage the digestive tract in others. Some insects and other animals consuming acorns have evolved counter adaptations to this defense, but blue jays lack a physiological means of dealing with tannins. Nonetheless, the birds consume large numbers of acorns each year. The question arises: how do blue jays deal with the harmful tannins and what, if anything, does this mean to forest ecology?

The answer seems to involve the presence of weevil larvae, which infest much of the acorn crop each year. The larvae burrow in the acorns, often preventing germination and the establishment of seedlings. Blue jays, of course, consume large numbers of the larvae when eating acorns, so perhaps the larvae somehow offset the adverse effects of the tannins. This relationship was suggested when ecologists fed blue jays diets of acorns with and without weevil larvae. Without the larvae, the birds lost weight on a diet of acorns; in fact, the weevil-free meals produced weight losses similar to jays that were fed nothing at all. With a 5 g (0.176 ounce) daily supplement of larvae, however, the blue jays stabilized their weight on a diet of acorns, thereby indicating that the weevil larvae indeed counteract the effects of acorn tannins. More important, the birds thus nourished also continue caching other acorns during the autumn.

At first glance, weevil larvae would seem detrimental to the reproduction of oak by ruining a part of the acorn crop; however, it may actually contribute to the far-flung dispersal of acorns and favor regeneration of the oak forest. Blue jays and oaks appear to have developed an ecological partnership that hinges on the presence of an insect once considered only in a negative light. For their part, blue jays benefit by safely feeding on tannin-laden foods and by caching a supply of uninfested acorns. The oaks benefit by having the birds disperse weevil-free seeds, some of which will later germinate and establish seedlings.

Deer yards

White-tailed deer occur throughout the Eastern Deciduous Forest. In northern areas within their range, however, deer are forced to conserve energy during periods of severe winter weather. Deer respond by concentrating in "yards," sheltered sites generally located in cedar swamps or other pockets of evergreen vegetation in the deciduous forest.

Conifers, particularly spruce, fir, and hemlock, offer good protection from cold winds and help reduce snow depth, whereas deciduous trees lacking winter foliage provide almost no shelter. However, the needles and twigs of spruce and other conifers are unpalatable and

(a)   (b)   (c)

Figure 4.9 White-tailed deer in the northern part of their range concentrate in relatively small wintering areas known as deer yards. (a) Overpopulation of deer soon leads to a browse line, where the best forage (northern white cedar) has been stripped within reach of all but the largest animals. (b) Only low-quality forage such as balsam fir remains at ground level, but even these marginal foods are eventually consumed during a hard winter. (c) An emaciated carcass bears witness to the diminished carrying capacity of an overbrowsed deer yard in Maine. Reproduced by permission of Gerald R. Lavigne, Maine Department of Inland Fisheries and Wildlife.

provide virtually no nutrition. Desirable sources of winter food (e.g., browse from maples and birches) are limited and are soon exhausted in most deer yards, especially when cold weather persists. Deer accordingly face a difficult choice: adequate food or ample cover. Cover typically "wins," but the result in some winters may be starvation and the death of many deer. Because deer establish traditions by returning to the same yards year after year, severe overbrowsing may create well-defined browse lines (Fig. 4.9). Some deer yards in the Adirondack Mountains of New York have remained in continuous use since the early 1800s. In such cases, the palatable vegetation is slow to recover and may eventually disappear. These situations grow worse, of course, when deer populations reach high levels. After unusually severe winters, carcasses of emaciated deer may litter overbrowsed yards, a situation that sometimes triggers hasty and ill-advised public responses (e.g., artificial feeding programs or restrictive hunting policies).

In comparison with spruce and fir, stands of northern white cedar offer deer better sources of winter food and cover. Deer also establish yards on south-facing slopes where exposure to the sun is greatest. Wherever they may be located, deer yards are key sites in the conservation and management of white-tailed deer populations.

Kirtland's warblers and fire

Jack pines offer an additional example (see *Longleaf pine forests*, this chapter) of a fire-dependent species whose role likewise concerns an endangered species of bird. Kirtland's warblers, or "jack pine warblers," build their nests on the ground immediately beneath the overhanging, lower branches of jack pines just where the tips touch the ground. Jack pines are widely distributed across North America, but become an important species in second-growth forests along the northern edge of the Eastern Deciduous Forest in the Great Lakes region. Nonetheless, Kirtland's warblers are extremely rare; in 1987 their numbers dipped to about 170 pairs, all of which nest in just a few counties in central Michigan and only in association with jack pines.

The special nesting niche – beneath the tips of ground-level branches – for Kirtland's warblers requires jack pines about the size of most Christmas trees. However, as the trees continue growing, their lower branches die and drop off, leaving the birds without nesting cover. If fires are suppressed jack pines grow too large to provide the essential nesting habitat for Kirtland's warblers, whereas periodic burning maintains the forest at a stage of development that meets these needs. Jack pines are themselves well adapted for frequent burning. Young trees produce cones which persist until exposed to heat;

they then open and disperse their seeds, quickly re-vegetating a burned-over forest.

Today, managers regularly burn blocks of jack pine forests within the limited breeding range of Kirtland's warblers to ensure that prime nesting habitat remains available. Approximately 60,700 ha (150,000 acres) of public lands in Michigan are managed for Kirtland's warblers, with about 12,140 ha (30,000 acres) constantly maintained as suitable nesting habitat. A nest was discovered in Wisconsin in 2006, the first record of nesting outside of Michigan since the 1940s. The number of Kirtland's warblers reached a record high in 2010 with 1733 singing males – indicators of nesting pairs – recorded in Michigan, 23 in Wisconsin, and 3 in Ontario.

Franklin's lost tree

On 1 October 1765, a discovery by John (1699–1777) and William Bartram (1739–1832), father and son and colonial America's foremost explorer-botanists, established a still unresolved botanical mystery. On the floodplain of Georgia's Altamaha River, they encountered "several curious shrubs," all new species. One of these, with large white flowers, was later named in honor of Benjamin Franklin. By 1803 however, fewer than 10 of the small trees remained at the original site – the only place where the species had ever been located – and none of these or any others were ever seen again in the wild.

Fortunately, the Bartrams had collected some seeds, some of which they later planted in their botanical garden which is still an attraction in Philadelphia. Other seeds were sent overseas to the wealthy Englishmen who sponsored the Bartrams' field trips. Today, the species survives only in gardens and nurseries in North America and Europe, where it is known as Franklin's lost tree.

Cicadas: buzz in the forest

Worldwide, at least 2500 species (and perhaps 3000) of cicadas occur in habitat ranging from deserts to forests. Cicadas, included in the insect order Hemiptera, are often incorrectly called "locusts"; instead, they are a type of grasshopper (Orthoptera) whose populations may become so numerous as to inflict serious crop damage (the misnomer likely originated from the biblical reference to a "plague of locusts"). The life-cycles of virtually all species of cicadas are relatively short and, because some individuals of these species emerge every year, they are collectively known as "annual cicadas."

In marked contrast, a single genus (*Magicicada*) of seven species with far longer life cycles occurs only in the Eastern Deciduous Forest of North America. These are the "periodical cicadas," including four largely southern and mid-western species with 13-year cycles and three primarily northern species with 17-year cycles. Uniquely,

these species develop in synchrony and therefore emerge *en masse* at predicable intervals. Each of the seven species are similar and share in common black bodies, red eyes, and a small black "W" formed by the veins near the tip of their forewings (Fig. 4.10, inset). Habitat preferences vary by species, with some found primarily in upland forests of walnut and hickory and others in floodplain forests of ash, elm, and oak, but species with the largest geographical distributions show less specificity and simply favor the canopies of mature forests. Like other cicadas, the males of each species have a distinctive "song" for attracting females.

Not all of the periodical cicadas with the same cycle emerge in the same year. Instead, each species follows staggered schedules of emergence within their respective geographical ranges. Each emergence represents an age class known as a "brood" and is distinguished by a Roman numeral (i.e., the broods of each species emerge in a cycle of the same length but in different years, not unlike students in, say, the class of 2014 that represent the same age group as those who will graduate in 2017). The broods are reproductively isolated from each other, and their appearance is so regular that the emergence of each can be accurately predicted many years in advance. Each brood has its own geographical range, varying in size from a few counties to several states (the largest covers 15 states). For reasons that are unclear, some broods are now extinct (e.g., Brood XI, once found in the Connecticut River Valley, was last recorded in 1954). The life history of periodical cicadas is essentially the same for all seven species (and, except for its duration, about the same for those annual species found in eastern North America).

Figure 4.10 The wings of adult periodical cicada show a darkly colored "W" in the venation of their wing tips (inset), which is absent in species with an annual cycle. An adult periodical cicada completes its molt after spending 17 years underground, leaving behind its nymphal exoskeleton (exuvia) as a tell-tale sign of this cyclic event. Adult cicada reproduced by permission of the Bob Rabaglia and the USDA Forest Service.

After mating, females lay their eggs in nests cut into the twigs of woody vegetation; each producing about 500–600 eggs distributed in lots of about 20 per nest. After 6–10 weeks, the first-stage nymphs hatch and drop to the ground where they burrow to depths of about 60 cm (24 inches) in search of succulent roots from which they extract xylem fluids as food. The nymphs, after four more stages of development and an interval of either 13 or 17 years, emerge to the surface in an amazing display of synchronization (soil temperatures of about 17°C or 64°F trigger their emergence). The nymphs climb the nearest woody stem – usually a tree trunk – then shed their exoskeletons (easily recognized husks that often remain attached to the bark; Fig. 4.10); now adults, the cicadas continue climbing upward where they perch until their wings harden. They remain active only for 4–6 weeks before dying.

Cicadas produce their noisy songs by flexing muscles attached to a tymbal, an elastic plate located on each side of the first abdominal segment. Each oval-shaped tymbal resembles a miniature washboard that bends when the muscles contract. These movements generate a series of popping sounds fired so rapidly as to be indistinguishable from each other. The sound is amplified by resonating in air sacs surrounding the apparatus. Other muscles alter the shape of the sacs and thus change the pitch of the sound. The species-specific songs are further modified by changing the rate at which the main muscles are contracted and by the synchrony of the two tymbals. Only males have tymbals, but both sexes have the tympana – the functional equivalent of eardrums – to hear the sounds.

The immensity of cicada populations is estimated by counting the holes in the ground from which the nymphs emerge. Based on this technique, 3.7 million ha^{-1} (1.5 million per acre) were recorded in Illinois, but most estimates for plots of the same sizes vary from several thousands to tens or hundreds of thousands. The evolutionary strategy seems to reduce the impact of predation. Periodical cicadas, in contrast to annual cicadas, are slower, less flighty, and easier to capture; there are therefore easy prey for many kinds of birds and mammals, even other insects. Indeed, when a brood emerges, the diet of these predators switches almost entirely to cicadas, hence potentially threatening the brood's survival. As a countermeasure, periodical cicadas "flood the market" in what is known as **predator satiation**, a strategy allowing predators to eat their fill without significantly reducing the breeding success of the prey population. Additionally, by emerging so infrequently the cicada populations cannot be matched by a corresponding increase in predator numbers. Because 13 and 17 are each prime numbers, cycles based on these intervals are mathematically less likely to be adopted in the life history of other species.

Ecological challenges

Declines of neotropical migrants

Recent decades have witnessed new and global concerns about the integrity of ecological systems. Among these is an awareness of the heightened rapidity with which species diminish in numbers and sometimes disappear. In the Western Hemisphere, declining populations of migrant birds illustrate the current threat to **biological diversity**. A group of birds known collectively as Neotropical migrants – among them wood thrushes and several species of wood warblers – forms a focus of attention. After breeding during the spring and summer months in the forests of North America, these birds migrate to Central and South America and the Caribbean islands (the Neotropics) for the winter months. Because they pass through a series of environments, most Neotropical migrants unusually remain vulnerable to wetland drainage, deforestation, and other anthropogenic disturbances of the landscape. The loss or degradation of any one type of habitat becomes the weak link in the environmental chain traveled each year by the migrating birds.

The Eastern Deciduous Forest is of particular significance as habitat for these birds. Neotropical migrants comprise almost 70% of all pairs of birds breeding in the deciduous forest associations of eastern North America, where they may reach a density of 587 pairs km^{-2}. Alarming reductions in songbird populations were first noted in the 1970s and continue today. At one site, for example, the numbers of Neotropical and other species, as well as their densities, dropped steadily from 325 pairs per 100 ha (247 pairs per acre) in 1947 to 273 pairs per 100 ha in 1983. Population data for 62 species of Neotropical migrants breeding in eastern North America revealed similar declines over a wide area, so the downward slides were not just local events.

In eastern North America, forest fragmentation (the process of cutting large blocks of forest into smaller units) seems to explain a good deal of the plight befalling Neotropical migrants (Fig. 4.11). Two factors are involved. First, some species are "area sensitive," which means that they avoid sites smaller than some minimum size, even if these places otherwise offer adequate food and cover. Wood thrushes, for example, seldom nest in less than 10 ha (25 acres) of suitable habitat. Area-sensitive species require blocks of breeding habitat whose size remains above a critical threshold.

Second, fragmentation increases the amount of relative linear edge per unit of area (Fig. 4.12). In other words, a small block of forest has proportionately more of its area exposed to external influences, which poses at least two hazards to birds nesting within the block. First, nests in small blocks of forest may be more easily located and destroyed by predators, which often travel along edges,

Figure 4.11 Forest fragmentation is widespread across much of eastern North America, once occupied by Eastern Deciduous Forest, as shown in this aerial view. Reproduced by permission of Craig A. Harper.

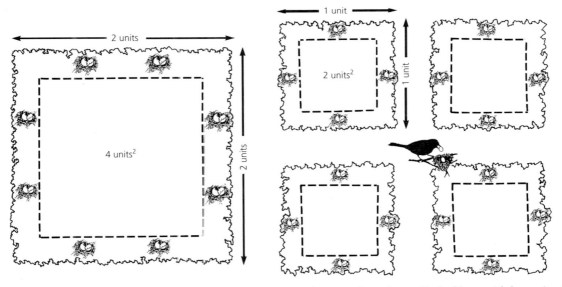

Figure 4.12 Forest fragmentation increases the amount of edge per unit of area. As shown here, a block of forest with four units of area (left) has eight nests at risk along its edges, whereas the same total area arranged in smaller blocks (right) jeopardizes twice as many nests to brown-headed cowbirds and predators. Dashed lines symbolize the limit to which brown-headed cowbirds penetrate into a block of forest when searching for nests; those nests farther inside the block experience relatively little parasitism and predation. Illustrated by Tamara R. Sayre.

whereas nests in the interior of larger blocks may avoid predation. For example, predation reached 95% on nests in some small woodlots but was only 2% on a large tract of forest in the Great Smoky Mountains.

A second and even greater hazard associated with increased edge concerns the nesting habits of brown-headed cowbirds. As nest parasites, these birds lay their eggs in the nests of other birds – often a species of Neotropical migrant – that raise the young cowbirds, usually at the expense of their own nestlings. Brown-headed cowbirds prefer open areas and seldom venture far into forests, but when forests are fragmented the resulting increase in edge offers cowbirds greater access to the nests of other birds. Forest fragmentation concurrently diminishes populations of forest-nesting songbirds as those of cowbirds increase.

The destruction of habitats elsewhere, particularly the forests along the Gulf Coast where these birds rest and refuel during migration and tropical forests where the birds overwinter, adds further concerns for the future of Neotropical migrants. Recent estimates suggest that tropical forests might be reduced to tiny fragments by 2030.

Migrant songbirds are important predators of leaf-chewing insects. In a telling experiment, the lack of birds doubled the insect populations on white oaks, causing up to 34% loss of leaf area and effects that carried over into the following years. Clearly, continued declines in North American populations of insectivorous birds may reduce forest productivity because of increased insect damage. A relationship in which the absence of predators (here birds, at the third trophic level) unleashes a herbivore population to the point of damaging primary producers (here oaks, the first trophic level) represents a **trophic cascade**. Such phenomena were once associated only with aquatic communities, but the experimental interactions between songbirds and the vigor of oak trees suggest a wider application of the concept. Forest managers clearly should encourage the conservation of insectivorous birds with practices that promote their diversity and abundance.

Forest destruction by exotic organisms

American chestnut once formed a major component in a large area of Eastern Deciduous Forest, but today the species is largely a memory, the result of a fungal disease detected in 1904 (Fig. 4.13). Known as chestnut blight, the disease immigrated on Chinese chestnut trees and immediately attacked the American species, which lacked immunity. By 1930, the disease had spread throughout the range of the American chestnut – most of the United States east of the Mississippi – eventually killing more than 4 billion trees. The windborne spores disperse rapidly and enter the bark in cracks (e.g., crotches of limbs) or lesions made by wood-boring insects. The fungus kills the vital layer of cambium, and wounds known as cankers eventually girdle the trunk, interrupting the flow of water and nutrients and killing the tree above the girdle (the roots and stumps remain alive, but the shoots they produce later die from the fungus). New crops of spores are produced from fruiting bodies extruding through the diseased bark. The disease is unusual because it virtually eliminated its host. Generally, pathogenic attacks ebb as the density of their victims diminishes, and the host population rebuilds before another attack occurs.

American chestnut was once one of the most commercially important species in Appalachian forests. A mature tree typically grew straight and branch-free for 15–30 m (50–100 feet) before reaching heights up to 60 m (200

Figure 4.13 American chestnut was once a primary species in some regions of the Eastern Deciduous Forest. Accidental introduction of the deadly fungal disease chestnut blight early in the 20th century completely eradicated mature trees such as those shown here in the Great Smoky Mountains. Reproduced by permission of Forest History Society, Durham, North Carolina.

feet). Its durable wood was valued for uses ranging from railroad ties to coffins, and the bark provided tannic acid for treating leather. The disappearance of the species also reduced the mast available to wild turkeys and other wildlife, but the spiny burr surrounding each cluster of three nuts probably evolved as a defense against foraging squirrels.

Chestnut blight likely represents the greatest botanical disaster befalling the forests of North America, a misfortune perhaps exceeded only by the extinction of passenger pigeons. Unlike the passenger pigeon, however, ongoing research offers some promise for restoring the America chestnut to its former status in the Eastern Deciduous Forest.

Whereas chestnut blight was accidentally introduced, gypsy moths were brought from Europe to North America with the idea of improving silk production. Not long after their introduction in 1869, however, some of the European moths escaped near Boston. Since then, more money has been spent to combat gypsy moths than to control all other forest insects combined. The caterpillars eat leaf tissues, beginning with spring foliage and continuing until the leaves mature later in the summer. Outbreaks defoliate hickories, maples, and especially oaks. Repeated outbreaks in consecutive years may kill half of the trees in some oak forests. Birches are also attacked, and some caterpillars eat the foliage of pines and other conifers.

Gypsy moths have attacked large areas of the northeastern United States for many years, and they reached northern Virginia in 1989. The greatest populations now occur in the southern Appalachians, the Ozark Mountains, and the northern Lake States and west into Minnesota. A second strain of gypsy moths, this one from Asia, landed in Washington and Oregon in 1991 and in North Carolina in 1993. The Asian strain, if permanently established, may pose an even greater threat to forest communities because it feeds on a wider range of species than the European strain.

Besides the direct damage they cause to trees, gypsy moths may indirectly influence the nesting ecology of songbirds. Defoliation reduces the number of nesting sites for canopy-nesting birds (e.g., scarlet tanagers). Similarly, reduced cover may increase nest predation, including species nesting on or near the ground (e.g., indigo buntings). With massive defoliation, as much as 73% more solar radiation may reach streams, raising water temperatures by nearly 4°C (almost 7°F) or enough to alter the composition of aquatic communities. Defoliation also kills a large percentage of oaks and other trees serving as winter dens for black bears, which may affect their reproduction and overwinter mortality.

Acorn production also is curtailed by 50–100%, which in turn affects gray squirrels and other mast-eating wildlife. Gypsy moth attacks reduce the acorn crop by: (a) consuming flowers; (b) lowering the carbohydrate supply (i.e., loss of foliage), which aborts immature acorns; and (c) decreasing the initiation of flower buds. When outbreaks are severe enough to kill large areas of forest, oaks are replaced by other trees to the further detriment of wildlife dependent on acorn mast.

DDT was once the accepted "medicine" for gypsy moths and large areas of forest were sprayed before the insecticide was banned in 1972. Today, attacks of these pests are countered with **integrated pest management** (IPM), including **biological controls** as well as chemical insecticides (e.g., diflubenzuron). Another treatment involves spraying a bacterium, *Bacillus thuringiensis* var. *kurstaki* (popularly known as BTK) which infects the caterpillars. The treatment also attacks other kinds of caterpillars, but with relatively few side effects. BTK apparently poses no harm to soil, water, vegetation, or vertebrate animals, and so seems safe for continuing the war against gypsy moths.

These examples – one a fungus, the other an insect – illustrate the ecological dangers presented by exotic species, those organisms finding their way to North America from other lands. These, of course, are not the only hard lessons learned from these situations. Fire ants, exotic earthworms, European starlings, wild pigs, and a host of exotic plants (e.g., kudzu) have their own tales to tell (see Infobox 4.3).

Reintroduction of red wolves

When the first settlers ventured into the relatively unspoiled Eastern Deciduous Forest, red wolves prowled the region from central Maine to eastern Texas, a range they had occupied since the end of the Wisconsin glacial period. They were especially abundant in higher-elevation forests in the mountains of eastern North America and in the dense bottomland forests and wetlands along rivers in the southeastern United States. Soon, however, populations of red wolves declined in the face of habitat fragmentation and widespread predator control. By the 1960s, biologists realized that red wolves were nearing extinction. These concerns increased when red wolves started to hybridize with coyotes, thereby losing their full genetic identity and leaving ever-fewer "pure" wolves in the population. In fact, the gene pool of red wolves was genetically "swamped" by coyote genes – a process known as **introgression** – with the result that much the wolf population acquired the heritable traits of coyotes. Eventually, the only genetically intact population of red wolves persisted in what seemed to be marginally suitable habitat: the prairies, marshes, and bottomlands of coastal Louisiana and Texas. Meanwhile, coyotes were expanding their range eastward and undoubtedly would soon invade even this last stronghold. The genetic clock was ticking rapidly for the few remaining red wolves.

After red wolves were listed as endangered in both the 1966 and 1973 versions of the Endangered Species Act, biologists located and captured the last 17 red wolves remaining in eastern Texas and western Louisiana. Of these, 14 were selected (after screening to eliminate those with coyote traits) for a captive breeding program. After the initial attempts to establish populations with captive-bred offspring met with little or no success, four pairs were released at Alligator River National Wildlife Refuge, an immense, coyote-free area of coastal lowland in northeastern North Carolina. By 2010, and bolstered by supplemental releases, the population on the refuge and surrounding federal lands had grown to more than 70 individuals organized into 26 packs that included 11 breeding pairs. However, the secretive nature of red wolves hinders their census, and some biologists estimate the population in North Carolina to be nearer 130 individuals. An additional 200 red wolves remain in captive breeding facilities at more than 30 locations throughout the United States.

Infobox 4.3 An invasion of earthworms

Charles Darwin (1809–1882) gained lasting fame for his *Origin of Species* (1859) but another book, *The Formation of Vegetable Mould Through the Action of Worms, with Observations of Their Habits* (1881), also attracted considerable attention. Indeed, "*Worms*" sold more copies in its first year of publication than "*Origin*." In general, earthworms are recognized as indicators of good soil; their functions enhance natural fertilization and the decomposition of leaves and other organic matter, and improve of soil aeration and drainage. Consequently, ecologists often regard earthworms as "ecosystem engineers" because of their ability to alter environments. However, invasive species of earthworms in the northern hardwood forests of North America may threaten plant and soil interactions and, in some cases, animal life.

When glaciers retreated from an area of North America now covered with hardwood forests (e.g., Michigan) some 12,000 years ago, they left behind soils barren of whatever **fossorial** fauna that may have existed before the huge ice sheets smothered the terrain. In the following millennia, the landscape and its biota steadily developed but remained free of earthworms. Eventually, however, this empty niche filled when Euro-Americans settled the region. They imported plants and dumped soil (used as ships ballast) containing worms or their egg cases. Still later, more foreign worms were introduced when anglers discarded unwanted bait, including those known as "night crawlers." In all, at least 15 species of exotic earthworms eventually invaded the soils in northern forests.

Exotic earthworms remove the thick litter of leaves and other organic matter accumulating on the forest floor. This alters both the flow of nutrients and the composition of the microfloral community in the soil, thereby affecting food webs and the above-ground life dependent on the litter. Among the more important results is a reduction in the mycorrhizal fungi necessary for the growth of many kinds of vascular vegetation. The spread of invasive earthworms across a forest can be traced by a visible leading edge marked by reductions in litter depth of 10 cm (4 inches) to none at all. Concurrently, increased leaching of nitrogen and phosphorous may reduce the availability of these nutrients to those horizons where fine roots are concentrated.

These effects alter the composition of the understory by replacing native herbaceous vegetation (e.g., violets and trillium) with monotypic communities of other species that are often themselves exotic species (e.g., buckthorn). By reducing the thickness of the organic horizon (O), one species of exotic earthworm, *Lumbricus rubellus*, eliminated the understory of herbaceous plants, reduced the density and survival of tree seedlings, and extirpated a rare fern (*Botrychium mormo*).

By removing leaf litter, exotic earthworms also reduce the density of ground-nesting songbirds such as ovenbirds and hermit thrushes. Likewise, the successful fledging of ovenbird nestlings appears lessened where invasive earthworms reduced litter depth, decreased ground cover, and diminished nest concealment, which increased losses to predators. Similarly, the abundance of woodland salamanders (e.g., the eastern red-backed salamander) declined exponentially as the volume of leaf litter diminished in hardwood forests invaded by exotic earthworms. The lack of arthropod prey contributed to the negative effects on salamander and bird populations.

While the effects described above are associated with forested areas devoid of native earthworms, the question arises about the resistance of these species elsewhere to invasions of exotic earthworms. Studies addressing this issue indicate that native species are not able to deter these invasions and, in fact, often coexist with exotic earthworms. **Competitive exclusion**, while difficult to determine in earthworms, does not seem to prevail as an ecological limitation and deterrent, if any, may be more related to habitat conditions than to biological interactions.

Coyotes eventually invaded northeastern North Carolina where they threaten to hybridize with the red wolf population established at Alligator River. To meet this challenge, the US Fish and Wildlife Service captures, sterilizes, and returns coyotes to their place of capture where they act as "placeholders." This process relies on their territorial behavior, which is unaffected by sterilization, to keep out other coyotes, limiting further growth in the resident coyote population and hybridization with red wolves. For the most part, the natural process of interspecific competition should enable red wolves to displace the placeholder coyotes and allow continued expansion of an endangered species with its genetic integrity fully intact.

Readings and references

Eastern Deciduous Forest Biome

Bailey, R.G. 1995. *Description of the Ecoregions of the United States.* Second edition. USDA Forest Service Miscellaneous Publication 1391, Washington, DC.

Brooks, M. 1965. *The Appalachians.* Houghton Mifflin, Boston, MA.

Davis, M.B. 1976. Pleistocene biogeography of temperate deciduous forests. In: *Geoscience and Man. Vol 13. Ecology of the Pleistocene* (R.C. West and W. G. Haag, eds). Louisiana State University Press, Baton Rouge, LA, pp. 13–26.

Peattie, D.C. 1950. *A Natural History of Trees of Eastern and Central North America.* Houghton Mifflin, Boston, MA.

Rohrig, E. and B. Ulrich (eds). 1991. *Ecosystems of the World. Vol. 7. Temperate Deciduous Forests*. Elsevier, New York, NY.

Climatic boundaries and soils

Küchler, A.W. 1964. Potential natural vegetation of the conterminous United States. American Geographical Society Special Publication 136: 1–116.

USDA Soil Conservation Service. 1975. Soil taxonomy: a basic system of soil classification for making and interpreting soil surveys. USDA Soil Conservation Service, Agricultural Handbook 436, Washington, DC.

Yahner, R. H. 1995. *Eastern Deciduous Forest Ecology and Wildlife Conservation*. University of Minnesota Press, Minneapolis, MN.

Features and adaptations

Evans, K.E. and R.N. Conner. 1979. Snag management. In: *Proceedings of the Workshop on Management of North Central and Northeastern Forests for Nongame Birds*. USDA Forest Service, North Central Forest Experimental Station, St Paul, MN, pp. 214–225.

Harris, W.F., R.S. Kinerson, Jr., and N.T. Edwards. 1977. Comparison of below ground biomass of natural deciduous forest and loblolly pine plantations Pedobiologia 17: 369–381.

Nye, P.H. 1966. The effect of nutrient intensity and buffering power of a soil, and the absorbing power, size and root hairs of a root, on nutrient absorption by diffusion. Plant Soil 25: 81–105.

Rudolph, R.R. and C.G. Hunter. 1964. Green trees and greenheads. In *Waterfowl Tomorrow* (J.P. Linduska, ed.) US Fish and Wildlife Service, Washington, DC, pp. 611–618.

The forest primeval

Berry, T., P. Beasley, J. Clements, and W. Dunbar. 2006. *The Forgotten Expedition, 1804–1805: The Louisiana Purchase Journals of Dunbar and Hunter*. Louisiana State University Press, Baton Rouge, LA.

Cox, T.R., R.S. Maxwell, P.D. Thomas, and J.J. Malone. 1985. *This Well-Wooded Land: Americans and their Forests from Colonial Times to the Present*. University of Nebraska Press, Lincoln, NE.

Cronon, W. 1983. *Changes in the Land: Indians, Colonists, and the Ecology of New England*. Hill and Wang, New York, NY.

Day, G.M. 1953. The Indian as an ecological factor in the northeastern forest. Ecology 34: 329–346.

Dyer, J.M. 2001. Using witness trees to assess forest change in southeastern Ohio. Canadian Journal of Forest Research 31: 1708–1718.

Feduccia, A. (ed.) 1985. *Catesby's Birds of Colonial America*. University of North Carolina Press, Chapel Hill, NC.

Flader, S.L. (ed.). 1983. *The Great Lakes Forest, an Environmental and Social History*. University of Minnesota Press, Minneapolis, MN. (Including *The Forest the Settlers Saw*)

Frick, G.F. and R.P. Stearns. 1961. *Mark Catesby: The Colonial Audubon*. University of Illinois Press, Urbana, IL.

Lawson, J. 1709. *A New Voyage to Carolina*. (H.T. Lefler, ed.) University of North Carolina Press, Chapel Hill, NC.

Moulton, G.E. (ed.). 1968. *The Journals of the Lewis & Clark Expedition*. Vol. 2. University of Nebraska Press, Lincoln, NE.

Nash, R. 1982. *Wilderness and the American Mind*. Third edition. Yale University Press, New Haven CT.

Russell, E.W.B. 1983. Indian-set fires in the forests of the northeastern United States. Ecology 64: 78–88.

Williams, M. 1989. *Americans and their Forests: A Historical Geography*. Cambridge University Press, New York, NY.

The layered forest

Braun, E.L. 1950. *Deciduous Forests of Eastern North America*. Blackburn Press, Caldwell, NJ. (Reprint of first edition.)

Smith, T.M. and R.L. Smith. 2009. *Elements of Ecology*. Seventh edition. Pearson Education, Inc., San Francisco, CA.

Autumn leaves

Kozlowski, T.T. (ed.). 1973. *Shedding of Plant Parts*. Academic Press, New York, NY.

Ground and leaf litter

Bray, J.R. and E. Gorham. 1964. Litter production in forests of the world. Advances in Ecological Research 2: 101–157.

Dixon, K.R. 1976. Analysis of seasonal leaf fall in north temperate deciduous forests. Oikos 27: 300–306.

Gosz, J.R., G.E. Likens, and F.H. Bormann. 1972. Nutrient content of litter fall on the Hubbard Brook, Experimental Forest, New Hampshire. Ecology 53: 769–784.

Lang, G.E. 1974. Litter dynamics in a mixed oak forest on the New Jersey Piedmont. Bulletin of the Torrey Botanical Club 101: 277–286.

Maser, C. and J.M. Trappe (eds.). 1984. The seen and unseen world of the fallen tree. General Technical report PNW-164, USDA Forest Service, Portland OR.

Maser, C. and J.R. Sedell. 1994. *From the Forest to the Sea: The Ecology of Wood in Streams, Rivers, Estuaries, and Oceans*. St Lucie Press, Delray Beach, FL.

McBrayer, J.F. and K. Comrack. 1980. Effect of snow-pack on oak-litter breakdown and nutrient release in a Minnesota forest. Pedobiologia 20: 47–54.

Seastedt, T.R. and D.A. Crossley. 1981. Microarthropod response following cable logging and clear-cutting in the southern Appalachians. Ecology 62: 126–135.

Swift, M.J., I.N. Healey, J.K. Hibberd, et al. 1976. The decomposition of branchwood in the canopy and floor of a mixed deciduous woodland. Oecologia 26: 139–149.

Mast

Christisen, D.M. and L.J. Korschgen. 1955. Acorn yields and wildlife usage in Missouri. Transaction of the North America Wildlife Conference 20: 337–357.

Cypert, E., and B.S. Webster. 1948. Yield and use by wildlife of acorns of water and willow oaks. Journal of Wildlife Management 12: 227–231.

Dalke, P.D. 1953. Yields of seeds and mast in second growth hardwood forest, southcentral Missouri. Journal of Wildlife Management 17: 378–380.

Duvendeck, J.P. 1962. The value of acorns in the diet of Michigan deer. Journal of Wildlife Management 26: 371–379.

Flyger, V. 1969. The 1968 squirrel "migration" in the eastern United States. Proceedings of the Northeast Fish and Wildlife Conference 26: 69–79.

Goodrum, P.D., V.H. Reid, and C.E. Boyd. 1971. Acorn yields, characteristics, and management criteria of oaks for wildlife. Journal of Wildlife Management 35: 520–532.

Janzen, D.H. 1971. Seed predation by animals. Annual Review of Ecology and Systematics 2: 465–492. (Includes theoretical aspects of "mast years.")

Korschgen, L.J. 1967. Feeding habits and foods. In: *The Wild Turkey and its Management* (O.H. Hewitt, ed.) The Wildlife Society, Washington, DC, pp. 137–198.

McQuilkin, R.A. and R.A. Musbach. 1977. Pin oak acorn production on green tree reservoirs in southeastern Missouri. Journal of Wildlife Management 41: 218–225.

McShea, W.J. and G. Schwede. 1993. Variable acorn crops: responses of white-tailed deer and other mast consumers. Journal of Mammalogy 74: 999–1006.

Schorger, A.W. 1955. *The Passenger Pigeon*. University of Wisconsin Press, Madison, WI.

Silverton, J.W. 1980. The evolutionary ecology of mast seeding in trees. Biological Journal of the Linnean Society 14: 235–250.

Sork, V.L., J. Bramble, and O. Sexton. 1993. Ecology of mast-fruiting in three species of North American deciduous oaks. Ecology 74: 528–541.

Biotic associations

Agrawal, A. and S.L. Stephenson. 1995. Recent successional changes in a former chestnut-dominated forest in southwestern Virginia. Castanea 60: 107–113.

Braun, E.L. 1950. *Deciduous Forests of Eastern North America*. Blackburn Press, Caldwell, NJ. (Reprint of first edition.)

Cain, S.A. 1943. The Tertiary character of the cove hardwood forests of the Great Smoky Mountains National Park. Bulletin of the Torrey Botanical Club 70: 213–235.

Daubenmire, R.F. 1936. The "Big Woods" of Minnesota: its structure and relation to climate, fires, and soils. Ecological Monographs 6: 233–268.

Davis, J.H. 1930. Vegetation of the Black Mountains of North Carolina: an ecological study. Journal of the Elisha Mitchell Scientific Society 45: 291–318.

Dyer, J.M. 2006. Revisiting the deciduous forests of eastern North America. BioScience 56: 341–352.

Eggler, W.A. 1938. The maple-basswood forest type in Washburn County, Wisconsin. Ecology 19: 243–263.

Ford, W.M., P.E. Hale, R.H. Odom, and B.R. Chapman. 2000. Stand-age, stand characteristics, and landform effects on understory herbaceous communities in southern Appalachian cove-hardwoods. Biological Conservation 93: 237–246.

Grimm, E.C. 1984. Fire and other factors controlling the Big Woods vegetation of Minnesota in the mid-nineteenth century. Ecological Monographs 54: 291–311.

Keever, C. 1953. Present composition of some stands of the former oak-chestnut forest in the southern Blue Ridge Mountains. Ecology 34: 44–54.

Mackay, H.E., Jr. and N. Sivec. 1973. The present composition of a former oak-chestnut forest in the Allegheny Mountains of western Pennsylvania. Ecology 54: 915–919.

Monk, C.D. 1965. Southern mixed hardwood forest in north-central Florida. Ecological Monographs 35: 335–354.

Nichols, G.E. 1935. The hemlock-white pine-northern hardwood region of eastern North America. Ecology 16: 403–422.

Quarterman, E. and C. Keever, 1962. Southern mixed hardwood forest: climax in the southeastern Coastal Plain, USA. Ecological Monographs 32: 167–185.

Shanks, R.E. 1953. Forest composition and species association in the beech-maple forest region of western Ohio. Ecology 34: 455–466.

Stahle, M.K., M.K. Cleavland, and J.G. Hehr. 1988. North Carolina climate changes reconstructed from tree rings: AD 372 to 1985. Science 240: 1517–1519.

Stephenson, S.L. 1985. Changes in a former chestnut-dominated forest after a half century of succession. American Midland Naturalist 116: 173–179.

Wells, B.W. 1928. Plant communities of the coastal plain of North Carolina and their successional relations. Ecology 9: 230–242.

Whitney, G.G. 1990. The history and status of the hemlock-hardwood forests of the Allegheny Plateau. Journal of Ecology 78: 443–458.

Whittaker, R.H. 1956. Vegetation of the Great Smoky Mountains. Ecological Monographs 26: 1–80.

Williams, A.B. 1936. The composition and dynamics of a beech-maple climax community. Ecological Monographs 6: 318–408.

Some associated communities

Longleaf pine forests

Chapman, H.H. 1932. Is the longleaf type a climax? Ecology 13: 328–334.

Chapman, H.H. 1936. Effects of fire in the propogation of seedbed for longleaf pine seedlings. Journal of Forestry 34: 852–854.

Chapman, H.H. 1944. Fire and pines. American Forests 50: 62–64, 91–93.

Christensen, N.L. 1977. Fire and soil-plant relations in a pine-wiregrass savannah on the coastal plain of North Carolina. Oecologia 31: 27–44.

Clewell, A.F. 1989. Natural history of wiregrass (*Aristida stricta* Michx., Graminae). Natural Areas Journal 9: 223–233.

Conner, R.N. and B.A. Locke. 1982. Fungi and red-cockaded woodpecker cavity trees. Wilson Bulletin 94: 64–70.

Cooper, R.W. 1975. Prescribed burning. Journal of Forestry 73: 776–780.

Croker, T.C., Jr. 1979. The longleaf pine story. Journal of Forest History 23: 32–43.

Earley, L.S. 2004. *Looking for Longleaf: The Fall and Rise of an American Forest*. University of North Carolina Press, Chapel Hill, NC.

Glitzenstein, J.S., W.J. Platt, and D.R. Streng. 1995. Effects of fire regime and habitat in tree dynamics in north Florida longleaf pine savannahs. Ecological Monographs 65: 441–476.

Jackson, J.A. 1986. Biopolitics, management of federal lands, and the conservation of the red-cockaded woodpecker. American Birds 40: 1162–1168.

Jackson, J.A. 1994. Red-cockaded woodpecker. The Birds of North America 85: 1–20.

Jose, S., E.J. Jokela, D. Miller, and D.L. Miller. 2006. *The Longleaf Pine Ecosystem: Ecology, Silviculture, and Restoration*. Springer, New York, NY.

Lemon, P.C. 1949. Successional responses of herbs in longleaf-slash pine forest after fire. Ecology 30: 135–145.

McFarlane, R.W. 1992. *A Stillness in the Pines, the Ecology of the Red-Cockaded woodpecker*. W.W. Norton & Company, New York, NY.

Outcalt, K.W. 1994. Seed production of wiregrass in central Florida following growing season prescribed burns. International Journal of Wildland Fire 4: 123–124.

Outland, R.B. 2004. *Tapping the Pines: The Naval Stores Industry in the American South*. Louisiana State University Press, Baton Rouge, LA.

Pietropaolo, J. and P. Pietropoalo. 2005. *Carnivorous Plants of the World*. Timber Press, Portland, OR. (Pages 15–24 are devoted to the Venus flytrap.)

Pyne, S.J. 1982. *Fire in America, a Cultural History of Wildland and Rural Fire*. Princeton University Press, Princeton, NJ.

Rudolph, D.C., H. Kyle, and R.N. Conner. 1990. Red-cockaded woodpecker vs rat snakes: the effectiveness of the resin barrier. Wilson Bulletin 102: 14–22.

Schnell, D.E. 2009. *Carnivorous Plants of the United States and Canada*. Timber Press, Portland OR.

Varner, J.M. and J.S. Kush. 2004. Remnant old-growth longleaf pine (*Pinus palustris* Mill.) savannas and forests of the southeastern USA: status and threats. Natural Areas Journal 24: 141–149.

Walker, S., and R.K. Peet. 1983. Composition and species diversity of pine-wiregrass savannas of the Green Swamp, North Carolina. Vegetation 55: 163–179.

Wright, H.A. and A.W. Bailey. 1982. *Fire Ecology, United States and Southern Canada*. John Wiley & Sons, New York NY.

New Jersey Pine Barrens

Andresen, J.W. 1959. A study of pseudo-nanism in *Pinus rigida* Mill. Ecological Monographs 29: 309–332.

Boyd, H.P. 1991. *A Field Guide to the Pine Barrens of New Jersey*. Plexus Publications, Medford, NJ.

Boyd, H.P. 2008. *The Ecological Pine Barrens of New Jersey: An Ecosystem Threatened by Fragmentation*. Plexus Publications, Medford, NJ.

Buchholz, D. and R.A. Zampella. 1987. A 30-year fire history of the New Jersey Pine Barrens plains. Bulletin of the New Jersey Academy of Science 32: 61–69.

Buell, M.F. and J.E. Cantlon. 1950. A study of two communities of the New Jersey Pine Barrens and a comparison of methods. Ecology 31: 567–586.

Collins, B.R. and K.H. Anderson. 1994. *Plant Communities of New Jersey*. Rutgers University Press, New Brunswick, NJ.

Forman, R.T.T. (ed.) 1998. *Pine Barrens: Ecosystem and Landscape*. Academic Press, New York, NY.

Harshberger, J.W. 1916. *The Vegetation of the New Jersey Pine-Barrens*. Christopher Sower Company. Philadelphia, PA. (Reprinted in 1970 by Dover Publications, New York, NY.)

Lutz, H.L. 1934. Concerning a geological explanation of the origin and present distribution of the New Jersey pine barren vegetation. Ecology 15: 399–406.

McCormick, J. and M.F. Buell. 1968. The plains: pygmy forests of the New Jersey Pine Barrens, a review and annotated bibliography. Bulletin of the New Jersey Academy of Science 13: 20–34.

Pierce, A.D. 1957. *Iron in the Pines*. Rutgers University Press, New Brunswick, NJ.

Carolina bays

Clark, M.K., D.S. Lee, and J.B. Funderberg. 1985. The mammal fauna of Carolina bays, pocosins, and associated communities in North Carolina: an overview. Brimleyana 11: 1–38.

Frey, D.G. 1951. The fishes of North Carolina's bay lakes and their interspecific variation. Journal of the Elisha Mitchell Scientific Society 67: 1–44.

Lee, D.S. 1986. Pocosin breeding bird fauna. American Birds 40: 1263–1273.

Melton, F.A. 1950. The Carolina "bays." Journal of Geology 58: 128–134.

Melton, F.A. and W. Schriever. 1933. The Carolina "bays"—are they meteoric scars? Journal of Geology 41: 52–66.

Prouty, W.F. 1952. Carolina bays and their origin. Geological Society of America Bulletin 63: 167–224.

Sharitz, R.R. and J.W. Gibbons. 1982. The ecology of southeastern shrub bogs (pocosins) and Carolina bays: a community profile. FWS/OBS-82-04, US Fish and Wildlife Service, Washington, DC.

Wells, B.W. and S.G. Boyce. 1953. Carolina bays: additional data on their origin, age, and history. Journal of the Elisha Mitchell Scientific Society 69: 117–141.

Highlights

Acorns and blue jays

Bosemma, I. 1979. Jays and oaks: an eco-ethological study of a symbiosis. Behavior 70: 1–117.

Darley-Hill, S. and W.C. Johnson. 1981. Acorn dispersal by the blue jay (*Cyanocitta cristata* L.) Oecologia 50: 231–232.

Gomez, J.M. 2003. Spatial patterns in long-distance dispersal of *Quercus ilex* acorns by jays in a heterogeneous landscape. Ecography 26: 573–584.

Johnson, W.C., and C.S. Adkisson. 1985. Dispersal of beechnuts by blue jays in fragmented landscapes. American Midland Naturalist 113: 319–324.

Johnson, W.C., and T. Webb III. 1989. The role of blue jays (*Cyanocitta cristata*) in the post-glacial dispersal of fagaceous trees in eastern North America. Journal of Biogeography 16: 561–571.

Johnson, W.C., L. Thomas, and C.S. Adkisson. 1993. Dietary circumvention of acorn tannins by blue jays. Oecologia 94: 159–164.

Scarlett, T.L. and K.G. Smith. 1991. Acorn preference of urban blue jays (*Cyanocitta cristata*) during fall and spring in northwest Arkansas. Condor 93: 438–442.

Thoreau, H.D. 1860. The succession of forest trees. In: *Henry David Thoreau, the Natural History Essays (with an introduction by R. Sattlemeyer)*. 1980. Pergrine Smith, Salt Lake City, UT, pp. 72–92.

Deer yards

Brown, D.T. and G.J. Doucet. 1991. Temporal changes in winter diet selections by white-tailed deer in a northern deer yard. Journal of Wildlife Management 55: 361–376.

Gill, J.D. 1957. Review of deer-yard management, 1956. Game Division Bulletin 5, Maine Department of Inland Fish and game, Augusta, ME.

Hurst, J.E. and W.F. Porter. 2008. Evaluation of shifts in white-tailed deer winter yards in the Adirondack region of New York. Journal of Wildlife Management 72: 367–375.

Oyer, A.M. and W.F. Porter. 2004. Localized management of white-tailed deer in the central Adirondack Mountains, New York. Journal of Wildlife Management 68: 257–265.

Verme, L.J. 1965. Swamp conifer deeryards in northern Michigan. Journal of Forestry 63: 523–529.

Kirtland's warblers and fire

Corace, R.G., III, P.C. Goebel, and D.L. McCormick. 2010. Kirtland's warbler habitat management and multispecies bird conservation: considerations for planning and management across jack pine (*Pinus banksiana* Lamb.) habitat types. Natural Areas Journal 30: 174–190.

Donner, D.M., J.R. Probst, and C.A. Ribic. 2008. Influence of habitat amount, arrangement, and use on population trend estimates of male Kirtland's warblers. Landscape Ecology 23: 467–480.

Donner, D.M., C.A. Ribic, and J.R. Probst. 2009. Male Kirtland's warblers' patch level response to landscape structure during periods of varying population size and habitat amounts. Forest Ecology and Management 258: 1093–1101.

Donner, D.M., C.A. Ribic, and J.R. Probst. 2010. Patch dynamics and the timing of colonization-abandonment events by male Kirtland's warblers in an early successional environment. Biological Conservation 143: 1159–1167.

Mayfield, H.F. 1960. *The Kirtland's Warbler.* Cranbrook Institute Science Bulletin 40, Bloomfield Hills, MI. (The basic reference for the natural history of this species.)

Franklin's lost tree

Berkeley, E. and D.S. Berkeley. 1982. *The Life and Travels of John Bartram.* University Presses of Florida, Tallahassee, FL.

Harper, F. (ed.) 1958. *The Travels of William Bartram, Naturalist's Edition.* Yale University Press, New Haven CT.

Thompson, K.S. 1990. Benjamin Franklin's lost tree. American Scientist 78: 203–206.

Cicadas: buzz in the forest

Alexander, R.D. and T.E. Moore. 1962. The evolutionary relationships of 17-year and 13-year cicadas, and three new species (Homoptera, Cicadidae, *Magicicada*). Miscellaneous Publication 121, Museum of Zoology, University of Michigan.

Bear, J.A., Jr. and L.C. Stanton (eds). 1997. *Jefferson's Memorandum Books, Accounts, with Legal Records and Miscellany, 1767–1826.* Volume 1. Princeton University Press, Princeton, New Jersey. (Regarding cicadas, see page 388.)

Cooley, J.R. and D.C. Marshall. 2001. Sexual signaling in periodical cicadas. Behaviour 138: 827–855.

Cooley, J.R., C. Simon, and D. C. Marshall. 2003. Temporal separation and speciation in periodical cicadas. BioScience 53: 151–157.

Cooley, J.R., G. Kritsky, M.J. Edwards, et al. 2009. The distribution of periodical cicada Brood X. American Entomologist 55: 106–112.

Dybas, H.S. 1969. The 17-year cicada: a four-year "mistake?" Bulletin of the Field Museum of Natural History 40 (August): 10–12. (Reports 133,000 to 1.5 million emergence holes per acre.)

Marshall, D. C. 2001. Periodical cicada (Homoptera:Cicadidae): life-cycle variations, the historical emergence record, and the geographic stability of brood distributions. Annals of the Entomological Society of America 94: 386–399.

Sorenson, C. 2011. How do little cicadas make such loud noises? Wildlife in North Carolina 75 (May): 39.

Williams, K.S. and C. Simon. 1995. The ecology, behavior, and evolution of periodical cicadas. Annual Review of Entomology 40: 269–295.

Ecological challenges

Declines in neotropical migrants

Askins, R.A. Hostile andscapes and the decline of migratory songbirds. Science 267: 1956–1957.

Brittingham, M.C. and S. A. Temple. 1983. Have cowbirds caused forest songbirds to decline? BioScience 33: 31–35.

Franzreb, K.E. 2005 The effects of timber harvesting on Neotropical migrants in cove hardwood forests in the southern Appalachian Mountains. USDA Forest Service General Technical Report PSW-GTR-191: 301–310.

Friesen, L.E., P.F.J. Eagles, and R.J. MacKay. 1995. Effects of residential development on forest-dwelling Neotropical migrant songbirds. Conservation Biology 9: 1408–1414.

Hall, G.A. 1984. Population decline of Neotropical migrants in an Appalachian forest. American Birds 38: 14–18.

Harris, L.D. 1984. *The Fragmented Forest, Island Biogeography Theory and the Preservation of Biotic Diversity.* University of Chicago Press, Chicago IL.

Holmes, R.T. and T.W. Sherry. 2001. Thirty-year bird population trends in an unfragmented temperate deciduous forest: importance of habitat change. Auk 118: 589–609.

Keller, G.S. and R.H. Yahner. 2006. Declines of migratory songbirds: evidence for wintering ground causes. Northeastern Naturalist 13: 83–92.

Lichstein, J., T.R. Simons, and K.E. Franzreb. 2002. Effects of landscape composition on songbird habitat use in managed southern Appalachian forests. Ecological Applications 12: 836–857.

Marquis, R.J. and C.J. Whelan. 1994. Insectivorous birds increase growth of white oak through consumption of leaf-chewing insects. Ecology 75: 2007–2114.

Price, J.T. and T.L. Root. 2005. Potential impacts of climate change on Neotropical migrants: management implications. USDA Forest Service General Technical Report PSW-GTR-191: 1123–1128.

Robbins, C.S., J.R. Sauer, R.S. Greenberg, and S. Droege. 1989. Population declines n North American birds that migrate to the Neotropics. Proceedings of the National Academy of Science 86: 7658–7652.

Terborgh, J. 1989. *Where Have All the Birds Gone?* Princeton University Press, Princeton, NJ.

Terborgh, J. 1992. Perspectives on the conservation of Neotropical migrant landbirds. In: *Ecology and Conservation of Neotropical Migrant Landbirds* (J.M. Hagen and D.W. Johnston, eds). Smithsonian Institution Press, Washington, DC, pp. 7–12.

Terborgh, J. 1992. Why American songbirds are vanishing. Scientific American 266: 98–104.

Wilcove, D.S. 1985. Nest predation in forest tracts and the decline of migratory songbirds. Ecology 66: 1211–1214.

Forest destruction by exotic organisms

Anagnostakis, S.L. 1992. Measuring resistance of chestnut trees to chestnut blight. Canadian Journal of Forest Research 22: 568–571.

Anderson, R.F. 1960. *Forest and Shade Tree Entomology.* John Wiley & Sons, New York, NY.

Anderson, R.F. 1973. A summary of white pine blister rust research in the Lake States. USDA Forest Service general Technical Report NC-6, North Central Forest Experiment Station, St. Paul, MN.

Beattie, R.K. and J.D. Diller. 1954. Fifty years of chestnut blight in America. Journal of Forestry 52: 323–329.

Brewer, L.G. 1995. Ecology of survival and recovery from blight in American chestnut trees (*Castanea dentate* [Marsh.] Borrkh) in Michigan. Bulletin of the Torrey Botanical Club 122: 40–57.

Britton, K.O. 1993. Anthracnose infection of dogwood seedlings exposed to natural inoculums in western North Carolina. Plant Diseases 77: 34–37.

Brown, J.H., D.B. Halliwell, and W.P. Gould. 1979. Gypsy moth defoliation: impact in Rhode Island forests. Journal of Forestry 77: 30–32.

Campbell, R.W. and R.J. Sloan. 1977. Forest stand responses to defoliation by the gypsy moth. Forest Science Monographs 19: 1–34.

Chen, B., G.H. Choi, and D.L. Nuss. 1994. Attenuation of fungal virulence by synthetic infectious hypovirus transcripts. Science 264: 1762–1763.

Choi, G.H. and D.L. Nuss. 1992. Hypovirulence of chestnut blight fungus conferred by infectious viral cDNA. Science 257: 800–803.

DeGraaf, R.M. 1987. Breeding birds and gypsy moth defoliation: short-term responses of species and guilds. Wildlife Society Bulletin 15: 217–221.

Dickey, R.D. and R.B. Clapper. 1965. A progress report on attempts to bring back the chestnut tree in the eastern United States, 1954–1964. Journal of Forestry 63: 186–188.

Dunlap, T.R. 1980. The gypsy moth: a study in science and public policy. Journal of Forest History 24: 116–126.

Gottschalk, K.W. 1989. Gypsy moth effects on mast production. In: *Proceedings of a Workshop, Southern Appalachian Mast Management*. USDA Forest Service and University of Tennessee, pp. 42–50.

Grace, J.R. 1986. The influence of gypsy moths on the composition and nutrient content of litter fall in a Pennsylvania oak forest. Forest Science 32: 855–870.

Hepting, G.H. 1974. Death of the American chestnut tree. Journal of Forest History 18: 60–67.

Kasbohm, J.W., M.R. Vaughn, and J.G. Kraus. 1996. Black bear denning during a gypsy moth infestation. Wildlife Society Bulletin 24: 62–70.

Kegg, J.D. 1973. Oak mortality caused by repeated gypsy moth defoliation in New Jersey. Journal of Forestry 69: 852–854.

Laycock, G. 1966. *The Alien Animals*. Natural History Press, Garden City, NY.

Reardon, R.C. 1991. Appalachian gypsy-moth integrated pest-management project. Forest Ecology and Management 39: 107–112.

Schultz, J.C. and I.T. Baldwin. 1982. Oak leaf quality declines in response to defoliation by gypsy moth larvae. Science 217: 149–151.

Sheath, R.G., J.M. Burkholder, M.O. Morriscn, et al. 1986. Effect of canopy removal by gypsy moth larvae on the macroalgae of a Rhode Island USA headwater stream. Journal of Phycology 22: 567–570.

Smith, H.R. 1985. Wildlife and the gypsy moth. Wildlife Society Bulletin 13: 166–174.

Stalter, R. and J. Serrao. 1983. The impact of defoliation by gypsy moths on the oak forest at Greenbrook Sanctuary, New Jersey. Bulletin of the Torrey Botanical Club 110: 526–529.

Thurber, D.K., W.R. McClain, and R.C. Whitmore. 1994. Indirect effects of gypsy moth defoliation on nest predation. Journal of Wildlife Management 58: 493–500.

Valentine, H.T. and D.R. Houston. 1984. Identifying mixed-oak stand susceptibility to gypsy moth defoliation: an update. Forest Science 30: 270–271.

Wilcove, D. 1988. Changes in the avifauna of the Great Smoky Mountains: 1947–1983. Wilson Bulletin 100: 256–271.

Williams, D.W., R.W. Fuester, W.W. Metterhouse, et al. 1991. Oak defoliation and population density relationships for the gypsy moth (Lepidoptera: Lymantriidae). Journal of Economic Entomology 84: 1508–1514.

Reintroduction of red wolves

Fredrickson, R. J. and P. W. Hedrick. 2006. Dynamics of hybridization and introgression in red wolves and coyotes. Conservation Biology 20: 1272–1283.

Goldman, E.A. 1947. The wolves of North America. Journal of Mammalogy 18: 37–45.

Nowak, R.M. 1992. The red wolf is not a hybrid. Conservation Biology 6: 593–595.

Nowak, R.M. 2002. The original status of wolves in North America. Southeastern Naturalist 1: 95–130.

Philips, M.K. and V.G. Henry. 1992. Comments on red wolf taxonomy. Conservation Biology 6: 596–599.

Philips, M.K., B. Kelly, and G. Henry. 2007. Restoration of the red wolf. In: *Wolves: Behavior, Ecology and Conservation* (D. Mech and L. Boitani, eds). University of Chicago Press, Chicago, IL, pp. 272–288.

Rhymer, J. M. and D. Simberloff. 1996. Extinction by hybridization and introgression. Annual Review of Ecology and Systematics 27: 83–109.

US Fish and Wildlife Service. 2013. Red wolf recovery program, 1st quarterly report FY13, October–December 2012. Southeast Region, Atlanta, GA. 8 pp. (Describes program to sterilize coyotes.)

Infobox 4.1. E. Lucy Braun (1889–1971), diva of forest ecology

Bolgiano, C. 1998. E. Lucy Braun, grandmother of eastern old-growth studies. Wild Earth 8(3): 84–86.

Stein, L.K. 1988. The sisters Braun: uncommon dedication. Cincinnati Museum of Natural History Quarterly 21(2): 9–13.

Stuckey, R.N. 1973. E. Lucy Braun (1889–1971), Outstanding botanist and conservationist. Michigan Botanist 12(2): 83–106.

Stuckey, R.N. 1997. Emma Lucy Braun (1889–1971). In: *Women in the Biological Sciences: A Bibliographic Sourcebook* (L.S. Grinstein, C.A. Biermann, and R.K. Rose, eds). Greenwood Press, Westport, CT, pp. 44–50.

Infobox 4.2. Red maple, a species on the move

Abrams, M.D. 1992. Fire and the development of oak forests. BioScience 42: 346–353.

Abrams, M.D. 1998. What explains the widespread expansion of red maple in eastern forests? BioScience 48: 355–364.

Christensen, N.L. 1977. Changes in structure, pattern, and diversity associated with forest maturation in Piedmont, North Carolina. American Midland Naturalist 97: 176–188.

Clark, J.S., C. Fastie, G. Hurtt, et al. 1998. Reid's paradox of rapid plant migration. BioScience 48: 13–24.

Gugger, P.F., J.S. McLachlan, P.S. Manos, and J.S. Clark. 2008. Inferring long-distance dispersal and topographic barrier during post-glacial colonization from the genetic structure of red maple (*Acer rubrum* L.) in New England. Journal of Biogeography 35: 1665–1673.

Larsen, J.A. 1959. A study of an invasion by red maple of an oak woods in southern Wisconsin. American Midland Naturalist 49: 908–914.

Lorimer, C.G. 1984. Development of the red maple understory in northeastern oak forests. Forest Science 30: 3–22.

Infobox 4.3. An invasion of earthworms

Blakemore, R.J. 2008. *American Earthworms (Oligochaeta) from North of the Rio Grande: A Species Checklist*. Yokohama National University, Yokohama, Japan. (Lists 183 species in 12 families for United States and Canada.)

Darwin, C.R. 1881. *The Formation of Vegetable Mould through the Action of Worms, with Observations of their Habits*. John Murray, London.

Eisenhauer, N., S. Partsch, D. Parkinson, and S. Scheu. 2007. Invasion of a deciduous forest by earthworms: changes in soil chemistry, microflora, microarthropods, and vegetation. Soil Biology and Biochemistry 39: 1099–1110.

Frelich, L.E., C.M. Hale, S. Scheu, et al. 2006. Earthworm invasion into previously earthworm-free temperate and boreal forests. Biological Invasions 8: 1235–1245. (See for changes in soils and vegetation.)

Gundale, M.J. 2002. Influence of exotic earthworms on the soil organic horizon and the rare fern *Botrychium mormo*. Conservation Biology 16: 1555–1561.

Hale, C.M., L.E. Frelich, and P.B. Reich. 2005. Exotic European earthworm invasion dynamics in northern hardwood forests of Minnesota, USA. Ecological Applications 15: 848–860. (Cites rates of spread and diminished thickness of leaf litter.)

Hale, C.M., L.E. Frelich, and P.B. Reich. 2006. Changes in hardwood forest understory plant communities in response to European earthworm invasions. Ecology 87: 1637–1649.

Hendrix, P.F., G.H. Baker, M.A. Callaham, Jr., et al. 2006. Invasion of exotic earthworms Into ecosystems inhabited by native earthworms. Biological Invasions 8: 1287–1300.

Holdsworth, A.R., L.E. Frelich, and P.B. Reich. 2007. Effects of earthworm invasion on plant species richness in northern hardwood forests. Conservation Biology 21: 997–1008.

Keller, R.P., A.N. Cox, C. Van Loon, et al. 2007. From bait shops to the forest floor: earthworm use and disposal by anglers. American Midland Naturalist 158: 321–328.

Loss, S.R. and R.B. Blair. 2011. Reduced density and nest survival of ground-resting songbirds relative to earthworm invasions in northern hardwood forests. Conservation Biology 25: 983–992.

Maerz, J.C., V.A. Nuzzo and B. Blossey. 2009. Declines in woodland salamander abundance associated with non-native earthworm and plant abundance. Conservation Biology 23: 975–981.

CHAPTER 5

Grasslands: Plains and Prairies

Grasslands challenge our senses, calling on us to open our eyes to impossibly broad horizons and then, in the very next breath, to focus on some impossibly tiny creature hidden in the grass.

Candace Savage

The western edge of the Eastern Deciduous Forest gives way gradually to a grassland known as the tallgrass prairie or, as it is sometimes known, the true prairie (Fig. 5.1). Conditions here are drier, and the frequency of fires plays a major role in the maintenance of vegetation in this region (or at least did so in the past). Farther westward are plains of shorter grasses and a land often gripped by drought. In addition to the vast area of prairie and plains once covering the continental interior, grasslands also developed at other locations in North America (e.g., Palouse Prairie, Chapter 6). The term prairie itself originates from the French word for "meadow;" apparently, it was the best word available to European explorers for the extensive grasslands stretching to the western horizon. Similar communities are absent from western Europe; as a result, their presence in North America understandably astonished those Euro-Americans who first ventured beyond the eastern forests.

The early years of the 19th century were punctuated with intrepid explorations of the American West, best known by the remarkable journey (1804–1806) of Lewis and Clark. Somewhat later, in 1819–1820, Major Stephen H. Long (1784–1864) searched for the headwaters of the Arkansas, Platte, and Red rivers and, in doing so, crossed the Great Plains to the Rocky Mountains. Long, however, was not impressed with the plains stretching before him: "… I do not hesitate in giving the opinion, that (this extensive region) is almost wholly unfit for cultivation, and of course, uninhabitable by a people depending on agriculture for their substance." Moreover, the plains would "… serve as a barrier to prevent too great an expansion of our population westward." As a final straw, on his official map Long labeled much of this region *The Great Desert*, later embellished as *The Great American Desert* in books and maps of the day.

Major Long's misguided ecological determination prevailed for years afterward, becoming something of a psychological barrier to westward expansion after the Civil War.

Nonetheless, 110 years later, prolonged drought in the Great Plains produced conditions not altogether unlike a desert, prompting one historian to declare, "The spectre of the Dust Bowl haunts the West even today for it proved, for all time, that there lurks behind the myth of the Great American Desert of the 1820's a frightening amount of reality."

Major associations

Tallgrass prairie

This is the easternmost association in the complex of North American grasslands (Fig. 5.2). Indiangrass, big bluestem, and switchgrass are the renowned species "reaching belly high to a horse." Big bluestem, in particular, is often designated as the indicator species of the tallgrass prairie. Because of the distinctive shape of its inflorescence, big bluestem is sometimes known as "turkeyfoot." Another tall species, slough grass, occurs on wetter sites and was a favorite of pioneers for thatching roofs and for twisting into cords.

The tallgrass prairie is the richest in diversity among the grassland associations in North America. A wealth of shorter grasses and other plants complement the taller species mentioned above. Forbs – herbaceous, flowering plants other than grasses – add a delightful palate of colors to the panorama of grasses and are particularly conspicuous components of this and other grasslands. Many species of **composites**, including daisies, asters, goldenrod, and thistle, are well represented as are buffalo pea, wild indigo, purple prairie clover, and other species in the **legume** family. Another legume, lead plant, is one of the few shrubs in the tallgrass prairie. Native Americans used the leaves of lead plant for making tea, and its occurrence indicates a prairie in healthy condition.

A projection known as the prairie peninsula originally extended eastward into the Eastern Deciduous Forest. Expressed in terms of geography, the prairie peninsula extended eastward from the Mississippi River across

Ecology of North America, Second Edition. Brian R. Chapman and Eric G. Bolen.
© 2015 John Wiley & Sons, Ltd. Published 2015 by John Wiley & Sons, Ltd.

Figure 5.1 Wild flowers add a palate of color to the green carpet of prairie grasses. Here, sunflowers highlight the tallgrass prairie at Neal Smith National Wildlife Refuge in Iowa, perhaps suggesting the appearance of a native grassland system at the time of Euro-American settlement. Reproduced by permission of Sara Hollerich and the US Fish and Wildlife Service.

Illinois and into Indiana, with isolated segments reaching Ohio. Like the tallgrass prairies west of the Mississippi River, the prairie peninsula is rich in forbs. One mid-19th century traveler counted 80 species of plants in flower on the Illinois prairie in July. In addition to big bluestem and other characteristic vegetation, the Plains garter snake represents the faunal element associated with ponds, woody draws, and other moist areas within the prairie peninsula (Fig. 5.3a). Some fishes also show affinities with the prairie peninsula. Suckermouth minnows, once occurring almost entirely west of the Mississippi River, moved eastward after the Civil War as farming increased the turbidity of streams in the prairie peninsula. Conversely, siltation eliminated the harelip sucker from its range which included the prairie peninsula; unfortunately, since the species has not been recorded anywhere since 1900 it is now considered extinct.

Virtually all of the tallgrass prairie is gone, the victim of its own rich soils and the rapid march of the plow. In this region, currently devoted almost entirely to agricultural production, corn is now the dominant tallgrass. Precious few sites of tallgrass prairie remain; "postage stamp prairies," small relicts in old cemeteries and abandoned railroad rights-of-way, offer glimpses of the once-magnificent blanket of grassland. In 1966, however, a long quest to preserve a block of pristine grassland reached fruition with the establishment of Tallgrass Prairie National Preserve in the Flint Hills of southeastern Kansas.

Two somewhat isolated areas of tallgrass prairie occur elsewhere. One of these, the Nebraska Sandhills, developed where soil moisture is sufficient to nurture big bluestem and other deeply rooted bunchgrasses. Ranching, rather than farming, is the basis for the regional economy. Therefore, unlike its highly altered counterpart to the east, the ecological integrity of tallgrass prairie in north-central Nebraska remains largely unchanged. The second area parallels the Texas Gulf Coast, where seacoast bluestem and Pan American balsamscale are among the dominant tallgrasses on the region's sandy soils. Gulf cordgrass covers large areas with poor drainage, including those where saline conditions prevail and, at some sites, this species dominates in monotypic communities. Communities of these and other grasses in part form the vegetational unit known as the Gulf Coast Prairies and Marshes. Attwater's prairie chickens, now precariously near extinction, are closely associated with this region.

Midgrass prairie

Also known as mixed grass prairie or simply mixed prairie, this association covers a large region of interior North America (Fig. 5.1). These names arise from the intermediate size of the dominant grasses, which include little bluestem, some kinds of grama grass, and various species of needlegrass. Where conditions are favorable, big bluestem and the other tall grasses persist as subordinates in the midgrass prairie. Land-use practices greatly influence this and other kinds of prairie. At a carefully managed site in south-central Nebraska, for example, 17 years of light grazing and frequent burnings increased the composition of tall grasses more than fivefold in what was previously considered a midgrass prairie. Westward, however, moisture steadily diminishes and the eventual disappearance of tall grasses marks the western edge of the Midgrass Prairie Association. This limit coincides with the point where soil moisture no longer exceeds a depth of 61 cm (24 inches) during the growing season.

Shortgrass prairie

Grasslands better adapted to arid conditions lie west of the midgrass prairie, ending where the Rocky Mountains rise from the plains (Fig. 5.3b). Buffalo grass and blue grama are the principal species in much of the shortgrass prairie, a region also widely identified as "plains" in comparison to regions where taller grasses prevail. Grasses in this region have two primary characteristics: their short stature, often little more than lawns in height, and sod-forming, an important adaptation for aridity especially when accompanied by significant grazing pressure. Some ecologists maintain that this region, if freed from grazing, would emerge in a climax of midgrasses. This conclusion,

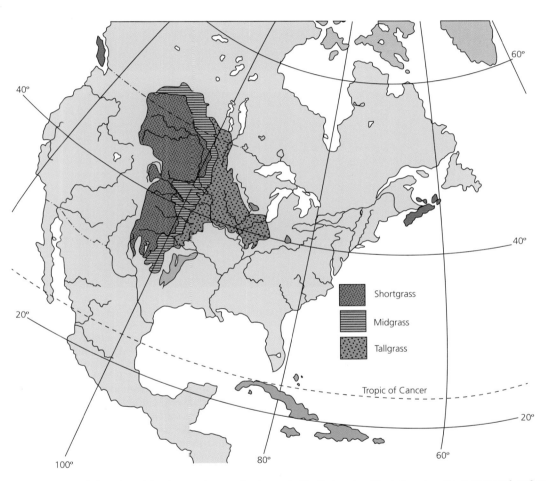

Figure 5.2 Approximate delineation of the three major grassland areas in North America. The tallgrass region is commonly referred to as true prairie, whereas the more arid midgrass and shortgrass areas together are often known as the Great Plains. Some ecologists refer to the midgrass region as mixed prairie. Illustrated by Tamara R. Sayre.

(a)

(b)

Figure 5.3 (a) In the Flint Hills south of Manhattan, Kansas, several grasses, especially big bluestem and numerous forbs, command the broad sweep of the tallgrass prairie landscape. Woody vegetation often develops in draws where mesic conditions prevail. (b) Scattered yuccas, in flower, highlight the shortgrass prairie in western Kansas. This site (Cimarron National Grassland) is public land managed by the USDA Forest Service. Reproduced by permission of Mike Blair, Kansas Department of Wildlife and Parks.

in part, results from observations of short grasses pro-tected with exclosures (fenced plots designed to keep out livestock). Exclosures remove the influence of cattle and other large herbivores on the vegetation inside the plot; that is, when protected with exclosures, midgrasses replace short grasses. Shortgrass prairie is therefore some-times considered as a **disclimax**, a disturbance climax developing *after* livestock began to graze on fenced range-lands. Fences confined cattle and other livestock, in comparison to free-ranging bison, and overgrazing there-after removed midgrasses from the community. A con-trary view recognizes shortgrass prairie as a true climax that evolved *with* the influence of grazing animals, namely the herds of bison that once roamed the western plains.

In any event, the short grasses form a sod remarkably adapted to exploit limited moisture. Ecologists have quipped that these grasses "are short only aboveground," a reference to the extensive root systems they develop in

response to semiaric conditions. The tough and wiry roots of buffalo grass, although thread-sized, form a thick mat spreading outward from each plant. Roots of the same threadlike structure also extend downward in excess of 1 m (4–6 feet). In all, buffalo grass produces an immensely abundant root system that fully occupies the soil around and beneath each plant. Early settlers on the western plains preferred the tough sod of buffalo grass as the "bricks" for their houses. The root system of blue grama, the codominant of the shortgrass prairie, is quite similar to buffalo grass. In places, the sod of buffalo grass and blue grama may form a patchwork with bare soil, but the roots of these shortgrasses thread through the soil just beneath the surface in these bare places. Buffalo grass and blue grama therefore gain moisture from even modest amounts of rainfall.

The region known as the Great Plains, although gener-ally regarded as including both the midgrass and shortgrass

Infobox 5.1 Aldo Leopold (1887–1948): conservationist, forester, and philosopher

A native of Iowa, Leopold graduated from Yale, earning an undergraduate degree in 1908 capped by a master's degree in forestry the following year. A career with the fledgling US Forest Service followed, beginning with field duties in the then territories of New Mexico and Arizona, and ending in 1927 as Assistant Director of the Forest Products Laboratory in Madison, Wisconsin. After two years of private consulting, he was appointed professor of wildlife management – the first curriculum of its kind in the nation – at the University of Wisconsin, where he taught until his death. His textbook *Game Management*, published in 1933, was also the first of its kind and marked the beginning of a new era of dealing with wildlife and their habitat as renewable resources.

Aldo Leopold shaped conservation in North America like few others. While in the US Forest Service, he planned what later became the Gila Wilderness Area, another first of its kind in the nation. In 1935, Leopold helped organize The Wilderness Society and, in 1937, co-founded The Wildlife Society. He served as president for the latter in 1939 and likewise for the Ecological Society of America in 1947.

In a collection of essays entitled *A Sand County Almanac,* published post-humously in 1949, Leopold emerged as a philosopher who established the land ethic, that is, recognition that land is a community of interlocking parts whose health is vital and not a mere commodity for economic exploitation. Humans, he stressed, must integrate their activities for the combined welfare of all components of communities. In some cases, this requires proactive scientific management with "axe, plow, cow, fire, and gun" – the same tools that, when used indiscriminately, had destroyed wildlife – whereas others call for full and unyielding protection. In one of his essays, *Thinking Like a Mountain*, Leopold rethinks his earlier attitude to the role of predators in natural communities. Once he believed a good wolf was a dead wolf, but he later came to realize how wolves – and other predators – helped maintain the ecological health of natural systems, and he argued for including their role as much as for any other trophic level when managing wildlife. In his memorable words, "…just as a deer herd lives in mortal fear of its wolves, so does a mountain live in mortal fear of its deer," clear reference to the overbrowsed vegetation resulting from too many deer on a "wolfless mountain." Today *A Sand County Almanac* is considered as essential reading for amateur and professional conservationist alike.

Many of his thoughts and much of his writing took shape at "the shack," a chicken coop on an abandoned, depression-era farmstead he used as a get-away-from-it-all and showcase for restoring worn out land with native vegetation. After renovation, the shack became a centerpiece for family activities, among them banding birds, hunting, and picking berries. One of the chickadees he banded became the subject of an essay; he followed the fate of #65290 for 5 years, learning about the rigors of chickadee life.

In his journals, Leopold compared natural communities to the workings of a fine watch, citing "To keep every cog and wheel is the first precaution of intelligent tinkering," which not only champions biodiversity but also heralds the protection of endangered species. In this view, each species has its own intrinsic value quite apart from any tangible or other value it may have for humans. Species therefore lack positive, neutral, or negative values and instead attain their worth as parts of a healthy ecosystem.

The Wildlife Society honors his memory with its highest tribute, the Aldo Leopold Award, a medal presented annually to a person who has demonstrated distinguished service to conservation. The US Forest Service established the Aldo Leopold Wilderness Institute at the University of Montana, and both a network of hiking trails in Wisconsin and a national wilderness area in New Mexico bear his name.

prairies, actually lacks firm geographical delineation because of intervening areas such as the Nebraska Sandhills. *Great Plains* consequently may better reflect a socioeconomic region instead of representing a well-defined ecological unit. In some classifications, grasslands known as *steppes* are identified with a modifier, the best known of these being the shortgrass steppes of present-day Ukraine. Today, these regions are heavily cultivated and form the "breadbaskets" of their respective continents.

Transition zones

The transition from eastern forest to grassland is not an abrupt ecotone. Instead, trees in the transition zone are clumped, dotting the intervening grasslands with diminishing frequency until the prairie clearly dominates the physiognomy of the landscape. Stands of bur oak are

among the vestiges of forest along the northern edges of the zone. Aldo Leopold (Infobox 5.1) described bur oaks as "the shock troops sent by the invading forest to storm the prairie; fire is what they had to fight," clear reference to the tension that fire renders to the dynamics of grassland vegetation. To the south, blackjack oak and post oak are typical of the transition zone; both species reach their western limits in central Oklahoma (Fig. 5.4). Black hickory is particularly tolerant of dry conditions and may be associated with the two oaks in this region.

Aspen parklands

To the north, a zone of vegetation known as aspen parkland separates the Boreal Forest from the expanse of North America's interior grasslands (Fig. 5.5). Aspen parkland lies almost entirely in Canada, with small areas occurring in a few places just south of the border (e.g., Glacier County, Montana). The physiognomy is

Figure 5.4 Scattered stands of blackjack and post oaks are typical of the ecotone between the Eastern Deciduous Forest and tallgrass prairie in southern locations, as shown here in Oklahoma. Bur oaks characterize the northern segment of the same ecotone. Reproduced by permission of Brian R. Chapman.

distinctive: groves of medium-sized trees dot a grassland panorama. Bearded wheatgrass and rough fescue are among the several species of grasses in this region of the North American prairie. The principal species of tree – quaking aspen – gains a foothold in moist areas, often after fires have swept across the landscape (Fig. 5.6). Much has been written about the poetry of aspen foliage, whose leaves dance and flash with the faintest breeze and sing in restless motion. Initially aspen establishes itself with seeds and then extends into the prairie with long root systems from which clones sprout, often at considerable distances from the parent plants. Frequent burning, together with grazing by bison, once kept aspen in check and favored the persistence of the grassland communities, but fire suppression and new grazing patterns later increased the encroachment of aspen within the parkland region. Frequent droughts reverse the expansion of aspen into the surrounding prairie.

A related species, balsam poplar, mixes with quaking aspen on poorly drained sites; on north-facing slopes and at higher elevations, paper birch may be included in the aspen groves. The groves otherwise are essentially pure stands of quaking aspen.

Quaking aspen grows rapidly, but the trees seldom exceed 60 years of age, with heights of 18 m (60 feet) and diameters of about 40 cm (16 inches). When mature, the lower branches die, leaving only the top third of the tree in foliage. A white bloom covers the south side of the trunk providing protection from sun scald in late winter and early spring, when temperatures on the sunny side rise above freezing and the north and shady side remains frozen. Beaver relish aspen for both food and structural material, and the lure of beaver pelts had much to do with the early exploration of the parkland

region. Aspen buds are a staple in the winter diet of ruffed grouse. Snowshoe hares likewise depend heavily on the bark and twigs of young aspen for winter food. At peak numbers, the hares frequently girdle 90% or more of the aspen seedlings.

Although aspen produces heavy crops of seeds, competition from an intact blanket of grasses rarely permits successful germination; aspen seeds require bare soil, and their viability is short-lived. Richardson ground squirrels, whose range includes all of the parkland region, and other burrowing animals (e.g., badgers) provide local sites of bare ground. In many cases, these sites of bare soil initially favor snowberry, a pioneer species, dubbed locally as "badger willow" or "buck brush." Sharp-tailed grouse dust in such places and, because these birds pass viable seeds of snowberry fruits, thickets of snowberry develop locally on bare areas in the prairie. In turn, foxes and coyotes den in the snowberry thickets and create even more bare soil. Aspen seeds consequently find these sites suitable for germinating, and the developing grove eventually overcomes the snowberries. Thus established, each aspen grove expands using an extensive system of rhizomes without further dependence on seedlings. Smaller groves eventually coalesce into larger areas of aspen. As predicted by the concept of **island biogeography**, the richness (i.e., number of species) of the biota increases with the size of the groves (= islands), as has been shown for their bird communities. Hence, larger groves may have greater importance as refuges, although smaller groves serve as "stepping stones" for birds migrating across the parkland ecotone.

Fire damages snowberry shrubs, so repeated burning of the prairie indirectly retards the development of aspen groves. Aspen itself also readily burns, but typically

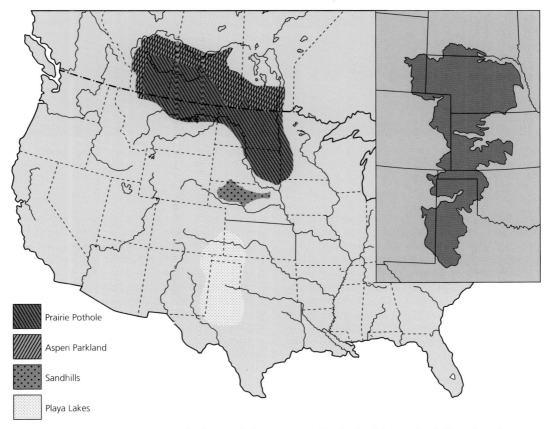

Prairie Pothole

Aspen Parkland

Sandhills

Playa Lakes

Figure 5.5 Regions of interest within the Grassland Biome include aspen parklands, the Nebraska Sandhills, and two large areas dotted with wetlands. Each spring, wetlands known as prairie potholes provide crucial nesting habitat for millions of ducks. The Ogallala Aquifer (inset) forms one of the world's largest deposits of subterranean freshwater. It supplies drinking water for 2.3 million people as well as irrigation water for an eight-state region of the western Great Plains. However, because of its slow rate of recharge and heavy use, the aquifer is essentially "mined" and steadily depleted as a supply of water. Illustration by Tamara R. Sayre and Brian R. Chapman based on the National Atlas of the United States (2009).

Figure 5.6 Aspen parklands are named for aspen thickets dotting the northern edges of grassland areas in North America, as shown here in central Saskatchewan. Ecologists generally regard aspen parklands as an ecotone between Boreal Forest and grassland vegetation. Reproduced by permission of J. T. (Jim) Romo.

regenerates from suckers. Conversely, overgrazing reduces grass cover and increases the incidence of bare soil, which in turn enhances snowberry thickets and, ultimately, more groves of aspen. In times gone by, soil laid bare by wallowing bison initiated the same sequence.

Ecologists are not in full agreement as to whether aspen parkland represents a true ecotone. At its northern edge, the parkland is nearly solid aspen with only a few patches of prairie intervening. Southward, the incidence of prairie increases and the cover of woody vegetation decreases until aspen occupies only small, isolated groves where the moisture regime is favorable (i.e., depressions and north-facing slopes). The issue of transition may be overly academic and, in any case, should not overshadow the fundamentals of northern prairie ecology: grassland and woodland vegetation interacting dynamically under influences of fire, drought, and grazing. Regrettably, and like prairie communities elsewhere, few areas of undisturbed aspen parkland remain available for either study or aesthetic appreciation.

Cross Timbers

Fingers of deciduous forest arranged on a north–south axis cross the prairie in northeastern Texas. These alternating belts of grassland and forest are known as the Cross Timbers. The distinctive pattern of this vegetation became a milepost in the westward march across North America. As pioneers left the Eastern Deciduous Forest and entered the prairie, they encountered additional forested areas – the Cross Timbers – before reaching the grasslands that extended to the Rocky Mountains.

The Cross Timbers forest features post oak and black-jack oak, which are present in a ratio of roughly 2:1 (Fig. 5.7). The stands of oak were often park-like, in which the trees are widely spaced in a savanna. Little bluestem represented the predominant grass, but tall-grasses such as big bluestem and Indiangrass and several species of midgrasses (e.g., sideoats grama) also were components of the Cross Timbers prairies. Sandy soils, likely ancient beaches, support the fingers of oaks, whereas the prairie developed on the intervening loams of greater fertility. Grazing, farming, and other disturbances have eliminated most of the original Cross Timbers prairie.

Western transition

At its western edges, the prairie slowly yields to more desert-like conditions. Aridity increases as the plains fall under the full impact of rain shadow from the Rocky Mountains. In the north, a community of sagebrush and wheatgrass marks the transition in the vegetation; similarly, sage grouse begin to replace prairie chickens in the avifauna. Fires are less frequent here than in the grasslands to the east, where the vegetation provides more fuel in comparison with widely spaced vegetation and greater areas of exposed ground in the transition zone. Fires occasionally occur, especially where grazing is limited and fuel accumulates. Such a fire occurred in the sagebrush–wheatgrass community at Little Bighorn National Battleground in 1983, thereby exposing a

Figure 5.7 Blackjack oak and post oak are the principal species in the Cross Timbers, which extend as fingers of park-like forest into the grasslands of northeastern Texas. Reproduced by permission of Gary Valentine, Natural Resources Conservation Service, Temple, Texas.

trove of artifacts from Custer's fateful "Last Stand" on 25 June 1876.

Along the Front Range in Colorado, the shortgrass prairie gives way in the foothills to mountain mahogany and sparse pine forests, although other grasses occur at these sites (e.g., mountain muhly). Farther southward, juniper woodlands border the prairie, and farther yet, a brushland of mesquite prevails along with the beginnings of desert grasslands (see *Tamaulipan Mezquital*, Chapter 6). With the rise of the Rockies the Great Plains reach their western limit, replaced by desert communities and the distinctive zones of montane vegetation.

Features and adaptations

Seasonal grasses

Grasses (Infobox 5.2) fall into two groups based on the season when they produce new growth. These distinctions are related to different types of photosynthetic activities and reflect either a northern or southern origin of each species. Some grasses therefore thrive in warm, dry conditions, others in cool, moist settings. Cool-season grasses mature and produce seed by early summer and then remain semi-dormant during hot weather. They resume growth in the cool months of autumn and remain green despite frosts, thereby considerably extending the availability of forage for grazing animals. Conversely, warm-season grasses produce much of their foliage during midsummer, when they also flower and set seed. Seed production extends well into autumn in some species, but warm-season grasses seldom continue growing after summer weather ends.

The mix of warm- and cool-season grasses is an efficient means of using water and other resources in a prairie ecosystem. Green forage is also available to grazing animals throughout all of the coldest months. Big bluestem, blue grama and buffalo grass are warm-season grasses. Examples of cool-season grasses include needlegrasses, wheatgrasses, and others that "green up" the prairie in the spring.

With the exception to their tolerance to fire, described later, perhaps the most significant ecological feature of grasses lies below ground: their root systems. Grasses have fibrous roots that represent a large part of each plant's total biomass. Based on detailed fieldwork in eastern Nebraska, for example, the roots and other underground parts of big bluestem represent a biomass of almost 9100 kg ha^{-1} (4.1 tons per acre) in just the top 10 cm (4 inches) of soil. This network of roots extends well down into the soil profile and explains the drought-resistance of many grasses.

Two kinds of specialized stems grow from the bases of some grasses; each functions in asexual (= vegetative) reproduction. Stolons creep above ground, taking root where the microhabitat is suitable. Rhizomes spread below ground from the "mother" plant, sending shoots upward at intervals. Stolons and rhizomes provide grasses with an effective means of reproduction during droughts, when germination and seedling survival are often poor.

Sod forms from the thick mass of stolons and/or rhizomes, their developing shoots, and the upper part of the root system and may eventually cover large areas (e.g., buffalo grass). Other grasses grow in well-defined clumps, hence are called bunchgrasses (e.g., fescue). New shoots, known as tillers, typically grow from the bases of bunchgrasses, thereby adding more plant material to the "bunch" each growing season. Some species may develop either as a bunchgrass or a sod-former, depending on soil moisture or other local site conditions.

Infobox 5.2 Poaceae, the grass family

Agrostology, the scientific study of grasses, is the only branch of botany based on a single family of plants. The grass family, once known as Gramineae, was renamed Poaceae when taxonomists formally adopted -aceae as the standard ending for the names of plant families. Grasses are distinguished by hollow stems, known as culms, that are plugged at intervals called nodes. Leaves consist of two parts: a basal sheath that wraps around the culm resembling a split tube, and the blade which is typically long, narrow, and with parallel veins. Poaceae alone has leaves with these features. Additionally, a membranous appendage (often a fringe of hairs) known as a ligule marks the junction of the sheath and blade. Grass leaves appear alternately on the side of culms. Importantly, the foliage emerges from a growing point at the base of the blade. This feature evolved as a result of frequent grazing, but also allows grasses to survive and regrow after fires (or mowing). Some plants, mainly sedges and rushes, superficially resemble true grasses but in fact are unrelated; the term graminoid ("grasslike") characterizes these plants (e.g., a swale of graminoid vegetation).

The flowers of grasses are complex structures of bracts, scales, and florets typically arranged in spikelets. The spikelets in turn are arranged in one of several types of inflorescences, of which spikes, panicles, and racemes are examples. The fruit is a caryopsis in which the seed coat is inseparably fused with the fruit; grass fruits are known collectively as grains. Grasses rely on wind for pollination, and therefore lack colorful flowers, scents, or honey to attract pollinators. Their seeds are dispersed by wind or animals, in the latter case either by attaching to hair or fur or by surviving in droppings. The fibrous root systems of grasses prevent soil erosion, including those that anchor coastal dunes (e.g., sea oats).

Recent analyses of dinosaur coprolites ("fossilized droppings") revealed tissues of plants similar to modern bamboo and rice. Grasses apparently originated at least 65 million years ago, probably in the Southern Hemisphere, diversifying later when drier climates created vast open areas in forest vegetation. Poaceae includes about 700 genera and 11,000 species, of which 139 genera and 906 species occur in North America north of Mexico. Grasses grow in virtually all terrestrial environments on every continent (a single species of hairgrass is one of just two vascular plants in Antarctica) and humans have purposely or accidently introduced grasses into new regions.

Many kinds of wildlife coevolved with grassland vegetation which supported vast herds of bison, antelope, and other grazing species. Giant pandas are entirely dependent on a single type of grass – bamboo – for their food. Grazing animals typically have specialized digestive systems to accommodate the tough fibrous tissues of grasses. Carnivores, of course, rely on herbivores to exist, and the vital role of grasses in terrestrial food chains was understood long ago when the prophet Isaiah (40: 6) declared that "All flesh is grass."

The family is not the largest in the plant kingdom, but it easily is the most important to human economies. Indeed, entire cultures were founded on rice, wheat, or corn. Several species of grasses serve as lawn turf or ornamental vegetation; others provide sugar and forage for domestic animals. Flushed with eloquence, Kansas Senator John J. Ingalls (1833–1900) paid homage to this force of nature when he opined that "Forests decay, harvests perish, flowers vanish, but grass is immortal."

Range managers recognize three groups of grasses and other plants based on the response of each species to grazing pressure (Fig. 5.8). **Decreasers** are those plants in the climax community that are preferred by livestock; these diminish under heavy grazing pressure or even disappear when excessive grazing continues unabated. Some key species of decreasers are popularly known as "ice-cream" plants because they are the first to disappear as a result of grazing. In contrast, **increasers** become more abundant with heavy grazing; such species are less palatable than decreasers, although other factors sometimes influence their abundance on rangelands. Finally, plants known as **invaders** appear under extreme grazing pressure. These differ from either decreasers or increasers because they are not a part of the climax vegetation, and many are exotic species (e.g., Russian thistle, popularly known as "tumbleweed"). Invaders are unpalatable and often represent pioneer or other species associated with the early stages of succession. Prickly pear is a typical invader on some overgrazed rangelands, and decreasers survive only where the spiny clumps of cactus protect these grasses from livestock.

The relative abundance of decreasers, increasers and invaders enables range managers to assess the current condition of rangeland. Ranges in excellent condition, for example, have an abundance of decreasers, few increasers, and virtually no invaders. Range condition indicates the potential of each site – its state of health – but is not a measure of the immediate availability of green forage. Appraisals of range condition are a practical application of concepts developed for plant succession and climax vegetation (Table 5.1).

Fortunately for humans, several kinds of grasses respond to artificial selection, a process analogous to natural selection and the foundation of agriculture. Rice, wheat, and corn are grasses on which

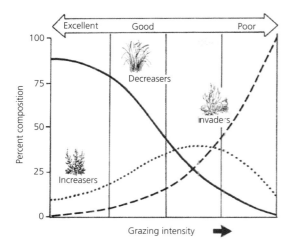

Figure 5.8 Heavy grazing changes the composition of native grasslands. Climax grasses, which livestock typically prefer as forage, become scarce as grazing intensity increases; hence these plants are known as "decreasers." Meanwhile, less palatable plants ("increasers") steadily become more abundant; with continued overgrazing however, these give way to low-quality invaders, often woody species and cacti. Range condition, shown at the top of the diagram, is indicated by the percentage of decreasers, increasers, and invaders in the vegetation at grazed sites. Illustrated by Tamara E. Sayre.

Table 5.1 Range condition can be assessed by comparing the composition of the climax community with the composition of the current vegetation. Composition is based on percent cover of each species or group of plants, which may be itemized in greater detail than shown in this example. Based on a concept developed by Dyksterhuis (1949).

Species or group	Climax vegetation[1]	Current vegetation	Current proportion of climax
Sideoats grama	100	10	10
Other grasses	10	20	10
Annuals	0	35	0
Forb increasers	5	15	5
Woody increasers	5	20	5
TOTAL	–	100	30[2]

[1] Maximum percentage expected in the absence of disturbance.
[2] Range condition is assigned to one of four classes: excellent (75–100% of the climax vegetation); good (50–75%); fair (25–50%); and poop (0–25%). In this example, range condition is at the low end of the fair class.

civilizations were built, but others include oats, barley, millet, and sugarcane. Cattle and most other meat-producing animals depend on diets dominated by grasses.

Many species of prairie forbs bear flowers that add a palate of colors to grasslands during the growing season. Insects or birds typically pollinate such plants. Forbs also have unique adaptations for surviving in grassland communities. Among these is the effective seed dispersal of Russian thistle – the "tumbleweed" heralded in ballad and narrative of the "Old West" – but which actually first appeared in South Dakota in 1873. Tumbleweeds are pioneers and scatter their abundant seeds on disturbed soils as they roll along, driven by the wind. Each plant may yield 250,000 seeds, many of which are loosely attached and fall free when the plant breaks free and begins its tumbling journey. Other seeds are embedded in stems and branches; these germinate when the plant lodges and becomes partially buried under drifting soil. Tumbleweeds – each a ball of branches 1–1.5 m (3–5 feet) in diameter – may travel considerable distances on the open plains, thereby distributing their seeds in virtually all sites suitable for germination. Even when tumbleweeds pile up against fences, a shifting wind often frees the plants for a journey in another direction.

Soils

Large areas of grassland developed on deposits of loess, fine mineral particles carried by wind and deposited where they form rich topsoils. Glaciers prompted loess deposits in North America: vast plains of exposed mud remained when the ice sheets melted and, as the mud dried, winds blew the silty particles to their present location in midcontinent (e.g., Iowa). Loess is not prevalent northward, however; here glaciers influenced soil formation by leaving behind deposits known as glacial till on which grasslands also developed (e.g., some parts of southern Canada).

Until 1960, when a newer system of soil classification was developed, soils in the grassland region of the North American interior were known as (from east to west): (a) prairie; (b) chernozem; (c) chestnut; and (d) brown soils in a progression representing increased aridity. Grassland soils in North America are now part of the soil order known as Mollisols. In part, these developed under tallgrass prairie in a relatively humid, temperate climate. Dark and extremely rich, these soils are typical of the loess deposits in North America, and their distribution today corresponds closely with the "Corn Belt."

Chernozem soils (Russian for 'black earth") developed under the somewhat drier, western area of the tallgrass prairie and adjacent parts of the midgrass prairie. The profile of these soils, which are also classified as Mollisols, includes a hard layer of calcium carbonate. This "hardpan" forms beneath the surface at about 1.5 m (6 feet) and indicates the depth where leaching ends (Fig. 5.9). The amount of precipitation in this region is sufficient to

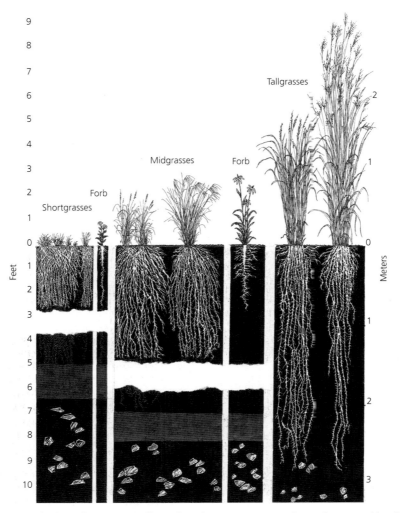

Figure 5.9 Big bluestem and other tallgrasses typically produce deep root systems, whereas in more arid regions the roots of midgrasses and shortgrasses develop in progressively shallower soils. As an adaptation to aridity, shortgrasses such as buffalo grass produce unusually thick mats of roots near the soil surface. Grasses have fibrous root systems, but forbs commonly have tap roots. The white zone in the soil profile indicates a layer of calcium carbonate (popularly known as a hardpan or *caliche*) which can hinder root penetration. Illustrated by Tamara R. Sayre.

transport calcium carbonate to this depth and no further. At that point, the calcareous solution hardens into a cement-like layer, which sometime lessens the penetration of roots deeper into the soil profile. Moreover, soils beneath the hardpan are deprived of moisture from the surface.

Chestnut soils are characteristic of semiarid regions where they are associated with the midgrass prairie, eventually grading into the brown soils farther to the west. The depth of the hardpan becomes correspondingly shallower with the reduction in annual precipitation. The "Wheat Belt" occupies most of the region characterized by chestnut soils.

Brown soils develop in the driest region of the plains, where shortgrass prairie is the dominant type of vegetation. The limited precipitation forms a hardpan at a depth of 60 cm (2 feet) or less. These soils are not as productive as the others, and farming may be risky unless crops are irrigated. Consequently, large areas of brown soils are grazed.

Role of fire

Fire is a dominant force in maintaining the integrity of grasslands, and its repeated occurrence greatly discourages most kinds of woody plants. Three phenomena are involved.

1 The growing points of most grasses lie near or below ground level, where they are protected from all but the severest fires. In comparison, the growing points of woody plants are well above ground – typically at the tips of twigs – and therefore exposed to fire.
2 For grasses, fires seldom remove more than 1 year of aboveground growth, whereas many years of growth are lost when trees burn to ground level. Grasses replace their lost tissues quickly – often within weeks – but severely burned woody plants may require years of regrowth to regain their former biomass.
3 Because grasses grow so rapidly after fires, they also quickly regain their maturity and produce seeds shortly thereafter, commonly within the same growing season. Trees and shrubs, by comparison, require several years to produce seeds. Fires therefore seldom inhibit reproduction in grasses, but may seriously limit seed production in woody plants.

Repeated burning clearly favors the development of grasslands by retarding growth of woody vegetation. The corollary, of course, is that trees and shrubs may invade grasslands when fire is suppressed. Wildfire suppression often invokes images of sleek fire trucks, specially equipped aircraft, and smoke jumpers, but fires are sometimes suppressed more passively. Before settlement, flames spread without check until reaching a creek, gully, or perhaps an area where a previous fire had temporarily exhausted the fuel. Large prairie dog towns may also have acted as fire breaks and rainfall from thunderstorms may limit the spread of lightning fires. Now, roadways serve as potential firebreaks and plowed land is even more effective at halting prairie fires. Author John Madson quotes a source in 1857 who noted that "... furrows were plowed all around the settlement to stop the burning of the prairies, whereupon the timber quickly grows up."

The frequency of fires in grasslands is difficult to determine in comparison to forests. Cross-sections of trees reveal fire scars among growth rings, which can be enumerated in terms of their occurrence, but grasses are obviously ill-suited for the same purpose. Studies of lightning strikes consequently offer the primary means of discovering just how often prairies might burn without the additional influence of anthropogenic fires. On the midgrass prairies of North Dakota, for example, lightning started fires with a frequency of up to 25 per year per 10,000 km^2 (3861 square miles). Most such fires occur during July and August, a period of high temperatures, dry fuel, and active thunderstorms.

Fires occurred naturally on the prairie and, long before Euro-Americans added sparks from chimneys, muzzle-loading guns, or cinder-belching engines, Native Americans ignited the prairie and did so intentionally. Such fires were an effective hunting tool, used to drive game over cliffs or toward hidden bowmen. Some grasses become rank and unpalatable as they mature, but their tender regrowth is protein-rich and savored by grazing animals, which move to recently burned ranges. The succulent regrowth that quickly followed a prairie fire also ensured fresh pasturage for their ponies and attracted game. Fires also consume the mulch of dead material, making the new green shoots more accessible to grazing animals. On 6 March 1805, Lewis and Clark wrote in their journal "... smokey all day from the burning of the plains which was set on fire by the (Indians) for an early crop of grass as an inducement for the buffalo to feed on ...." Similar comments by early travelers prompts debate about the influence of tribal-set fires on the ecological see-saw between prairie and forest in North America: did forests once extend farther into the grasslands, only to be pushed back by centuries of repeated and intentional burning by Native Americans?

Prairie streams

Prairies are relatively flat landscapes, and the gentle relief produces low-gradient streams. The resulting flow is sluggish and without the turbulence of falls and rapids. Because water flows to the nearest point of lower elevation, prairie streams typically meander across the landscape, greatly adding to their length; that is, the beds of prairie streams loop back and forth in response to relatively small changes in elevation. In time, new channels form as the necks steadily erode between the loops. The loops isolated from the stream remain as landforms known as oxbow lakes.

Overflow from periodic flooding keeps oxbow lakes filled and, although sediments and organic materials slowly fill their basins, they retain their identity almost indefinitely. Depending on their age (i.e., length of isolation from the main stream), oxbow lakes may be in various stages of succession. Those that are still young will feature aquatic communities, but older oxbows may develop mesic communities, including thickets and small stands of trees. These sites therefore add considerable diversity to the prairie biota, and at least one large oxbow – Horseshoe Lake State Fish and Wildlife Area – serves as a recreational area in Illinois.

Streams and rivers coursing the North American grasslands also carry heavy loads of sediment. Indeed, the mightiest of the prairie rivers, the Missouri, was characterized as "too thick to drink, too thin to plow." Such streams develop bars from accumulations of silt, sand, and gravel at points where the sediment load becomes excessive in relation to the flow of water. As more sandbars form and raise the stream bed – a process known as aggradation – the channel steadily widens and becomes progressively shallower in cross-section. Aggradation continually adds new sandbars, which form an ever-changing network in the streambed known as a braided channel.

Prairie wetlands and waterfowl

A large segment of the waterfowl population in North America depends on prairie wetlands for breeding, migrating, and wintering habitat. For breeding waterfowl, however, prairie potholes are without peer (Fig. 5.10). These small wetlands, glacial in origin, extend across a broad area from western Minnesota through the Dakotas and eastern Montana and in Prairie Canada (southern Manitoba, Saskatchewan, and Alberta).

Prairie potholes represent just 10% of the breeding habitat for waterfowl in North America, but produce 50% of the continental duck crop each year (and even more when conditions are exceptional). These wetlands – popularly called "duck factories" – are indeed the greatest duck-production habitat in the world. Northern pintails, redheads, mallards, and canvasbacks are among the several species of ducks that breed in the prairie pothole region.

Snowmelt is the primary source of water for prairie potholes, bur precipitation varies greatly from year to year and droughts are a normal climatic phenomenon in grassland ecosystems. In response to drought, northern pintails – one of the more common species nesting in the prairie pothole region – move farther north in search of breeding areas in the sub-Arctic, where wetlands are more permanent but less suitable as breeding habitat. Production is low under these conditions, but it is high enough to ensure that a core population survives to reoccupy the highly productive pothole wetlands when the droughts end.

In contrast to pintails, lesser scaup on drought-stricken prairies often leave without making any attempt at nesting. Of those that remain, most simply loaf in groups of pairs. The reproductive organs of the females steadily regress; drought inhibits their reproductive efforts. Lesser scaup is a species better adapted for nesting in wetlands less prone to drought.

Overall, the regular occurrence of a drought produces a 'boom-or-bust' system in which the prairie potholes offer optimal nesting habitat for waterfowl only for three years of each decade. Year-to-year swings in waterfowl production are dramatic. At one site, for example, just 23 broods were produced in a drought year, but when better water conditions prevailed, production at the same area jumped to 557 broods.

Periods of drought, while seemingly harsh, actually contribute to the high productivity of prairie potholes and other wetlands. During wet cycles, when the potholes brim with water, dead vegetation and other organic matter

Figure 5.10 A profusion of prairie potholes once covered the northern plains and prairies, but large numbers of these wetlands were drained after World War II to increase agricultural production. Prairie potholes are popularly called "duck factories" because of their importance as nesting habitat for several species of waterfowl. Reproduced by permission of Karen Bataille of the Missouri Department of Conservation.

accumulate on the bottom. Decomposition is slow underwater because oxygen is limited – an **anaerobic** condition – and relatively few nutrients are released into the aquatic system. When a pothole dries, however, the organic matter on the bottom is exposed to oxygen – an **aerobic** situation – and decomposition progresses rapidly. Nutrients are consequently released into the soil, thereby becoming accessible to plant and animal life when the drought ends and the potholes refill. A boom-or-bust regime in nutrient cycling therefore establishes the foundation for corresponding swings in abundance for virtually all components in the food chains of prairie potholes.

Superficially similar to prairie potholes, wetlands known as playa lakes provide key habitat for waterfowl wintering on the southern plains of North America. *Playas* – "beach" in Spanish – are round, shallow depressions filled by local rainfall; hence many basins may be dry at any given time (Fig. 5.11). While crossing the Texas Panhandle in search of the mythical cities of gold in 1541, Coronado (1510–1554) was the first European to encounter playa lakes which he recorded as, "Round like plates, a stone's throw or more across." A relatively impermeable layer of clay seals the floor of each basin and keeps most of the water from percolating any deeper; the playas dry primarily by evaporation. Geologists debate the origin of the saucer-like basins. One widely held idea proposes that the basins are etched by the steady winds coursing across the plains; another

(a)   (b)

Figure 5.11 (a) Shallow, saucer-like playa lakes dot the flat terrain of the southern high plains of Texas. These wetlands depend on surface runoff and quickly dry without periodic rainfall. When filled, however, playas provide winter habitat for a million or more waterfowl. (b) Note intensive agriculture, made possible by irrigation, in what was once a semiarid region of shortgrass prairie in Kansas. Reproduced by permission of (a) Eric G. Bolen and (b) William C. Johnson.

far more romantic, but less plausible, idea suggests that the playas were the result of generations of bison dusting in "wallows." Whatever their origin, some 25,000 playas dot the region known as the *Llano Estacado*, or Staked Plains. With favorable conditions, these circular wetlands serve as winter habitat for a million or more ducks each year.

Regrettably, large numbers of prairie potholes are no longer viable wetlands. Drainage, largely from agricultural activities, claimed more than half of the 2.8 million ha (6.9 million acres) originally covered with potholes in North and South Dakota, and 3.6 million ha (about 8.9 million acres) are gone in Minnesota. The situation is slightly better in Prairie Canada, but the loss of these valuable wetlands continues on both sides of the border. Playas also experienced reduced capacity as habitat for waterfowl and other wildlife. Many are now modified sources of irrigation water; excavations dug into their otherwise shallow basins drain and concentrate the surface water in deep pits.

Pleistocene extinctions

The Pleistocene fauna of North America featured about 100 species of large mammals. Many of these ranged over the grasslands then developing in the continent's midlands. Known as megafauna because their adult body weight exceeded 44 kg (100 lb), this rich collection of native mammals was diverse and included relatives of modern-day elephants, sloths, large rodents, horses, and camels, among others. Also included, of course, were the predators of these herbivores, of which the saber-toothed cat is no doubt the most familiar. The megafauna also included giant bears. Presumably abundant as well as diverse, this assemblage of animals probably resembled the herds still roaming the plains of Africa. Nonetheless, two-thirds of the Pleistocene megafauna of North America shared a common fate: extinction.

Just how extinction befell the megafauna remains a mystery. Some propose that Paleolithic hunters simply killed too many animals. The fossil record reveals that the megafauna disappeared rather soon after humans immigrated from Siberia into North America across the land bridge known as Beringia. Briefly stated, the megafauna in the New World evolved independently of humans and therefore developed no suitable means of defense against this new and potent predator. The result was overkill. In this view, humans, with their hunting technology, either quickly exterminated large animals in a fashion likened to blitzkrieg warfare or, with less intensive killing, so disrupted animal populations as to initiate their decline beyond recovery (e.g., upset long-standing equilibria between herbivores and carnivores).

Climatic changes are chief among the other explanations for the extinction of the North American megafauna. For example, in the past 10 million years the mammalian fauna in North America experienced six episodes of extinction, of which losses in the late Pleistocene are simply the most recent. According to this view, each of these episodes corresponds to a glacial cycle when climatic extremes and environmental instability were at their greatest. The issue, while unsettled, represents a fine example of the interdisciplinary nature of paleoecology, replete with contributions from other disciplines (e.g., archeology, geochemistry, and climatology).

One result of the Pleistocene extinctions was the emergence of the modern grazing animals. Competition diminished when extinction eliminated mammoths and most of the other large grazing animals from the megafauna. A greatly expanded niche therefore remained available for the surviving species. Chief among these were bison, which increased on the grasslands of North America to perhaps 60 million.

Selected prairie mammals

Bison

No other animal, except perhaps the prairie dog, symbolizes grasslands as much as the bison. In his landmark publication *The Great Plains*, Walter Prescott Webb (1888–1963) wrote of the bison's historic importance: "In the Plains lived one animal that came nearer to dominating the life and shaping the institutions of a human race than any other in the land, if not the world – the buffalo." The "human race" was the Plains Indians – Lakota (Sioux), Comanche, Pawnee, and Cheyenne, among others – who depended on a buffalo economy.

Popularly known as "buffalo," bison once grazed the grasslands of North America in seemingly endless numbers; the immense herds reached the threshold of extinction in the two decades following the Civil War, however. Bison were killed for their well-furred hides, which made fine carriage robes; for meat, but often only for the delicacy of their tongues; and sometimes simply for bloodlust. After the herds were vastly reduced, some sources claimed that the slaughter was also a tactical means of destroying tribal resistance to the encroachment of Euro-Americans. General "Little Phil" Sheridan (1831–1888) declared that hide hunters had helped more in two years to "settle the vexing Indian question" than the entire army had done in 30 years. The discovery of bison leather's elastic qualities hastened the killing; it was perfect for running the pulleys, shafts, and wheels of industry. Industrial uses required green hides and, because these were heavy to transport, the destruction of bison often followed the ever-growing network of railroads crossing the plains.

Estimates for the bison population west of the Mississippi vary greatly from 28 million to 60 million at the close of the Civil War. Whatever the actual number, however, by 1889 fewer than 1000 bison of the millions once on the western plains survived the relentless slaughter. Fortunately, a restoration program spearheaded by William T. Hornaday (1854–1937) and other conservationists averted the pending disaster of extinction. Bison herds were slowly reestablished from a nucleus of animals at Yellowstone National Park and a small group in the care of the New York Zoological Garden.

Bison are naturally efficient grazers, converting grass to meat at a ratio exceeding domestic livestock. The immense herds of bison moved unfettered across the plains and, although grazing pressure was undoubtedly heavy for the moment at a given location, the continual movement of the herds permitted recovery of the grasses before the herds revisited previously grazed areas. To rub their shaggy coats where there were no trees and to rid themselves of vermin – and perhaps just to frolic – generations of bison dusted in the sod, eventually wearing circular depressions into the landscape. Because the shallow depressions were repeatedly disturbed and collected moisture, buffalo wallows contributed to the diversity of the plains flora.

Buffalo chips, of which hundreds of millions must have existed at any given time, were also local sites of biological activity. Each chip provided microhabitat for a succession of insect communities and, of course, richly fertilized the grasslands. Homesteaders, lacking sources of wood on the western plains, found a handy, if pungent, fuel in the once-endless supply of dried chips. So numerous were the pats of buffalo dung that the official report of the Long Expedition noted how the quantity of rafting dung almost obscured the surface of a flooded creek. Even today, westerners often refer to a torrent of rain as a "chip floater."

Because of their size and somewhat ill-tempered nature, bison were prey only for wolves which likely favored the weaker adults or calves. Native Americans, afoot and armed with stone-tipped weapons, initially killed relatively few bison, but larger numbers were slaughtered where tribes could drive herds over steep cliffs at sites known as "buffalo jumps" (Fig. 5.12). Later, after horses and rifles were available, Native Americans on the plains effectively adopted both for hunting bison, but this situation existed for less than 200 years. Bison populations nonetheless fluctuated from other influences. Unchecked prairie fires surely killed many bison and stampeded others into ravines.

Prairie dogs

Five species of prairie dogs occur in North America, three with white-tipped tails (Utah, Gunnison, and white-tailed prairie dogs) and two with black-tipped tails (Mexican and black-tailed prairie dogs). Perhaps no more than 6000 Utah prairie dogs remain, making them an endangered species. Although relatively abundant, white-tailed prairie dogs are generally found at higher elevations where they hibernate in response to the long winters. In comparison, black-tailed prairie dogs (Fig. 5.13a) remain active in all but the severest winter weather; they are also by far the most numerous and widely distributed of the five species. Explorers beginning with Coronado took note of prairie dogs, and Lewis and Clark shipped a live black-tailed prairie dog to President Jefferson as evidence of the West's remarkable creatures.

In the recent past, those who wished for better grazing lands (i.e., more forage for livestock) relentlessly

Figure 5.12 Native Americans once stampeded bison, often killing hundreds at a time, over cliffs known as "buffalo jumps." The one shown here, Head-Smashed-In Buffalo Jump in Alberta, Canada and a UNESCO World Heritage Site, served as a hunting location for nearly 6000 years. Reproduced by permission of Kenneth D. Thomas (www.KenThomas.us).

(a)

(b)

Figure 5.13 (a) Prairie dogs and burrowing owls live in close association in many grassland areas of western North America. (b) Prairie chickens court females on display grounds known as leks. As shown here, males defend small territories within a lek against other males. Note erected neck feathers (pinnae) on the male at right. Females watching these displays from the perimeter eventually mate with males holding territories near the center of the lek. Reproduced by permission of (a) Dan Flores and (b) Brian R. Chapman.

persecuted prairie dogs, usually with poisons. Estimates are difficult to verify, but some authorities believe fewer than 10% of the original population remains today. Like the slaughter of bison, the attempt to exterminate prairie dogs attacked one of the most abundant mammals in North America. An exceptionally large town of black-tailed prairie dogs once covered an area of 6475 km² (2500 square miles) near present-day Lubbock in the Texas Panhandle. A government biologist estimated that this town alone housed a population of 400 million prairie dogs, about half of their total number in Texas. Another estimate suggested a 19th century population of 5 billion black-tailed prairie dogs for all of North America.

Like most plains animals, prairie dogs depend heavily on vision for protection, and they continually clip vegetation within their colonies for an unobstructed view of the surrounding terrain. The mounds around the entrances to their dens serve as vantage points and also keep out floodwaters. Black-tailed prairie dogs communicate with a variety of calls, including an alarm bark warning of predators as well as an "all clear" signal. Large towns are divided into smaller units, "wards" of 2–4 ha (5–10 acres), within which social units known as coteries live. About eight prairie dogs form each coterie and defend a territory against members of other coteries.

Some ecologists regard prairie dogs as keystone species in shortgrass ecosystems because their burrow systems provide important ecosystem functions, including soil mixing and aeration, nutrient cycling, and modified vegetation. Additionally, their burrow systems offer food resources, nesting habitats, and shelter for many invertebrates (mainly beetles and spiders) and for more than 100 species of vertebrates, including tiger salamanders, burrowing owls, mountain plovers, killdeer, and the critically endangered black-footed ferret. Prairie dogs themselves provide prey for ferruginous hawks, bobcats, coyotes, black-footed ferrets, and many other avian and mammalian predators. All told, suppression or eradication of prairie dogs populations may alter the vegetation and associated fauna of some grassland areas.

Pronghorns

Hooved herbivores, popularly called "antelope," shared much of the western plains with bison. In reality, the species is not closely related to the true antelope of Africa and Asia. This is the pronghorn, a species endemic to North America and assigned to a monotypic family (Antilocapridae) in most taxonomic schemes.

Pronghorns have large and distinctive patches of white hair on their rumps. When the hairs in these patches are erect, the patches serve as an effective means of intraspecific signaling across the open plains. Their horns are also distinctive and unlike either the horns of bison or African antelopes or the antlers of caribou, elk, or other members of the deer family. Each horn consists of an outer, branched sheath over a permanent inner core. Moreover, the sheaths are uniquely formed of fused hairs instead of bone; they develop on both sexes, although those of females are relatively small. Male pronghorns shed the outer covering each year after breeding, but the females shed their horns with less regularity. Pronghorns, like other grassland ungulates throughout the world, are hoofed, swift runners. In comparison with other ungulates, however, pronghorn have disproportionally wide trachea, so their lungs receive large volumes of air.

This physical adaptation coincides with their behavior of running open-mouthed with extended tongues, thereby increasing the flow of air entering their enlarged trachea.

Pronghorns are more browsers than grazers, especially in the winter, but forbs also represent a significant part of their diet. Sagebrush is a major food, which is surprisingly rich in protein and carbohydrates. Only 1% of the pronghorn diet consists of grasses, so ranges overgrazed by livestock can recover despite the presence of large pronghorn herds. This feature also suggests how, in pre-settlement times, medium and large herbivores (pronghorn and bison) could share the plains without competing for forage.

More than 62% of the pronghorn population is associated with grasslands; 41% with shortgrass prairie and 21% with midgrass habitats. Most of the remainder (32%) occurs in bunchgrass–sagebrush plains. Wooded areas represent effective barriers, and the riparian woodlands along the Missouri River kept pronghorns from moving eastward at the latitude of Nebraska. However, farther north along the river (e.g., the Dakotas) where the riparian woods diminished, pronghorns crossed the river and extended their range eastward into the tallgrass prairie. In other words, pronghorns moved around the woodland barrier.

Curiously, pronghorns do not jump over obstacles, and even relatively low fences present serious barriers. Pronghorns safely passed through or under the strands of barbed wire fences but sheep require woven-net fences, which effectively kept pronghorns from moving across the plains to forage or seek winter cover. Large numbers of pronghorns died, especially in hard weather, when sheep-proof fences imprisoned entire herds. Drought reduced a pronghorn population by 40% largely because net fences confined pronghorns to overgrazed livestock ranges where they were forced to eat low-quality or toxic vegetation. These and other situations eventually led to recommendations for types of fencing that confine sheep and yet permit pronghorn movements.

Pronghorns were heavily hunted in the past, and stringent protection was necessary early in the last century. Some estimates suggest a pronghorn population of 35 million before Euro-American settlement, but fewer than 20,000 remained by 1924. The species eventually recovered, and well-regulated sport hunting of pronghorns is permitted in several western states.

Selected prairie birds

Burrowing owls

A small but long-legged species of owl is closely associated with prairie dog colonies on the western plains of North America. In some cases, burrowing owls adopt the dens of ground squirrels, but most find refuge and nesting sites in tunnels constructed by prairie dogs. Burrowing owls are unusual because of their fossorial nesting habits, but they also evolved a unique means of defense. These birds are active during the day and are easily observed at the entrance to their burrows (Fig. 5.13a). The long legs of burrowing owls assist with thermoregulation, precluding overheating when the birds stand on bare ground in hot environments. Burrowing owls are mostly insectivorous but on occasion eat other foods (e.g., rodents, especially in winter). In most species of owls, females are larger than males, but this relationship is strangely absent from burrowing owls, and it remains inviting to include their fossorial habits in any hypothesis that might be forthcoming.

Burrowing owls line their tunnels and underground chambers with dried dung, which may serve as a means of camouflaging their scent from predators. Livestock provide the dung for this purpose today, but bison undoubtedly were the primary source of dung when this interesting habit evolved. Burrowing owls will replace the dung if it is removed and, during the nesting season, the presence of a well-maintained lining indicates an active burrow. Moreover, insects living in the dung linings may be important local sources of food for burrowing owls.

When disturbed in their nests, young burrowing owls employ an extraordinary behavioral adaptation: the nestlings mimic the buzz of rattlesnakes, a defense mechanism of obvious value for helpless birds confined in the small space of an underground burrow. The similarity is remarkable; when compared electronically, the buzz of prairie rattlesnakes proved little different from the vocalizations of young burrowing owls.

Controls aimed at reducing prairie dog and ground squirrel populations on rangelands indirectly reduced the numbers of burrowing owls. Because fewer burrows are available, conservation agencies at times express concern about the welfare of burrowing owl populations. On the other hand, because of the endangered status of black-footed ferrets – another species closely associated with prairie dog colonies – habitat protection for the ferrets coincidently protects the burrowing owls. Where burrows are not available as nesting sites, the owls sometimes accept artificial structures designed for the purpose. Curiously, an eastern subspecies of burrowing owl occurs in southern Florida, often living in urbanized areas without the benefit of burrowing mammals.

Prairie chickens

Three races of greater prairie chickens occurred in North America at the time of European settlement. One of these, the heath hen, occupied the coastal plain from Virginia to Massachusetts, where overhunting and habitat loss quickly reduced the population to precarious numbers. Extinction eventually claimed the heath hen,

but two other races persisted on the grasslands lying west of the eastern woodlands.

Settlers pushing westward quickly relied on greater prairie chickens as a ready source of fresh meat. Flocks flushed before the wagons, and even novice hunters downed enough for an evening meal. Hunting pressure remained heavy after settlement and, in 1878, Iowa enacted the first "bag limit" for sport hunting in the United States: a legal kill of no more than 25 prairie chicken per day! Immense numbers were also killed for market, a practice significantly furthered by the western expansion of the railroads. The plow ultimately finished what ill-advised shooting began, and the birds diminished as grain fields replaced the tallgrass prairie. Only a small fraction of the former population remains today, and these are hunted only in limited numbers.

Attwater's prairie chicken represents the third race, which clings tenuously to existence in the tallgrass prairies bordering the Texas coast. Where up to 1 million Attwater's prairie chickens thrived more than a century ago, a census in 2005 revealed only 40. Captive breeding programs maintain more than 200 birds at various locations, but despite habitat on a federal refuge and full legal protection, their future in the wild seems bleak indeed.

Prairie chickens display during courtship at sites known as "booming grounds" or **leks**, which are areas of sparse cover as much as 150 m (500 feet) in width. Leks are used year after year, with some sites having histories of several decades of continual use. Here, a dozen or more males from the region gather at dawn each spring to display for about 90 minutes; they return again at sunset for about the same period of time. Each male displays on a small part of the lek – a **territory** – which he defends against the intrusion of other males. Competition is keen for the innermost locations within the lek, and dominant males hold these positions against subordinates.

The courtship behavior of prairie chickens is engaging (5.13b). Several males usually display at one time, although each performs separately for dominance and for the attention of females. The display begins with a short run, which ends abruptly, followed by rapid foot-stamping in one spot. While stamping, the male often pivots but always erects a collar of elongated neck feathers called pinnae (some biologists thus identify prairie chickens as *pinnated grouse*). Concurrently, each bird: (a) inflates an air sac on each side of his neck, which resembles a small orange in both size and color; (b) initiates a booming vocalization somewhat resembling blowing across the open neck of a bottle; and (c) droops his wings and fans his tail. More is still to come: a male then jumps upward about 30 cm (1 feet) and twists in mid-air, often facing in the opposite direction upon landing. Should a male feel challenged by another male on an adjoining territory, he lowers his head and rushes toward his opponent in what is usually a bluff but sometimes results in brief, physical contact. Female prairie chickens gather at the center of the lek, where they mate with the innermost males. Less-dominant males at the perimeter of the lek seldom acquire mates. Perhaps no more than 10–20% of the males at each lek complete most of the matings, a fine example of natural selection based on social dominance.

Lek behavior, which evolved as a group-based visual performance, clearly befits open environments and thereby contrasts with the courtship behavior of ruffed grouse (i.e., isolated individuals drumming in forests, an acoustical performance befitting the limited visibility in thickly wooded terrain). As might be expected, the ceremonial dances of Omaha and other plains tribes mimicked the courtship displays of prairie chickens (e.g., foot stamping).

Highlights

Riparian forests

Even in pre-settlement times, North American grasslands did not exist as an uninterrupted "sea of grasses." Mesic sites support woody vegetation of which the best examples are riparian forests, ribbons of deciduous trees following the course of rivers and streams crossing the prairie. The width of these woodlands varies, with those in favorable sites often representing forests of significant proportions. As shown earlier, riparian forests along some parts of the Missouri River were wide enough to prevent pronghorns from moving eastward. Conversely, white-tailed deer extend their range into grassland regions along riparian forests, which in this case act as corridors. Woodchucks also moved westward into prairie areas by following the riparian corridors on the tributaries of the Kansas River. In sum, riparian forests extend like fingers into the prairie, where they act as barriers or corridors for the local or regional fauna.

The Platte River

Two headwater branches of the Platte River arise in the Rocky Mountains but eventually unite for a prairie journey across Nebraska to the Missouri River. The Platte supplied the vital artery for westward migration during the 1840s. The Oregon Trail paralleled its southern bank, eventually guiding the plodding traffic of immigrants to the continental divide at South Pass and beyond. Wagon ruts still cut deeply into the prairie along the route. Along the north bank, the Mormon Trail stretched westward from Omaha. The Pony Express also followed the river, and the level floor of the Platte valley served as the roadbed for the Union Pacific Railroad; the larger trees in the riparian forest were felled for cross ties. Of more immediate ecological interest, however, is that the Platte: (a) is one of the few large rivers that crosses the prairies and therefore represents a major source of water in a

semiarid region; and (b) features a braided channel (as described in *Prairie streams*).

Various schemes (canals, ditches, and dams) eventually harnessed the flow of water in the Platte. These turned the semiarid plains into irrigated fields, but they also tamed the Platte's seasonal floods. The Platte River system today is contained by 15 reservoirs and, with the other alterations, no more than 30% of the river's historical flow still courses its braided channels. Similarly, the river bed has shrunk dramatically. Today, the channel at places is about 10–20% of its former width.

The reduced flow produced significant ecological impacts, among which is the degradation of habitat for about 80% of North America's population of lesser sandhill cranes. Each year, during their spring migration from the Texas Panhandle, as many as 500,000 sandhill cranes linger for a few weeks in Nebraska, roosting at night in the Platte's riverbed and representing the largest gathering of cranes in the world. Whooping cranes also visit the Platte during migration, and bald eagles are common winter residents. Wet meadows nearby offer abundant sources of food for sandhill cranes and, at night, the river's twisted network of bars and sandy islands isolate the cranes from predators. Their diet, enriched by corn, helps condition the cranes for the remainder of their long flight. Other birds also utilize habitat on the Platte, among them least terns and piping plovers – each a federally protected species – that nest on sandy sites just above the waterline.

Because of their importance to migratory birds, stopping points as significant as the Platte are identified as staging areas. In addition to providing crucial resources, these locations are etched into the traditions of migratory birds so alternate locations, even if available, may not be adopted when age-old **staging areas** are destroyed or altered. Today, the Platte River provides far less sanctuary for migratory birds. Because of the dams, spring floods no longer scour the Platte's shallow riverbed or enrich the surrounding meadows. More importantly, without the water's scouring action each year, vegetation gains a foothold on the sandbars and smaller islands. Cottonwood, willows, and other vegetation currently choke much of the Platte, and with this overgrowth, key habitat has disappeared for future generations of sandhill cranes and other birds.

Conversely, the riparian forests developing along the Platte in the last century enabled the westward dispersal of several woodland birds. Among these are blue jays and brown thrashers, species that colonized the Front Range of the Rocky Mountains in Colorado after following the Platte westward across the Great Plains. In fact, the majority of the avifauna today breeding on the shortgrass prairie in eastern Colorado was absent in 1900. These additions are the likely consequence of a westward march along the Platte River.

Nebraska Sandhills

The watershed of the Platte drains about 75% of the Nebraska Sandhills which, at 51,800 km² (24,000 square miles), is the largest dune system in North America. The dunes reach heights of 30–90 m (100–295 feet) and extend for as much as 16 km (10 miles), but average 1.2 km (0.75 miles) in length. Within the dune system, topographic position plays an important role in the distribution of upland vegetation, which changes in composition between north- and south-facing slopes as well as in the intervening valleys. Some depressions between the hills often lie below the water table; here numerous lakes provide oases of wetland habitat for waterfowl and other aquatic life in this semiarid region.

The origin of the sands is debated, but the soil-forming parent material is geologically young; a wind-generated regime formed the dunes about 7000 years ago. Because of favorable moisture conditions in the sandy soils, a large pocket of tallgrass prairie, instead of a midgrass community otherwise typical of this region, developed on the Nebraska Sandhills. The tallgrasses in the Nebraska Sandhills include the same species as those in the true prairie region father eastward (e.g., big bluestem and Indiangrass). Closely related to big bluestem is sand bluestem, whose creeping rhizomes undoubtedly represent an adaptation associated with the regional soils. Unlike other tallgrass areas, however, the plow has not converted the Nebraska Sandhills into croplands (the sandy soils quickly "blow out" when deprived of their cover of grasses) and ranching remains the dominant use of the land. Nonetheless, when various agents expose local areas of soil, the aptly named blowout grass is foremost among the pioneer species that gradually revegetate these sites. Blowout grass spreads across the exposed sand with tough, coarse rhizomes, which may extend for 12 m (40 feet), and the flexible blades of this grass endure wind-whipping for long periods without injury. A legume, lance-leaved psoralea, is also a major pioneer species on blowouts or other disturbed sites in the Nebraska Sandhills.

Sand reedgrass is the most characteristic species of the Nebraska Sandhills. On favorable sites, such as south-facing slopes of dunes, the stems of this reedgrass reach 1.8 m (6 feet) in height, but elsewhere the stems may be less than half as tall. The rhizomes of sand reedgrass are strong and scaly, and their creeping nature binds the sandy soil. Hairy grama and prairie dropseed are among the midgrasses associated with sand reedgrass in the Nebraska Sandhills. Livestock graze these and other grasses on larger ranches. Indeed, because the sandy soils are unsuitable for farming, the size of allotments authorized by the Homestead Act (1862) was increased fourfold in this area by the Kinkaid Act (1904), a measure enacted expressly to encourage ranching in the Nebraska Sandhills.

Ants

Grasslands are key habitat for numerous insects, the majority of which consume foliage. Grasshoppers are obvious examples, but many other taxa are represented in the insect fauna on healthy grasslands. In fact, insects may reflect range condition; the diversity of the insect community decreases and abundance increases on overgrazed rangelands.

Other insects are granivorous. Chief among these are harvester ants, which consume seeds in arid and semi-arid regions throughout the world. Their foraging choices, in fact, may influence the composition of plant communities. The ants effectively remove seeds they prefer as food, thereby limiting the reproduction and occurrence of certain plants, whereas plants produced from unpalatable seeds remain in greater abundance. On a shortgrass prairie in Colorado harvester ants foraged on 39 species of seeds, but their forage also included litter from other plants as well as dead insects.

Individual colonies of red harvester ants reach a stable size of about 12,000 workers after five years and remain active for 15–20 years. Red harvester ants also clear all vegetation around the entrance to their hills, exposing large disks of bare soil in vegetated areas. In a shortgrass community in Texas, the cleared area around each hill averaged about 1.4 m² (15 square feet). These occurred at a density of about 21 hills per ha (8.5 per acre); harvester ants are therefore a visible presence in grassland communities. Even larger disks may occur, sometimes in greater densities, in other grassland communities. The cleared areas apparently serve two functions: as fire breaks around the ant hills and as root-free zones. Roots otherwise could penetrate the underground galleries and, after dying, leave channels along which water could saturate the soil. When foraging, harvester ants fan out along trunk trails worn into the surrounding vegetation, but they end their searches for seeds by midday because of elevated soil temperatures.

One or two species of three-awn grasses typically ring the disks of harvester ants in Texas. Consequently, it was widely believed that harvester ants cultivated the grasses as crops for additional sources of seed. In reality, however, these grasses are the result of seeds that sprouted while in storage in the ant's underground granaries. Because they are no longer suitable as food, the ants remove and carry the sprouting seeds to the edge of the disk, where they take root and eventually encircle the nest. The seeds of three-awn grasses are a common and favorite food; as a result, these species predominate in the vegetation at the rim of the disk. Based on samples collected in Texas, each disk surrounding an active ant hill contained more than 22,500 seeds, most of which were discards; even larger numbers of seeds were collected and stored as food.

Isolation and contact on the plains

An interesting example of speciation concerns two closely related species of grassland birds. Despite their remarkable similarity in appearance, eastern and western meadowlarks represent "good species" (i.e., well-separated gene pools and not races of the same species). On close examination these two birds show slight differences in plumage, but they differ significantly in their respective songs. The song of the western meadowlark is often described as flutelike, whereas that of the eastern meadowlark resembles a slurred whistle.

How might this situation develop? The best explanation seems to be that glaciers split a single ancestral population during the Pleistocene. The wedge need not have been the ice sheet itself; cold along the glacial front created severe conditions well beyond the ice's southernmost advance, and these circumstances undoubtedly produced an effective east–west barrier for many organisms. Isolated on either side of this barrier, meadowlarks eventually developed species-specific vocalizations but retained similar plumages. Millennia later, after the glaciers melted, the two species came into contact (assisted by human settlement), but their distinctive calls continued to isolate the two populations just as effectively as had the glacial barriers; the calls of one species did not attract the opposite sex of the other, thwarting interspecific courtship. Meadowlarks therefore provide an example of how physical isolation eventually produced behavioral barriers to the exchange of genes. Natural selection, in fact, works against the formation of interspecific pairs, as well as against hybrids themselves which are seldom fertile. Such evidence reflects the strong genetic integrity of eastern and western meadowlarks and affirms their designation as separate species.

Settlement eliminated much of the prairie, but humans also created new habitats deep within the grasslands of North America. Among these were strips of trees known as shelterbelts or windbreaks, planted to reduce the sweep of wind across the open landscape.

Shelterbelts also provided woodland habitat for birds that were otherwise excluded from grassland environments. Mississippi kites, for example, readily nest in shelterbelts, resulting in a geographic change in their distribution in North America. Even more significantly, some tree-dwelling birds living east and west of the grasslands came into even greater contact as a result of shelterbelt planting. One – the northern flicker – is of interest because it represents a "species pair," with eastern and western forms. The northern flicker population east of the grasslands is distinguished by yellow-shafted wing and tail feathers and, on the male, a black facial "moustache," whereas western flickers have red-shafted wing and tail feathers and a red moustache. With shelterbelts, however, the eastern and western counterparts of flickers came into greater contact than previously possible, and the two forms readily hybridized, producing offspring with intermediate coloration. A similar situation occurs with other species-pairs, including eastern (Baltimore) and western (Bullock) forms of the northern oriole. However, ready hybridization in these birds clearly suggests that they lack the same degree of reproductive isolation as occurs in eastern and western meadowlarks.

Grassland settlement

The full measure of western expansion was not long in coming after European claims to North America were finally resolved. Settlers trekked westward beyond the Appalachians through the Cumberland Gap and, after 1825, commercial traffic with the Ohio frontier busily plied the Erie Canal. Upon reaching the edge of the eastern forests, however, settlers faced a new landscape: the grasslands. Western Europe lacked prairies or plains, and the descendants of Old World immigrants brought with them no legacy for subsistence on the immense treeless area of the North American interior. For them, the prairie seemingly marked the end of soil fertility.

At first, settlers remained near the wooded fringes of the eastern prairie, often near streams where wood was plentiful. Ironically, settlers sometimes expended considerable effort clearing riparian forests for fields instead of cultivating the adjacent grasslands, which for a time remained intact as untilled grazing lands. Eventually, however, the discovery of coal, the introduction of barbed wire, and the coming of the railroads hastened occupation of the prairies. Another development – the steel plow – forever altered the ecology of the American prairie.

From an economic point of view, settlement of the prairies became a war on roots. Whereas the land sold cheaply, more capital was often required to plow the thick sod of the true prairie. The work sometimes required five yokes of oxen – two to a yoke – an investment exceeding the means of most settlers. The fibrous grass roots, much like a network of wire, stymied horse-drawn plows. Some forbs, particularly red root, added significantly to the problem. Moreover, the heavy prairie soils balled up on the iron plows of the day and required frequent cleanings. An alternative was so-called sod corn, planted by striking the turf with an axe and dropping the seed into the cut, but the yield was predictably marginal.

In 1837 John Deere (1804–1886), a blacksmith from Vermont, changed the direction of prairie farming when, in the frontier town of Grand Detour, Illinois, he fashioned a plow from the hardened steel of a saw blade. The plow's smooth surface no longer choked with soil, and its sharp edge sliced the network of roots without the labor of oxen. The tallgrass prairie was therefore "broken" – jargon for cleaving the original sod but no less applicable for breaking the backbone of grassland communities in much of North America. *Sodbusters* eventually became the less than complimentary name for farmers on the western plains in what was regarded, at least by some, as cattle country. For Plains Indians, the plow produced a world where "Grass no good upside down."

Tallgrass prairie did not last long in the expansion that followed the Civil War. Ornithologist Robert Ridgway (1850–1929) visited prairie locations in his native Illinois in 1871. When he returned 12 years later, he found changes "almost beyond belief;" except for an unfenced area of 64 ha (9160 acres), the original prairie had been replaced by fields, orchards, and buildings. "As a consequence we searched in vain for the characteristic prairie birds," finding representatives only in the small plot still remaining. The ornithologist left with a "sad heart" and despaired of again finding unspoiled prairie in Illinois. Today, North America's inland sea of plains and prairie is largely gone. In its place is a landscape of domestic grasses: great belts of corn and wheat.

Another development also forever transformed the western grasslands of North America: barbed wire initiated an era of fenced rangelands that continues today. Patented in 1874, barbed wire quickly triggered range wars, but eventually the new fencing became a means of regulating grazing pressure and a silent tool of range management. With barbed wire, livestock could be held in one place and then moved elsewhere to maintain the health of the grasses, patterns of movement known as grazing systems. Poor management could of course lead to overgrazing and the attendant features of degraded vegetation and soil erosion. Barbed wire presented an industrial-age analog for mesquite and other thorned vegetation on which loggerhead and northern shrikes impaled their prey (e.g., insects and small vertebrates). However, the fencing also became a source of mortality for birds (e.g. whooping cranes) and mammals as large as deer.

In the 1930s, a combination of forces – natural and anthropogenic – produced what some have called one of the greatest environmental disasters the world has known. The Dust Bowl was one of any number of draughty periods on the Great Plains, but none had previously struck when the land lay so bare. Huge areas of land, much of which was marginally suited for farming, were tilled to satiate the national glut for wheat. As was now painfully clear, the plow had long before turned under the sod-forming shortgrasses that protected the soil against drought (Infobox 5.3). Dark blizzards of soil swept across the plains during the "Dirty Thirties." At least one soil storm reached Washington, DC, thereby bringing the calamity of the Dust Bowl to the direct attention of Congress.

The Dust Bowl coincided with the Great Depression, which together ended further settlement of the North American plains. The misery of this era echoes in the words and music of John Steinbeck and Woody Guthrie:

stories and ballads of dusty landscapes and a failed economy. Among the ecological consequences was an alarming plunge in waterfowl populations (the prairie potholes were no more than parched depressions), but these events triggered major initiatives for wildlife conservation (e.g., sales of "duck stamps" began in 1934, with the revenues committed to wetland acquisition). With hard times, however, came countermeasures, including the Prairie States Forest project which halted wind erosion by planting some 128 million trees in shelterbelts.

Beneath the Great Plains lies the Ogallala Aquifer, a vast underground water deposit that extends some 450,000 km² (174,000 square miles) beneath parts of eight states (Fig. 5.4). The aquifer is primarily paleowater, deposits which accumulated during the last ice age or earlier, and receives limited recharge as seepage from playa lakes. The upper surface of water-bearing formation lies about 30–60 m (100–200 feet) below the grassland in the southern extent of the deposit and almost 120 m (400 feet) deep in the north. About 100 years ago, wells began tapping the Ogallala Aquifer for irrigation and municipal use. Relatively easy access to this water likely accelerated the conversion of grassland to farmlands, especially in the more arid central and western plains. About 95% of the water currently pumped from the aquifer irrigates farmlands. Consequently, the water level in the Ogallala has dropped, with the remaining supply perhaps lasting only another 25 years at present rates of consumption and recharge. This forecast has prompted conversion to crops and farming methods that require less irrigation.

The Taylor Grazing Act, also initiated in 1934, closed what remained of public land from further settlement under the Homestead Act. Instead, these lands (mostly on the semiarid plains) would henceforth be dedicated to grazing and managed in districts by local ranchers acting under government supervision. Moreover, the Land Utilization Project purchased badly degraded private lands and added these to the public domain. Some 4.6 million ha (11.3 million acres) were eventually acquired, mostly in the plains, and designated as National Grasslands. These units still exist, scattered from the panhandles of Texas and Oklahoma (the heartland of the Dust Bowl) northward to Montana.

Prairie preservation

Prairies are difficult communities to preserve for at least three reasons: (a) most areas of tallgrass prairie have been cultivated for many years, few of the remaining areas are of great size, and even fewer are "virgin"; (b) grasslands seldom instill a sense of majesty as might a redwood forest, and public outcries for the protection of prairies are therefore infrequent; and (c) prairie

Infobox 5.3 John E. Weaver (1884–1966), father of grassland ecology

Born in Iowa and a long-time ecologist stationed in Nebraska, Weaver was a true son of the tallgrass prairie. The plains and prairies of North America remained his calling throughout a career that, after one year as an instructor of botany in Washington State, spanned decades (1917–1952) at the University of Nebraska at Lincoln. He earned a PhD in 1916 at the University of Minnesota, studying under Frederic E. Clements (1874–1945).

With Clements as his coauthor, Weaver published the first American textbook dealing with plant ecology in 1929; a new edition appeared in 1938. A second book of note, *North American Prairie*, appeared in 1954, followed in 1956 by *Grasslands of the Great Plains: Their Nature and Use* written with F. W. Albertson. *Native Vegetation of Nebraska* (1965) and a post-humous publication, *Prairie Plants and Their Environment, a Fifty-year Study in the Midwest* (1968), rounded out his list of books. He also authored or coauthored scores of journal papers and two major studies concerning the effects of extended drought on prairies and pastures.

Root development was among Weaver's primary interests and he made numerous, pains-taking studies of the extensive systems that anchored grassland vegetation. He examined the root systems of prairie plants either by digging deep trenches to expose a cross section of roots or by extracting entire blocks of sod (root monoliths) from which the soil was washed away. In either case, the roots, stolons, and rhizomes were measured and mapped to scale on graph paper. In 1917–1918, Weaver dug more than 100 pits by hand and excavated about 1150 individual plants of some 140 species to learn how their roots systems knit together and responded to climatic conditions. By literally "working in the trenches," he learned of the great underground biomass and root-depth of the tallgrasses of the true prairie and the equally large mass but lateral spread of the shortgrasses that characterized the drier plains of western North America. In one study, he recorded an average of nearly 17 m (55 feet) of rhizomes per square foot of bluestem sod, or more than 644 km (400 mile) per acre.

His field work spanned the devastating years of the Dust Bowl, during which he turned out more students who studied the ravages of long-term drought on grassland vegetation than anyone else. Nearly a decade later, when the drought ended, he initiated follow-up work to reveal how the prairie responded to the return of favorable precipitation. All told, his research covered the most severe cycle of drought and recovery in the recorded history of North America. Weaver once believed that the prairie, because of the stability and toughness it acquired after centuries of evolution, "approaches the eternal." But after witnessing years of human abuse to the prairie, abetted by the Dust Bowl, Weaver reached a painful conclusion: "Once destroyed it can never be replaced...." Wistfully, he noted, "To prairie sod, only the plow is lethal."

In 1930, Weaver's peers acknowledged his stature when he was elected president of the Ecological Society of America and, in 1950, as honorary president of the International Botanical Congress held in Stockholm, Sweden. His biography was included as one of the 100 most highly recognized scientists among the thousands of others listed in the then-current edition of American Men of Science.

For John Weaver, the prairies represented the heart and enduring strength of North America. Thoughtfully and prophetically, he wrote "Nature is an open book for those who care to read. Each grass-covered hillside is a page on which is written the history of the past, conditions of the present, and predictions of the future. Some see without understanding; but let us look closely and understandingly, and act wisely, and in time bring our methods of land use and conservation activities into close harmony with the dictates of nature."

preserves must be burned regularly, particularly in moister regions, yet many people regard fire as too dangerous to employ as a management tool.

Despite these difficulties, several prairie preserves are now established in North America. One of these, the Konza Prairie Research Natural Area, is a 3488 ha (8618 acre) tallgrass site in the Flint Hills region of northwestern Kansas, where it was once a part of a cattle empire. "The Konza," now owned by The Nature Conservancy, is managed for ecological research by Kansas State University. The vegetation, dominated by big bluestem but including more than 550 other species of plants, is burned at selected intervals in keeping with a long-term research plan. A herd of 210–220 bison currently graze 950 ha (2345 acres) all year round without supplemental feeding. The herd produces 60–80 calves each spring; to maintain a constant size, some of the young and low-vigor older adults are culled to simulate predation. The bison area contains more than 50 small exclosures, thus permitting detailed comparisons between grazed and ungrazed vegetation.

One of the most imaginative – and certainly controversial – proposals for the restoration of grasslands is popularly known as the "Buffalo Commons." In 1893, the historian Frederick Jackson Turner (1861–1932) eloquently noted the passing of the American frontier, which ended when settlement exceeded a density of 0.7 people km^{-2} (2 people per square mile). Almost a century later, Frank and Deborah Popper discovered that more than 140 counties in the 48 contiguous United States had lost enough population to qualify once again as "frontier." Of the "frontier counties," the majority lie within the Great Plains (e.g., western Nebraska), well within the former range of bison. Accordingly, the Poppers proposed that the region might better be managed as a plains ecosystem, that is, a national reserve replete with bison herds and a mecca for tourists. Unlike wheat farming, ecotourism does not demand massive infusions of energy each year; with the return of the native vegetation, regional economies will no longer depend on annual rainfall and soil moisture, nor will tourists deplete basic resources (e.g., most anthropogenic soil erosion should end). Buffalo ranching offers still other economic opportunities. Bison are far more efficient than cattle at converting grass into meat and, because they are well adapted to the plains environment, they seldom require food supplements, the attention of veterinarians, or other extraordinary care. A Buffalo Commons may also alleviate some degree of the socioeconomic ailments experienced in the region.

Lawyer-turned-artist George Catlin (1796–1872) long ago lamented the day when bison and Native Americans might no longer occupy "these vast and idle plains." As the Poppers themselves eventually learned, Catlin had proposed a "nation's park, containing man and beast, in all the wild and freshness of their nature's beauty," established by a "great protecting policy of government." In such a place, he wrote, "the world could see for ages to come, the native Indian . . . amid the fleeting herds of elks and buffaloes."

Readings and references

Introduction

Bensen, M. (ed.) 1988. *From Pittsburg to the Rocky Mountains, Major Stephen Long's Expedition 1819–1820*. Fulcrum, Golden, CO. (The occurrence of a rain-swollen creek blanketed with floating buffalo chips is among the observations.)

Dillon, R.H. 1967. Stephen Long's Great American Desert. Proceedings of the American Philosophical Society 111: 93–108.

Major associations

Tallgrass prairie

Carr, W.R. 1981. Vascular plants of Bigelow (Chuckery) Cemetery State Nature Preserve in northern Madison County, Ohio. In: *The Prairie Peninsula: In the "Shadow" of Transeau* (D.L. Stucky and K.J. Reese, eds). Proceedings of the Sixth North American Prairie Conference, Ohio Biological Survey Notes Number 15, Ohio State University, Columbus, OH, pp. 128–130. (An example of how old cemeteries are useful for studies of prairie flora.)

Chamrad, A.D. and T.W. Box. 1965. Drought-associated mortality of range grasses in south Texas. Ecology 46: 780–785.

Collins, S.L. and D.E. Adams. 1983. Succession in grasslands: 32 years of change in a central Oklahoma tallgrass prairie. Vegetation 51: 181–190.

Dyksterhuis, E.J. 1949. Condition and management of range land based on quantitative ecology. Journal of Range Management 2: 104–115.

Hart, R.H. and J.A. Hart. 1997. Rangelands of the Great Plains before European settlement. Rangelands 19: 4–11.

Joern, A. and K.H. Keeler. (eds) 1995. *The Changing Prairie: North American Grasslands*. Oxford University Press, New York, NY.

Knapp, A.K., J.M. Briggs, D.C. Hartnett, and S.L. Collins. (eds) 1998. *Grassland Dynamics: Long-Term Ecological Research in Tallgrass Prairie*. Oxford University Press, New York, NY.

Knapp, A.K., J.M. Blair, J.M. Briggs, et al. 1999. The keystone role of bison in North American tallgrass prairie. BioScience 49: 39–50. (Suggests that bison movements, grazing, and wallowing maintained habitat for many species.)

Livingston, R.B. 1952. Relict true prairie communities in central Colorado. Ecology 33: 72–86.

Madson, J. 1982. *Where the Sky Began: Land of the Tallgrass Prairie*. Sierra Club Books, San Francisco, CA.

Malin, J.C. 1984. *History and Ecology, Studies of the Grassland*. University of Nebraska Press, Lincoln, NE.

Manning, R. 1995. *Grassland, the History, Biology, Politics, and Promise of the American Prairie*. Viking, New York, NY.

Reichman, O.J. 1991. *Konza Prairie: a Tallgrass Natural History*. University of Kansas Press, Lawrence, KS.

Risser, P.G., E.C. Birney, H.D. Blocker, et al. 1981. *The True Prairie Ecosystem*. Hutchinson Ross Publishing Company, New York, NY.

Wood, W.R. and M. Liberty. (eds) *Anthropology on the Great Plains*. University of Nebraska Press, Lincoln, NE.

Zimmerman, J.L. 1993. *The Birds of Konza*. University Press of Kansas. Lawrence, KN.

Midgrass prairie

Coupland, R.T. 1950. Ecology of mixed prairie in Canada. Ecological Monographs 20: 271–315.

Hanson, H.C. and W. Whitman. 1938. Characteristics of major grassland types in western North Dakota. Ecological Monographs 8: 57–114.

Weaver, J.E. 1943. Replacement of true prairie by mixed prairie in eastern Nebraska and Kansas. Ecology 24: 421–434.

Weaver, J.E. and W.E. Bruner. 1954. Nature and place of transition from true prairie to mixed prairie. Ecology 35: 117–126.

Shortgrass prairie

Brockaway, D.G., R.G. Gatewood, and R.B. Paris. 2002. Restoring fires as an ecological process in shortgrass prairie ecosystems: initial effects of prescribed burning during the dormant and growing seasons. Journal of Environmental Management 65: 135–152.

Larson, F. 1940. The role of bison in maintaining the short grass plains. Ecology 21: 113–121.

Peden, D.G., G.M. Van Dyne, R.W. Rice, and R.M. Hansen. 1974. The trophic ecology of *Bison bison* L. on shortgrass plains. Journal of Applied Ecology 11: 489–498.

Transition zones

Aspen parklands

Aikman, J.M. and A.W. Smelser. 1938. The structure and environment of forest communities in central Iowa. Ecology 19: 141–150.

Bird, R.D. 1930. Biotic communities of the aspen parkland of central Canada. Ecology 11: 356–442.

Bird, R.D. 1961. *Ecology of the Aspen Parkland of Western Canada in Addition to Land Use*. Canadian Department of Agriculture, Publication 1066, Ottawa, ONT.

Coupland, R.T. and T.C. Brayshaw. 1953. The fescue grassland in Saskatchewan. Ecology 34: 386–405.

Daubenmire, R.F. 1936. The "Big Woods" of Minnesota: its structure, and relation to climate, fire, and soils. Ecological Monographs 6: 233–268.

Howell, D.L. and C.L. Kucera. 1956. Composition of presettlement forests in three counties in Missouri. Bulletin of the Torrey Botanical Club 83: 207–217.

Johns, B.W. 1993. The influence of grove size on bird species richness in aspen parklands. Wilson Bulletin 105: 256–264.

Kiel, W.H., Jr, A.S. Hawkins, and N.G. Perret. 1972. Waterfowl Habitat Trends In The Aspen Parklands Of Manitoba. Canadian Wildlife Service, Report Series 18. Ottawa, ONT.

Leopold, A. 1949. *A Sand County Almanac and Sketches Here and There*. Oxford University Press, New York, NY.

Little, E.L., Jr. 1939. The vegetation of Caddo County canyons, Oklahoma. Ecology 20: 1–10.

Lynch, D. 1955. Ecology of the aspen groveland in Glacier County, Montana. Ecological Monographs 25: 321–344.

Moss, E.H. 1932. The poplar association and related vegetation of central Alberta. Journal of Ecology 20: 380–415.

Pelton, J. 1953. Studies on the life history of *Symphoricarpos occidentalis* Hook in Minnesota. Ecological Monographs 23: 17–39.

Pool. R.J., J.E. Weaver, and F.C. Jean. 1918. Further studies in the ecotone between prairie and woodland. University of Nebraska Studies 18: 1–47.

Telfer, E.S. and G.W. Scotter. 1975. Potential for game ranching in boreal aspen forests of western Canada. Journal of Range Management 28: 172–180.

Cross Timbers

Dyksterhuis, E.J. 1948. The vegetation of the western cross timbers. Ecological Monographs 18: 325–376.

Francaviglia, R.V. 2000. *The Cast Iron Forest: A Natural and Cultural History of the North American Cross Timbers*. University of Texas Press, Austin, TX.

Johnson, S.L. and P.G. Risser. 1975. A quantitative comparison between an oak forest and an oak savanna in Central Oklahoma. Southwestern Naturalist 20: 75–84.

Shutler, A. and B.W. Hoagland. 2004. Vegetation in the cross timbers, Carter County Oklahoma. Proceedings of the Oklahoma Academy of Science 84: 19–26.

Western transition

Buffington, L.C. and C.H. Herbel. 1965. Vegetational changes on a semidesert grassland range from 1858 to 1963. Ecological Monographs 35: 139–164.

Hennessy, J.T., R.P. Gibbens, J.M. Tromble, and M. Cardenas. 1983. Vegetation changes from 1935 to 1980 in mesquite dunelands and former grasslands of southern New Mexico. Journal of Range Management 36: 370–374.

Wright, H.A. 1972. Fire as a tool to manage tobosa grasslands. Proceedings of the Tall Timbers Fire Ecology Conference 12: 153–167.

Features and adaptations

Dahlman, R.C. and C.L. Kucera. 1965. Root productivity and turnover in native prairie. Ecology 46: 84–89.

Johnsgard, P.A. 1979. *Birds of the Great Plains: Breeding Species and their Distribution*. University of Nebraska Press, Lincoln, NE.

Jones, J.K., Jr, D.M. Armstrong, R.S. Hoffman, and C. Jones. 1983. *Mammals of the Northern Great Plains*. University of Nebraska Press, Lincoln, NE.

Knopf, F.L. and F.B. Sampson. 2010. *Ecology and Conservation of Great Plains Vertebrates*. Ecological Studies 125. Springer, New York, NY.

Weaver, J.E. 1954. *North American Prairie*. Johnsen Publications, Lincoln, NE.

Weaver, J.E. 1958. Summary and interpretation of underground development in natural grassland communities. Ecological Monographs 28: 55–78.

Weaver, J.E. and R.W. Darland. 1949. Soil-root relationships of certain native grasses in various soil types. Ecological Monographs 19: 303–338.

Seasonal grasses

Borchert, J.R. 1950. The climate of the central North American grassland. Annals of the Association of American Geographers 40: 1–39.

Carpenter, J.R. 1940. The grassland biome. Ecological Monographs 10: 617–684.

Costello, D.F. 1969. *The Prairie World*. Crowell, New York, NY.

Dort, W., Jr and J.K. Jones, Jr. (eds) 1970. *Pleistocene and Recent Environments of the Central Great Plains*. Special Publication 3, Department of Geology, University of Kansas Press, Lawrence, KA.

Weaver, J.E. 1968. *Prairie Plants and their Environment, a Fifty-Year Study in the Midwest*. University of Nebraska Press. Lincoln, NE.

Weaver, J.E. and T.J. Fitzpatrick. 1934. The prairie. Ecological Monographs 4: 109–295.

Weaver, J.E. and F.W. Albertson. 1956. *Grasslands of the Great Plains*. Johnsen Publishers, Lincoln, NE.

Soils

Box, T.W. 1961. Relationships between plants and soils of four major range plant communities in south Texas. Ecology 42: 794–810.

Boul, S.W., R.J. Southard, R.C. Graham, and P.A. McDaniel. 2011. *Soil Genesis and Classification*, sixth edition. John Wiley & Sons, New York, NY.

Brady, N.C. and R.R. Weil. 1999. *The Nature and Properties of Soils*. 14th edition. Prentice-Hall. Upper Saddle River, NJ.

Role of fire

Abrams, M.D. 1985. Fire history of oak gallery forest in a northeast Kansas tallgrass prairie. American Midland Naturalist 114: 118–191.

Anderson, K.L., E.F. Smith, and C.F. Owensby. 1970. Burning bluestem range. Journal of Range Management 23: 81–92.

Bragg, T.B. and L.C. Hulbert. 1976. Wood plant invasion of unburned Kansas bluestem prairie. Journal of Range Management 29: 19–24.

Brockway, D.G., R.G. Gatewood, and R.B. Paris. 2002. Restoring fire as an ecological process in shortgrass prairie ecosystems: initial effects of prescribed burning during the dormant and growing seasons. Journal of Environmental Management 65: 135–152.

Collins, S.L. and L.L. Wallace. 1990. *Fire in North American Tallgrass Prairies*. University of Oklahoma Press, Norman, OK.

Cooper, C.F. 1961. The ecology of fire. Scientific American 204: 150–160.

Coppock, D.L. and J.K. Detling. 1986. Alteration of bison and black-tailed prairie dog grazing interaction by prescribed burning. Journal of Wildlife Management 50: 452–455.

Daubenmire, R. 1968. Ecology of fire in the grasslands. Advanced Ecological Research 5: 209–66.

Higgins, K.F. 1984. Lightning fires in North Dakota grasslands and in pine-savanna lands of South Dakota and Montana. Journal of Range Management 37: 100–103. (Provides data for frequency of lightning-initiated fires.)

Hulbert, L.C. 1969. Fire and litter effects in undisturbed bluestem prairie in Kansas. Ecology 50: 874–877.

Kucera, C.L. 1960. Forest encroachment in native prairie. Iowa State Journal of Science 43: 635–640.

Nelson, J.G. and R.E. England. 1971. Some comments on the causes and effects of fire in the northern grasslands area of Canada and the nearby United States, ca 1750–1900. Canadian Geography 15: 295–306.

Parmenter, R.R. 2008. Long-term effects of a summer fire on desert grassland plants demographics in New Mexico. Rangeland Ecology and Management 61: 156–168.

Peet, M., R.C. Anderson, and M.S. Adams. 1975. Effect of fire on big bluestem production. American Midland Naturalist 94: 15–24.

Pyne, S.J. 1982. *Fire in America, a Cultural History of Wildland and Rural Fire*. Princeton University Press, Princeton, NJ.

Raby, S. Prairie fires in the northwest. Saskatchewan History 19: 81–99.

Scheintaub, M.R., J.D. Demer, E.F. Kelly, and A.K. Knapp. 2009. Response of the shortgrass steppe plant community to fire. Journal of Arid Environments 73: 1136–1143.

Scott, D.D., R.A. Fox, Jr, M.A. Conner, and D. Harmon. 1989. *Archaeological Perspectives on the Battle of Little Bighorn*. University of Oklahoma Press, Norman, OK. (A grass fire exposed artifacts long obscured by thick vegetation.)

Towne, G. and C. Owensby. 1984. Long-term effects of annual burning at different dates on ungrazed Kansas tallgrass prairie. Journal of Range Management 37: 392–397.

Wolfe, C.W. 1973. Effects of fire on a sandhills grassland environment. Proceedings of the Tall Timbers Fire Ecology Conference 12: 241–255.

Wright, H.A. and A.W. Bailey. 1982. *Fire Ecology: United States and Southern Canada*. John Wiley & Sons, New York, NY.

Prairie streams

Allan, J.D. and M.M. Castillo. 2007. *Stream Ecology: Structure and Function of Running Waters*, second edition. Springer, Dordrecht, The Netherlands. (Describes stream channel changes during flooding that result in isolated oxbows.)

Dodds, W.K., K. Gido, M.R. Wiles, K.M. Fritz, and W.J. Matthews. 2004. Life on the edge: the ecology of Great Plains prairie streams. BioScience 54: 205–216.

Matthews, W.J. 1988. North American prairie streams as systems for ecological study. Journal of the North American Benthological Society 7: 387–409.

Prairie wetlands and waterfowl

Bolen, E.G., G.A. Baldassarre, and F.S. Guthrie. 1989. Playa lakes. In: *Habitat Management for Migrating and Wintering Waterfowl in North America* (L.M. Smith, R.L. Pederson, and R.M. Kaminiski, eds). Texas Tech University Press, Lubbock, TX, pp. 341–356.

Bolen, E.G., L.M. Smith, and H.L. Schramm, Jr. 1989. Playa lakes: prairie wetlands on the Southern High Plains. BioScience 39: 615–623.

Duebbert, H.F. and A.M. Frank. 1984. Value of prairie wetlands to duck broods. Wildlife Society Bulletin 12: 27–24.

Haukos, D.A. and L.M. Smith 1994. The importance of playa wetlands to biodiversity of the Southern High Plains. Landscape and Urban Planning 28: 83–98.

Henny, C.J. 1973. Drought displaced movement of North American pintails into Siberia. Journal of Wildlife Management 37: 23–29.

Hestbeck, J.B. 1995. Response of northern pintail breeding populations to drought, 1961–1962. Journal of Wildlife Management 59: 9–15.

Lynch, J.J., C.D. Evans, and V.C. Conover. 1963. Inventory of waterfowl environments in prairie Canada. Transactions of the North American Wildlife and Natural Resources Conference 28: 93–109.

Pederson, R.L., D.G. Jorde, and S.G. Simpson. 1989. Northern Great Plains. In: *Habitat Management for Migrating and Wintering Waterfowl in North America* (L.M. Smith, R.L. Pederson, and R.M. Kaminski, eds). Texas Tech University Press, Lubbock, TX, pp. 281–310.

Rogers, J.P. 1964. Effect of drought on reproduction in the lesser scaup. Journal of Wildlife Management 28: 213–222.

Sargeant, A.B., S.H. Allen, and R.T. Eberhardt. 1984. Red fox predation on breeding ducks in midcontinent North America. Wildlife Monographs 89: 1–41.

Smith, A.G., J.H. Stout, and J.B. Gollop. 1964. Prairie potholes and marshes. In: *Waterfowl Tomorrow* (J.P. Linduska, ed.). US Department of Interior, Washington, DC, pp. 39–50.

Smith, R.I. 1970. Response of pintail breeding populations to drought. Journal of Wildlife Management 34: 943–946.

Stewart, R.E. and H.A. Kantrud. 1974. Breeding waterfowl populations in the prairie pothole region of North Dakota. Condor 76: 70–79.

Tiner, R.W., Jr. 1984. *Wetlands of the United States: Current Status and Recent Trends*. National Wetlands Inventory. US Fish and Wildlife Service, Washington, DC. (Review of prairie potholes and other wetlands of importance to waterfowl.)

van der Valk, A. (ed.) 1988. Northern prairie wetlands. Iowa State University Press, Ames, IA. (Primary reference for ecological features of prairie potholes.)

Pleistocene extinctions

Grayson, D.K. 1991. Late Pleistocene mammalian extinction in North America: taxonomy, chronology and explanations. Journal of World Prehistory 5: 193–231.

Martin, P.S. 1975. Vanishings, and future, of the prairie. Geoscience and Man 10: 39–49. (Discusses grassland evolution in conjunction with grazing mammals.)

Martin, P.S., and R.G. Klein. (eds) 1984. *Quaternary Extinctions, A Prehistoric Revolution*. University of Arizona Press, Tucson, AZ.

Selected prairie mammals

Bison

Barsness, L. 1985. *Heads, Hides, & Horns, the Complete Buffalo Book*. Texas Christian University Press, Fort Worth, TX.

Callenbach, E. 1996. *Bring Back the Buffalo! A Sustainable Future for America's Great Plains.* Island Press, Washington, DC.

Collins, S.L. and G.E. Uno. 1983. The effect of early spring burning on vegetation in buffalo wallows. Bulletin of the Torrey Botanical Club 110: 474–481.

England, R.E. and A. DeVos. 1969. Influence of animals on pristine conditions on the Canadian grasslands. Journal of Range Management 22: 87–94.

Flores, D. 1991. Bison ecology and bison diplomacy: the southern plains from 1800 to 1850. Journal of American History 78: 465–485.

Freese, C.H., K.E. Aune, D.P. Boyd, et al. 2007. Second chance for the Great Plains bison. Biological Conservation 136: 175–184.

Haley, J.E. 1936. *Charles Goodnight, Cowman and Plainsman.* Houghton Mifflin, New York, NY.

Hornaday, W.T. 1889. *The Extermination of the American Bison.* Annual Report of the Board of Regents of the Smithsonian Institution, Washington, DC.

Larson, F. 1940. The role of bison in maintaining the short grass plains. Ecology 21: 113–121.

Malouf, C. and S. Conner. (eds) 1962. *Symposium on Buffalo Jumps.* Montana Archaeological Society Memoir, Number 1. Helena, MT.

McDaniel, B. and E.U. Balspaugh, Jr. 1968. Bovine manure as an overwintering medium for Coleoptera in South Dakota. Annals of the Entomological Society of America 61: 765–768.

McHugh, T. 1958. Social behavior of the American buffalo (*Bison bison bison*). Zoologica 43: 1–40.

Peden, D.G. 1976. Botanical composition of bison diets on shortgrass plains. American Midland Naturalist 96: 225–229.

Pfeiffer, K.E. and D.C. Hartnett. 1995. Bison selectivity and grazing response of little bluestem in tallgrass prairie. Journal of Range Management 48: 26–31.

Polley, H.W. and S.L. Collins. 1984. Relationships of vegetation and environment in buffalo wallows. American Midland Naturalist 112: 178–186.

Polley, H.W. and L.L. Wallace. 1986. The relationships of plant species heterogeneity to soil variation in buffalo wallows. Southwestern Naturalist 31: 493–501.

Roe, F.G. 1970. *The North American Buffalo, a Critical Study of the Species in its Wild State.* Second edition. University of Toronto Press, Toronto, ONT.

Shaw, J.H. 1995. How many bison originally populated western rangelands? Rangelands 17: 148–150.

Smits, D.D. 1994. The frontier army and the destruction of the buffalo, 1865–1883. Western History Quarterly 25: 313–388. (Supports the position that General Sheridan adopted a formal policy for eliminating bison to subjugate the Native Americans of the plains.)

Vinton, M.A. and D.C. Harnett. 1992. Effects of bison grazing on *Andropogon gerardii* and *Panicum virgatum* in burned and unburned tallgrass prairie. Oecologia 90: 374–382.

Webb, W.P. 1931. *The Great Plains.* Grosset & Dunlap. New York, NY.

Prairie dogs

Augustine, D.J. and B.W. Baker. 2013. Associations of grassland bird communities with black-tailed prairie dogs in the North American Great Plains. Conservation Biology 27: 324–334.

Bailey, V. 1905. Biological survey of Texas. North American Fauna 25: 1–222. (Includes a population estimate for what was likely the largest single prairie-dog town.)

Gedeon, C.I., L.C. Drickamer, and A.J. Sanchez-Meador. 2012. Importance of burrow-entrance mounds of Gunnison's prairie dogs (*Cynomys gunnisoni*) for vigilance and mixing of soil. Southwestern Naturalist 57: 100–104.

Hansen, R.M. and I.K. Gold. 1977. Blacktail prairie dogs, desert cottontails, and cattle trophic relations on the shortgrass prairie. Journal of Range Management 30: 210–214.

Hoogland, J.L. 1995. *The Black-Tailed Prairie Dog: Social Life of a Burrowing Animal.* University of Chicago Press, Chicago, IL.

King, J.A. 1955. *Social Behavior, Social Organization, and Population Dynamics in a Black-Tailed Prairie Dog Town in the Black Hills of South Dakota.* Contribution 67, University of Michigan Laboratory of Vertebrate Biology, Ann Arbor, MI.

Koford, C.B. 1958. Prairie dogs, whitefaces, and blue grama. Wildlife Monographs 3: 1–78.

Kotliar, N.B., B.J. Miller, R.P. Reading, and T.W. Clark. 2005. The prairie dog as a keystone species. In: *Conservation of the Black-Tailed Prairie Dog: Saving North America's Western Grasslands* (J.L. Hoogland, ed.). Island Press, Washington, DC, pp. 53–64.

Johnsgard, P.A. 2005. *Prairie Dog Empire: A Saga of the Shortgrass Prairie.* University of Nebraska Press, Lincoln, NE.

Odell, E.A., F.M. Pusateri, and G.C. White. 2008. Estimation of occupied and unoccupied black-tailed prairie dog acreage in Colorado. Journal of Wildlife Management 72: 1311–1317.

Seton, E.T. 1928. *Lives of Game Animals.* Doubleday, Page and Company, Garden City, NY. Volume 4, Part 1. (Provides an estimate for the prairie dog population in North America.)

Smith, G.A. and M.V. Lomolino. 2004. Black-tailed prairie dogs and the structure of avian communities on the shortgrass plains. Oecologia 133: 592–602.

Snell, G.P. and B.D. Hlavachick. 1980. Control of prairie dogs – the easy way. Rangelands 2: 239–240.

Whicker, A.D. and J.K. Detling. 1988. Ecological consequences of prairie dog disturbances: prairie dogs alter grassland patch structure, nutrient cycling, and feeding site selection by other herbivores. BioScience 38: 778–785.

Witmer, G., M. Pipas, and T. Linder. 2006. Animal use of black-tailed prairie dog burrows: preliminary findings. Proceedings of the 22nd Vertebrate Pest Conference 22: 195–197.

Pronghorns

Buechner, H.K. 1950. Life history, ecology and range use of the pronghorn antelope in Trans-Pecos Texas. American Midland Naturalist 43: 257–354.

Gross, B.D., J.L. Holechek, D. Hallford, and R.D. Piper. 1983. Effectiveness of antelope pass structures in restriction of livestock. Journal of Range Management 36: 22–24.

Hailey, T.L., J.W. Thomas, and R.M. Robinson. 1966. Pronghorn die-off in Trans-Pecos Texas. Journal of Wildlife Management 30: 488–496.

Hoover, R.L., C.E. Till, and S. Ogilive. 1959. The antelope of Colorado: a research and management study. Colorado Department of Fish and Game, Technical Bulletin 4, Denver, CO.

Jones, J.K., Jr, D.M. Armstrong, R.S. Hoffman, and C. Jones. 1983. *Mammals of the Northern Great Plains.* University of Nebraska Press, Lincoln, NE.

Spillett, J.J., J.B. Low, and D. Sill. 1967. *Livestock Fences: How they Influence Pronghorn Antelope Movements*. Utah Agricultural Experiment Station, Bulletin 470, Logan UT.

Selected prairie birds

Burrowing owls

Butts, K.O. and J.C. Lewis. 1982. The importance of prairie dog towns to burrowing owls in Oklahoma. Proceedings of the Oklahoma Academy of Science 62: 46–52.

Collins, C.T. and R.E. Landry. 1977. Artificial burrows for burrowing owls. North American Bird Bander 2: 151–154.

Coulombe, H.N. 1970. Physiological and physical aspects of temperature regulation in the burrowing owl, *Speotyto cunicularia*. Comparative Biochemistry and Physiology 35: 304–335.

Earhart, C.M., and N.K. Johnson. 1970. Size dimorphism and food habits of North American owls. Condor 72: 251–264.

Klute, D.S., L.W. Ayers, M.T. Green, et al. 2003. Status assessment and conservation plan for the western burrowing owl in the United States. Fish and Wildlife Service Biological Technical Publication FWS/BTP-R6001-2003. US Department of Interior, Washington, DC.

Levey, D.J., R.S. Duncan, and C.F. Levins. 2004. Use of dung as a tool by burrowing owls. Nature 431(7004): 39.

Lutz, R.S. and D.L. Plimpton. 1999. Philopatry and nest site reuse by burrowing owls: implications for productivity. Journal of Raptor Research 33: 149–153.

Martin, D.J. 1973. Selected aspects of burrowing owl ecology and behavior. Condor 75: 446–456.

Martin, D.J. 1973. A spectrographic analysis of burrowing owl vocalizations. Auk 90: 564–578.

Meuller, H.C. 1986. The evolution of reversed sexual dimorphism in owls: an empirical analysis of possible selective factors. Wilson Bulletin 98: 387–406.

Rowe, M.P., R.G. Gross, and D.H. Owings. 1986. Rattlesnake rattles and burrowing owl hisses: a case of acoustic Batesian mimicry. Ethology 72: 53–71.

Wasemann, T. and M. Rowe. 1987. Factors influencing the distribution and abundance of burrowing owls in Cape Coral, Florida. In: *Integrating Man and Nature in the Metropolitan Environment* (L.W. Adams and D.L. Leedy, eds). National Institute of Urban Wildlife, Columbia, MD, pp. 129–137.

Prairie chickens

Ballad, W.B. and R.J. Robel. 1974. Reproductive importance of dominant male greater prairie chickens. Auk 91: 75–85.

Heth, C. (ed.) 1992. *Native American Dance: Ceremonies and Social Traditions*. Smithsonian Institution, Washington, DC. 196 pp. (Notes the incorporation of animal behavior and symbols into the ceremonial dancing of Native Americans.)

Johnsgard, P.A. 1973. *Grouse and Quails of North America*. University of Nebraska Press, Lincoln, NE.

Lehmann, V.W. 1941. Attwater's prairie chicker, its life history and management. North American Fauna 57: 1–65.

Schroeder, M.A. and C.E. Braun. 1992. Greater prairie-chicken attendance at leks and stability of leks in Colorado. Wilson Bulletin 104: 273–284.

Westemeier, R.L., J.D. Brawn, S.A. Simpson, et al. 1998. Tracking the long-term decline and recovery of an isolated population. Science 282(5394): 1695–1698.

Wiley, R.H. 1974. Evolution of social organization and life history patterns among grouse (Aves: Tetraornidae). Quarterly Review of Biology 49: 201–227. (Discussion of leks and other social behaviors.)

Highlights

Riparian forests

Abrams, M.D. 1986. Historical development of gallery forest in northeast Kansas. Vegetation 65: 29–37.

Choate, J.R. and K.M. Reed. 1986. Historical biogeography of the woodchuck in Kansas. Prairie Naturalist 18: 37–42. (Riparian forests acted as corridors for the westward expansion of an eastern species.)

Jones, J.K., Jr. 1964. *Distribution and Taxonomy of Mammals of Nebraska*. University of Kansas Publication 16, Number 1. Museum of Natural History, University of Kansas, Lawrence, KA. (Riparian forests acted as barriers for pronghorns in eastern Nebraska.)

The Platte River

Currier, P.J. 1984. Woody vegetation clearing on the Platte River: restoration of sandhill crane roosting habitat (Nebraska). Restoration Management 2: 30.

Faanes, C.A. 1982. Aspects of the nesting ecology of least terns and piping plovers in central Nebraska. Prairie Naturalist 15: 145–154.

Johnsgard, P.A. 1984. *The Platte: Channels in Time*. University of Nebraska Press, Lincoln, NE.

Knopf, F.L. 1992. Faunal mixing, faunal integrity, and the biopolitical template for diversity conservation. Transactions of the North American Wildlife and Natural Resources Conference 57: 330–342.

Knopf, F.L. and M.L. Scott. 1990. Altered flows and created landscapes in the Platte River headwaters, 1840–1990. In: *Management of Dynamic Ecosystems* (J.M. Sweeney, ed.). North Central Section, The Wildlife Society, West Lafayette, IN, pp. 47–70.

Krapu, G.L. (ed.) 1981. The Platte River ecology study. Special Research Report, US Fish and Wildlife Service, North Prairie Wildlife Research Center, Jamestown, ND.

Mattes, M.J. 1969. *The Great Platte River Road: The Covered Wagon Mainline via Fort Kearny to Fort Laramie*. University of Nebraska Press, Lincoln, NE.

McDonald, P.M. and J.G. Sidle. 1992. Habitat changes above and below water projects on the North Platte and South Platte rivers in Nebraska. Prairie Naturalist 24: 149–158.

Sidle, J.G., E.D. Miller, and P.J. Currier. 1989. Changing habitats in the Platte River valley of Nebraska. Prairie Naturalist 21: 91–104.

Ziewitz, J.W., J.G. Sidle, and J.J. Dinan. 1992. Habitat conservation for nesting least terns and piping plovers on the Platte River, Nebraska. Prairie Naturalist 24: 1–20.

Nebraska Sandhills

Hibbard, B.H. 1965. *A History of Public Land Policies*. University of Wisconsin Press, Madison, WI.

Johnsgard, P.A. 1995. *This Fragile Land, a Natural History of the Nebraska Sandhills*. University of Nebraska Press, Lincoln, NE.

Novacek, J.M. 1989. The water and wetland resources of the Nebraska Sandhills. In: *Northern Prairie Wetlands* (A. van der Valk, ed.). Iowa State University Press, Ames, IA, pp. 340–384.

Schacht, W.H., J.D. Volesky, D. Bauer, et al. 2000. Plant community patterns on upland prairie in the eastern Nebraska Sandhills. Prairie Naturalist 32: 43–58. (See for the influence of topography.)

Smith, H.T.U. 1965. Dune morphology and chronology in central and western Nebraska. Journal of Geology 73: 557–578.

Warren, A. 1976. Morphology and sediments of the Nebraska Sandhills in relation to Pleistocene winds and the development of aeolian bedforms. Journal of Geology 84: 685–700.

Ants

Box, T.W. 1960. Notes on the harvester ant, *Pogonmyrmex barbatus* var. *molefacieus*, in south Texas. Ecology 41: 381–382.

Cole, B.J. 1994. Nest architecture in the western harvester ant, *Pogonmyrmex occidentalis* (Cresson). Insectes Sociaux 41: 401–410.

Gordon, D.M. 1995. The development of an ant colony's foraging range. Animal Behavior 49: 649–659.

Holldobler, B. and E.O. Wilson. 1990. *The Ants*. Harvard University Press, Cambridge University Press, Cambridge, MA.

Rogers, L.E. 1974. Foraging activity of the western harvester ant in the shortgrass plains ecosystem. Environmental Entomology 3: 420–424.

Isolation and contact on the plains

Lanyon, W.E. 1956. Ecological aspects of the sympatric distribution of meadowlarks in the north-central states. Ecology 37: 98–108.

Lanyon, W.E. 1957. The comparative biology of the meadowlarks (*Sturnella*) in Wisconsin. Publications of the Nuttall Ornithological Club 1: 1–67.

Lanyon, W.E. 1966. Hybridization in meadowlarks. Bulletin of the American Museum of Natural History 134: 1–25.

Love, D. and F.L. Knopf. 1978. The utilization of tree plantings by Mississippi kites in Oklahoma and Kansas. In: *Trees, A Valuable Great Plains Multiple Use Source*. Proceedings of the 13th Annual Meeting of the Forestry Commission, Great Plains Agricultural Council Publication 87, Lincoln, NE, pp. 70–74.

Mengel, R.M. 1970. The North American central plains as an isolating agent in bird speciation. In: *Pleistocene and Recent Environments of the Central Great Plains* (W. Dort and J.K. Jones, Jr, eds). Special Publication 3, Department of Geology, University of Kansas, University of Kansas Press, Lawrence, KA, pp. 379–340.

Rising, J.D. 1983. The Great Plains hybrid zones. Current Ornithology 1: 131–157.

Short, L.L. 1965. Hybridization of the flickers (*Colaptes*) of North America. American Museum of Natural History Bulletin 129: 307–428.

Silbey, C.G. and L.L. Short. 1964. Hybridization in the orioles of the Great Plains. Condor 66: 130–150.

Szijj, L.J. 1966. Hybridization and the nature of the isolating mechanism in sympatric populations of meadowlarks (*Sturnella*) in Ontario. Zeitschrift für Tierpsychologie 23: 677–690.

Grassland settlement

Agrow, K.A. 1962. Our National Grasslands: dustland to grassland. American Forests 68: 10–12, 48, 50.

Allen, G.T. and P. Ramirez. 1990. A review of bird deaths on barbed wire fences. Wilson Bulletin 102: 553–558.

Bogue, A.G. 1963. *From Prairie to Corn Belt*. University of Chicago Press, Chicago, IL.

Broehl, W.G., Jr. 1984. *John Deere's Company, a History of Deere & Company and its Times*. Doubleday & Company, New York, NY.

Cade, T.J. 1967. Ecological and behavioral aspects of predation by the northern shrike. Living Bird 6: 43–86.

Hewes, L. 1950. Some features of early woodland and prairie settlement in a central Iowa county. Annals of the Association of American Geographers 40: 40–57.

Hudson, J.C. 1994. *Making the Corn Belt: A Geographical History of Middle-Western Agriculture*. University of Indiana Press, Bloomington, In.

Kendall, E.C. 1959. John Deere's steel plow. Bulletin of the American Museum of Natural History 218: 15–25.

McCallam, H.D. and F.T. McCallam. 1965. *The Wire that Fenced the West*. University of Oklahoma Press, Norman, OK.

McGuire, V.L. 2013. Water-level and storage changes in the High Plains aquifer, predevelopment to 2011 and 2009–11. US Geological Survey Scientific Investigations Report 2012-5291.

Ridgway, R. 1889. *The Ornithology of Illinois. Part I. Descriptive Catalog*. Natural History Survey, Illinois State Laboratory of Natural History.

Sloan, S.A. 1991. The shrike's display advertising. Natural History 6/91: 32–38.

Vestal, A.G. 1939. Why the Illinois settlers chose forest lands. Illinois State Academy of Science Transactions 32: 85–87. (Lists reasons why settlers initially avoided farming tallgrass prairies.)

Watkins, T.H. 1993. *The Great Depression: America in the 1930s*. Little, Brown. Boston, MA.

Watkins, T.H. and C.S. Watson, Jr. 1975. *The Lands No One Knows: America and the Public Domain*. Sierra Club Books, San Francisco, CA.

Weeks, J.B., E.D. Gutentag, F.J. Heimes, and R.R. Lucky. 1988. Summary of the High Plains regional aquifer-system analysis in parts of Colorado, Kansas, Nebraska, New Mexico, Oklahoma, South Dakota, Texas, and Wyoming. US Geological Survey Professional Paper 1400-A. 30 pp.

Worster, D. 1979. *Dust Bowl, the Southern Plains in the 1930s*. Oxford University Press, New York, NY.

Prairie preservation

Anonymous. 1994. Buffalo-ranching: back to the frontier. Economist 331(7861): 30–31.

Catlin, G. 1841. Letters and notes on the manners, customs, and conditions of the North American Indians. 2 volumes, reprinted in 1973 by Dover Publications, New York, NY.

Freese, C.H., K.E. Aune, D.P. Boyd, et al. 2007. Second chance for the Great Plains bison. Biological Conservation 136: 175–184.

Hamilton, R.G. 1996. Using fire and bison to restore a functional tallgrass prairie landscape. Transactions of the North American Wildlife and Natural Resources Conference 61: 208–215.

Marchello, M.J., W.D. Slanger, D.B. Milne, et al. 1989. Nutrient composition of raw and cooked *Bison bison*. Journal of Food Compositions and Analysis 2: 177–185.

Matthews, A. 1992. *Where the Buffalo Roam, the Storm over the Revolutionary Plan to Restore America's Great Plains*. Grove Weidenfeld, New York, NY. (A close view of the Poppers and their proposal for a Buffalo Commons.)

Nagel, H.G., R.A. Nicholson, and A.A. Steuter. 1994. Management effects on Willa Cather Prairie after 17 years. Prairie Naturalist 26: 241–250.

Packard, S. and C.F. Mutel. (eds) 1997. *The Tallgrass Restoration Handbook for Prairies, Savannas, and Woodlands* Island Press, Washington, DC.

Peden, D.G., G.M Van Dyne, R.W. Rice, and R.M. Hansen. 1974. The trophic ecology of *Bison bison* L. on shortgrass plains. Journal of Applied Ecology 11: 489–497. (Suggests bison subsist better than cattle on plains vegetation.)

Popper, D.E. and F.J. Popper. 1987. The Great Plains: from dust to dust. Planning 53(12): 12–18.

Reichman, O.J. 1991. *Konza Prairie: A Tallgrass Natural History*. University of Kansas Press, Lawrence, KS.

Samson, F.B. and F.L. Knopf. (eds) 1996. *Prairie Conservation: Preserving North America's Most Endangered Ecosystem*. Island Press, Washington, DC.

Shirley, S. 1994. *Restoring the Tallgrass Prairie an Illustrated Manual for Iowa and the Upper Midwest*. University of Iowa Press, IA.

Vassar, J.W., G.A. Hence, and C. Blakely. 1981. Prairie restoration in north-central Missouri. In: *The Prairie Peninsula – in the "Shadow" of Transeau*. Proceedings of the Sixth North American Prairie Conference, Ohio State Biological Survey Biological Notes 15, Ohio State University, Columbus, OH, pp. 197–199.

Infobox 5.1. Aldo Leopold (1887–1948): conservationist, forester, and philosopher

Flader, S.L. 1974. *Thinking Like a Mountain: Aldo Leopold and the Evolution of an Ecological Attitude toward Deer, Wolves, and Forests*. University of Nebraska Press, Lincoln, NE.

Leopold, A. 1933. *Game Management*. Chas. Scribner's Sons, New York, NY.

Leopold, A. 1949. *A Sand County Almanac*. Oxford University Press, New York, NY.

Leopold, L.B. 1953. *Round River, from the Journals of Aldo Leopold*. Oxford University Press, New York, NY. (Source for the quote about "tinkering.")

Meine, C. 1988. *Aldo Leopold, his Life and Work*. University of Wisconsin Press, Madison, WI.

Infobox 5.2. Poaceae, the grass family

Barkworth, M.E., L.K. Anderson, K.M. Capels, S. Long, and M.B. Piep. (eds) 2007. *Manual of Grasses for North America North of Mexico*. Utah State University Press, Logan, UT.

Piperno, D.R. and H.D. Sues. 2005. Dinosaurs dined on grass. Science 310(5751): 112.

Prasad, V., C.A. Stroemberg, H. Alimohammadian, and A. Sahni. 2005. Dinosaur coprolites and the early evolution of grasses and grazers. Science 310(5751): 1177–1180.

Stefferud, A. (ed.) *Grass, the Yearbook of Agriculture 1948*. US Department of Agriculture, Washington, DC.

Infobox 5.3. John E. Weaver (1884–1966), father of grassland ecology

Tobey, R.C. 1981. *Saving the Prairies: The Life Cycle of the Founding School of American Plant Ecology, 1895–1955*. University of California Press, Berkeley, CA.

Weaver, J.E. 1944. North American prairie. The American Scholar 13: 329–339.

Weaver, J.E. 1954. *North American Prairie*. Johnsen Publishing Company, Lincoln, NE. (Source of the quote, "Nature is an open book…".)

Weaver, J.E. 1963. The wonderful prairie sod. Journal of Range Management 16: 165–171. (Features rhizome measurements for several prairie grasses.)

Weaver, J.E. 1965. *Native Vegetation of Nebraska*. University of Nebraska Press, Lincoln, NE.

Weaver, J.E. 1968. *Prairie Plants and their Environment: A Fifty-Year Study in the Midwest*. University of Nebraska Press, Lincoln, NE.

Weaver, J.E. and F.E. Clements. 1938. *Plant Ecology*, second edition. McGraw-Hill Book Company, New York, NY.

Weaver, J.E. and F.W. Albertson. 1956. *Grasslands of the Great Plains, their Nature and Use*. Johnsen Publishing Company, Lincoln, NE.

CHAPTER 6
Regional Grasslands and Related Areas

What a thousand acres of compass plant looked like when they tickled the bellies of the buffalo is a question never again to be answered, and perhaps not even asked.

Aldo Leopold

A few distinctive grassland areas are either peripheral to, or disjunct from, the midcontinent prairies and plains (Fig. 6.1). These include regional communities that are highly altered by agriculture and encroaching urbanization and therefore poorly represented today by tracts of native biota, whereas large areas of others remain less disturbed by anthropogenic activities. Desert grasslands often are something of an anomaly because they occur in arid zones where the physical and ecological distinctions between desert and grassland are not always clear.

Regional associations

Palouse prairie

A grassland once covered much of eastern parts of Washington and Oregon and extended into western Idaho. The fertile soils in this region are exceptionally deep and support a thick carpet of bunchgrasses. Settlers, impressed with the plush vegetation, applied the name *Palouse* (French for lawn). Bluebunch wheatgrass and Idaho fescue represent the dominant grasses along with western wheatgrass, wild rye, and Sandberg bluegrass.

The Palouse Prairie developed after the Cascade Mountains formed in the Pliocene Epoch. The mountains interrupted the moisture-rich air masses flowing eastward from the Pacific Ocean. Thereafter, the forests east of the mountains gave way to grassland vegetation as the region dried.

Some ecologists recognize two zones within the Palouse Prairie. The first of these is represented in relatively dry sites, dominated by bluebunch wheatgrass and Sandberg bluegrass. Forbs are scarce, and the wheatgrasses develop in clumps to form the archetypical model of a bunchgrass prairie. The second zone is more mesic, and perennial forbs are more abundant. Idaho fescue replaces Sandberg bluegrass as co-dominant with bluebunch wheatgrass. In this zone, the wheatgrasses are more rhizomatous, producing sod and therefore lessening the appearance of a clumped grassland.

Rabbitbrush is a conspicuous but widely scattered shrub in the Palouse Prairie. Snowberry and rose are two other common shrubs. Several forbs, among them squaw weed and balsam root, fill in the spaces between the bunchgrasses. In temporary wetlands, the starchy bulbs of camas lilies once provided food for Native Americans. They established seasonal camps at these sites, known locally as "camas prairies," where continual digging for the bulbs may have been an important ecological influence.

The Palouse Prairie is discontinuous, and other types of vegetation occupy the intervening sites. Some mountain valleys in western Montana support stands of Palouse Prairie. An isolated community of bluebunch wheatgrass was also discovered in northern Utah, well removed climatically as well as geographically from the region more typically associated with Palouse Prairie.

Thick deposits of Pleistocene **loess** occur in much of the region occupied by Palouse Prairie. These soils are exceptionally deep, often extending to depths of more than 100 m (325 feet). In places, volcanic ash deposited from an eruption of Mount Mazama some 6500 years ago further enriches the loess in this region. These soils retain winter moisture (mostly from snowfall) well into the following summer, and plant growth therefore continues with little or no summer precipitation.

The climate associated with Palouse Prairie features wet winters and dry summers. The vegetation accordingly is green in winter and golden brown in summer. Fortunately, the winter rains are gentle (almost mist-like), thereby preventing serious soil erosion on farmed hillsides. Winter temperatures are cool but may become quite cold when masses of polar air spill westward over the Rocky Mountains. Summers are mild and, on a climatic gradient, Palouse Prairie falls between shrub steppe and coniferous forest.

Ecology of North America, Second Edition. Brian R. Chapman and Eric G. Bolen.
© 2015 John Wiley & Sons, Ltd. Published 2015 by John Wiley & Sons, Ltd.

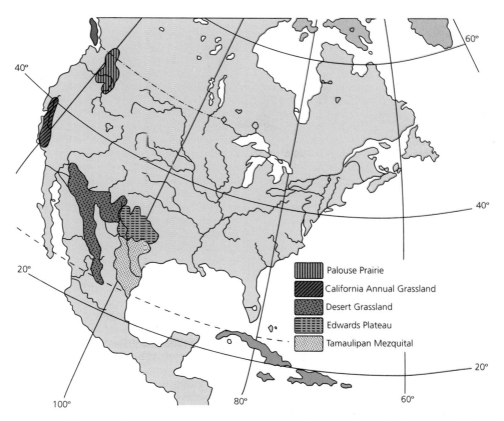

Figure 6.1 Approximate delineation of Palouse Prairie, Edwards Plateau, and other regional grassland areas in North America. Illustrated by Tamara R. Sayre.

Virtually all of the Palouse Prairie is now under cultivation. Wheat, typically alternated with peas, is the predominant crop. The reserve of winter moisture in the soil ensures production, and crop failures are less likely in this region in comparison to cultivated prairies elsewhere (e.g., wheat belt regions in western Kansas). Yet, because of the soil's great depth, wasteful erosion is often tolerated. Conservation measures such as contour plowing are seldom practiced, even where steep hillsides require caterpillar tractors for operating farm equipment.

Livestock also graze in the Palouse Prairie, particularly where stony soils prevent cultivation. However, apparently because native grazing animals exerted little influence on the evolution of the Palouse Prairie, many plants are relatively intolerant to grazing pressure. The great herds of bison, while once common on the plains east of the Bitterroot Mountains, were little in evidence on the grasslands west of the mountains. Early explorers commented on the scarcity of game on the Palouse Prairie and, after they acquired horses, Native Americans in this region made seasonal trips eastward across the northern Rockies in search of big game. Hence, at least until the arrival of Euro-Americans and their livestock, the Palouse Prairie seems to have remained largely free of grazing animals. The moisture pattern in this region, with its pronounced summer aridity, undoubtedly contributes to the considerable vulnerability of native grasses to grazing during the warm months of the year.

Since settlement, overgrazing also helped spread an introduced annual, cheatgrass, over this and other western rangelands. Cheatgrass is a winter annual, which makes it especially competitive with bluebunch wheatgrass on heavily grazed ranges. The seeds of cheatgrass germinate in autumn, so the seedlings have established well-developed root systems by the following spring. In contrast, the seedlings of the native wheatgrasses begin to grow in spring when their young roots cannot compete with cheatgrass for soil moisture. Kentucky bluegrass and rabbitbrush also increase under heavy grazing pressure. Today, large areas in this region are shrub steppe, which may represent the northward invasion of sagebrush from the Great Basin into sites once covered by Palouse Prairie.

California Annual Grasslands

The Central Valley of California supports grasslands dominated by annual grasses. The grassland communities extend into the foothills of the Sierra Nevada and Coast Range on either side of the Central Valley, where they mix with scattered trees (e.g., blue oak). A Mediterranean climate, that is, cool, moist winters and hot, dry summers, is typical of the region. Much of the annual vegetation has adapted to this regime by initiating growth when the rains begin in autumn, maturing in the following spring, and then dying when the soil

moisture is depleted. Similar grasslands also developed on the Channel Islands off the southern California coast (Infobox 6.1).

California condors, today one of the rarest birds in North America, once foraged across much of this region. After years of precarious existence, the last of the remaining condors was captured in 1987 to establish a captive breeding population and the eventual release of their offspring. About 240 birds currently live in the wild at five locations with the largest population centered around Los Padres National Forest in southern California, an area with cliffs where the condors perch and nest, as well as grasslands where the birds search for carrion. Unfortunately, condors at times ingested bullet fragments when feeding on the remains of deer and other hunter-killed wildlife and thereby suffered from lead poisoning. The problem ended in 2008 when the California State Assembly required hunters to use lead-free ammunition in areas occupied by condors.

Virtually all of California Annual Grasslands have been altered by human activities, and it is now difficult to determine the exact nature of the original vegetation. Introduced plants date to the Mission Period (1769–1824), and heavy grazing of the native grasses began with Spanish land grants. Interestingly, studies of plant materials in the adobe bricks from missions provide rather accurate dates for some of the earliest introductions. Large herds of Tule elk and pronghorns once inhabited these grasslands, but both species were extirpated during the Mission Period. However, recent reintroductions of Tule elk and pronghorn create a more natural grassland ecosystem in parts of the Central Valley.

A community of perennial bunchgrasses – principally purple needlegrass, nodding needlegrass, and two species of fescue – is representative of the original climax vegetation. Some ecologists refer to this community as "California Prairie." However, these perennials have given way to annual grasses, several of which came from the Mediterranean area of Europe where the climate is similar to the Central Valley. The seedheads of one invader, barb goatgrass, injure livestock, pets, and wildlife (e.g., deer), so animals may avoid walking through mature stands of this vegetation. Because their barbed seeds easily hook onto clothing and animal skins, the plants disperse widely and eventually comprise as much as 70% of the prairie vegetation. Wild oat, slender oat, common ryegrass, and soft chess – all annuals – are among the other dominant species in these grasslands. Most authorities consider the transformation from perennial to annual vegetation as a permanent change.

Two species of filaree and Napa thistle, also annuals, are among the many forbs in the community. Fully 94% of the cover in the San Joaquin arm of the Central Valley is annual vegetation, and 63% of the cover is derived from exotic species. The exotic vegetation effectively

Infobox 6.1 The island fox

The smallest of its kind in North America, the island fox inhabits six of the eight islands in the Channel Island archipelago off the southern California coast. After a colonization event that occurred between 10,000 and 16,000 years ago, the island fox descended from ancestral gray foxes that probably reached the northern islands in the archipelago on drifting logs, a dispersal process biogeographers call rafting. Thereafter the foxes spread to the southern islands about 6400 years ago, perhaps facilitated by the original Native American settlers, the Chumash.

Island foxes, although somewhat darker in overall coloration, appear similar to gray foxes in most aspects except size. They are about 25% smaller than their ancestors (or about the size of a large domestic housecat) and evolved into a distinct species through an evolutionary process resulting in **insular dwarfism**. This process acts over many generations favoring smaller individuals that can survive on small territories during periods when resources, especially the size and availability of prey, are reduced. Only those foxes that successfully endure these periods of hardship contribute their genes to subsequent generations.

Six islands (San Miguel, Santa Rosa, Santa Cruz, San Nicolas, Santa Catalina, and San Clemente) support populations of island fox, but these vary substantially in their size, topography, vegetative composition, and current and past human impacts. The climate, often characterized as "Mediterranean," is typified by hot, dry summers and mild, damp winters. The major vegetative communities on most islands include coastal scrub, chaparral, and grasslands, but small pine forests develop on islands with higher elevations.

An ocean barrier of at least 4.8 km (3 miles) separates each island from its nearest neighbor, which prevents genetic interchange among the populations. Over time, the foxes on each island developed their own unique genetic and morphological distinctions (e.g., the number of tail vertebrae differs from one population to another). Six subspecies of island fox, each native to a specific island, are now recognized.

Each of the Channel Islands has experienced a long history of environmental changes wrought by modern humans. On some, overgrazing by domestic livestock (sheep, burros, goats, pigs, and cattle) and agricultural activities contributed to erosion. On others, human habitation, roads, military activities, feral animals, and introductions of exotic plants and animals (e.g., bison, mule deer, Roosevelt elk, and blackbuck antelope) altered native plant communities.

Feral dogs and cats compete with island foxes for food and introduced diseases and parasites for which the foxes lacked immunity. Inevitably, the collective impact of these pressures reduced the numbers of island foxes, but the severest of all occurred in the 1990s when biologists discovered that golden eagles were preying on the diminutive foxes. The birds steadily reduced the island fox population from 2000 on Santa Cruz Island in 1994 to fewer than 135 in 2000. Similar reductions occurred on other islands. Previously, bald eagles dominated the islands, but they were decimated after decades of exposure to DDT; their aggressive behavior once seemed to keep the islands free of golden eagles. Moreover, a fish-eating predator was replaced by one that favors small mammals.

Four of the six island fox subspecies are federally listed as endangered; the other two subspecies are considered "threatened" by California Department of Fish and Game. The National Park Service, which protects all or parts of several Channel Islands, leads a broad-based research and management program for restoring island foxes, beginning with the removal of many domestic and feral animals. Efforts also include relocating golden eagles to mainland areas, reintroducing bald eagles, and supplemented fox populations by annual releases from those reared in captive breeding programs. Whereas the island fox still faces a somewhat uncertain future, a note of optimism now prevails for the smallest fox in North America.

resists re-establishment of the original bunchgrass community and, because of their permanent status, the annuals have been called "new natives." Nonetheless, because of fire, grazing, and swings in precipitation, the composition may change greatly in just 2–3 years. Fire, for example, can increase cover from forbs from 11% to 53% and reduce grasses from 89% to 47%.

Despite the overwhelming invasion and persistence of exotic vegetation in California Annual Grasslands, native plants still thrive in seasonal wetlands known as vernal pools (Fig. 6.2). These shallow depressions generally overlay clay soils; hence, vernal pools initially collect and hold winter rainfall and then slowly dry from evaporation and percolation during the spring and summer months. The pools have a depth range of 10–60 cm (4–24 inches), with the larger of these about 50 m (165 feet)

across. Vernal pools retain water long enough for growth and reproduction of their biota, but not long enough to develop into typical wetland systems (e.g., marshes).

The native biota associated with vernal pools includes 15 grasses and wildflowers as well as four crustaceans, a beetle, and a salamander, all currently listed as threatened or endangered species (e.g., hairy Orcutt grass, vernal pool tadpole shrimp, California tiger salamander, and delta green ground beetle). The pool habitats have existed for many thousands of years, as suggested by the number or endemic species. Curiously, about 10–15 of the 130 taxa associated with vernal pools display an unusual distributional pattern known as **amphitropical disjunction**; that is, the same or similar taxa occur in temperate or polar areas of both North and South America but are absent from the intervening tropical

Figure 6.2 Vernal pools are threatened habitats scattered across parts of the California Annual Grasslands, as shown here in San Diego County. The pools fill with runoff each spring but dry completely by midsummer. Many species, such as San Diego fairy shrimp and hairy Orcutt grass, are among the 21 endangered species adapted to the wet–dry regime associated with vernal pools. Photo courtesy of Richard W. Halsey of The California Chaparral Institute.

Figure 6.3 A fine example of desert grassland in southern Arizona. Bush muhly and other grasses cover the ground between scattered cacti and mesquite, but desert grasslands elsewhere more often are badly degraded, usually leading to the dominance of shrubs. Photo courtesy of Mitchel P. McClaran.

regions. Such a pattern can be explained only by: (a) the existence of a former, contiguous distribution that was later interrupted in the tropical zone; or (b) the long-distance dispersal of propagules that "hop-scotched" over the tropics (e.g., seeds carried by migratory birds to and from north temperate and south temperate latitudes).

More than 95% of the vernal pools in the Central Valley are already gone because of urban development, agricultural activities, and projects designed to manipulate surface water. Many of the remaining vernal pools

persist on terraces at the base of the Sierra Nevada foothills on the east side of the Central Valley.

Southwestern desert grasslands

Across the southwestern United States and northern Mexico are large but scattered areas of desert grasslands (Fig. 6.3). Most of these lie at elevations above the Sonoran and other warm deserts (Chapter 7) and below the Pinyon–Juniper Woodlands at the lower elevations of Montane Forests (Chapter 8), locations at 1100–1800

Infobox 6.2 Ladies-only lizards

While strolling through a southwestern desert grassland, a hiker might be startled by the rustle of dry grass and a fleeting glimpse of a speedy reptile. Often the source of the disturbance is a small, dark brown lizard adorned dorsally with six yellow- to cream-colored stripes extending to the base of its long and bluish-green tail. The slim lizard moves about in quick bursts, rooting under bushes and at the base of rocks for insects and other small prey. This is the desert grassland whiptail lizard, one of about 23 species of whiptail lizards that inhabit sparse grasslands of southwestern North America.

Biologists were perplexed for many years with why museum collections of whiptail lizards consisted entirely of females. Why were there no males? Eventually, the first clue leading to the answer came from a study of parthenogenic fishes.

Parthenogenesis is a form of asexual reproduction in which eggs produced by a female develop into embryos without fertilization, and later into fully-formed adult females. This naturally occurring process was long known in plants, but only more recently discovered in a number of invertebrates, fish, amphibians, reptiles, and a few birds. About one-third of the 10 genera and 230 species of whiptail lizards (family Teiidae) distributed widely in the Americas are parthenogenetic. Of these, nine female-only species occur in desert grasslands of the western United States and northern Mexico.

The parthenogenetic process begins when paired chromosomes duplicate within an ovum. The tetraploid egg cell then divides by normal meiosis, producing egg cells with the same chromosome number as the female parent. A **gravid** female lizard deposits a clutch of 1–4 unfertilized eggs in mid-summer. The young which hatch about two months later are clones, genetically identical to the mother.

Parthenogenic species of whiptail lizards generally arise from hybridization between two closely related, sexually reproducing species. For example, the New Mexico whiptail lizard is the product of successful matings between the little striped whiptail and the tiger whiptail. Because parthenogenetic species arise by hybridization, the genotypes of different lineages may vary considerably, but individuals reflect the genetic diversity within a mother's lineage.

Most all-female hybrids occur where the ranges of their parental stocks overlap in **ecotones** or ecologically disturbed habitats. The establishment and perpetuation of a l-female species depends on the existence of such habitats, but the habitats vary in composition among the parthenogenic species. Because of their relationship with an unsettled environment, parthenogenic whiptails are sometimes likened to "weeds."

The cytogenetic mechanism that initiates embryonic development in parthenogenic lizards remains a mystery. However, whiptails and several other all-female lizards exhibit a rather strange behavior in which one female behaves as a male and attempts to mate with another female. This behavior (pseudocopulation) has been observed in nature, but not commonly. Under laboratory situations, captive females being courted are almost always preovulatory and females making male-like advances are either oögenetically inactive or have already laid eggs. The courtship behavior stimulates hormone production and, in turn, ovulation in the courted female. Thus, courted females respond to the courtship behavior as though they had mated sexually, and the pseudocopulation enhances the rate at which the courted females produce and lay eggs.

The evolutionary significance of parthenogenesis remains unclear. Some biologists suggest that all-female populations of whiptail lizards may be ephemeral because all members of the species are essentially clones with limited genetic variability. Others wonder why unisexual reproduction in vertebrates is so rare. More than just startling rustles in dry grass, whiptail lizards represent a marvel of an evolutionary strategy in which males no longer have much to do – or even exist!

m (3600–5900 feet) above sea level with extremes as high as 2500 m (8200 feet) in parts of Mexico. Many of these communities blanket the broad basins lying between isolated mountain ranges, landforms geologists identify as the Basin and Range Province. Because of differences in elevation and exposure within these regions, desert grasslands often form a mosaic with desert and woodland vegetation (Infobox 6.2).

Desert grasslands in the southwestern United States generally receive between 230 and 460 mm (9–18 inches) of precipitation, but rainfall may be greater at some sites in Mexico. Aridity in these areas is intensified because of excessive solar radiation, which increases evaporation and therefore reduces the moisture available to the vegetation. Moreover, the ground heats rapidly during the day and then cools swiftly at night,

which produces strong winds and further depletes the availability of moisture. Desert grasslands accordingly acquire moisture deficits two or more times greater than precipitation and it is these shortages, rather than temperature extremes, that limit plant growth.

The soils of desert grasslands are quite unlike those of other grasslands in North America. Because of limited rainfall, few materials leach from the upper horizons and the resulting soil profile is not well differentiated. With limited plant production and organic matter, the soils are light-colored unlike the rich, dark coloration typical of topsoils developing in grasslands in more humid regions. Calcium carbonate accumulates near the surface, forming a hard layer known in the southwest as caliche.

Unlike tallgrass prairie and other grasslands that have disappeared under the plow, large expanses of desert grassland remain much in evidence today. Grazing, for the most part, is the primary anthropogenic influence affecting the ecological features of desert grasslands, albeit their condition and composition now vary greatly. In some locations, however, irrigated croplands have completely replaced these communities.

Grasses such as bush muhly, curly mesquite, and alkali sacaton are characteristic of southwestern desert grasslands. Several species of grama grasses are also common and, among these, black grama is especially representative. Black grama, however, does not tolerate heavy grazing, and other species quickly increase when grazing pressure exceeds the somewhat limited capacity of these rangelands to support livestock. Some annual species of grama grass (six-weeks grass, for example) are "increasers," as are various woody plants. Among the latter are mesquite and creosote bush, as well as succulent plants such as yucca and some species of cacti. Desert grasslands, once overgrazed, deteriorate into brush-infested ranges that are extremely difficult to revitalize.

Tobosa grass is another indicator species of southwestern desert grasslands. This grass is quite coarse, almost straw-like, when mature and lacks much forage value. However, when mature stands of tobosa grass are burned, livestock favor the succulent, protein-rich regrowth. Regular treatments with prescribed burning also effectively fight the encroachment of mesquite on tobosa rangelands.

A bunchgrass known as sacaton provides sleeping and resting covers for collared peccaries in the desert grasslands of southeastern Arizona. During late summer, peccaries also forage for succulent roots in the sacaton community. Periodic fires, including management with prescribed burning, help maintain desert grasslands, including sacaton communities. To benefit collared peccaries, however, sacaton should be burned in blocks, thereby producing a mosaic of stands in various stages of post-fire regrowth.

Masked bobwhites are closely associated with the desert grasslands in south-central Arizona and northwestern Mexico. They are a subspecies of the more familiar and widely distributed northern bobwhite, one of the most prized game birds in North America. The masked bobwhite males are distinctively marked with black head plumage and a red breast, but the females resemble female northern bobwhites. Masked bobwhites are susceptible to changes wrought by overgrazing and, where rangelands have been degraded, the birds declined as grasses disappeared and shrubs increased. The birds were extirpated from Arizona by 1900, and fears of their extinction were expressed in the early 1950s. Fortunately, a wild population was discovered in Mexico in the following decade, but the birds remain listed as an endangered species.

Efforts to restore masked bobwhite included the release of hand-reared broods under the foster-parent care of male northern bobwhites; the latter were vasectomized to prevent hybridization. Successful restoration of masked bobwhite populations, of course, requires the availability of suitable habitat for the released birds and their future offspring. A limited amount of ideal grassland for the species came under federal protection in 1985 with the establishment of Buenos Aires National Wildlife Refuge in Arizona. As many as 300–500 masked bobwhites now occupy the 47,700 ha (118,000 acre) refuge where management practices restore native grasses and prohibit grazing and hunting.

Edwards Plateau

A large region of highly dissected tableland in south-central Texas is popularly known as the "Hill Country." This is the Edwards Plateau, which lacks the usual physiognomy of a grassland but is geographically included in the Great Plains (Fig. 6.4). The southern and eastern borders of the plateau are sharply defined by the Balcones Escarpment, a geologic fault extending in a wide arc across central Texas from Del Rio to the Red River. Soils are shallow and of limestone origin, but a centrally located area in the Edwards Plateau features granite-based soils (*Granite outcrops and inselbergs*, Chapter 12).

No natural lakes occur in the Edwards Plateau, but many creeks and rivers contribute water to caves, springs, and the large Edwards Aquifer underlying the eastern portion of the plateau. Numerous caves riddling the limestone bedrock are used by some of the world's largest populations of cave-dwelling bats – primarily Mexican free-tailed bats – as maternity and summer roosts. The caves, springs, and aquifer also host a large number of endemic invertebrates and salamanders, many of which are federally protected as threatened or endangered species.

Before settlement, much of the Edwards Plateau was a fire-maintained, live-oak savanna. Both shortgrasses (e.g., Texas grama and muhly grass) and midgrasses (e.g., little bluestem and hairy grama) occur in the Edwards Plateau, and taller grasses are common on hillsides. A prominent overstory of oaks and junipers is present throughout the community. Thicker stands of juniper are known locally "cedar breaks."

A small population of golden-cheeked warblers winter in northern Mexico and Central America, but nesting occurs only in a limited area within the Edwards Plateau. This is one of the most restricted breeding ranges in North America, and the few remaining birds accordingly represent an endangered species. Golden-cheeked warblers are inexorably linked with Ashe juniper, one of several species forming the cedar breaks in the Hill Country (Fig.6.5). The birds construct their nests only from the bark of Ashe juniper, and only mature trees provide strips long enough for this purpose. The bark from other junipers – three other species occur in the Edwards Plateau – does not satisfy this crucial necessity.

Figure 6.4 The hilly terrain of the Edwards Plateau in west-central Texas lies along the southern edge of the Great Plains. The soils, typically thin and rocky, support a grassland with a scattered overstory of oaks, mesquite, and junipers. Photo courtesy of Brian R. Chapman.

Figure 6.5 The golden-cheeked warbler, a Neotropical migrant and an endangered species, nests only in the Edwards Plateau of Texas. Males have a prominent yellow face with a black stripe running through the eye. For their nesting materials, golden-cheeked warblers depend solely on the shredding bark of mature Ashe juniper. Habitat fragmentation threatens the species, as does cutting mature Ashe juniper for fence posts. Photo courtesy of Steve Maslowski and the US Fish and Wildlife Service.

Unfortunately, cedar breaks often disappear in the course of range improvements and housing developments; urban sprawl around San Antonio and Austin steadily constricts the remaining habitat. Ashe juniper is also harvested for fence posts. All told, habitat destruction is foremost among the reasons for the imperiled status of golden-cheeked warblers. Fortunately, scout camps, military bases, and other uncleared lands act as refuges, which supplement a few sites specifically set aside to protect breeding habitat for these birds.

A basin of Precambrian strata lies within the Edwards Plateau. The granite-gravel soils in much of this basin, the Central Mineral Region, contrast with the limestone soils of the surrounding area. Curiously, a condition popularly known as "velvet horn" occurs in the deer herd living on these soils. Instead of hardened antlers, as many as 9% of the bucks of white-tailed deer in the Central Mineral Region retain velvet-covered antlers during the fall and winter months. The animals also have small, dysfunctional testicles, and their social behavior is subordinate to other deer of either sex. The exact cause of velvet horn in unknown, but the association with granite-gravel soils seems clear. Nonetheless, the disorder (hypogonadism) does not impair the overall reproductive performance of the deer herd in the Central Mineral Region (e.g., fawn production is the same as in other habitat types).

Unfortunately, only about 2% of the remaining habitat in the Edwards Plateau remains intact. Overgrazing and extensive soil loss underlie the concurrent loss of native grasslands and expansion of woody vegetation.

Tamaulipan Mezquital

A savannah-like region of live oaks, interspersed among clumps of honey mesquite, and an extensive groundcover of curly mesquite grass once characterized about 200,000 km² (77,220 square miles) of northeastern Tamaulipas and parts of southern Texas. This vegetation (native to a region known as the Tamaulipan Mezquital) uniquely blends plants with western desert, northern, coastal, and tropical affinities. During the last 100 years, however, heavy grazing and conversion to urban and agricultural uses greatly altered the original vegetation. What remains is largely a mesquite-dominated "mezquital," a thornscrub-grassland where plant communities are sorted by soil type and the impacts of livestock grazing (Fig. 6.6). Several species of *Acacia* form a large part of the thornscrub community that favors drier sites, whereas honey mesquite, prickly pear, and a

Figure 6.6 Overgrazing encourages mesquite and prickly pear while replacing the understory of native grasses with coarse vegetation of little value as forage for livestock. White prickly poppies, protected by toxic sap and sharp spines, represent the majority of understory vegetation at this overgrazed site in the Tamaulipan Mezquital. Photo courtesy of Brian R. Chapman.

drought-resistant understory form dense stands on heavy clay soils. Fires may have been frequent in the region, contributing to the development of thornscrub communities.

From the Gulf Coastal Plains of Texas and Tamaulipas, the altitude of the Tamaulipan Mezquital rises to about 620 m (2000 feet) along the eastern flank of the Sierra Madre Oriental. Flatlands and gently undulating slopes interspersed with hills up to 50 m higher than the surrounding terrain are typical of the region. Average rainfall varies from 38 to 76 cm (15–30 inches) per year, but the irregular distribution of rainfall often leaves little moisture for plant growth. In some months, for example, a single thunderstorm provides all of the moisture, much of which is absorbed by a layer of soft, gravel-like limestone ("caliche") lying just beneath the rather shallow soils. Streams are therefore uncommon and their beds are typically dry gullies ("arroyos") bordered by strips of dense brush known as "ramanderos." Woody plants in the ramanderos (e.g., granjeno, retama, huisache, brazil, and mesquite) withstand periodic flooding; their dense cover provides important travel corridors for wildlife.

The Rio Grande, which bisects the Tamaulipan Mezquital, was once lined by riparian thickets with trees as high as 21 m (66 feet). From the Gulf of Mexico inland for about 130 km (80 miles), a wide palm forest adjoined the river (*Palm forest*, Chapter 12). Farther upstream, palms were replaced with bottomland hardwood forests composed of cedar elm, Berlandier ash, sugar hackberry, Texas ebony, honey mesquite, and granjeno. The early Spanish called these riparian woodlands "bosques," a term still used in places along the Rio Grande. Landscapes on either side of the lower Rio Grande feature resacas, former river beds that remain dry unless inundated by heavy rainfall from a tropical storm, hurricane, or an especially wet winter. In some deeper resacas, rainwater may persist in oxbows through the dry summer season and, although semi-permanent, provide critical water supplies for wildlife. An impenetrable undergrowth of thorny shrubs and vines often surrounds these sites.

The morning air near water-filled resacas resounds with the challenge calls of male plain chachalacas, deep-toned songs of white-winged doves, and the soft cooing of mourning doves. Chachalacas (sometimes called "Mexican pheasants") are related to the guans and currasows of Central and South America but, unlike typical ground-nesting gallinaceous birds, they nest as high as 10 m (32 feet) above ground in vines or the crotches of tree branches (Fig. 6.7). When moving through the trees, chachalacas hop from branch to branch, usually without using their wings. When disturbed, they drop silently to the ground and disappear in the tangled underbrush.

Except for their namesake markings, white-winged doves resemble the more familiar and slightly smaller mourning dove. Unlike mourning doves, however, white-winged doves are colonial nesters that also congregate while roosting and feeding. Historically, white-wings nested in the riparian forests and thornscrub bordering the lower Rio Grande, and an estimated population of about 12 million white-winged doves once nested in the south Texas portion of the Tamaulipan Mezquital. Extensive brush removal, largely for agricultural development, eliminated about 200,000 ha (493,827 acres) of nesting habitat in Texas

Figure 6.7 Plain chachalacas are common but seldom-seen birds of dense thickets in the Tamaulipan Mezquital. When foraging or disturbed, chachalacas hop quietly from branch to branch in tangled vegetation. The name of these large, chicken-like birds is derived from their loud calls that punctuate the dawn hours. Photo courtesy of Brian R. Chapman.

and, by 1939, the whitewing population fell to about 500,000. Similar development in Tamaulipas likewise removed about 1,067,000 ha (2,635,000 acres) of native scrubland. Fortunately, the birds adapted to some of the ecological changes and expanded their breeding range well into central Texas. Breeding colonies of white-winged doves are now well established in citrus groves in the Lower Rio Grande Valley of Texas and in urban areas with mature trees as far north as San Antonio.

Ocelot and jaguarundi still roam the dense thorn forests and brushlands of the Tamaulipan Mezquital. Both are shy, secretive cats that hunt rabbits, small rodents, and birds. Ocelots prowl the underbrush at night and spend the day resting in nearly impenetrable thickets. Jaguarundis also hunt at night, but sometimes climb trees from which they spring to catch prey. Both species

travel in densely vegetated corridors along resacas and arroyos, but clearing prevents such travel and thus negatively impact these cats, both of which are federally listed as endangered species. Fewer than 40 ocelots still occur in Texas, but jaguarundis have not been seen since 1969. The current status of their populations in Mexico is unknown, but these are presumably more numerous than in Texas.

Santa Ana National Wildlife Refuge (NWR) in Hidalgo County, Texas, preserves one of the largest remaining tracts of subtropical riparian forest and native brushland in the Tamaulipan Mezquital. Although surrounded by extensive areas of cleared farmland, Santa Ana NWR secures habitat for more threatened and endangered species than any other refuge in the United States; overall, the refuge protects more than 400 species of birds, 30 species of mammals, 50 species of reptiles and amphibians, 260 species of butterflies, and 450 species of plants. Many species, such as the Altimira oriole, gray hawk, ringed kingfisher, blue spiny lizard, southern yellow bat, and Mexican burrowing toad reach their northern limits on or near the refuge.

Highlights

Rodents and vegetation

An abundance of small mammals occupy the desert grasslands and associated communities (e.g., oak woodlands) in the arid regions of southwestern North America. Among these are banner-tailed kangaroo rats, which exert considerable ecological influence on their immediate habitat. On black grama rangelands in New Mexico, for example, the denuded patches around their dens represented almost 11% of the total area. Banner-tailed kangaroo rats also store large quantities of seeds and other plant materials in their dens (e.g., about 10 kg ha^{-1} or 9.2 lb/acre). In desert and other grassland types (e.g., California Annual Grasslands), **granivorous** rodents may consume up to 95% of the annual seed production of some plants.

Because of their influence on the communities in which they dwell, some rodents may represent keystone species. Such a relationship was illustrated on a grassland-shrub ecotone in the Chihuahuan Desert. In this case, three species of kangaroo rats produced the keystone effect by forming a **guild**, a group of organisms with similar habits living in the same community (e.g., the member species in a guild have similar diets or breed in specialized sites such as tree cavities). The guild of kangaroo rats exerted two ecological influences: they consumed large seeds and they disturbed the soil with their burrows and runways. Dramatic changes developed after the guild was removed from plots of shrubland for 12 years: a grassland replaced the shrub-dominated vegetation; the density of grasses increased several-fold; and the dominant large-seeded species, no longer consumed by kangaroo rats, also gained a competitive advantage over the plants producing smaller seeds.

Significant changes also occurred in the animal community. The dense cover of grasses and forbs reduced the area of bare ground, which limited the foraging habitat for seed-eating birds. Six species of grassland rodents colonized the plots in the absence of the kangaroo rats. The presence of these species, however, did not compensate for the absence of kangaroo rats; that is, after the six colonizing species arrived, the grassland vegetation did not revert to shrub and thus demonstrated that they did not play the same keystone role shown by the guild of kangaroo rats. These results were likely enhanced because the plots were located in a grassland-shrub ecotone, which may be more sensitive than "pure" desert communities to small shifts in abiotic and biotic conditions. Nonetheless, the presence or absence of kangaroo rats profoundly influenced the structure and composition of this site.

The hoarding behavior of woodrats is of special interest to climatologists, plant geographers, and paleoecologists. Popularly known as "packrats" because of their collecting habits, woodrats amass piles of debris atop their living chambers. The materials collected by packrats include discarded food, non-edible plant trash, the bone-laden droppings of predators, and the dung of large herbivores. The fossil materials in these ancient middens of woodrat debris – some are more than 40,000 years old – yield troves of ecological information, as shown by those of white-throated woodrats for large areas of Texas, New Mexico, Arizona, and northern Mexico.

Woodrat middens are found in fissures, in rocky outcrops, or in thickets of cactus or brush. Aridity and shelter help to preserve middens of course, but their contents may be especially well protected because urine from generations of woodrats eventually crystallizes, forming an amber-like casing around fragments of plant and animal materials. Middens preserved in this way resemble glistening masses of pitch. Some doughty but naïve "forty-niners" *en route* to the California goldfields thought they had discovered a natural source of candy, which they found "sweetish but sickish" and were "troubled with nausea afterward." Just over a century later, and coincident with the development of radiocarbon dating, the scientific value of woodrat middens was realized – reconstructions of ancient communities and windows into the past.

Careful analysis of woodrat middens reveals clear evidence of woodland communities as sites currently occupied by vegetation adapted to more arid conditions. Glaciers never reached southwestern North America, yet they created relatively cool, moist climatic regimes far removed from the ice itself. In turn, these conditions produced Ice Age communities of pinyon pine, juniper, and shrub oak, vegetation only growing today at higher elevations.

Later, as the influence of the last glacial advance diminished, the woodland vegetation gradually shifted, giving way to desert grassland. The fossil evidence collected from packrat middens further suggests that these communities remained more or less similar to modern grasslands only for the last 4000 years. Desert grasslands are therefore relatively recent communities and, for most of the Pleistocene, woodland communities dominated the vegetation of southwestern North America.

Channeled Scablands

A large area on the Columbian Plateau in eastern Washington is distinctively etched by surface erosion resulting from ancient floods. Indeed, the Channeled Scablands form some of the most distinctive erosional features on our planet. They were formed by floodwaters scouring a bed of volcanic rock at or near the surface of a region encircled by the Spokane, Columbia, and Snake rivers. Eastward lie the hills of the Palouse Prairie.

A shrub steppe of sagebrush and bunchgrasses characterizes most of the natural vegetation in this region. Idaho fescue and bluebunch wheatgrass – the same species typical of the Palouse Prairie – dominate the grassland aspect, but these form separate vegetation units within the shrub steppe (e.g., big sagebrush-bluebunch wheatgrass). Overall, the shrub steppe in this region generally resembles the cold-desert regime found in much of the Great Basin (Chapter 7). However, in more mesic sites in the Channeled Scablands, the shrub steppe community gives way to forest steppe in which ponderosa pine replaces sagebrush. Riparian and wetland vegetation develops in association with the channels, where potholes, lakes, and intermittent streams occur.

Catastrophic forces produced the unique geological features of the Channeled Scablands. At first, geologists believed a single, giant flood produced the network of channels, but later work suggested that several huge floods scoured the region. The last of the floods, which occurred during and following the Pleistocene, was apparently the largest and destroyed most of the evidence of previous flooding. These events, known as the Spokane Floods, resulted from the release of immense volumes of glacial meltwater trapped behind an ice dam. The impounded water, known as Lake Missoula, contained 2084 km³ (500 cubic miles) of water, which drained in just a few days when the ice dam broke. At its peak, the estimated flow was ten times greater than the current flow of all of Earth's rivers combined. Such a force lies beyond imagination but the numerous channels, including Grand Coulee, bear clear witness to the tremendous energy once unleashed across this landscape.

Snake River Birds of Prey Conservation Area

Conservation reached a measure of maturity when key habitat along the Snake River in southern Idaho became a federal sanctuary for hawks and other raptors. The site,

Figure 6.8 Although the origin of the "Mima mounds" remains unknown many hypotheses, some quite fanciful, have been advanced. These regularly spaced mounds near Puget Sound in Thurston County, Washington, cover a vast area. Photos courtesy of Roberta Davenport and the Washington State Department of Natural Resources.

covering an area of nearly 196,000 ha (485,000 acres), lies within the Columbian Plateau, a shrub-steppe of bunchgrasses and sagebrush resembling the characteristic vegetation of the Channeled Scablands. More than 700 pairs of 15 species of raptors nest at the site, which includes lofty cliffs bordering the river. Another nine species regularly visit as migrants or winter residents. In addition to the nesting areas on the cliffs, the large area of shrub-steppe provides habitat for rabbits and numerous rodents necessary as a prey base for the raptor population. All told, the Snake River Birds of Prey National Conservation Area protects an ecosystem supporting the densest and most diverse population of eagles, hawks, falcons, and owls ever recorded.

Mima mounds

Viewed from above, the Mima Prairie in Thurston County, Washington, looks like the vastly enlarged surface of a basketball with its dots of circular, domelike mounds typically about 2 m (6.5 feet) high and 10–20 m (33–66 feet) in diameter. Known as "Mima mounds," these domes consist of sandy or loamy soils, often mixed with gravel or small stones, but not heavy clays. Charcoal residues in the soil were added when aboriginal inhabitants of North America burned the grasslands to promote fresh growths of edible vegetation. In northwestern North America, Mima mounds sit atop a substrate with a shallow basement layer composed of bedrock, densely bedded gravel, or dense clay. In these areas, the basement layer impedes drainage, and the soil beneath the mounds remains waterlogged for prolonged periods.

The largest contiguous expanse of Mima mounds occurs in Thurston County where the 258 ha (637 acre)

Mima Mounds Natural Area Preserve protects mounds at one of the last native prairies remaining in the Puget Sound region (Fig. 6.8). Mima mounds occur in three other major locations: from south-central Oregon to northwestern Baja California; from central Wyoming south to north-central New Mexico; and from southern Missouri to parts of western Louisiana and eastern Texas. Small patches of Mima mounds also exist in Iowa, eastern North Dakota, and northwestern Minnesota.

The origin of Mima mounds remains a long-standing mystery despite their widespread occurrence in North America west of the Mississippi. More than a dozen hypotheses have attempted to explain their formation, but none has met with full acceptance. An early concept proposed their construction by Native Americans, but lacks supporting archaeological evidence. Another suggests that pocket gophers piled the soil in mounds to keep their tunnels above the water table. Other oft-cited hypotheses concern geological phenomena, including earthquakes shaking the ground, shrinking and swelling of clay basement layers, or wind-blown sediments accumulating around clumps of vegetation. In a lighter vein, folklorists suggested the mounds represent spots where Paul Bunyan deposited shovelfuls of soil while digging out Puget Sound. In part, the lack of a suitable explanation arises from the occurrence of Mima-like mounds in different ecosystems elsewhere (e.g., Kenya, Australia, and China).

Mima mounds often lie adjacent to vernal pools, especially in the vicinity of San Diego, California. An impermeable basement layer beneath Mima mounds in this region impedes drainage, allowing water to collect in the depressions between mounds. The faunal and floral diversity supported by the mounds and vernal pools add even more to the uniqueness of the California Annual Grasslands, described earlier.

Desertification

Many grasslands in North America were converted long ago into croplands. Others, while still somewhat intact, are highly vulnerable to degradation known as desertification, the spread or intensification of desert conditions, leading to: (a) lessened biological productivity; (b) increased soil deterioration; and (c) systems no longer sustaining human livelihood. Desertification may trigger images of sand dunes creeping over richer soils. While this sometimes happens, desertification more often occurs when drought strikes previously abused arid and semiarid lands (e.g., overgrazed grasslands). Such lands thereafter degrade permanently into deserts. High rates of soil erosion, reduced water resources, salinization, and the elimination of native vegetation are the paramount symptoms of desertification.

On a global scale, much of the area bordering the Mediterranean Sea perhaps serves as the best illustration of desertification. Centuries of overgrazing, as well as thoughtless deforestation, ruined the once-fertile soils in this region, which is now mostly a landscape of rocky soils, scrubby vegetation, and considerable poverty. *Mediterraneanization* accordingly joined the lexicon of terms that describe the ways natural resources might be ruined. We should remind ourselves of this disaster whenever the remaining grasslands in North America are included among those sites too quickly judged as "expendable." The Dust Bowl, previously described in Chapter 5, offers harsh testimony concerning the past and, hopefully, future management of the Great Plains and the regional grasslands in North America. Unfortunately, desertification remains pervasive in the southwestern United States and northern Mexico, particularly in areas where the human condition is already far too impoverished.

Readings and references

Regional associations

Palouse prairie

Daubenmire, R. 1942. An ecological study of the vegetation of southeastern Washington and adjacent Idaho. Ecological Monographs 12: 53–79.

Duffin, A.P. 2005. Vanishing earth: soil erosion in the Palouse, 1930–1945. Agricultural History 79: 173–192.

Hanson, T., Y. Sanchez de Leon, J. Johnson-Maynard, and S. Brunsfeld. 2008. The influence of soil and site characteristics on Palouse Prairie plant communities. Western North American Naturalist 68: 231–240.

Harris, G.A. 1967. Some competitive relationships between *Agropyron spicatum* and *Bromus tectorum*. Ecological Monographs 37: 89–111.

Johnsgard, P.A. and W.H. Rickard. 1957. The relation of spring bird distribution to a vegetation mosaic in southeastern Washington. Ecology 38: 171–174.

Mack, R.N. 1981. The invasion of *Bromus tectorum* L. into western North America: an ecological chronicle. Agro-ecosystems 7: 145–165.

Rickard, W.H. 1960. The distribution of small mammals in relation to the climax vegetation mosaic in eastern Washington and northern Idaho. Ecology 41: 99–106.

Stoddart, L.A. 1941. The Palouse grassland association in northern Utah. Ecology 22: 158–163.

Tisdale, E.W. 1961. Ecologic changes in the Palouse. Northwest Science 35: 134–138.

Weddell, B. and J. Lichthardt. 1998. Identification of conservation priorities for and threats to Palouse grassland and Canyon grassland remnants in Idaho, Washington, and Oregon. Idaho Bureau of Land Management Technical Bulletin 98(13): 1–57.

California Annual Grasslands

Alexander, D.G. and R. Syrdahl. 1992. Invertebrate biodiversity in vernal pools. Northwest Environmental Journal 8: 161–163.

Bacon, C.R. 1983. Eruptive history of Mount Mazama and Crater Lake Caldera, Cascade Range, USA. Journal of Volcanology and Geothermal Research 18: 57–115.

Barry, S. 1995. Vernal pools on California's annual grasslands. Rangelands 17: 173–175.

Bauder, E.T. 2000. Inundation effects on small-scale plant distributions in San Diego, California vernal pools. Aquatic Ecology 34: 43–61.

Bauder, E.T. 2005. The effects of unpredictable precipitation regime in vernal pool hydrology. Freshwater Biology 50: 2129–2135.

Bauder, E.T. and S. McMillan. 1998. Current distribution and historical extent of vernal pools in southern California and northern Baja California, Mexico. In: *Ecology, Conservation, and Management of Vernal Pool Ecosystems* (C.W. Witham, E.T. Bauder, D. Belk, and W.R. Ferren, Jr, eds). California Native Plant Society, Sacramento, CA pp. 56–70.

Bauder, E.T., D.A. Kreager, and S.C. McMillan. 1998. *Vernal Pools of Southern California Draft Recovery Plan*. US Fish and Wildlife Service, Region 1, Portland OR.

Biswell, H.H. 1956. Ecology of California's grasslands. Journal of Range Management 9: 19–24.

Burcham, L.T. 1961. Cattle and range forage in California: 1770–1880. Agricultural History 35: 140–149.

Heady, H.F. 1958. Vegetational changes in the California annual type. Ecology 39: 402–416.

Hendry, G.W. and M.K. Bellue. 1936. An approach to southwestern agricultural history through adobe brick analysis. In: *Symposium on Prehistoric Agriculture*. University of New Mexico Bulletin 296, Anthropology Series 1, No. 5. Albuquerque, NM, pp. 65–72. (Records the plants, fungi, insects in adobe bricks to establish periods when exotic species entered the local biota.)

Hervey, D.F. 1949. Reaction of a California annual-plant community to fire. Journal of Range Management 2: 116–121.

Holland, R. and S. Jain. 1977. Vernal pools. In: *Terrestrial Vegetation of California* (M.G. Barbour and J. Major, eds). John Wiley & Sons, New York, NY, pp. 515–533.

Kelly, T.R., P.H. Bloom, S.G. Torres, et al. 2011. Impact of the California lead ammunition ban on reducing lead exposures in golden eagles and turkey vultures. PLoS One 6(4): e17656.

Koford, C.B. 1953. *The California Condor.* Research Report 4, National Audubon Society, New York, NY.

Menke, J.W. 1992. Grazing and fire management for native perennial grass restoration in California grasslands. Fremontia 20: 22–25.

Nielsen, J. 2006. *Condor: To the Brink and Back – The Life and Times of One Giant Bird.* Harper Perennial, New York, NY.

Peters, A., D.E. Johnson, and M.R. George. 1996. Barb goatgrass: a threat to California's rangelands. Rangelands 18: 8–10.

Snyder, N. and H. Snyder. 2000. *The California Condor: A Saga of Natural History and Conservation.* Princeton University Press, Princeton, NJ.

Solomeshch, A.L., M.G. Barbour, and R.F. Holland. 2007. Vernal pools. In: *Terrestrial Vegetation of California,* third edition. (M.G. Barbour, T. Keeler-Wolf, and A.A. Schoenherr, eds). University of California Press, Berkeley, CA, pp. 394–424.

Stromberg, M.R., J.D. Corbin and C.M. D'Antonio. (eds) 2007. *California Grasslands: Ecology and Management.* University of California Press, Berkeley, CA.

Thacker, P.D., N. Lubick, R. Renner, et al. 2006. Condors are shot full of lead. Environmental Science & Technology 40: 6143–6150.

US Fish and Wildlife Service. 2005. Recovery plan for vernal pool ecosystems of California and southern Oregon. Region 1, US Fish and Wildlife service, Portland, OR. (Includes details for 20 threatened and endangered species.)

White, K.L. 1967. Native bunchgrass (*Stipa pulchra*) on Hastings Reservation, California. Ecology 48: 949–955.

Zedler, P.H. 1987. The ecology of southern California vernal pools: a community profile. US Fish and Wildlife Service, Biological Report 85(7.11).

Southwestern desert grasslands

Banks, R.C. 1975. Plumage variation in the masked bobwhite. Condor 77: 486–487.

Bock, C.E. and J.H. Bock. 1979. Relationship of the collard peccary to sacaton grassland. Journal of Wildlife Management 43: 813–816.

Brown, A.L. 1950. Shrub invasion of southern Arizona desert grassland. Journal of Range Management 3: 172–177.

Buffington, L.C. and C.H. Herbel. 1965. Vegetational changes on a semidesert grassland range from 1858 to 1963. Ecological Monographs 35: 139–164.

Cable, D.R. 1967. Fire effects on semidesert grasses and shrubs. Journal of Range Management 20: 170–176.

Cox, J.R., H.L. Morton, T.N. Johnson, Jr, et al. 1982. Vegetation restoration in the Chihuahuan and Sonoran deserts of North America. Agricultural Reviews and Manuals ARM-W-28, Agricultural Research Service, US Department of Agriculture, Oakland, CA.

Ellis, D.H., S.T. Dobrott, and J.G. Goodwin, Jr. 1977. Reintroduction techniques for masked bobwhites. In: *Endangered Birds, Management Techniques for Preserving Threatened Species* (S.A. Temple, ed.). University of Wisconsin Press, Madison, WI, pp. 345–354.

Hennessey, J.T., R.P. Gibbens, J.M. Tromble, and M.Cardenas. 1983. Vegetation changes from 1935 to 1980 in mesquite dunelands and former grasslands of southern New Mexico. Journal of Range Management 36: 370–374.

Hernandez, F., W.P. Kuvlesky, Jr, R.W. DeYoung, et al. 2006. Recovery of a rare species: case study of the masked bobwhite. Journal of Wildlife Management 70: 617–631.

Humphrey, R.R. 1958. The desert grassland, a history of vegetational change and an analysis of causes. Botanical Review 25: 193–252.

McClaran, M.P. and T.R. Van Devender. (eds) 1995. *The Desert Grassland.* University of Arizona Press, Tucson, AZ.

Tomlinson, R.E. 1972. Current status of the endangered masked bobwhite quail. Transactions of the North American Wildlife and Natural Resources Conference 37: 294–311.

Wood, J.E. 1969. Rodent populations and their impact on desert rangelands. New Mexico State University Agricultural Experiment Station Bulletin 555, Las Cruces, NM.

Wright, H.A. 1972. Fire as a tool to manage tobosa grasslands. Tall Timbers Fire Ecology Conference 12: 153–167.

Edwards Plateau

Amos, B.B. and F.R. Gehlbach. (eds) 1998. *Edwards Plateau Vegetation: Plant Ecological Studies in Central Texas.* Baylor University Press, Waco, TX.

Anders, A.D. and D.C. Dearborn. 2004. Population trends of the endangered golden-cheeked warbler at Fort Hood, Texas, from 1992–2001. Southwestern Naturalist 49: 39–47.

Bryant, F.C. 1991. Managed habitats for deer in juniper woodland of west Texas. In: *Wildlife and Habitats in Managed Landscapes* (J.E. Rodiek and E.G. Bolen, eds). Island Press, Washington, DC, pp. 59–75.

Fowler, N.L. and D.W. Dunlap. 1986. Grassland vegetation of the eastern Edwards Plateau. American Midland Naturalist 115: 146–155.

Graber, A.E., C.A. Davis, and D.M. Leslie, Jr. 2006. Golden-cheeked warbler males participate in nest-site selection. Wilson Journal of Ornithology 118: 247–251.

Pulich, W.M. 1976. *The Golden-Cheeked Warbler, a Bioecological Study.* Texas Parks and Wildlife Department, Austin, TX.

Rappole, J.H., D.I. King, and P. Leimgruber. 2000. Winter habitat and distribution of the endangered golden-cheeked warbler (*Dendroica chrysoparia*). Animal Conservation 3: 45–59.

Thomas, J.W., R.M. Robinson, and R.G. Marburger. 1964. Hypogonadism in white-tailed deer of the Central Mineral Region in Texas. Transactions of the North American Wildlife and Natural Resources Conference 29: 225–236. (A description of "velvet horn.")

Vallentine, J.F. 1960. Live oak and shin oak as desirable plants on Edwards Plateau ranges. Ecology 41: 545–548.

Wink, R.L. and H.A. Wright. 1976. Effects of fire on an Ashe juniper community. Journal of Range Management 26: 326–329.

Tamaulipan Mezquital

Bogush, E.R. 1952. Brush invasion in the Rio Grande Plain of Texas. Texas Journal of Science 4: 85–91.

Box, T.W., J. Powell, and D.L. Drawe. 1967. Influence of fire on south Texas chaparral. Ecology 48: 955–961.

Collier, B.A., K.L. Skow, S.R. Kremer, et al. 2012. Distribution and derivation of white-winged dove harvests in Texas. Wildlife Society Bulletin 36: 304–312. (Documents the northward range expansion in Texas.)

Cottam, C. and J.B. Trefethen. 1968. *Whitewings, the Life History, Status, and Management of the White-Winged Dove*. D. Van Nostrand Company, Inc., Princeton, NJ.

Gehlbach, F.R. 1981. *Mountain Islands and Desert Seas: A Natural History of the US–Mexican Borderlands*. Texas A&M University Press, College Station, TX.

Jahrsdoerfer, S.E. and D.M. Leslie, Jr. 1988. Tamaulipan brushland of the Lower Rio Grande Valley of South Texas: description, human impacts, and management options. US Fish and Wildlife Service, Biological Report 88(36), US Department of Interior, Washington, DC.

Johnson, Y.S., F. Hernandez, D.G. Hewitt, et al. 2009. Status of white-winged dove nesting colonies in Tamaulipas, Mexico. Wilson Journal of Ornithology 121: 338–346.

Johnston, M.C. 1962. Past and present grasslands of southern Texas and northern Mexico. Ecology 44: 456–466.

Lehman, V.W. 1969. *Forgotten Legions: Sheep in the Rio Grande Plain of Texas*. Texas Western Press, El Paso, TX.

Longoria, A. 1997. *Adios to the Brushlands*. Texas A&M University Press, College Station, TX.

Marion, W.R. and R.J. Fleetwood. 1978. Nesting ecology of the plain chachalaca in south Texas. Wilson Bulletin 90: 386–395.

Rappole, J.H., A.S. Pine, D.A. Swanson, and G.L. Waggerman. 2007. Conservation and management of migratory birds; insights from population data and theory in the case of the white-winged dove. In: *Wildlife Science: Linking Ecological Theory and Management Applications* (T.E. Fulbright and D.G. Hewitt, eds). CRC Press, Boca Raton, FL, pp. 3–20.

Reid, N., D.M. Stafford-Smith, P. Beyer-Münzel, and J. Marroguín. 1990. Floristic and structural variation in the Tamaulipan thornscrub, northeastern Mexico. Journal of Vegetation Science 1: 529–538.

Shindle, D.B. and M.E. Tewes. 1998. Woody species composition of habitats used by ocelots (*Leopardus pardalis*) in the Tamaulipan Biotic Province. Southwestern Naturalist 43: 273–279.

Tewes, M.E. and D.D. Everett. 1986. Status and distribution of the endangered ocelot and jaguarundi in Texas. In: *Cats of the World: Biology, Conservation, and Management* (S.D. Miller and D.D. Everett, eds). National Wildlife Federation, Washington, DC, pp. 147–158.

Highlights

Rodents and vegetation

Betancourt, J.L., T.R. Van Devender, and P.S. Martin. (eds) 1990. *Packrat Middens: The Last 40,000 Years of Biotic Change*. University of Arizona Press, Tucson, AZ.

Betancourt, J.L., K.A. Rylander, C. Peñalba, and J.L. McVickar. 2001. Quaternary vegetation history of Rough Canyon, south-central New Mexico. Paleogeography, Paleoclimatology, and Paleoecology 165: 7195.

Borchert, M.I. and S.K. Jain. 1978. The effect of rodent seed predation on four species of alifornia annual grasses. Oecologia 33: 101–113.

Brown, J.H. and E.J. Heske. 1990. Control of a desert-grassland transition by a keystone rodent guild. Science 250: 1705–1707.

Gray, S.T., J.L. Betancourt, S.T. Jackson, and R.G. Eddy. 2006. Role of multidecadal climate variability in a range extension of pinyon pine. Ecology 87: 1124–1130.

Hall, S.A. 2000. Was the High Plains a pine-spruce forest? Rangelands 22: 3–5.

Hall, S.A. 2005. Ice Age vegetation and flora of New Mexico. In: *New Mexico's Ice Ages* (S.G. Lucas, G.S. Morgan, and K.E. Zeigler, eds). New Mexico Museum of Natural History Science Bulletin 28. Albuquerque, NM, pp. 171–184.

Hall, S.S. and D.H. Riskind. 2010. Palynology, radiocarbon dating, and woodrat middens: new applications at Hueco Tanks, Trans-Pecos, USA. Journal of Arid Environments 74: 725–730.

Heske, E.J., J.H. Brown, and Q. Guo. 1993. Effects of kangaroo rat exclusion on vegetation structure and plant species diversity in the Chihuahuan Desert. Oecologia 95: 520–524.

Johnson, L. and J. Johnson. (eds) 1987. *Escape from Death Valley*. University of Nevada Press, Reno, NV. (Includes the story of candy-hungry forty-niners sampling the sticky contents of packrat middens.)

Scholt, L.F. 1973. Consumption of primary production by a population of kangaroo rats (*Dipodomys merriami*) in the Mojave Desert. Ecological Monographs 43: 357–376.

Stones, R.C. and C.L. Howard. 1968. Natural history of the desert woodrat, *Neotoma lepida*. American Midland Naturalist 80: 458–476.

Van Devender, T.R. 1995. Desert grassland history. In: *Desert Grasslands* (M.P. McClaren and T.R. Van Devender, eds). University of Arizona Press, Tucson, AZ, pp. 68–99.

Wells, P.V. and C.D. Jorgensen. 1964. Pleistocene wood rat middens and climatic change in Mohave Desert – a record of juniper woodland. Science 143: 1171–1174.

Channeled Scablands

Allen, J.E., M. Burns and S. Burns. 2009. *Cataclysms on the Columbia*, second edition. Ooligan Press, Portland OR.

Bjornstad, B. 2006. *On the Trail of the Ice Age floods: A Geological Field Guide to the Mid-Columbian Basin*. Keokee Books, Sandpoint, ID.

Bretz, J.H., H.T.U. Smith, and G.E. Neff. 1956. Channeled Scabland of Washington – new data and interpretations. Geological Society of America Bulletin 67: 957–1049.

Daubenmire, R. 1970. *Steppe Vegetation of Washington*. Washington Agricultural Experiment Station Technical Bulletin 62. Washington State University, Pullman, WA.

Daubenmire, R. 1972. Annual cycle of soil moisture and temperature as related to grass development in the steppe of eastern Washington. Ecology 53: 419–424.

Soennichsen, J. 2009. *Bretz's Flood: The Remarkable Story of a Rebel Geologist and the World's Greatest Flood*. Sasquatch Books, Seattle WA.

Weis, P.L. and W.L. Newman. 1989. *The Channeled Scablands of Eastern Washington: The Geologic Story of the Spokane Flood*. Eastern Washington University Press, Cheney, WA.

Snake River Birds of Prey Conservation Area

US Bureau of Land Management. 1980. Final environmental statement, Snake River Birds of Prey National Conservation Area. University of Michigan Library, Ann Arbor, MI.

US Bureau of Land Management. 2008. Snake River Birds of Prey National Conservation Area: proposed resource management plan and final environmental impact statement. US Bureau of Land Management, Publication ID-111-20060EIS-1740, Boise, ID.

Mima mounds

Berg, A.W. 1990. Formation of Mima mounds, a seismic hypothesis. Geology 18: 281–285.

Cox, G.W. 1984. The distribution and origin of Mima mound grasslands in San Diego County, California. Ecology 65: 1397–1405. (See also Cox, G.W. 1984. Mounds of mystery. Natural History 93(6): 36–45.)

Reifner, R.E., S. Boyd, and R.J. Shelmon. 2007. Notes on native vascular plants from Mima mound-vernal pool terrain and the importance of preserving coastal terraces in Orange County, California. Aliso 24: 19–28.

Veatch, A.C. 1906. On the human origin of the small mounds of the lower Mississippi Valley and Texas. Science 21: 310–311.

Walter, D. and A. Bryant. 2001. Mima mounds of Thurston County: a study of evapotranspiration, geologic history & myths. Environmental Analysis Program, The Evergreen State College. Available at: http://archives.evergreen.edu/webpages/curricular/2000-2001/ENVANA/Mima%20Mounds/Instrumentation.htm (accessed 8 January 2015).

Desertification

Arnalds, O. and S. Archer. 1999. *Rangeland Desertification*. Kluwer Academic Publishers, Dordrecht, The Netherlands.

Cloudsley-Thompson, J.L. 1977. Animal life and desertification. Environmental Conservation 4: 199–204.

Dregne, H.E. 1983. *Desertification of Arid Lands*. Harwood Academic Publisher, New York, NY.

Geist, H. 2005. *The Causes and Progression of Desertification*. Ashgate Publishing Company, Burlington, VT.

Sheridan, D. 1981. *Desertification of the United States*. Council on Environmental Quality, US Government Printing Office, Washington, DC.

Infobox 6.1. The island fox

Collins, P.W. 1991. Interaction between the island foxes (*Urocyon littoralis*) and Indians on the islands off the coast of southern California. I. Morphologic and archaeological evidence of human assisted dispersal. Journal of Ethnobiology 11: 51–82.

Conover, A. and A. Curry. 2004. Fighting for foxes. Smithsonian 35(7): 66–71.

Coonan, T.J. and M. Dennis. 2006. Island fox recovery program: 2005 annual report. Channel Islands National Park Technical Report 06-02: 1–84.

Courchamp, F., R. Woodroffe, and G. Roemer. 2003. Removing protected populations to save endangered species. Science 302(5650): 1532. (Describes removal of golden eagles from the Channel Islands.)

George, S.B. and R.K. Wayne. 1991. Island foxes: a model for conservation genetics. Terra 30: 18–23.

Gilbert, D.A., N. Lehman, S.J. O'Brien, and R.K. Wayne. 1990. Genetic fingerprinting reflects population differentiation in the Channel Island fox. Nature 344: 764–767.

Moore, C.M. and P.W. Collins. 1995. Urocyon littoralis. Mammalian Species 489: 1–7.

Philbrick, R.N. and J.R. Haller. 1977. The Southern California islands. In: *Terrestrial Vegetation of California* (M.G. Barbour and J. Major, eds). John Wiley & Sons, NY, pp. 893–906.

Roemer, G.W., C.J. Donlan, and F. Courchamp. 2002. Golden eagles, feral pigs, and insular carnivores: how exotic species turn native predators into prey. National Academy of Science 99: 791–796.

Runyan, C. 2009. Outfoxed: island restoration brings record recovery. Nature Conservancy Magazine 59: 10–11.

Van Vuren, D.H., M.L. Johnson, and L. Bowen. 2001. Impacts of feral livestock on island watersheds. Pacific Science 55: 285–289.

Infobox 6.2. Ladies-only lizards

Cole, C.J., L.M. Hardy, H.C. Dressaeur, et al. 2010. Laboratory hybridization among North American whiptail lizards, including *Aspidoscelis inornata arizonae* x *A. tigris marmorata* (Squamata: Teiidae), ancestors of unisexual clones in nature. American Museum Novitates 3698: 1–43.

Crews, D. and K.T. Fitzgerald. 1980. "Sexual" behavior in parthenogenic lizards (*Cnemidorphorus*). Proceedings of the National Academy of Science 77: 499–502.

Crews, D., M. Grassman, and J. Lindzey. 1986. Behavioral facilitation of reproduction in sexual and unisexual whiptail lizards. Proceedings of the National Academy of Science 83: 9547–9550.

Grassman, M. and D. Crews. 1987. Dominance and reproduction in a parthenogenic lizard. Behavioral Ecology and Sociobiology 21: 141–147.

Hubbs, C.L. and L.C. Hubbs. 1932. Apparent parthenogenesis in nature, in a form of fish of hybrid origin. Science 76: 628–630.

Lowe, C.H. and J.W. Wright. 1966. Evolution of parthenogenic species of *Cnemidophorus* (whiptail lizards) in western North America. Journal of the Arizona Academy of Science 4: 81–87.

Moore, M.C., J.M. Whittier, A.J. Billy, and D. Crews. 1985. Male-like behavior in an all-female lizard: relationship to ovarian cycle. Animal Behavior 33: 284–289.

Reeder, T.W., C.J. Cole, and H.C. Dressaeur. 2002. Phylogenetic relationships of whiptail lizards of the genus *Cnemidophorus* (Squamata: Teiidae): a test of monophyly, reevaluation of karyotypic evolution, and a review of hybrid origins. American Museum Novitates 3365: 1–64.

Vitt, L.J. and J.P. Caldwell. 2009. *Herpetology*, third edition. Elsevier, Inc., Burlington, MA.

Wright, J.W. and C.H. Lowe. 1968. Weeds, polyploids, parthenogenesis, and the geographical and ecological distribution of all-female species of *Cnemidophorus*. Copiea 1968: 128–138.

CHAPTER 7

Deserts

The desert has a beauty all its own, a rustic scene of something to behold. So next, when you are at the desert skirt, look passed, see more than just a bunch of dirt.

From a poem by Lady Kathleen

North America includes four large areas assigned to the Desert Biome (Fig. 7.1). The "hot deserts" – Sonoran, Mojave, and Chihuahuan – receive most of their limited precipitation as winter rainfall, whereas snowfall accounts for the majority of precipitation in the Great Basin Desert, a "cold desert." Although united by aridity each of these deserts features a distinguishable biota, often highlighted by an indicator species (e.g., giant saguaro of the Sonoran Desert). Successional developments are not readily apparent in desert communities, although degraded sites are all too common in the wake of heavy-handed human activities. At first glance deserts seem hard and durable, but they are in fact easily damaged and slow to recover from anthropogenic activities. Heavy grazing and other distresses, and a view that deserts are as suitable for dumps as they are for lizards and cacti, are among the agents that degrade desert environments. Fortunately, state and national park systems include several outstanding desert areas, and laws protect several species of desert biota.

Physical geography

Why deserts are dry

Deserts result from phenomena related to the dramatic reduction of airborne moisture, but these forces vary in scope; some are global, but others are regional. A simple principle governs the mechanism in each case: warm air can hold a good deal of moisture, whereas cold air cannot.

1 At the equator, where solar radiation is greatest, warm air rises then moves to the north or south over each hemisphere. This air cools rapidly as it moves upward and releases immense quantities of moisture, creating lush zones of tropical vegetation on either side of the equator. Now greatly cooled and depleted of moisture, the air sinks to Earth's surface at latitudes between 15° and 35° north and south of the equator. As it sinks, the already dry air undergoes **adiabatic** warming, rendering it incapable of producing precipitation. In consequence, great deserts such as the Sahara lie in those zones where masses of dry air return to Earth's surface (i.e., these subtropical deserts result from global patterns of air circulation).

2 Moisture-laden air moves across cold ocean water, which often originates from upwellings and currents along the western margins of some continents. Thus cooled, the air is greatly depleted of moisture offshore and the little that remains condenses into coastal fogs. In short, the air has dried by the time it reaches land. The cold California Current (see Currents and Climates, Chapter 11) acts in this fashion and contributes to desert conditions in parts of North America (e.g., Baja California). Elsewhere, cold offshore currents have formed large coastal deserts in Peru and southwestern Africa, both among the driest locations on Earth.

3 Deserts also form in continental interiors, especially in high plateaus surrounded by high mountains and far removed from sources of moisture. Winds in these places travel long distances across land, eventually drying to the point where precipitation is nearly impossible. The Great Basin Desert, to some degree, represents a continental desert, but the Gobi Desert in Asia is a better example.

4 Air also cools as it rises over mountains. This occurs in the western United States when air moves upward from sea level, eventually crossing the high Sierra Nevada. Heavy precipitation, including deep snows, occurs on Sierra's western slopes, but little moisture remains when the air descends across their eastern slopes. These conditions produce arid environments on the lee sides of mountain ranges, which lie in so-called **rain shadows** (Fig. 7.2). The Mojave Desert developed in a rain shadow, and much of the aridity in the Great Basin Desert results from the same influence.

Ecology of North America, Second Edition. Brian R. Chapman and Eric G. Bolen.
© 2015 John Wiley & Sons, Ltd. Published 2015 by John Wiley & Sons, Ltd.

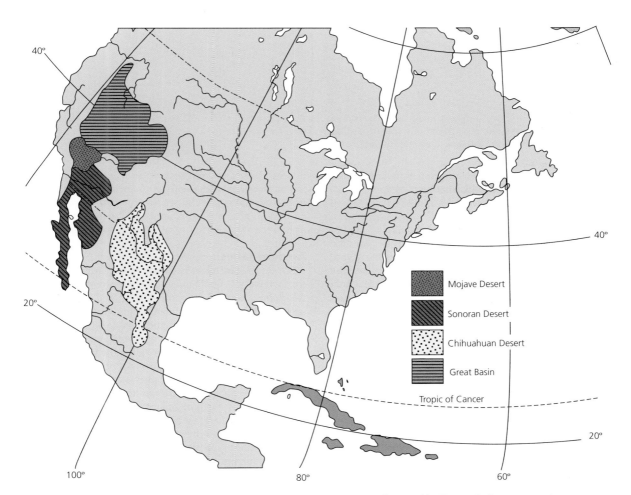

Figure 7.1 Approximate delineation of the four major deserts in North America. Illustrated by Tamara R. Sayre.

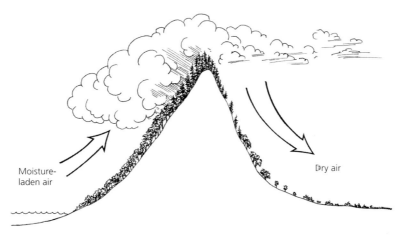

Figure 7.2 Air cools as it rises, dropping moisture on the windward side of mountains such as the Sierra Nevada, but sites in the rain shadow on the opposite side are comparatively arid. Rain shadows, which are produced by large mountain ranges, are a major reason for the development of deserts. Illustrated by Tamara R. Sayre.

Desert mountains and bajadas

Mountains often rise from the floor of deserts and, when these are tall enough, mesic conditions support communities of woodland vegetation including animals such as white-tailed deer, porcupines, and gray-footed chipmunks. In the Trans-Pecos area of Texas, for example, the Chisos and Davis mountains (well within the borders of the Chihuahuan Desert) include communities with oak and pine dominants (e.g., pinyon and ponderosa pines). Douglas fir occurs at the highest elevations in the Chisos Mountains. These and similar communities developing in deserts elsewhere create mountaintop "islands" amidst a desert "sea."

Bajadas (Spanish for "slopes") are alluvial fans typically forming from outwash at the mouths of canyons in mountainous deserts. Surges from infrequent but powerful thunderstorms produce the discharge of sediments. Bajadas typically expand intermittently, remaining unchanged for long intervals. They expand unevenly however, because the deposition shifts across the fan in keeping with the changing elevations. As one part builds up, the flow is directed to lower areas. Some bajadas spread over large areas of desert floor, producing sites where distinctive communities may develop. In particular, coarse soils occur on the upper slopes of bajadas, but these gradually become finer towards the outer edge of the fan. Such changes in soil texture in turn affect the availability of soil moisture. Less moisture is available to plants in fine-textured soils because of the higher surface tension at these sites (i.e., roots can less easily absorb water at the outer edge of the bajada). Soil salinity also follows a "downstream" gradient, becoming steadily more saline in the lower bajada. Because of these features, the vegetation develops greater diversity on the upper bajadas where plant cover is also greater.

Similarly, plant forms with a range of features such as leaf shape and size, stem height, root system, succulence, and woodiness are more diversely represented in the vegetation established on the upper slopes of bajadas.

Ancient lakes

Basin-and-range terrain characterizes many areas of the southwestern United States and northern Mexico. When flat-floored desert valleys are surrounded by mountains or hills, the runoff from rainfall and snowmelt drains into basins with no outlet. In these areas, called **"bolsons"** by geologists, temporary lakes form then gradually dry, leaving a crusty salt pan. Death Valley, the lowest, hottest, and driest area in North America, may be the best-known example of a bolson. Encompassing a total area of about 7800 km² (3000 square miles) in the Mojave Desert, Death Valley is surrounded to the south by Owlshead Mountains, to the north by the Sylvania Mountains, the Amargosa Range on the east, and the Panamint Range to the west. The basin occasionally receives enough water from the surrounding mountains to temporarily hold a small, salty lake.

Two enormous prehistoric lakes once covered large areas of the Great Basin Desert. Glacial meltwater at the end of the Pleistocene formed Lake Bonneville and Lake Lahontan centered in Utah and Nevada, respectively. Lake Bonneville was larger than modern Lake Huron but it gradually dried, leaving behind smaller lakes of which Great Salt Lake is the largest remnant (4400 km² or 1700 square miles in size during an average year). At its peak, the surface of Lake Bonneville extended some 305 m (1000 feet) above the present level of Great Salt Lake and covered nearly 58,000 km² (22,400 square miles). Lake Lahontan, once about the size of present Lake Erie, also evaporated away; among its remains are Pyramid Lake and two shallow depressions, the Carson and Humboldt sinks. Today, prominent benches on the mountainsides are reminders of the ancient shorelines of these lakes.

Water levels in the Great Salt Lake change annually in keeping with the year-to-year runoff of snowmelt from adjacent mountain ranges. Between 1983 and 1988, the level of the Great Salt Lake rose about 9 m (30 feet) and inundated large areas of the surrounding terrain. Lacking any outlet, the waters flowing into Great Salt Lake are trapped and, as they evaporate, they leave behind their burden of salts. Salinities at times are sevenfold those of ocean water. No fishes survive in Great Salt Lake, although an enormous population of brine shrimp flourishes in this inland sea. The brine shrimp, along with the plankton on which they filter feed, normally form a simple community in the lake. Predators are absent, likely because they cannot survive the extreme salinity of the lake water. Brine shrimp can tolerate salinities less than those of seawater, but they are seldom abundant under such conditions, perhaps because predators are then able to invade the community and diminish the shrimp population. In fact, when Great Salt Lake flooded with fresh water in the late 1980s, the reduced salinity allowed invasions of a predaceous insect, water boatman, which preyed on the smaller larval stages of brine shrimp. Sharp reductions in the biomass of brine shrimp followed thereafter, along with other changes in the community; for instance, water clarity diminished fourfold because plankton numbers increased. In sum, a predator temporarily initiated a tropic cascade that ended when Great Salt Lake returned to pre-flood levels and salinities again precluded the presence of water boatman.

Despite the salinity of the lake itself, vast marshes formed at the mouths of rivers flowing into the eastern rim of Great Salt Lake. In Lovelock Cave, adjacent to desert wetlands in Humboldt Sink, Nevada, archeologists discovered decoys fashioned from rushes and bearing a remarkable likeness to canvasback ducks. A now-vanished civilization of desert natives hunted with these at least 1000 years ago. The devastating disease botulism occasionally strikes ducks and other birds in these marshes, at times killing tens of thousands in epizootics of unparalleled proportions. Many of these wetlands, renown as habitat for large numbers of waterfowl, are now protected and managed by wildlife agencies (e.g., Bear River Migratory Bird Refuge). The impact of botulism lessened after dikes built on the refuges stabilized water levels in the marshes bordering Great Salt Lake. Other marshes in the Great Basin Desert arise from springs. In nearby Carson Sink, both state and federal agencies manage the Stillwater Marshes – the largest wetland complex in Nevada – for waterfowl and other wildlife.

Features and adaptations

Desert soils and surfaces

Soils in most desert regions of North America are classified as aridisols, but in earlier taxonomies these were known as gray or red desert soils. Aridisols form under the influence of high temperatures, low but often torrential rainfall, and wind. As might be expected because of the sparse vegetation in deserts, little organic matter occurs in the upper profile of aridisols, resulting in little biological activity for long periods of time. Materials deeper within the profile of desert soils may be cemented into "caliche," a hardpan of calcium carbonate. Undeveloped strata composed of little more than rocks – lithosols – are common on slopes and other desert areas subject to heavy erosion. Mollisols, mentioned in Chapter 5, characterize some of the grassier areas in the northern Great Basin. Dunes dominate some sites, of which the White Sands area in the Chihuahuan Desert is the best known (*Chihuahuan Desert*, this chapter).

On certain types of soils in arid regions, wind erosion removes fine particles leaving behind hardened surfaces known as a desert pavement. Stones and pebbles in desert pavement thereafter join together in a fashion resembling a floor of small tiles. Other processes also form desert pavement, particularly where the soils develop from coarser materials. Because of these tight and hardened surfaces, the soil profiles absorb little rainfall from heavy thunderstorms that instead runs off in immense volumes. Flash floods are therefore routine in deserts and, where the soil is vulnerable, the runoff often cuts deep gullies known as arroyos. Several types of plants take advantage of the generally moister soil conditions near arroyos. These include trees (desert-willow, smoke tree, and large mesquite are typical) which form a thin riparian woodland that traces the meandering of the arroyo across the desert. An exotic species, saltcedar, dominates the riparian vegetation in many places, at times forming monotypic stands. Willows and saltcedar are good examples of **phreatophytes**, plants whose deep root systems penetrate into permanent sources of water. Saltcedar is widely regarded as an undesirable species, in part because it so effectively translocates large volumes of water into the atmosphere.

Runoff from desert pavement also fills shallow depressions in the desert, forming temporary ponds where animal life flourishes until evaporation again dries the surface. A crust of salts usually forms as the water evaporates, on which concentric rings of vegetation develop in a target-like pattern. Moving toward the center, each ring indicates plants with progressively greater tolerances to salinity; in some cases, the center of the depression is entirely barren. Salt-tolerant plants are known as **halophytes**, of which various species of glasswort are good examples in North America.

Another type of crust may form on the surface of desert soils. However, unlike the surface layer of salt in depressions, these cryptogamic crusts are dark, thin, and delicate, with a brittle, sponge-like structure that crumbles when disturbed. Formation of these crusts is concentrated in the top 3 mm (0.12 inch) of the soil and results from the biological activities of spore-bearing **cryptogams** – algae, fungi, and lichens – and microorganisms, many of which are cyanobacteria. These crusts help stabilize the surface of desert soils against erosion. Perhaps of greater importance, cryptogamic crusts incorporate atmospheric nitrogen into the biological systems of the desert. This nitrogen in the air is changed into a form that is soluble in the soil and thereafter available to higher plants, a process known as nitrogen fixation. Cryptogamic communities also grow and die, and their decay contributes to soil fertility.

Obviously, disturbances that harm cryptogamic crusts also impair important ecological processes in deserts. Anthropogenic disturbances, including foot, vehicular, and livestock traffic, readily compromise the integrity of these crusts. When their structure is damaged, for example, even normal wind speeds can erode desert soils. Because these crusts remain all year round and offer ground cover during droughts, they provide stability that may otherwise be lacking on the soil surface. Ground cover is especially important because desert plants are widely spaced and little else is available to hold the soil. When damaged, the crusts may not recover for decades or even longer.

Another phenomenon known as "rock varnish" also forms on the surface of desert soils, including the surfaces of boulders and other rocky surfaces. Rock varnish forms from a mixture of clay minerals with iron and manganese oxides, which covers rocky surfaces in desert regions with a rich patina. Manganese oxides, which are concentrated by bacterial action, produce dark-colored varnish, whereas the film is bright orange when iron oxides predominate. The blackish or reddish stains often seen on cliff surfaces are good examples of rock varnish.

In ways somewhat similar to the Tundra Biome, soils in deserts are easily damaged. Off-road vehicles break the surface crusts, increase compaction, and greatly disturb soil–plant interactions (see *The wheeled menace*, this chapter). Ill-considered irrigation schemes increase soil salinity to levels harmful to most plants, and overgrazing contributes to the deterioration of desert soils. As with the Tundra, damaged areas in the desert are extremely slow to recover and, unless carefully protected from further misuse, large areas of desert will suffer long-lasting deterioration across a full range of soil–plant–animal relationships.

Plant adaptations

Some ecologists catalog the ways desert plants cope with aridity. The first strategy is to escape droughts altogether. A large group of desert plants achieve their growth during those brief periods when water is available; that is, they complete their life cycle immediately after heavy rainfalls and before arid conditions return. Such plants typically survive long periods of drought as seeds or, in some cases, as modified underground stems (e.g., bulbs or corms).

Deserts are famous for "blooming" after heavy rains. When rains soak the dry soil, flowers of many kinds quickly color the desert and then vanish, sometimes for several years, until favorable conditions return again. After bursting forth, these plants temporarily represent the most abundant and eye-catching vegetation in the desert: a literal carpet of flowers. These plants are annuals which live and complete their life cycle in just one year. However, because these plants may skip several growing seasons, they are better described as **ephemerals**, plants that temporarily appear on an irregular basis rather than each year. Diversity in ephemeral communities may be quite high, with as many as 56 species recorded in small plots in the Mojave Desert.

To govern germination, the seeds of ephemeral plants are impregnated with chemical inhibitors that vary from species to species; these are known as "drought-escaping" plants. The germination inhibitors in the seeds of some species wash away easily. Others require heavy rainfall before the inhibitor is removed and, for these, a series of light rains will not trigger germination. Temperature also influences the germination of desert ephemerals. In the Sonoran Desert, where rain occurs in two peaks each year, the combination of temperature and rainfall produce two rather distinctive communities of ephemerals, one germinating in response to rain and cool conditions (November–March) and a second germinating in response to summer heat and thunderstorms (July and August).

Immense numbers of seeds are produced whenever communities of desert ephemerals erupt and flower. Sites in the Mojave Desert yielded densities of 8000 to 187,000 seeds per square meter, and even higher densities have been recorded in deserts elsewhere. These persist in the soil, forming what ecologists call a **seed bank**. Because of species-specific differences in germination requirements, seed banks normally contain a rich variety of species but not all of these germinate at one time. Seed banks also occur in prairies, forests, and other communities, but are seldom as rich in species as those in deserts.

Drought evasion is the second means of coping with aridity. Plants in this group restrict water loss, usually with one or more kinds of adaptations such as waxy cuticles, modified leaf structures, and other means of preventing or decreasing moisture loss from plant tissues. Drought-evading plants may also have extensive root systems. This category includes plants that endure droughts. These persist because their tissues can survive long periods without absorbing water. Cacti, the most familiar example of drought-enduring plants, store water in succulent tissues and slowly utilize these reserves.

When rainfall returns, the spongy tissues rapidly absorb large quantities of water, often enlarging the diameter of the trunk and stems.

Of these two categories, the drought-escaping plants form the most distinctive group. Many plants have additional adaptive features that permit survival in deserts. Most species of cactus are equipped with spines. These help protect their succulent tissues from grazing animals, although spines do not limit collared peccaries from feeding on prickly pear and desert rodents easily maneuver between the spines on at least the larger species of cacti. Other features of spines also have adaptive value for cacti; spines shade the remainder of the plant, do so without losing water, and diminish evaporation by obstructing the flow of air across the surface of cacti (Fig. 7.3).

Plants exchange carbon dioxide and oxygen through small pores called stomata, which are usually located on leaf surfaces. However, water vapor escapes through the same openings, and this loss requires constant replenishment. In response, the stomata of many desert plants develop at the base of small pits instead of flush with the leaf's surface. Thus located, evaporative water loss is reduced in desert plants because their sunken stomata are shielded from the dry air currents sweeping across the plants.

A desert tree, palo verde, is uniquely adapted with complementary features for continuing photosynthetic activities while coping with moisture loss. It sheds its small leaves when moisture is limited, thereby removing a major site of evaporation, but the plant continues to produce sugars because of its photosynthetic bark (hence the name *palo verde*, or "green wood").

A spiny shrub with whip-like stems, ocotillo also sheds its small leaves during dry periods but the foliage quickly regrows with rainfall. Embryonic leaves appear within 24 hours following a strong rain and reach their full development within 5 days (Fig. 7.4a). This cycle may be repeated as many as five or six times a year, but two or three leafy periods are more typical. Energy for the new leaves is derived from reserves in the stem instead of those in the root system. Leaves produced on the long shoots at the tips of ocotillo stems eventually drop their blades, but the remaining leaf stalks harden into spines. Hummingbirds are among the important pollinators of ocotillo, which bears scarlet flowers (Fig. 7.4b). Indeed, the flowering period of ocotillo coincides with the spring migration of hummingbirds across the geographical range of these plants.

Animal adaptations

Animals have evolved several means of dealing with water scarcity in desert environments. Some of these are relatively simple, such as the waterproof skins of reptiles that effectively hinder evaporation. Many species of both invertebrates and vertebrates avoid heat stress and therefore conserve water by limiting their activities to the cooler nights. These species typically spend the hot days in burrows, where water loss is curtailed because of the higher relative humidity.

Metabolic water represents a somewhat more sophisticated adaptation. Kangaroo rats, for example, do not drink but instead derive water from seeds in their diet. The water is produced internally from the oxidation of carbohydrates and fats, but protein metabolism may also contribute small amounts. The reaction, based on the oxidation of a simple sugar, glucose, is

$$C_6H_{12}O_6 + 6O_2 \rightarrow 6CO_2 + 6H_2O$$

In contrast, birds show few adaptations for coping with the scarcity of water. Gamble quail, which also eat seeds, rely on drinking for their water requirements and most species simply fly to water sources to meet their needs. Black-throated sparrows excrete highly concentrated urine and dry feces however, and they remain unusually independent of drinking water. These sparrows likely represent the most desert-adapted species of songbird in North America.

Figure 7.3 The dense array of spines on some cacti provides an almost impenetrable defense against herbivores. Spines also shade the soft tissues and restrict the flow of dry desert air across the plant's surface; the combined effects reduce water loss from evaporation. Reproduced by permission of Brian R. Chapman.

Salt balance is an important physiological function in desert animals, although of all the desert birds only the roadrunner seems to have salt glands. Conversely, many species of desert-dwelling reptiles have nasal salt glands, which excrete potassium and sodium salts without incurring much water loss. Salt excretion of this sort is pronounced in chuckwallas. One species of these relatively large lizards lives in both Mojave and Sonoran deserts where they eat succulent halophytes. To solve the problem of voiding salts without losing water, they expel salty fluids from their nostrils by sneezing, often leaving salt encrusted on their snouts. Chuckwallas also store water in tissues within sac-like folds in their skin.

During periods of rainfall, extracellular water accumulates in these tissues and the folds expand accordingly.

Perhaps no reptile is more popularly associated with North American deserts than horned lizards (Fig. 7.5). Several species, which are widely and erroneously called "horny toads" or "horned frogs," occur throughout the arid regions of western and southwestern North America. One of these, the Texas horned lizard, collects drinking water using a behavioral process called "rain-harvesting." When it rains these animals flatten their bodies, increasing the surface area on which raindrops fall. Water is captured even during light rains and is transported by capillary action through a network of minute

Figure 7.4 Sharp spines studding the long, whip-like stems of ocotillo, which appears dead for most of the year (inset), underscore the plant's popular name: the devil's walking stick. The leaves parch and drop off during dry periods and spines form from the leaf bases. However, within days after a rain, the seemingly dead ocotillo springs to life with masses of bright red flowers blooming at the tips of mature stems and clusters of small leaves along the length of each stalk (right). During their migration, hummingbirds seek nectar from the plant's tubular flowers. Reproduced by permission of Brian R. Chapman.

Figure 7.5 In addition to their defensive function, the spiny scales on horned lizards channel water from raindrops toward the lizard's mouth. The artificial background in this photograph of a Texas horned lizard allows better definition of the spines. Reproduced by permission of Brian R. Chapman.

grooves running between the scales on the lizard's skin, eventually reaching the mouth. The lizards drink by opening and closing their jaws, which is part of the behavioral response to the stimulus of rainfall (i.e., **stereotypic behavior**). In short, these lizards drink water collected and transported by their skin.

Texas horned lizards also demonstrate the unusual behavior of squirting blood from their eyes. Of the explanations proposed for this ability, recent evidence confirms that it serves as a defense against predators. More specifically, blood squirting is employed when Texas horned lizards are attacked by canids (dogs, foxes, and coyotes) but not by roadrunners or other predators. Kit foxes, for example, reacted adversely when squirted by horned lizards in laboratory tests and instead selected other foods. The active ingredients in this chemical defense are carried in the bloodstream of horned lizards and do not originate from glands near their eyes. Similarly, some factor in the blood of Texas horned lizards detoxifies the venom of harvester ants, the principal food of these interesting lizards.

Amphibians, because of their dependence on aquatic or at least moist habitats, are not well represented in desert faunas. Spadefoot toads nonetheless have adapted to the aridity of the Sonoran and other deserts in North America. Using a hardened projection – the "spade" – on their rear feet as shovels, these toads entomb themselves up to 90 cm (35 inches) below ground where they may spend many months in **estivation**, inactively awaiting the return of favorable climatic conditions (e.g., thunderstorms producing temporary ponds). Removed from the harsh conditions at the surface, some species form a cocoon of dead skin which retards the loss of body water; others encapsulate themselves in a chamber of mud and mucus. During this inactive period, spadefoot toads further counter the loss of body water by accumulating urea, a nitrogenous waste. The concentrated urea produces an osmotic gradient that moves water from the soil into the tissues of the buried toad instead of the reverse. The toads absorb water from the soil as long as the osmotic pressure of their body fluids exceeds the water tension in the soil (i.e., more urea is required in drier soils).

These and other physiological adaptations of spadefoot toads are complemented by those that affect the aquatic part of their life cycle. Adult toads emerge quickly and breed after heavy thunderstorms, laying eggs that hatch in less than 48 hours, and the tadpoles develop legs in about 10 days. Such rapid development counters the rapid evaporation that soon befalls the small temporary ponds where the toads breed. In contrast, the tadpoles of toads and frogs breeding where water is relatively permanent may not mature for several weeks or even months (e.g., tadpoles of the southern chorus frog, a species found in the Eastern Deciduous Forest, require 1.5–4 months to transform). Spadefoot toads respond to the sound of rainfall as a cue to emerge from their burrows. Moisture is not a necessary cue for emergence, as determined experimentally when toads emerged after water was sprinkled on soil covered with a plastic sheet. The sound travels through soil at low frequencies, and toads may sense a tactile rather than an auditory stimulus (i.e., vibrations absorbed by their bodies instead of their eardrums).

Spadefoot toads are not the only animals to take advantage of the temporary ponds formed by desert thunderstorms. Several kinds of invertebrates are among the explosion of life occurring after heavy rainfall temporarily fills long and dry depressions. Crustaceans are predominantly represented, especially those known collectively as phyllopods: fairy shrimps, clam shrimps, and tadpole shrimps. These animals, as for spadefoot toads, are faced with two challenges related to the unstable conditions in which they live. First, they must rapidly mature and reproduce after the ponds fill and complete these activities before the water evaporates. Second, they must survive extended periods of extreme aridity while awaiting the arrival of another thunderstorm (a wait that may span several years). However, unlike spadefoot toads, phyllopods address this challenge with **diapause** (arrested development), which these organisms accomplish by remaining as eggs until conditions favor continuation of their life cycle. As eggs, the organisms survive at the bare edge of metabolic activity. Despite limited nutrient reserves, their viability persists for many years (i.e., eggs in diapause lack complex tissues, so metabolic demands are much lower). During diapause the eggs also withstand extreme desiccation, succumbing only when virtually all their water content is exhausted.

Desert cicadas have evolved a unique means to effect evaporative cooling. These insects use their needlelike mouthparts to pierce the xylem – the water-conducting tissues – in desert plants. The cicadas thereafter extract water from this constant and abundant source. The ingested water quickly moves through the cicadas to pores on the abdominal and thoracic surfaces of their bodies, where it evaporates on contact with the hot external environment. Desert cicadas thereby cool themselves, in effect mimicking the familiar thermoregulatory mechanism of mammals of sweating. Using evaporative cooling, desert cicadas can maintain body temperatures up to 5°C (9°F) below ambient temperature.

To conclude, adaptations such as the formulation of metabolic water in kangaroo rats or the accumulation of urea in spadefoot toads fall within the realm of **physiological ecology**. Plants also exhibit physiological adaptations which are equally representative of this branch of ecology. Physiological ecology by no means is limited to the study of desert organisms. Recall, for example, the unique way wood frogs cope with long Arctic winters in the Tundra (Chapter 2). Nonetheless, in the austere laboratory of deserts we find an abundance of powerful examples that collectively focus on physiological ecology as one of the more incisive probes into the workings of nature.

The major deserts

Chihuahuan Desert

The Chihuahuan Desert extends on a northwest–southeast axis from Arizona and New Mexico into the states of Chihuahua and Coahuila in northern Mexico. Eastward, the Chihuahuan Desert is bordered by the Sierra Madre Oriental and westward by the Sierra Madre Occidental. Highlands of the Mexican Plateau form the southern border. The Big Bend area and other parts of the Trans-Pecos region in western Texas are also included. The Chihuahuan Desert today is completely isolated from other deserts in North America although, together with the Sonoran and Mojave deserts, it likely once formed a far larger region of aridity.

The soils of the Chihuahuan Desert contain more limestone than the other North American deserts. Because of the Chihuahuan Desert's relatively high elevations – up to 2000 m (6560 feet) above sea level – the climate is somewhat cooler and wetter than the other hot deserts in North America. These conditions, together with the limestone soils, may account for the better representation of grasses in the vegetation of the Chihuahuan Desert in comparison with either the Mojave or Sonoran deserts.

Gypsum dunes are a local feature and, indeed, the world's largest gypsum dune field is preserved at White Sands National Monument at the northern edge of the Chihuahuan Desert in New Mexico (Fig. 7.6a). These dunes form in the Tularosa Basin, where rainfall washes gypsum (a soluble mineral) from the surrounding mountains into the basin floor. Lacking any outlet the gypsum-laden waters evaporate, leaving behind an encrustation of minerals. The gypsum deposit, mostly in the form of soft crystals known as selenite, breaks into small particles that strong spring winds thereafter blow into dunes.

(a)

(b)   (c)

Figure 7.6 (a) Gypsum dunes creep across White Sands National Monument, a unique area of the Chihuahuan Desert in south-central New Mexico. (b) Some plants, including yuccas, adapt to the shifting dunes with elongated stems. (c) Others, including this skunkbush sumac, anchor compact pedestals of gypsum on which they persist after the dunes migrate elsewhere. Reproduced by permission of Tara Cuvelier and the National Park Service.

Several plants are endemic to gypsum-dominated soils in the Chihuahuan Desert (e.g., purple borage and a grass, gyp grama). Those living in dune fields must adapt to the moving sand or become buried. Soaptree yuccas avoid this problem with the rapid growth of their stems, which keeps the leafy part of these plants above the surface of the shifting dunes (Fig. 7.6b). The stems of this species may grow about 30 cm (12 inches) a year, in some cases extending 9 m (30 feet) downward through the dune. However, long-stemmed yuccas may collapse after the dune shifts away. Some dune plants deal with the moving sand by forming mounds known as pedestals. Aided by the plaster-of-Paris nature of gypsum, the plants compact part of the dune with their numerous stems and deep root systems. The result is pedestals of sand on which plants continue growing when the remainder of the dune travels away (Fig. 7.6c). Nonetheless, the heart of the dune field lacks vegetation because the sands move so rapidly and are so deep that plants cannot grow quickly enough to deal with these circumstances.

Key species in the Chihuahuan Desert include creosote bush, tarbush, sotol, ocotillo, and several yuccas and agaves. Lechuguilla, an agave, is widely regarded as an indicator species (Infobox 7.1). Creosote bush, however, also is common in the flora of other North American warm deserts, although its chromosome numbers apparently differ in each (diploid in the Chihuahuan Desert, tetraploid in the Sonoran Desert, and hexaploid in the Mojave Desert). Small- to medium-sized cacti are commonplace. Lechuguilla and sotol favor limestone areas, as does a rubber-producing plant, guyule. The understory includes a sparse cover of grasses (e.g., tobosa grass, bush muhly, and sacaton), but these may also occur in recognizable communities where grazing is curtailed and fire prevails with recurring frequency (see Chapter 6).

A natural wax can be rendered from the fleshy, reed-like stems of candelilla, and at times has been of economic importance. The wax forms a gray coating on the usually leafless stems of candelilla and reduces the evaporative loss of water. Apparently, more wax forms during periods of drought than in years with more abundant precipitation. When harvested, "wax weed" is pulled from the soil and bound in bundles for transport on burros to crude rendering vats. Normally, enough of the root system remains intact to restore the plants in about 20 years, but cropping may be excessive when prices are high and families harvesting this resource have to travel great distances to gather enough plants. Candelilla wax was of vital importance for waterproofing munitions during World War I and later as an ingredient for various kinds of polishes. At the local level, the plants provided building materials (e.g., roof thatching), a medicine for treating venereal diseases and, of course, wax for candles.

The Rio Grande is the most prominent river traversing the Chihuahuan Desert. In the Big Bend region of the Trans-Pecos in Texas, the river and its canyon-walled channel effectively hinder the dispersal of some mammals. Black-tailed prairie dogs, for example, occur on the north side of the Rio Grande, but their range does not extend southward across the river into Mexico in this region. Likewise, a population of kangaroo rats reaches its northern distributional limits on the Mexican side of the Rio Grande, but they are absent in Texas despite the availability of virtually identical habitat just north of the river. Several other species, whose ranges occur on both sides of the river, represent well-defined subspecies without intermediate forms; those on one side of the river are sharply unlike those on the opposite side. Among these are subspecies of the eastern cottontail, Ord's kangaroo rat, and two species of pocket gophers.

Sonoran Desert

In simple geographical terms, the Sonoran Desert surrounds the head of the Gulf of California (= Sea of Cortez). From south-central Arizona, it extends southward into the Mexican state of Sonora on the east side of the Gulf of California and throughout most of Baja California on the west side of the Gulf. In area, the Sonoran Desert covers about 310,300 km² (119,300 square miles), almost all which lies below 915 m (3000 feet) in elevation. The Colorado River bisects the Sonoran Desert, but the river exerts little direct influence on the desert vegetation. Nonetheless, immediately west of the river at the head of the Gulf of California lies a subregion known as the Colorado Desert, much of which exists in a basin near to or below sea level. The Salton Sea, which is in a bolson about 82 m (272 feet) below sea level, receives runoff from the Colorado Desert. Part of this area (Imperial Valley) is famous for its rich soils and agricultural production, but rocky and other poorly developed soils characterize most of the Colorado Desert. The San Andreas Fault runs through the Colorado Desert and gives rise to numerous springs and seeps along its route.

The Sonoran Desert is the least barren of the North American deserts and, indeed, boasts the richest floral diversity among the world's arid lands. One magnificent species of cactus, the giant saguaro, serves as a familiar indicator for much of the Sonoran Desert (Fig. 7.7). Despite their size and robust appearance however, saguaros are usually killed when exposed to more than 24 hours of freezing temperatures, which is foremost among the reasons why the species does not occur at elevations above 1400 m (4500 feet). Saguaro communities are also badly damaged by heavy grazing, which removes protective cover of grasses and other vegetation required by saguaro seedlings (see *Nurse trees*, this chapter).

Saguaros bloom in May and June and, while mature plants may subsequently produce 200 fruits each with 2000 seeds, only a few young plants survive from this wealth of propagules. Growth is exceedingly slow; the seedlings extend above ground by only 2 cm (less than 1 inch) after 10 years, and perhaps by about 1 m (39 inches)

Infobox 7.1 Agave, margaritas, and bats

Several legends surround the origin of the popular tequila-based drink known as a "Margarita," but most suggest that the refreshment was named after the first person to try it. The 1977 release of "Margaritaville," the chart-topping song written and performed by Jimmy Buffet, further enhanced the popularity of the lime-flavored cocktail. While many enjoy Margaritas, few know of the relationship between agave, tequila, and bats.

Tequila is made from sugars produced by blue agave, a large succulent species native to the sandy soils of the Chihuahuan Desert in Jalisco, Mexico. Sometimes called "century plants" because they require several years of growth before flowering, agaves produce a basal rosette of large, fleshy leaves armed with serrated margins of sharp teeth and a terminal spine. After 5–7 years, agaves bear tubular, yellow flowers atop a central flower stalk (*quiote*) that reaches heights of 5 m (16 feet). Mexican long-nosed bats pollinate the flowers during the single flowering season. Several thousand seeds are produced, and the plant dies soon afterward.

Agaves produce several vegetative shoots when they are about one year old. On commercial agave farms, these shoots are used in replanting because sugar production is maximized by cutting off the flower stalks shortly after they emerge, thus preluding seed production. After the flower stalks are removed, blue agave may live on for many years. In a blue agave's 12th year, the leaves are cut off near the heart (*piña*) which may weigh up to 90 kg (200 pounds); the latter is then heated to remove the sugary sap and tequila is distilled from the fermented sap.

Mexican long-nosed bats engage in long-distance seasonal migrations related to the blooming periods of agave and cacti in the deserts of southern New Mexico, western Texas, and northern Mexico. Roosting in caves along the way, the bats emerge at night to feed on agave pollen and nectar, sometimes traversing up to 39 km (24 miles) between roosting and feeding locations. In the process of feeding on flowers, the bats become one of the most significant pollinators of the tequila-producing blue agave. For their part, agaves provide the most important food resource for the bats during the winter months in central Mexico.

Most agaves used for tequila production are now cultivated and propagated vegetatively, but wild populations of agave still depend on long-nosed bats for cross-pollination. Because the wild populations of agave provide the only sources of new genetic varieties for commercial plants, both wild agaves and the bats that pollinate them seem important to the industries that produce tequila and other commodities derived from agave. Indeed, low genetic diversity for cultivated agave facilitated outbreaks of plant diseases and insect pests, both of which have threatened agave production in recent decades. Meanwhile, conversions of desert areas to croplands reduced the populations of wild agaves that support the bat populations. Because of their diminished numbers, Mexican long-nosed bats were listed as endangered in the United States in 1988 and in Mexico in 1991. Thus, as a social dimension of ecology, one might well ponder if the future of a popular distilled spirit depends on the continued survival of wild agaves and bats. Photo courtesy of Carlos Alejandro Luna Arangoré.

Figure 7.7 Giant saguaros are the most recognizable indicator species of the Sonoran Desert but, because of their distinctive features, these plants are also widely regarded as an icon representative of all North American deserts irrespective of their own characteristic vegetation. Reproduced by permission of Brian R. Chapman.

after 20–50 years. Some eventually tower as much as 9 m (30 feet) above the desert floor and may live 150–200 years. After dying, saguaros leave behind a tough, fibrous wood-like skeleton that may endure as a "log" for many years after falling.

Dead and dying cacti, particularly those at Saguaro National Park, Arizona, understandably produced more than a little anxiety about the welfare of the community. Many of the larger saguaros exhibited what seemed to be a disease that first killed sizable areas of tissue and eventually the entire plant. "Bacterial necrosis disease" soon became widely accepted as the responsible agent, which in turn stimulated a good deal of research aimed at finding the pathogen and, of course, a remedy. However, instead of infections from a bacterial pathogen, saguaros were experiencing an ecological "disease," namely freezing temperatures. The large stands of cacti at Saguaro National Park occur at the northern edge of the species' range, where unfavorable temperatures limit further expansion of saguaro. At this cold-limited margin, periods of severe cold kill even the largest plants, although death may not occur until several years later.

Saguaros are valued for residential landscaping and by cactus collectors. Consequently, a black market has developed despite laws protecting the species, and smaller saguaro (those 10–30 years old) are clandestinely removed from the desert for transplanting elsewhere.

The organ pipe cactus is also distinctive of the Sonoran Desert. This large cactus is common in Mexico but occurs rarely within the borders of the United States. Organ Pipe Cactus National Monument in southwestern Arizona protects a part of the Sonoran Desert in the United States where this plant is relatively abundant (see also Infobox 7.2).

Mojave Desert

This desert lacks the diversity of plant life seen in other North American deserts; for example, the flora of the Mojave Desert does not include many types of cacti. Creosote bush alone (or with just one or two other species) frequently covers large areas. Of these, the combination of creosote bush and white bursage form one of the most characteristic communities in the Mojave Desert. Despite these features, some ecologists regard the Mojave Desert as a transition between the Great Basin Desert to the north and the Sonoran Desert to the south.

Creosote bush thrives on well-drained, oxygenated soils free of salt concentrations. The shrubs produce clones, often forming rings of 2–9 satellite plants; in undisturbed communities, these may be centuries old. In places, creosote bush seems regularly spaced in a uniform pattern across the desert floor, which has triggered numerous explanations (Fig. 7.8). One of these is **allelopathy**, the harmful effects caused by one plant on the welfare of another as a result of adding chemicals to the environment. Typically the plant itself produces the chemicals harmful to another plant; for example, the leaves of some species contain toxins that prevent seedlings from growing within a certain radius of the allelopathic plant. When the leaves fall, the toxins released into the soil beneath the canopy eliminate would-be competitors, including the seedlings of the same species, and create regular spacing in the surrounding vegetation. While allelopathy was once accepted as the primary reason for

Infobox 7.2 Desert seeds and endangered whales

A shrub known as jojoba (pronounced ho-ho bah) is endemic to the Sonoran Desert. The lipids contained in the large seeds of jojoba are not the least of the plants unique features. A total of 40–50% of the jojoba seeds consist of a liquid wax (a fatty acid attached to a long-chain alcohol), a biochemical substance not produced by any other plant and distinctly different from a fat.

Native Americans long ago used jojoba seeds for medicinal purposes and, perhaps, employed the liquid wax for cooking. Interest in jojoba products later spread to Europeans settling in the Sonoran desert. Later in the 1930s came the discovery of the plant's unique chemistry, including the realization that jojoba's liquid wax closely paralleled the oil (spermaceti) of sperm whales. The similarity gained public attention during World War II which interrupted whaling and limited the supply of spermaceti, a high-grade lubricant for delicate machinery. Such oils were difficult to synthesize in commercial quantities at that time. Jojoba received even more attention when sperm whales were protected by the Endangered Species Conservation Act of 1969. The liquid wax of jojoba offered a substitute that might eliminate much of the overhunting which was then pushing sperm whales to extinction.

Jojoba was touted as a crop of economic potential and, given its native range and adaptations, might be cultivated on plantations in the Sonoran Desert. The conversion of agricultural fields from crops requiring irrigation to jojoba also frees up water to meet the ever-increasing needs of Tucson, Arizona. However, estimates suggest that as many as 10 years may be required before seed production reaches a commercial level, and the low yield of seeds never matched the demands. These drawbacks thwarted efforts to establish commercial-scale jojoba plantations, including those maintained on reservations for the benefit of Native Americans.

Luckily, the development of synthetic lubricants of equal quality eliminated any claims for resuming the harvest of sperm whales for their oil, and the species continues to recover from more than a century of exploitation. Today, an increased demand for jojoba oil skin care and shampoo products, cattle feed, and nutritional supplements is supplied by crops of high-yield clones grown in arid environments worldwide. The unique biochemical properties of jojoba illustrate the value of maintaining biological diversity, even in so-called desert wastelands.

Figure 7.8 The consistent distances between creosote bushes in the Chihuahuan Desert suggest a pattern caused by allelopathy, but other factors may also influence spacing between the bushes. Reproduced by permission of Brian R. Chapman.

the spacing in creosote bush, newer evidence indicates that moisture conditions beneath older, established shrubs may be limiting seedling survival.

Joshua trees are by far the most distinctive plant of the Mojave Desert (Fig. 7.9). Indeed, the distribution of this unusual yucca closely coincides with the area designated as Mojave Desert, but the plants are limited to higher elevations within the region. Unlike yuccas with the familiar basal rosette of long, stiff and sharply pointed leaves surrounding a head-high central stalk, Joshua trees reach heights of about 12 m (40 feet) with clusters of short, needle-like leaves extending their arm-like branches. The fruits borne by Joshua trees are also different from those developing on other species of yucca. In favorable sites, the plants reach densities of more than 100 ha^{-1} (40 per acre). The branches of Joshua trees seem to gesture, and legend holds that Mormons crossing the Mojave Desert were urged onward by the beckoning plants which the travelers named after the Old Testament patriarch who led the way into Cannan. Yucca night lizards, a small species barely 7 cm (3 inches) long, find key habitat in the dead stems and leaves of fallen Joshua

Figure 7.9 Joshua trees, a tall species of *Yucca*, serve as indicators of the Mojave Desert. Reproduced by permission of Wendy Hodgson.

trees, but otherwise they are not accurately named; although these "night" lizards forage after dusk, they remain fully active during the daytime.

Death Valley, the lowest point in North America, lies within the Mojave Desert. A temperature of 57°C (134°F), the hottest on record for the United States, occurred in Death Valley in 1913 and temperatures of 52°C (125°F) are not uncommon. Rainfall in Death Valley averages a scant 5 cm (2 inches) per year, and no rain fell at all in 1929 and 1953.

Great Basin Desert

The Great Basin contains the only "cold desert" in North America. Both the vegetation and physical setting in the Great Basin resemble cold deserts elsewhere in the Northern Hemisphere (e.g., the Gobi Desert in Asia). These deserts experience great extremes in daily and seasonal temperatures and they occur in large, enclosed basins favoring the development of saline soils and halophytic communities lacking species richness.

The Great Basin Desert, which lies in an extensive bolson between the Rocky Mountains and on the western rim the Sierra Nevada and the Cascade Mountains, includes most of Utah and Nevada; additional areas lie in bordering states (e.g., western Wyoming and much of southern Idaho). Mountain ranges within the Great Basin interrupt the presence of desert vegetation, and the taller of these support the Montane Forest (Chapter 9). Great Salt Lake is the largest body of water within the Great Basin, but it represents only a remnant of an immense lake once covering a far larger area (see *Ancient lakes*, this chapter). Most of the Great Basin Desert lies north of the 36th parallel, but a lobe – sometimes identified as the Painted Desert – extends southward into northwestern New Mexico and along the drainage of the Little Colorado River in northern Arizona. This region of the Great Basin Desert includes remarkable arch formations in southeastern Utah and a forest of petrified logs in northern Arizona. Lying scattered across the desert floor, the logs were pines that apparently drifted down an ancient river to their present location during the Triassic Period (circa 225 million years ago).

One or more species of sagebrush, saltbush, rabbitbrush, and greasewood are among the dominant plants in the Great Basin (Fig. 7.10). These and other shrubs typical of the desert are multi-branched and form relatively compact plants. They have evergreen leaves which are generally aromatic, but most lack spines. In many locations, largely because of soil salinity, large areas of vegetation may consist of essentially one species (e.g., shadscale). Sharp boundaries often mark the edges of these stands, producing clearly defined zones of vegetation. Rainfall is sparse during the growing season, and accordingly the shrubs are deep-rooted and experience limited growth during a short period each spring.

Creosote bush is notably absent in the Great Basin Desert, whereas this species occurs throughout the three "hot deserts" in North America. The northern limit of creosote bush separates the two types of desert (e.g., an ecological boundary where the Great Basin borders the Mojave Desert).

Major grasses include bluebunch wheatgrass and squirreltail, but several other species are also important (e.g., Indian ricegrass). In the northern Great Basin, grasses form a significant part of the sagebrush communities; some ecologists therefore recognize this vegetation as *shrub steppe*, a term also applied to rangelands associated with the Palouse Prairie (Chapter 6). However, grasses are uncommon on the southern edge of the Great Basin Desert, where sagebrush may grow to the virtual exclusion of grasses. Elsewhere, cheatgrass has invaded much of the Great Basin, especially in response to heavy livestock grazing. This Eurasian grass provides an abundance of highly flammable fuel in what was previously a barren understory, thereby increasing the frequency of fire in sagebrush communities. The latter are heavily damaged by fire and, because its roots do not resprout after fires, sagebrush is slow to reoccupy burned sites.

Figure 7.10 Rabbitbrush represents the predominant vegetation in this view across a valley floor in the Great Basin, western Nevada. Reproduced by permission of Brian R. Chapman.

With repeated fires, sagebrush communities may give way to rabbitbrush and other sprouting shrubs.

Sagebrush foliage is enriched with compounds known as volatile oils. These compounds burn readily and thereby contribute significantly to the susceptibility of sagebrush to fires. Their occurrence underscores the notion that sagebrush did not evolve in an environment regularly subject to fire. Instead, volatile oils perhaps evolved as a protection against grazing. The oils apparently interfere with the microbial activities that ruminants require for digesting plants materials, but the issue is not fully resolved. In any event, both domestic and native grazing animals generally avoid sagebrush.

Forbs are not abundant in the Great Basin Desert except locally, but globemallows are typical representatives. Halogeton is an exotic forb whose toxicity at times poisons large numbers of sheep. Russian thistle ("tumbleweed") has also invaded most of the Great Basin Desert. Cacti are not commonplace either in abundance or in species, and those that do occur are typically short or prostrate. Trees are absent from the Great Basin Desert, except for cottonwoods and willows along some of the better-watered stream beds.

Highlights

Nurse trees

Plants and animals have evolved numerous means for coping with harsh desert environments. One of these is an ecological association involving **nurse trees**, so-called because they provide the microhabitat necessary for the establishment and survival of other plants

(Fig. 7.11). George Engelmann (1809–1884), a botanist attached to the United States–Mexican boundary survey, described young saguaros as "…almost always found under the protecting shade of some shrub, especially 'green-barked acacia'…." Two species of nurse trees, palo verde (Engelmann's "green-barked acacia") and desert ironwood, are of particular importance to saguaro, the distinctive cactus of the Sonoran Desert. For example, the desert ironwood's shade cools the immediate environment, enhances soil moisture, and protects the underlying vegetation from harmful radiation. Beneath their branches, ironwood also produces "pools" of decomposing organic matter from their fallen leaves, thereby enriching and developing a soil profile unlike those barren mineral soils of the open desert.

Saguaro produce large numbers of seeds, but perhaps no more than 1 in 1000 yields a successful seedling. The seeds are dispersed by various agents, among them white-winged doves which often construct their nests in palo verde trees. Adult doves eat saguaro seeds but, when feeding their nestlings a regurgitated substance known as "pigeon milk," the birds at times also regurgitate some seeds which fall to the ground beneath the nest. Saguaro seeds dropped in this way then germinate and thrive in the immediate "care" of a palo verde tree. Similarly, the sowing and germinating of seeds may also occur under ironwood and mesquite trees, where white-winged doves also nest.

Desert ironwood is a legume and a **monotypic taxon**, the only species in its genus. About twice as many species of perennial plants grow in the shade of ironwood trees than occur in equivalent areas lying beyond their shade; all told, 31 species occur exclusively in the understory

(a)   (b)

Figure 7.11 (a) A young saguaro cactus develops under the protection of a "nurse tree". (b) Years later, evidence of the nurse tree will disappear, leaving this and other giant saguaros standing prominently in the Sonoran Desert. Note the cavity near the tip of the uppermost trunk. Reproduced by permission of Wendy Hodgson.

beneath ironwood. When measured in terms of the richness and abundance of plant life, the ecological conditions provided by ironwood at dry sites match those at sites watered by desert streams. Desert ironwood also offers cover for wildlife, which in turn often disperse the seeds of plants growing in the understory.

The flowers of desert ironwood are purplish, but seed pods are produced irregularly and seedlings may survive only a few times in several decades. Limited reproduction of this sort likely reflects the erratic occurrence of favorable moisture conditions. Bufflegrass, a species introduced from India and Africa, threatens ironwood and the welfare of an entire community today. Bufflegrass rapidly accumulates a layer of highly combustible litter, thereby providing fuel in an environment where few plants are adapted to fire. Moreover, ironwood burns readily, usually to the ground. Given these circumstances, fires eliminate desert ironwood along with its nourishing shade and instead produce a dry grassland where conditions no longer favor other kinds of life. Wood carvers in Mexico also prize desert ironwood; artisans harvest large quantities each year for their handicrafts. Fortunately, the carvers prefer wood from dead branches, so greater threats to desert ironwood result from its demand for firewood and from clearing land. Ironwood is not an endangered species, but its slow capacity for regeneration, susceptibility to fire, and key role in desert ecology suggest that the species be closely watched.

A type of **symbiosis** known as **commensalism** is the usual interpretation for an ecological association between nurse trees and, in this case, saguaro seedlings. Commensalism is defined as the relationship between two species in which one gains some benefit, whereas the other is neither helped nor harmed (i.e., symbolically, + and 0). In this case, saguaros (+) enjoy the protection of palo verde or other species of nurse trees (0). The interaction may be of another kind, however, involving the harsher aspects of interspecific competition (at least in the long term). As the saguaros grow, their shallow network of roots steadily expands and intercepts water before it can reach the deep tap roots of their nurse trees. As a result, the vigor of palo verde serving as a nurse tree is reduced as shown by the greater incidence of dead limbs and mortality compared with those not protecting saguaro cacti. Instead of a commensal relationship, the association therefore seems closer to predation in its basic features (i.e., + and −). In any case, the long-term influence of maturing saguaros on their nurse tree alters the structure of the local vegetation. Any given area within the Sonoran Desert may show a mosaic resulting from the long period of development of these species.

"Trees" for desert woodpeckers

Two species of woodpeckers excavate cavities in saguaros, which in this regard function as trees in much of the Sonoran Desert. These are the Gila woodpecker and the somewhat larger gilder flicker, a subspecies of the widely distributed northern flicker discussed in Chapter 5. The giant cacti respond to the excavations of these woodpeckers

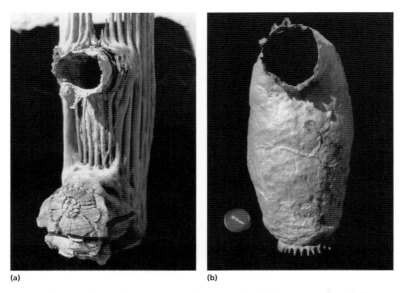

(a)   (b)

Figure 7.12 (a) The exposed "skeleton" of a giant saguaro shows the vertical rods that support the tall plants. A gilded flicker excavated the cavity by chiseling through the woody rods to the pith inside. (b) In response, the saguaro produces a "boot" of corky material, much like a cast, to seal off the damaged tissues inside the stem. Reproduced by permission of Wendy Hodgson.

by forming hard layers of cork material that seal the living tissues exposed by the excavation. The results are encasements known as "boots," which are durable and persist on the ground long after the saguaros in which they formed have fallen and decayed. Such boots provide ecologists with useful evidence concerning woodpecker cavities and the anatomy of saguaros (Fig. 7.12). Other birds (e.g., elf owl) and some bats also live in the cavities that woodpeckers excavate in saguaros. An exotic species, the European starling, recently invaded the Sonoran Desert and competes for residence in cavities occupied by Gila woodpeckers, but it does not compete with gilded flickers.

Gilded flickers usually chisel their nest cavities within 3 m (7 feet) of the tip of the main stem, whereas those of Gila woodpeckers are located over a broader range of heights but generally lower on the stem. However, these preferences for different parts of the stem are not the result of interspecific competition. Instead, the vertical position of the cavities is related to the anatomy of the cactus itself and to differences in the chiseling abilities of the two species.

An inner vertical skeleton of woody rods that are arranged in a circle support the soft, water-holding tissues of saguaros. These rods are thickest at the base of the plant – each is about the diameter of a large thumb – but taper to the size of a straw toward the apex of the main stem. An outer cortex of fleshy tissue surrounds this skeleton, which encloses an inner pith of additional soft tissue. The thickness of the cortex reaches a maximum of about 16 cm (6 inches) and steadily thins toward the tips of the plants.

Gila woodpeckers excavate small cavities which are typically located entirely within the cortex; the birds excavate only in fleshy tissues, halting when they encounter the skeleton of woody rods. By selecting the lower parts of saguaro stems, Gila woodpeckers avoid chiseling into hard materials. In comparison, the larger gilded flickers cannot limit their cavities to the cortex, which is not thick enough, and must instead chisel through the rods and into the pith to hollow out a cavity large enough for their needs. To accomplish this they excavate toward the tips of saguaros, where the rods are thin and relatively weak. Gilded flickers generally forage on the ground where they seek ants, but this type of feeding behavior limits the evolution of a bill fully suitable for chiseling into thick wood. Therefore, if gilded flickers excavated farther down the stems of saguaros, they would encounter rods too thick to penetrate.

The two woodpeckers differ in the damage they inflict on saguaros. Because Gila woodpeckers typically excavate entirely within the outer cortex, they produce little harm except for the loss of a small amount of soft tissue (i.e., relatively little loss of water-storing capacity and a meager reduction in photosynthetic activity). In contrast, gilded flickers cause greater damage. Vital structural support is lost when the rods are severed – they seldom regenerate – and stems above the cavities may fall. Decapitated and weakened, many of the plants become prematurely moribund. The rods also transport water, so the stems above gilded flicker cavities may shrink even when only some of the rods are cut. This in turn may produce unfavorable volume-to-surface area

(a)   (b)

Figure 7.13 (a) The unusual profile of the boojum tree resembles an upside-down carrot, but the name *cirio* is used in Mexico because the slender trees suggest the shape of altar candles. (b) Bark regularly peels from the swollen trunks of elephant trees. Both of these desert trees may be leafless for much of the year, and each has a limited distribution in North America. Reproduced by permission of Wendy Hodgson.

ratios in the upper stems, which can increase water loss as well as increase the chances of frost damage.

Boojums and elephants: unique trees

Two truly unique North American plants, the boojum tree and the elephant tree, are endemic to the Sonoran Desert. To some observers, "typical" boojum trees appear as upside-down carrots, although the tips of some develop whip-like branches (Fig. 7.13a). To Jesuit padres of the 18th century, the slender trees suggested altar candles; the name *cirio* (a taper) therefore emerged as a local name and is still commonly used in Mexico. In 1922 however, a botanist familiar with Lewis Carroll's nonsense poem *The Hunting of the Snark* recalled Carroll's appellation *boojum* for a mythical entity found in desolate far-off regions. On viewing the strange trees through a telescope, the botanist remarked to colleagues "Ho, ho, a boojum, definitely a boojum," and the name persisted thereafter.

Full-grown boojums may reach a height of about 15 m (50 feet), tapering upward from a base seldom more that about 50 cm (20 inches) in diameter. A few may have as many as 10 stems arising from ground level. Individual trees are often much contorted, and some bear branches dropping grotesquely toward the ground. Spines appear on new stem growth, but these wear away with age. Small leaves of two types – those borne singly or in clusters – cover the trunk (and branches, if any) at times, and panicles of highly aromatic flowers develop in season at

the tips. Boojum leaves may be shed during dry periods, and more than one crop may be produced per year. At locations near the Pacific coast, heavy coats of lichens sometimes suggest the presence of foliage on otherwise barren trees.

Two sites in the Sonoran Desert represent the entire range of the boojum tree. The larger area is a 400 km (250 miles) section forming the waist of Baja California. Boojums also occur on a smaller 48 × 5 km (30 × 3 miles) strip along the coast of Sonora. The two sites, about 120 km (75 miles) apart, are separated by the Gulf of California (= Sea of Cortez). The more extensive site in Baja California forms a subregion within the Sonoran Desert known as the Central Desert, and hosts what may be the world's richest variety of desert plants.

The disjunct distribution of boojum can be explained by one of two ideas. First, the mainland population of boojum trees – those occupying the small area of coastal Sonora – might represent a relict from an earlier time. Geologists agree that mainland Sonora and Baja California were united about 4 million years ago but thereafter parted along a geological fault, perhaps separating boojums into two populations. Second, storms may have blown boojum seeds from one site – presumably Baja California – to the mainland site in Sonora. Whatever the case, extreme aridity and other climatic features apparently limit boojum trees to a small area on the Sonoran mainland, whereas habitat conditions in Baja California allow a somewhat broader distribution.

Another unusual tree also endemic to the Central Desert region of Baja California is the elephant tree, so named because its swollen trunk resembles the thick legs of elephants (Fig. 7.13b). Elephant trees seldom reach heights much beyond 4.5 m (15 feet), yet their columnar trunks may be 1 m (3 feet) in diameter. The tree's swollen appearance is exaggerated because the massive branches that arise from the trunk taper rapidly. They have smooth, light-colored bark that peels repeatedly in a fashion somewhat like a paper birch. Like boojums, elephant trees may be leafless much of the year and develop in a contorted and somewhat grotesque fashion. Near the Pacific coast however, salt-bearing winds stunt their development; although the trees hug the ground like a dwarfed hedge, they still possess their elephantine trunks.

Yucca moths

Deserts, especially the Chihuahuan Desert, are rich with yuccas. Many species feature a basal cluster of dagger-like leaves and an elongated stem bearing a prominent inflorescence of whitish flowers. Other yuccas resemble trees whose trunks retain a thatch of dead leaves (e.g., Joshua tree). Among the ecological features of these plants is their intimate relationship with small white moths aptly known as yucca moths.

The mouth parts of female yucca moths include curled tentacles whose function is to collect pollen. Once laden with a mass of sticky pollen, the females seek another plant of the same species. The insects display consider-able ability to distinguish among the various species of yucca (of which there are more than 40, although not in the same region). Once at the second plant, the female first deposits her eggs in the ovary of a flower and then pollinates the same flower with the pollen from the first plant – a process that may be repeated several times. When the eggs hatch, the larvae feed on some – but not all – of the seeds developing in the ovary or on the ovarian tissue itself; they then crawl from the fruit and drop to the soil where they pupate.

Yuccas appear to depend on moths for pollination and, in turn, the moths require the developing fruits of yuccas as nurseries, another discovery (in 1872) by frontier botanist George Engelmann. Ecologists traditionally considered this tight relationship as the premier exem-plar of **obligate mutualism**, which in this instance is specifically known as brood place pollination. Obligate mutualism is rare in nature, presumably because: (a) either partner faces extinction if the other fails; and (b) the geographical distribution of one species is limited by the range of the other.

All told, the yuccas and yucca moths seemed wholly interdependent for successful reproduction until new evidence suggested a somewhat less-dependent asso-ciation than previously believed. In addition to the moths, other insects at times may pollinate the flowers of some yuccas, although none of these apparently depend on the plants for successful completion of their own life cycle. The ecological exchange between yuccas and yucca moths, although still widely practiced, may there-fore be more accurately regarded as **facultative mutu-alism**, a beneficial but not rigidly obligate relationship.

Desert fishes

Visions of deserts seldom include fishes or even aquatic sites suitable as fish habitat. Nonetheless, several fishes have evolved in some small, isolated, and often harsh aquatic sites in North American deserts. These fishes and their environments are noteworthy because they illustrate evolutionary processes in North America in ways parallel-ing those Charles Darwin witnessed among the finches in the Galapagos Islands. The welfare of these fishes also reflects the health of desert ecosystems, where large num-bers of other species depend on the same aquatic habitats for survival. In other words, the ecological importance of an aquatic site in a desert does not stop at the water's edge.

Pupfishes are among the better-known groups of desert-dwelling fishes. Some of these survive exclusively in Death Valley, one of the most stressful environments on Earth. In particular, pupfishes have evolved physiological mechanisms for coping with temperature extremes and high salinities. Water temperatures, for example, may vary from almost 0°C (32°F) in winter to about 36°C (97°F) in summer, although under some conditions these heat-tolerant fishes may survive water temperatures of 42–44°C (108–111°F).

The events leading to the endemic distribution of pup-fish and other desert fishes began in the Pleistocene Epoch. About 10,000–20,000 years ago, the environ-ment was cool and wet in Death Valley and other desert areas in North America, resulting in a network of lakes and streams lying within the intermountain basins. Fishes and other aquatic organisms moved freely through this network, which included contact with major river systems (e.g., Colorado River). A warmer, drier climate later produced arid conditions that diminished the aquatic system into isolated springs and small streams, which became refugia for remnants of the once wide-spread fish fauna. The water emitting from the springs in Death Valley exceeds the amount available from annual precipitation, so much of the current surface flow origi-nates from an ancient underground reservoir.

The remnant fish populations in Death Valley thus evolved in isolation. Today, for example, 20 pupfish populations persist in Death Valley, represented by five species of which one has further separated into six sub-species. The latter have been isolated for 400–4000 years and were interconnected at times during that period, whereas the others have been fully isolated for 10,000–20,000 years.

The Devils Hole pupfish is perhaps the most famous of the desert fishes in North America. With a length seldom exceeding 2.5 cm (1 inch), the species is the

Figure 7.14 A view into Devils Hole in Death Valley National Park, Nevada. The submerged limestone shelf in the shallow water represents the entire spawning habitat of the Devils Hole pupfish (inset), which may be the world's rarest fish. The equipment monitors water levels in the aquifer-fed pool. Reproduced by permission of Stan Shebs and the (inset) US Fish and Wildlife Service.

smallest of its kind and lacks pelvic fins. More unusual, the species is endemic to a single spring in the Death Valley region of southwestern Nevada. Its limited range of Devils Hole consists of about 19 m² (207 square feet) of surface water, the smallest of any species of vertebrate. In the past, irrigation pumping reduced the water level in the spring, which gradually exposed the surface of a limestone ledge where the pupfish feed and nest (Fig. 7.14). The population of Devils Hole pupfish declined to precarious numbers until court action modified the pumping, so that enough of the ledge remained underwater to maintain a stable population. Thereafter, the population grew to about 200 pupfish (based on spring counts which are somewhat lower than fall counts) until about 1995. For the last two decades, however, the population steadily declined, reaching an all-time low of 35 in the spring of 2013. Unfortunately restoration measures which established three additional populations failed, as did attempts at rearing the pupfish in captivity. Currently, Devils Hole pupfish again survive only at Devils Hole and the species remains endangered, a dire reflection for its continued existence.

Realm of reptiles

Reptiles are especially well represented in the fauna of North American deserts. Of these, the variety of snakes and lizards is particularly rich, but chelonians are represented primarily by just two species: Agassiz's desert tortoise and Morafka's desert tortoise. The full geographic range of these desert tortoises includes much of the Sonoran Desert in both the United States and Mexico, exclusive of Baja California. Agassiz's desert tortoise occurs west and north of the Colorado River (i.e., in Utah, Nevada, northern Arizona, and California), whereas the distribution of the recently described Morafka's desert tortoise includes southern Arizona south of the Colorado River and Mexico. Unfortunately, because of human disturbances (e.g., off-road vehicle traffic), Aggassiz's desert tortoise is currently on the federal list of threatened species in the Mojave Desert.

Like the three other species of tortoises in North America, Agassiz's and Morafka's desert tortoises are adapted for life in an arid habitat. Tortoises live for decades and take about 15 years to mature sexually. As adults, tortoises construct catchment basins and drink from the puddles in these during rains. Juveniles, however, may experience greater mortality during dry years, perhaps because they experience difficulty acquiring sufficient nutrients for growth and survival. Rainfall heightens the activities of desert tortoises, which seemingly appear relatively plentiful during storms but become far less noticeable at other times.

Livestock grazing is often implicated with the deterioration of habitat favored by desert tortoises in the western United States and Mexico. In this view, grazing alters the composition of the plant community, particularly leading to losses of perennial grasses, spring ephemerals, and other plants that the tortoises favor for food and cover, but the relationship is not clearly established. Overall, forage quality may be more important than quantity to the welfare of desert tortoises, but the nutritional requirements for the species remain undetermined. Desert tortoises have a curious habit of ingesting weathered bones, stones, and calcium-rich soils. This behavior suggests a nutritional demand for minerals, which the tortoises require to supplement their diet of herbaceous foods.

Gila monsters are one of the more notable reptiles in North America, particularly in the Sonoran Desert. These thick-bodied, fat-tailed lizards are venomous, a trait shared only with one other of the world's lizard fauna. Gila monsters are covered with bead-like scales forming a pattern of black markings on a yellow-to-orange background. Adults reach a length of 25–40 cm (10–16 inches).

Unlike rattlesnakes and other venomous snakes, the role of the delivery system in Gila monsters is primarily defensive and not used as a means of immobilizing prey. Gila monsters feed on the eggs of birds and other reptiles, as well as on small mammals. Such foods are simply seized, crushed, and swallowed, without employing venom.

Also unlike venomous snakes, Gila monsters deliver venom from their lower jaws where ducts from two venom glands carry the venom to the bases of 8–10 grooved teeth on the dentary bone. When provoked, Gila monsters lunge forward, or quickly turn their heads sideways, and bite with a tenacious "bulldog" grip. Venom is then "chewed" into the victim, not injected as is the case with the fangs of venomous snakes. The venom of Gila monsters is a neurotoxin – similar to that of cobras – and therefore quite dangerous. However, because they are not aggressive unless provoked, Gila monsters seldom bite humans. They become hazardous only when handled or when humans foolishly reach under rocks where one of these unusual lizards might be hiding. Gila monsters are protected in the United States (they are illegal as pet trade) but habitat deterioration remains a threat to their welfare.

A species of rattlesnake aptly known as the sidewinder is uniquely adapted to sandy areas of desert. These snakes move in a series of lateral looping movements in which only two parts of the body usually touch the sand at any time. These movements produce unconnected J-shaped tracks that angle diagonally forward along the route of travel. Sidewinders thus skip across the sand, although at times they move with the same undulating crawling motion typical of other snakes. Although sidewinding movements are an effective means for moving in sandy deserts, they may also provide some help with thermoregulation, that is, only two points are in contact with the sand which is extremely hot during the day and quite cool at night. Horn-like projections over their eyes distinguish sidewinders from any other snakes in the deserts of North America. No clearly defined function has been assigned to these projections, although they may protect the eyes of sidewinders from abrasion. Subspecies designations reflect the geographic affinities of each with major desert regions (e.g. Sonoran sidewinder and Mojave sidewinder).

Of soils and mice

The soils of some deserts are variously colored, in keeping with local or regional geological developments. These conditions are especially prevalent in parts of the Chihuahuan Desert, where black soils derived from lava are scattered among much lighter-colored soils. The contrasting nature of this "checkerboard" habitat became a laboratory for studies of adaptive coloration in small rodents. At least five taxa of rodents, for example, are markedly darker on the black lava beds of the Tularosa basin in south-central New Mexico, whereas pale-colored rodents are associated with the White Sands area in the same vicinity (e.g., a nearly white race of Apache pocket mouse, which is endemic to this unusual site). Black pigmentation indicates **melanism**, the opposite of **albinism**, and **polymorphic** populations result when the gene pool produces individuals with various amounts of black coloration. White-throated wood rats and rock pocket mice are among the rodents often showing clear evidence of "color matching" with the desert soils in their habitats. The bleached earless lizard also matches the light-colored gypsum soils at White Sands National Monument.

The relationship between the pelage and soil color readily illustrates the workings of natural selection. Such a correlation does not result from the conscious decision of the rodents themselves of course, but instead is the consequence of a steady "weeding out" (i.e., selection against) of individuals with traits that place them at some type of disadvantage. In this case, selection likely works against those individuals whose pelage colors offer little protection against detection by predators. Pale-colored rodents therefore gain an advantage on light-colored soils (i.e., camouflage) but are in greater jeopardy where lava beds formed darker backgrounds, with the reverse acting on rodents with dark pelage. Local populations of distinctively colored rodents therefore result from natural selection. Other research, conducted under laboratory conditions, later illustrated the effectiveness of background coloration as a means of reducing predation in similarly colored mice.

These circumstances echo the basic workings of evolution, namely that genes of individuals occur in a multitude of combinations; each of these combinations produces variations in physical, behavioral, and other traits, or pelage color in this case. Accordingly, some individuals are eliminated, whereas those with advantageous traits survive and pass on their genes to the next generation. Populations adapted to their environments steadily emerge. To summarize, natural selection operates on individuals, whereas populations are the manifestation of evolutionary success.

Deserts and predators

Desert fauna in North America, and indeed throughout the world, include a high percentage of predators. This relationship, which is not clearly understood, holds true for invertebrates as well as vertebrates. In forests for example, 25–33% of the invertebrate fauna are predators, but in semiarid and arid regions the proportion increases to 33–50%. About two-thirds of the avifauna in the Mojave Desert are partially or fully carnivorous. Moreover, a large percentage of the cavity-nesting birds in the Sonoran Desert select vegetation endowed with spines, which hints at the ecological importance of predators in desert environments. Desert waterholes represent sites where prey gather and seem especially vulnerable to predators (Infobox 7.3).

Infobox 7.3 *El tigré! El gran gato de las Americas*

Jaguars range over much of South and Central America, living primarily in rainforests and savannahs. At the northern edge of their distribution however, these secretive cats occur in desert and near-desert environments. Early records indicate jaguars extended into Arizona and New Mexico, and once roamed widely in Texas north to the Red River and perhaps eastward into Louisiana. According to John James Audubon (1785–1851), Texas Rangers reported jaguars as most commonly seen at watering-places frequented by mustangs and deer, and the Rangers prized the rosette-spotted skins as saddle blankets and holster covers. However, like wolves, grizzlies, and other large predators, jaguars were extirpated early in the course of Euro-American settlement and have not been recorded in Texas since 1946. More recent sightings in the United States, including those confirmed by photographs, are largely from mountainous areas in chaparral and desert grassland regions of Arizona and New Mexico. Most of these records originated within 100 km (50 miles) of the border with Mexico.

With adult males weighing up to 158 kg (348 pounds), jaguars are the third-largest felid on Earth and the largest in the Americas. Their bite is one of the strongest of all terrestrial predators; when hunting in aquatic habitat jaguars easily crush turtle shells and pierce the thick hides of caiman. Peccaries, armadillos, fish, and (in South America) capybaras are favored foods. Jaguars pose virtually no threat to humans and records to the contrary are rare or questionable. Some individuals appear totally black, but close inspection reveals the distinctive rosette pattern typical of all jaguars. The melanistic morph – so-called "black panthers" – is inherited as a monogenic dominant of the normal coloration and is not representative of a recessive trait as in leopards.

Much debate surrounds the status of jaguar populations in the United States. Like many predators, especially solitary species with large home ranges, jaguars are difficult to census, but most agree the population is quite limited. This limited population supports the interpretation that the relatively few occurrences – past and present – resulted from transients entering the United States from a "source" population in northern Mexico. Others believe jaguars in the American Southwest represent a resident but persistent breeding population steadily thinned by hunting and shrinking habitat.

Today jaguars are protected by the Endangered Species Act in the USA, and various laws in most other nations within their range offer some degree of legal oversight (Ecuador and Guyana being notable exceptions). In 2012, the US Fish and Wildlife Service proposed designating 3392 km² (1310 sqaure miles) – an area larger than Rhode Island – as **critical habitat** for jaguars. The designated area, later enlarged by 2%, includes mountain ranges in Arizona and New Mexico that serve as important corridors for jaguars crossing the US–Mexican border. This action also recognizes that even marginal habitat at the edge of a larger core area may provide a species' gene pool with individuals capable of adapting to environmental extremes. Some opinions differ however, and it has been suggested that resources are better directed at maintaining habitat for breeding populations south of the border, notably in Sonora, Mexico, if jaguars are to continue their presence in the American Southwest.

Gene flow via jaguars moving north from Mexico into the United States is among the controversies concerning construction of a fence along the border, as an impermeable fence would deter wildlife as well as human immigration. At present, such fences are not a deterrent for jaguars in the area designated as critical habitat, and none is anticipated at the site unless needed to counteract a threat to national security. Many conservationists nonetheless remain alarmed about the issue and its effect on borderland wildlife. Meanwhile, we say to *El tigre, "Bienvenido de nuevo a los Estados Unidos!"*

Pygmies of the sagebrush steppe

North America's smallest rabbit (adults weigh just 375–500 g or 13–17.6 ounces) occurs widely in the Great Basin and some adjacent intermountain areas from southwestern Montana to eastern California (Fig. 7.15). In 2003, an isolated population of pygmy rabbits in the Columbia Basin of Washington state was listed as endangered only to disappear in the wild the following year. The species is strongly associated with sagebrush, particularly a community often known as a sagebrush steppe in which bunchgrasses are well represented. Appropriately, the species is identified as a sagebrush specialist. Rabbitbrush and greasewood are among the other species of woody vegetation occurring in association with the habitat of pygmy rabbits.

Pygmy rabbits prefer sites on deep soils where tall, dense stands of sagebrush offer optimal food and cover. Sagebrush consists of nearly 100% of the rabbits' winter diet and about 50% in summer, with grasses and forbs making up the balance. Their liver functions allow pygmy rabbits to minimize the absorption and maximize the detoxification of terpenes and other volatile oils in sagebrush foliage. Unlike most other rabbits, they dig burrows which are used primarily for winter shelter and less often at other times. The burrows are often located at the base of sagebrush bushes and lead to tunnels that widen into chambers about 1 m (3.3 feet) below ground; most have 4–5 entrances but some may have as many as 10. Denning sites are likely selected based on **edaphic** features, particularly soil depth and the ease of digging. Where soils are unsuitable, pygmy rabbits den in stone walls, holes in volcanic rocks, as well as in burrows dug by other species or around abandoned buildings. They also tunnel extensively under deep snow in search of forage.

A variety of predators including foxes, coyotes, hawks, owls, and especially weasels feed on pygmy rabbits. In winter, the overall annual mortality for adults reaches 88%

Separated from other populations for at least 15,000 years, the Columbian Basin pygmy rabbit developed its own genetic signature. The long period of isolation, coupled with small numbers, significantly reduced its genetic diversity that in turn affected reproductive success, however (e.g., **inbreeding depression** reduced pregnancy rates). Unfortunately, pygmy rabbits throughout their range are threatened by habitat loss and fragmentation, particularly as a result of fires and heavy grazing. Sagebrush is extremely sensitive to fire, which is often used to remove this and other woody vegetation as a means of producing more grass on grazing lands, and a burned area may take decades to recover after a fire. Further, pygmy rabbits are poor dispersers and are reluctant to travel across open areas; to thrive, they therefore require large blocks of unaltered vegetation.

Recovery efforts in Washington focus on a captive breeding program using stock removed from the Columbia Basin population intercrossed with those from other populations in the Great Basin (e.g., Idaho). Regrettably, the source population available for this work was small, just 16 adults whose **effective population size** actually represented fewer than six founders. The goal of the program is to produce and release offspring that retain at least 75% of the genetic profile of the endangered population but which are also "energized" by genes from other populations. Fortunately, intercrossing improved

Figure 7.15 A pygmy rabbit, the smallest of North America's rabbits and hares, emerges from one of its many burrow entrances in the Great Basin Desert. Even as adults, the diminutive size of this endangered species is little more than a handful (inset). Reproduced by permission of Rod Sayler.

Figure 7.16 After tumbling from a nearby mountainside, these stones "sailed" across a dry lake bed known as Racetrack Playa in Death Valley National Park, Nevada. After winter storms deposit a film of water on the playa's fine clay soils and the water freezes overnight, strong wind gusts propel large sheets of ice across the playa surface after thawing begins the next morning. The moving ice sheets push the stones, creating furrows in the slick surface. Curvilinear tracks in the lake bed indicate changes in wind direction. Reproduced by permission of Paula Messina.

both reproductive rates and litter sizes, and several hundred juveniles have been released since 2011, some of which have successfully produced their own offspring.

Desert quail, rainfall, and vitamin A

Gambel's quail occur widely in the Mojave Desert and other warm deserts in North America. These birds must drink to survive and therefore visit springs and other sources of water on a daily basis. For their breeding success, however, Gambel's quail require rainfall. In this relationship, rain stimulates the production of green foliage, which in turn contains carotene, the precursor of vitamin A. In dry years desert plants yield little carotene, and the autumn populations of Gambel's quail may include as few as 19% juveniles. With abundant rainfall, however, the percentage of juveniles in autumn may jump to nearly 75%, strong evidence of successful breeding. An examination of the quail's reproductive system explains these differences: the organs fail to develop without adequate vitamin A.

By such means, the reproductive efforts of Gambel's quail are linked to years when young birds have the best chance of surviving. In other words, the adults do not reproduce when rainfall is insufficient, preventing wasteful expenditure of energy for fruitless results.

Sailing stones

In addition to its scorching heat and fabled aridity, Death Valley is home for two sites where rocks "sail" across smooth **playa** floors leaving long, shallow paths to mark

their passage (Fig. 7.16). One site, playfully named "Racetrack Playa," underscores these curious movements. The stones, which originate from surrounding outcrops, roll down the slopes and onto the flat, dry lakebeds where some later "sail" for varying distances (some tracks extend for 250 m or 820 feet) from their previous locations. These movements, which involve rocks weighing up to 36 kg (80 pounds), occur only during the winter months.

The forces propelling the rocks remain unclear despite more than a century of study. However, recent research which filmed stones mounted with GPS units now seems to explain these mysterious movements. The process begins when rainwater freezes overnight and forms a blanket of ice on the playa's shallow bed. Later, when the morning sun initiates melting, the ice breaks into thin sheets – some up to 16 m (50 feet) wide – that even gentle winds can push across the slick clay surface of the playa floor. The force provided by the moving ice sheets slides the stones on meltwater, sometimes quite rapidly, for considerable distances and etches a tell-trail along the way. While rare, the sequence of conditions necessary for "sailing" (rain; cold; sunshine; wind; slippery surface) create a unique event in North America's harshest desert.

Wheeled threats to deserts

Historically, human impacts on desert ecosystems largely resulted from livestock grazing in areas with accessible water. During the last century however, additional anthropogenic influences have altered many desert environments; these include introductions of exotic plants and animals and expanding urbanization (e.g., housing, malls, and golf courses). In particular, traffic from off-road vehicles (ORVs) has emerged as a major recreational activity in desert locations.

ORVs initiate several negative ecological effects, among them soil compaction, destruction of soil stabilizers (e.g., desert pavement, cryptogamic crusts, and lichens), increased wind and water erosion, reduced rates of water infiltration, crushed burrows, and direct damage to plants and animals. Soil disturbances are especially significant because, once damaged, desert soils may take 3000–10,000 years to develop anew. ORVs crush the stems, foliage, roots, and seedlings of plants, and small animals and their burrow systems are especially vulnerable to vehicular damage. For example, Agassiz's desert tortoises and their burrows were significantly more abundant on a control plot in California where ORVs were excluded when compared to a similar plot subject to ORV use. Furthermore, the body mass was larger for adult and sub-adult tortoises on the control plot.

Noise produced by some ORVs can reach 110 decibels, a level that can cause pain in humans and, presumably, some species of animals. Indeed, noise from ORVs in the Mojave Desert caused hearing loss in kangaroo rats, desert iguanas, and Mojave fringe-toed lizards as well as triggered the emergence of estivating spadefoot toads in lieu of the usual stimulus of rainfall. These effects are not without harm, as hearing loss increases the vulnerability of kangaroo rats to rattlesnakes, and the emergence of spadefoot toads when there are no rain-formed pools leads to their death by heat and dessication without any chance for successful reproduction.

Deserts disturbed by human activities may not recover for centuries, but banning ORVs from these (and other) vulnerable communities remains unlikely. Instead, it seems politically and socially possible to restrict their use to specified locations, excluding still-pristine environments, the ranges of protected species, and areas where restoration efforts are underway. Elsewhere, damage would likely be diminished if ORV traffic were limited to gentle slopes, sandy dunes, and areas naturally paved with gravel-mulch soils. Such considerations offer feasible options for managing sensitive and slow-to-recover desert environments.

Readings and references

Introduction

Barbour, M.G. and J. Major. (eds) 1977. *Terrestrial Vegetation of California*. John Wiley & Sons, New York, NY.

Brown, G.W. (ed.) 1968. *Desert Biology, Special Topics of the Physical and Biological Aspects of Arid Regions*. Vol. 1. Academic Press, New York, NY.

Larson, L. and P. Larson. 2000. *Deserts of the Southwest: A Sierra Club Naturalist's Guide*. Sierra Club Books, San Francisco, CA.

McMahon, J.A. and F.H. Wagner. 1985. The Mojave, Sonoran, and Chihuahuan deserts of North America. In: *Hot Deserts and Arid Shrublands* (M. Evenari, I. Noy-Meir, and D.W. Goodall, eds). Ecosystems of the World 12 A. Elsevier, New York, NY, pp. 105–202.

Whitford, W.G. 2002. *Ecology of Desert Systems*. Academic Press, New York, NY.

Physical geography
Why deserts are dry

Polis, G.A. (ed.) 1991. *The Ecology of Desert Communities*. University of Arizona Press, Tucson, AZ.

Shreve, F. 1942. The desert vegetation of North America. Botanical Review 8: 195–246.

Sowell, J.B. 2001. *Desert Ecology*. University of Utah Press, Provo, UT.

Ward, D. 2009. *The Biology of Deserts: Biology of Habitats*. Oxford University Press, New York, NY.

Yeaton, R.I. and M.L. Cody. 1979. The distribution of cacti along environmental gradients in the Sonora and Mojave deserts. Journal of Ecology 67: 529–541.

Desert mountains and bajadas

Bowers, M.A. and C.H. Lowe. 1986. Plant-form gradients on Sonoran Desert bajadas. Oikos 46: 284–291.

Bull, W.B. 1968. Alluvial fans. Journal of Geological Education 16: 101–106.

McAuliffe, J.R. 1994. Landscape evolution, soil formation, and ecological patterns and processes in Sonoran Desert bajadas. Ecological Monographs 64: 111–148.

Phillips, D.L. and J.A. McMahon. 1978. Gradient analysis of a Sonoran Desert bajada. Southwestern Naturalist 23: 669–679.

Vander Wall, S.B. and J.A. McMahon. 1984. Avian distribution patterns along a Sonoran desert bajada. Journal of Arid Environments 7: 59–74.

Ancient lakes

Allen, J.P. and S.S. Wilson. 1977. A bibliography of references to avian botulism. US Fish and Wildlife Service Special Science Report, Wildlife 206: 1–6.

Blackwelder, E. 1933. Lake Manly: an extinct lake of Death Valley. Geographical Review 23: 464–471.

Bolen, E.G. 1964. Plant ecology of spring-fed salt marshes in western Utah. Ecological Monographs 34: 143–166.

Loud, L.L. and M.R. Harrington. 1929. Lovelock Cave. University of California Publications in American Archeology and Ethnology 25: 1–183. (Discoveries include Indian decoys some 1000 years old.)

Nelson, N.F. 1954. *Factors in the Development and Restoration of Waterfowl Habitat at Ogden Bay Refuge*. Utah Department of Fish and Game, Publication 6. Salt Lake City, UT.

Sigler, W.F. and J.W. Sigler. 1996. *Fishes of Utah: A Natural History*. University of Utah Press, Salt Lake City, UT.

Smith, L.M. and J.A. Kadlec. 1986. Habitat management for wildlife in marshes of Great Salt Lake. Transactions of the North American Wildlife and Natural Resources Conference 51: 222–231.

Stephens, D.W. 1990. Changes in the lake levels, salinity, and the biological community of Great Salt Lake, USA (1847–1987). Hydrobiologia 197: 139–146.

Williams, C.S. and W.H. Marshall. 1938. Duck nesting studies, Bear River Migratory Bird Refuge, Utah. Journal of Wildlife Management 2: 29–48.

Wurtsbaugh, W.A. 1992. Food-web modification by an invertebrate predator on Great Salt Lake (USA). Oecologia 89: 168–175.

Wurtsbaugh, W.A. and T.S. Berry. 1990. Cascading effects of decreased salinity on the plankton, chemistry, and physics of the Great Salt Lake (Utah). Canadian Journal of Fishery and Aquatic Sciences 47: 100–109.

Features and adaptations

Desert soils and surfaces

Anderson, D.C., K.T. Harper, and R.C. Holmgren. 1982. Factors influencing development of cryptogamic soil crusts in Utah deserts. Journal of Range Management 35: 180–185.

Belnap, J. 2003. The world at your feet: desert biological soil crusts. Frontiers in Ecology and the Environment 1: 181–189.

Belnap, J. and D.A. Gillette. 1998. Vulnerability of desert biological soil crusts to wind erosion: the influence of crust development, soil texture, and disturbance. Journal of Arid Environments 39: 133–142.

Belnap, J., K.T. Harper, and S.D. Warren. 1994. Surface disturbance of cryptobiotic crusts: nitrogen activity, chlorophyll content, and chlorophyll degradation. Arid Soil Research and Rehabilitation 8: 1–8.

Bowker, M.A., M.E. Miller, J. Belnap, et al. 2008. Prioritizing conservation effort through the use of biological soil crusts as ecosystem function indicators in an arid region. Conservation Biology 22: 1533–1543.

Brotherson, J.D., S.R. Rushforth, and J.R. Johansen. 1983. Effects of long-term grazing on cryptogam crust cover in Navajo National Monument, Arizona. Journal of Range Management 36: 579–581.

Buol, S.W. 1965. Present soil-forming factors and processes in arid and semiarid regions. Soil Science 99: 45–49.

Cooke, R.U. 1970. Stone pavements in deserts. Annals of the Association of American Geographers 60: 560–577.

Davidson, E. and M. Fox. 1974. Effects of off-road motorcycle activity on Mojave Desert vegetation and soil. Madroño 22: 381–309.

Gile, H.F., F.F. Peterson, and R.B. Grossman. 1966. Morphological and genetic sequences of carbonate accumulation in desert soils. Soil Science 101: 347–360.

Hendricks, D.M. 1985. *Arizona Soils*. College of Agriculture, University of Arizona, Tucson, AZ.

McDonald, C.D. and G.H. Hughes. 1964. Studies of consumptive use of water by phreatophytes and hydrophytes near Yuma, Arizona. US Geological Survey, Professional Paper 486-F.

McFadden, L.D., S.G. Wells, and M.J. Jercinovich. 1987. Influences of eolian and paedogenic processes on the origin and evolution of desert pavements. Geology 15: 504–508.

Rychert, R.C. and J. Skujins. 1974. Nitrogen fixation by bluegreen algae-lichen crusts in the Great Basin Desert. Proceedings of the Soil Science Association of America 38: 768–771.

Snyder, J.M. and L.H. Wullstein. 1973. The role of desert cryptogams in nitrogen fixation. American Midland Naturalist 90: 257–265.

Vollmer, A.T., B.G. Maza, P.A. Medica, F.B. Turner, and S.A. Bamberg. 1976. The impact of off-road vehicles on a desert ecosystem. Environmental Management 1: 115–129.

Watchman, A. 2000. A review of the history of dating rock varnishes. Earth-Science Reviews 49: 261–277.

Plant adaptations

Beatley, J.C. 1974. Phenological events and their environmental triggers in Mojave Desert ecosystems. Ecology 55: 856–863.

Bowers, M.A. 1987. Precipitation and the relative abundance of desert winter annuals: a 6-year study in the northern Mojave Desert. Journal of Arid Environments 12: 141–149.

Branson, F.A., R.F. Miller, and I.S. McQueen. 1967. Geographic distribution and factors affecting the distribution of salt shrubs in the United States. Journal of Range Management 29: 287–296.

Brown, D.E. (ed.) 1982. Biotic communities of the American Southwest – United States and Mexico. Desert Plants 4: 1–342.

Darrow, R.A. 1943. Vegetative and floral growth of *Fouquieria splendens*. Ecology 24: 310–322.

Ebert, T.A. and P.H. Zedler. 1984. Decomposition of ocotillo (*Fouquieria splendens*) wood in the Colorado Desert of California. American Midland Naturalist 111: 143–147.

Evenari, M. 1949. Germination inhibitors. Botanical Review 15: 153–194.

Fitter, A. 2003. Making allelopathy respectable. Science 301: 1337–1338.

Hadley, N.F. and S.R. Szarek. 1981. Productivity of desert ecosystems. BioScience 31: 747–753.

Halvorson, W.L. and D.T. Patten. 1975. Productivity and flowering of winter ephemerals in relation to Sonoran Desert shrubs. American Midland Naturalist 93: 311–319.

Humphrey, R.R. 1975. Phenology of selected Sonoran Desert plants in Punta Cirio, Sonora, Mexico. Journal of the Arizona Academy of Science 10: 50–67.

Killingbeck, K.T. 1990. Leaf production can be decoupled from root activity in the desert shrub ocotillo (*Fouquieria splendens* (Englem.)). American Midland Naturalist 124: 124–129.

Nelson, J.F. and R.M. Chew. 1977. Factors affecting seed reserves in the soil of a Mojave Desert ecosystem, Rock Valley, Nye County, Nevada. American Midland Naturalist 97: 300–320.

Pake, C.E. and D.L. Venable. 1996. Seed banks in desert annuals: implications for persistence and coexistence in variable environments. Ecology 77: 1427–1435.

Reichman, O.J. 1984. Spatial and temporal variation of seed distributions in Sonoran Desert soils. Journal of Biogeography 11: 1–11.

Rice, E.L. 1974. *Allelopathy*. Academic Press, New York, NY.

Romeo, J.T. 2000. Raising the beam: moving beyond phytotoxicity. Journal of Chemical Ecology 26: 2011–2014.

Tevis, L. 1958. A population of desert ephemerals germinated by less than one inch of rain. Ecology 39: 688–695.

Waser, N.M. 1979. Pollinator availability as a determinant of flowering time in ocotillo (*Fouquieria splendens*). Oecologia 39: 107–121.

Yang, T.W. 1970. Major chromosome races of *Larrea divaricata* in North America. Journal of the Arizona Academy of Science 6: 41–45.

Animal adaptations

Bartholomew, G.A. and T.J. Cade. 1963. The water economy of land birds. Auk 80: 504–539.

Belk, D. and G.A. Cole. 1975. Adaptational biology of desert temporary-pond inhabitants. In: *Environmental Physiology of*

Desert Organisms (N.F. Hadley, ed.). Dowden, Hutchinson and Ross, Stroudsberg, PA, pp. 207–266.

Bentley, P.J. 1966. Adaptations of amphibians to arid environments. Science 152: 619–623.

Dimmitt, M.A. and R. Ruibal. 1980. Environmental correlates of emergence in spadefoot toads (*Scaphiopus*). Journal of Herpetology 14: 21–29.

Dorcas, M. and W. Gibbons. 2008. *Frogs and Toads of the Southeast*. University of Georgia Press, Athens, GA. (Source for tadpole development in southern chorus frogs.)

Hadley, N.F. 1972. Desert species and adaptation. American Scientist 60: 338–347.

Hadley, N.F, M.C. Quinlan, and M.L. Kennedy. 1991. Evaporative cooling in the desert cicada: thermal efficiency and water/metabolic costs. Journal of Experimental Biology 159: 269–283.

Heatwole, H. 1995. *Energetics of Desert Invertebrates*. Springer, New York, NY.

Louw, G.N. and M.K. Seely. 1982. *Ecology of Desert Organisms*. Longman, Inc., New York, NY. (Deals primarily with physiological adaptations.)

Mayhew, W.W. 1965. Adaptations of the amphibian, *Scaphiopus couchi*, to desert conditions. American Midland Naturalist 74: 95–109.

McClanahan, L. Jr. 1972. Changes in body fluids of burrowed spadefoot toads as a function of soil water potential. Copeia 1972: 209–216.

McClanahan, L., R. Ruibal, and V.H. Shoemaker. 1994. Frogs and toads in deserts. Scientific American 270(3): 82–88.

Middendorf, G.A., III and W.C. Sherbrooke. 1992. Canid elicitation of blood-squirting in a horned lizard (*Phrynosoma cornutum*). Copiea 1992: 519–527.

Nagy, K.A. 1972. Water and electrolyte budgets of a free-living desert lizard, *Sauromalus obesus*. Journal of Comparative Physiology 79: 39–62.

Nagy, K.A. 1988. Seasonal patterns of water and energy balance in desert vertebrates. Journal of Arid Environments 14: 201–210.

Newman, R.A. 1989. Developmental plasticity of *Scaphiopus couchi* tadpoles in an unpredictable environment. Ecology 70: 1775–1789.

Norris, K.S. and W.R. Dawson. 1964. Observations on the water economy and electrolyte excretion of chuckwallas (Lacertilia, *Sauromalus*). Copeia 1964: 638–646.

Omart, R.D. 1972. Physiological and ecological observations concerning the salt-excreting glands of the roadrunner. Comparative Biochemistry and Physiology 43A: 311–316.

Ruibal, R., L.Tevis, Jr, and V. Roig. 1969. The terrestrial ecology of the spadefoot toad, *Scaphiopus hammondii*. Copiea 1969: 571–584.

Schmidt, P.J., W.C. Sherbrooke, and J.O. Schmidt. 1989. The detoxification of ant (*Pogonomyrmex*) venom by a blood factor in horned lizards (*Phrynosoma*). Copeia 1989: 603–607.

Sherbrooke, W.C. 1990. Predatory behavior of captive greater roadrunners feeding on horned lizards. Wilson Bulletin 102: 171–174.

Sherbrooke, W.C. 1990. Rain-harvesting in the lizard, *Phrynosoma cornutum*: behavior and integumental morphology. Journal of Herpetology 24: 302–308.

Shoemaker, V.H. 1988. Physiological ecology of amphibians in arid environments. Journal of Arid Environments 14: 145–153.

Smyth, M. and G.A. Bartholomew. 1966. The water economy of the black-throated sparrow and the rock wren. Condor 68: 447–458.

Toolson, E.C. 1987. Water profligacy as an adaptation to hot deserts: water loss rates and evaporative cooling in the Sonoran Desert cicada, *Diceroprocta apache* (Homoptera: Cicadidae). Physiological Zoology 60: 379–385.

The major deserts

Chihuahuan Desert

Campbell, R.S. and I.F. Campbell. 1938. Vegetation on gypsum soil of the Jornada Plain, New Mexico. Ecology 19: 572–577.

Gardner, J.L. 1951. Vegetation of the creosote bush area of the Rio Grande Valley in New Mexico. Ecological Monographs 21: 379–403.

Johnston, M.C. 1974. Brief resume of botanical, including vegetational, features of the Chihuahuan Desert Region with special emphasis on their uniqueness. In: *Transactions of the Symposium on the Biological Resources of the Chihuahuan Desert Region, United States and Mexico* (R.H. Wauer and D.H. Riskind, eds). National Park Service, Transactions and Proceedings Series No. 3, Washington, DC, pp. 335–359.

Maxwell, R.A. 1968. *The Big Bend of the Rio Grande*. Guidebook 7, Bureau of Economic Geology, University of Texas, Austin, TX.

McKee, E.D. 1966. Structures of dunes at White Sands National Monument, New Mexico. Sedimentology 7: 1–69.

Morafka, D.J. 1977. *A Biogeographical Analysis of the Chihuahuan Desert Through its Herpetofauna*. W. Junk, The Hague, The Netherlands.

Parsons, R.F. 1976. Gypsophily in plants – a review. American Midland Naturalist 96: 1–20. (Deals with plants confined to gypsum soils.)

Raitt, R.J. and S.L. Pimm. 1976. Dynamics of bird communities in the Chihuahuan Desert, New Mexico. Condor 78: 427–442.

Schmidly, D.J. 1977. *The Mammals of the Trans-Pecos, Texas*. Texas A&M University Press, College Station, TX.

Schmidt, R.H., Jr. 1979. A climatic delineation of the "real" Chihuahuan Desert. Journal of Arid Environments 2: 243–250.

Schneider-Hector, D. 1993. *White Sands: The History of a National Monument*. University of New Mexico Press, Albuquerque,NM.

Smith, S.D. and J.A. Ludwig. 1978. The distribution and phytosociology of *Yucca elata* in southern New Mexico. American Midland Naturalist 100: 202–212.

Wauer, R.H. and D.H. Riskind. (eds) 1974. *Transactions of the Symposium on the Biological Resources of the Chihuahuan Desert Region, United States and Mexico*. National Park Service, Transactions and Proceedings Series No. 3, Washington, DC.

Welsh, R.G. and R.F. Beck. 1976. Some ecological relationships between creosote bush and bush muhly. Journal of Range Management 29: 472–475.

Yeaton, R.I. 1978. A cyclic relationship between *Larrea tridentata* and *Opuntia leptocaulis* in the northern Chihuahuan desert. Journal of Ecology 66: 651–656.

Sonoran Desert

Dunbier, R. 1968. *The Sonoran Desert, its Geography, Economy and People*. University of Arizona Press, Tucson, AZ.

Hensley, M.M. 1954. Ecological relations of the breeding bird population of the desert biome in Arizona. Ecological Monographs 24: 185–207.

Martin R.M.H., J.R. Cox, and F. Ibarra-F. 1995. Climatic effects on bufflegrass productivity in the Sonoran Desert. Journal of Range Management 48: 60–63.

McAuliffe, J.R., Jr. 1986. Herbivore-limited establishment of a Sonoran Desert tree, Cercidium microphyllum. Ecology 67: 276–280.

Phillips, S.J. and P.W. Comus. (eds) 1999. A Natural History of the Sonoran Desert. University of California Press, Berkeley, CA.

Shreve, F. and I.L. Wiggins. 1964. Vegetation and Flora of the Sonoran Desert. Vol. I. Stanford University Press, Stanford, CA.

Turner, R.M., J.E. Bowers, and T.L. Burgess. 1995. Sonoran Desert Plants, An Ecological Atlas. University of Arizona Press, Tucson, AZ.

Waser, N.M. and M.V. Price. 1981. Effects of grazing on diversity of annual plants in the Sonoran Desert. Oecologia 50: 407–411.

Wiggins, I.L. 1980. Flora of Baja California. Stanford University Press, Stanford, CA.

Mojave Desert

Miller, A.H. and R.C. Stebbins. 1964. The Lives of Desert Animals in Joshua Tree National Monument. University of California Press, Berkelely, CA.

Rundel, P.W. and A.C. Gibson. 1996. Ecological Communities and Processes in a Mojave Desert Ecosystem: Rock Valley, Nevada. Cambridge University Press, New York, NY.

Vasek, F.C. 1980. Creosote bush: long-lived clones in the Mojave Desert. American Journal of Botany 67: 246–255.

Great Basin Desert

Beatley, J.C. 1975. Climates and vegetation patterns across the Mojave/Great Basin transition of southern Nevada. American Midland Naturalist 93: 53070.

Chambers, J.C., B.A. Roundy, R.R. Blank, et al. 2007. What makes Great Basin sagebrush ecosystems invasible by Bromus tectorum? Ecological Monographs 77: 117–145.

Comstock, J.P. and J.R. Ehleringer. 1992. Plant adaptations in the Great Basin and Colorado Plateau. Great Basin Naturalist 52: 195–215.

Grayson, D.K. 1987. The biogeographic history of small mammals in the Great Basin: observations on the last 20,000 years. Journal of Mammalogy 68: 359–375.

Menakis, J.P., D. Osborne, and M. Miller. 2003. Mapping the cheatgrass-caused departure from historical natural fire regimes in the Great Basin, USA. USDA Forest Service Proceedings RMRS-P-29: 281–288.

Nagy, J.G., H.W. Steinhoff, and G.M. Ward. 1964. Effect of essential oils of sagebrush on deer rumen microbial functions. Journal of Wildlife Management 28: 785–790.

Ryser, F.A., Jr. 1985. Birds of the Great Basin: A Natural History. University of Nevada Press, Reno, NV.

Sigler, W.F. and J.W. Sigler. 1987. Fishes of the Great Basin: A Natural History. University of Nevada Press, Reno, NV.

Trimble, S. 1989. The Sagebrush Ocean: A Natural History of the Great Basin. University of Nevada Press, Reno, NV.

Young, J.A., A. Evans, and J. Mager. 1972. Alien plants in the Great Basin. Journal of Range Management 25: 199–201.

Highlights

Nurse trees

Engelmann, G. 1859. Cactaceae of the boundary. In: Report on the United States and Mexican Boundary Survey. US Department of Interior, Washington, DC.

Hutto, R.L., J.R. McAuliffe, and L. Hogan. 1986. Distributional associates of the saguaro (Carnegiea gigantea). Southwestern Naturalist 31: 469–476.

McAuliffe, J.R. 1984. Sahuaro-nurse tree associations in the Sonoran Desert: competitive effects of saguaros. Oecologia 64: 319–321. (Note alternate spelling for saguaro in the title.)

McAuliffe, J.R. 1990. Paloverdes, pocket mice, and bruchid beetles: interrelationships of seeds, seed dispersers, and seed predators. Southwestern Naturalist 35: 329–337.

Nabhan, G.P. and J.L. Carr. (eds) 1994. Ironwood: an ecological and cultural keystone in the Sonoran Desert. Occasional Paper No. 1, Conservation International, Washington, DC.

Olin, G., S.M. Alcorn, and J.M. Alcorn. 1989. Dispersal of viable saguaro seeds by white-winged doves (Zenaida asiatica). Southwestern Naturalist 34: 282–284.

Steenbergh, W.F. and C.H. Lowe. 1969. Critical factors during the first years of life of the saguaro (Cereus giganteus) at Saguaro National Monument, Arizona. Ecology 50: 825–834.

Steenbergh, W.F. and C.H. Lowe. 1977. Ecology of the Saguaro: II. Reproduction, Germination, Establishment, Growth, and Survival of the Young Plant. National Park Service Scientific Monograph Series 17(8). National Park Service, Washington, DC.

Suzán, H., D.T. Patten, and G.P. Nabhan. 1997. Exploitation and conservation of ironwood (Olneya tesota) in the Sonoran Desert. Ecological Applications 7: 948–957.

Turner, R.M., S.M. Alcorn, G. Olin, and J.A. Booth. 1966. The influence of shade, soil, and water on saguaro seedling establishment. Botanical Gazette 127: 95–102.

Vandermeer, J. 1980. Saguaro and nurse trees: a new hypothesis to account for population fluctuation. Southwestern Naturalist 25: 357–360.

"Trees" for desert woodpeckers

Hardy, P.C. and M.L. Morrison. 2001. Nest site selection by elf owls in the Sonoran Desert. Wilson Bulletin 113: 23–32.

Hensley, M.M. 1959. Notes on the nesting of selected species of birds of the Sonoran desert. Wilson Bulletin 71: 86–92.

Inouye, R.S., N.J. Huntly, and D.W. Inouye. 1981. Non-random orientation of Gila woodpecker nest entrances in saguaro cacti. Condor 83: 88–89.

Kerpez, T.A. and N.S. Smith. 1990. Competition between European starlings and native woodpeckers for nest cavities in saguaros. Auk 107: 367–375.

Kerpez, T.A. and N.S. Smith. 1990. Nest site selection and nest cavity characteristics of Gila woodpeckers and northern flickers. Auk 107: 193–198.

Korol, J.J. and R.L. Hutto. 1984. Factors affecting nest site location in Gila woodpeckers. Condor 86: 73–78.

McAuliffe, J.R. and P. Hendricks. 1988. Determinants of the vertical distributions of woodpecker nest cavities in the saguaro cactus. Condor 90: 791–801.

Boojums and elephants: unique trees

Felger, R. and M.B. Moser. 1985. People of the Desert and Sea: Ethnobiology of the Seri Indians. University of Arizona Press, Tucson, AZ.

Humphrey, R.R. 1974. The Boojum, Idria columnaris, Kellogg, and its Home and its Ecological Niche. University of Arizona Press, Tucson, AZ.

Mooney, H.A. and W.A. Emboden. 1968. The relationship of terpene composition, morphology, and distribution of populations of Bursera microphylla (Burseraceae). Brittonia 20: 44–51.

Yucca moths

Addicott, J.F. 1986. Variation in the costs and benefits of mutualism: the interaction between yuccas and yucca moths. Oecologia 71: 221–228.

Baker, H.G. 1986. Yuccas and yucca moths – a historical commentary. Annals of the Missouri Botanical Garden 73: 556–564.

Bolger, D.J., J.L. Neff, and B.B. Simpson. 1995. Multiple origins of the yucca-yucca moth association. Proceedings of the National Academy of Science 92: 6864–6867.

Dodd, R.J. and Y.B. Linhart. 1994. Reproductive consequences of interactions between *Yucca glauca* (Agavaceae) and *Tegeticula yuccasella* (Lepidoptera) in Colorado. American Journal of Botany 81: 815–825.

Englemann, G. 1872. The flower of yucca and its fertilization. Bulletin of the Torrey Botanical Club 3: 33. (Reveals the role of yucca moths.)

Pellmyr, O. 2003. Yuccas, yucca moths, and coevolution: a review. Annals of the Missouri Botanical Garden 90: 35–55.

Powell, J.A. 1992. Interrelationships of yuccas and yucca moths. Trends in Ecology and Evolution 7: 10–15.

Webber, J.M. 1953. *Yuccas of the Southwest*. US Department of Agriculture Monograph No. 17, Washington, DC.

Desert fishes

Baugh, T.M. and J.E. Deacon. 1988. Evaluation of the role of refugia in conservation efforts for the Devils Hole pupfish, *Cyprinodon diabolis*. Zoo Biology 7: 351–358.

Brown, J.H. and C.R. Feldmeth. 1971. Evolution in constant and fluctuating environments: thermal tolerances of desert pupfish (*Cyprinodon*). Evolution 25: 390–398.

Dudley, W.W., Jr and J.D. Larson. 1976. Effect of irrigation pumping on desert pupfish habitats in Ash Meadows, Nye County, Nevada. US Geological Survey Professional Paper 927: 1–52.

Miller, R.R., D.L. Soltz, and P.G. Sanchez. 1977. *Fishes and Aquatic Resources in the Death Valley System, California/Nevada: A Bibliography, 1878–1976*. National Park Service (Western Division), San Francisco, CA.

Minckley, W.L. 1991. Native fishes of arid lands: a dwindling resource of the desert Southwest. USDA Forest Service General Technical Report RM 206: 1–45.

Minckley, W.L. and J.E. Deacon. (eds) 1991. *Battle against Extinction, Native Fish Management in the American West*. University of Arizona Press, Tucson, AZ.

Naiman, R.J. and D.L. Soltz. (eds) 1981. *Fishes in North American Deserts*. John Wiley & Sons, New York, NY.

Naiman, R.J., S.D. Gerking, and T.D. Ratcliff. 1973. Thermal environment of a Death Valley pupfish. Copeia 1973: 366–369.

National Park Service. 2013. Devils Hole pupfish status remains precarious. Available at: http://www.nature.nps.gov/water/Homepage/devil_Hole.cfm (accessed 29 December 2014).

Pister, E.P. 1974. Desert fishes and their habitats. Transactions of the American Fisheries Society 103: 531–540.

Soltz, D.L. and R.J. Naiman. 1978. The natural history of native fishes in the Death Valley System. Natural History Museum of Los Angeles County, Science Series 30: 1–76.

Vrijenhoek, R.C., M.E. Douglas, and G.K. Meffe. 1985. Conservation genetics of endangered fish populations in Arizona. Science 229: 400–402.

Realm of reptiles

Beck, D.D. 1990. Ecology and behavior of the Gila monster in southwestern Utah. Journal of Herpetology 24: 54–68.

Beck, D.D., B.E. Martin, and C.H. Lowe. 2009. *Biology of Gila Monsters and Beaded Lizards*. University of California Press, Berkeley, CA.

Berry, K.H. 1978. Livestock grazing and the desert tortoise. Transactions of the North American Wildlife and Natural Resources Conference 43: 505–519.

Bogert, C.M. and R. Martin del Campo. 1956. The Gila monster and its allies. The relationships, habits, and behavior of the lizards of the family Helodermatidae. Bulletin of the American Museum of Natural History 109: 1–238.

Cohen, A.C. and B.C. Myres. 1970. A function of the horns (suprocular scales) in the sidewinder rattlesnake, *Crotalus cerastes*, with comments on other horned snakes. Copeia 1970: 574–575.

Corn, P.S. 1994. Recent trends of desert tortoise populations in the Mojave Desert. In: *Biology of North American Tortoises* (R.B. Bury and D.J. Germano, eds). Fish and Wildlife Research 13, National Biological Survey, Washington, DC, pp. 85–93.

Ernst, C.H. 1992. *Venomous Reptiles of North America*. Smithsonian Institution Press, Washington, DC.

Esque, T.C. and E.L. Peters. 1994. Ingestion of bones, stones, and soil by desert tortoises. In: *Biology of North American Tortoises* (R.B. Bury and D.J. Germano, eds). Fish and Wildlife Research 13, National Biological Survey, Washington, DC, pp. 105–111.

Hansen, R.N., M.K. Johnson, and T.R. Van Devender. 1976. Foods of the desert tortoise, *Gopherus agassizii*, in Arizona and Utah. Herpetologica 32: 247–251.

Jayne, B.C. 1986. Kinematics of terrestrial snake locomotion. Copeia 1986: 915–927.

Jones, K.B. 1983. Movement patterns and foraging ecology of Gila monsters (*Heloderma suspectum* Cope) in northwestern Arizona. Herpetologica 39: 247–253.

Longshore, K.M., J.R. Jaeger, and J.M. Sappington. 2003. Desert tortoise (*Gopherus agassizii*) survival at two Mojave Desert sites: death by short-term drought? Journal of Herpetology 37: 169–177.

Marlow, R.W. and K. Tollestrup. 1982. Mining and exploitation of natural mineral deposits by the desert tortoise, *Gopherus agassizii*. Animal Behavior 32: 475–478.

Medica, P.A., R.B. Bury, and R.A. Luckenbach. 1980. Drinking and construction of water catchments by the desert tortoise in the Mojave Desert. Herpetologica 36: 301–304.

Murphy, R., K. Berry, T. Edwards, et al. 2011. The dazed and confused identity of Agassiz's land tortoise, *Gopherus agassizii* (Testudines: Testudinidae), with descriptions of a new species and its consequences for conservation. ZooKeys 113: 39–71.

Nagy, K.A. and P.A. Medica. 1986. Physiological ecology of desert tortoises. Herpetologica 42: 73–92.

Oftedal, O.T., S. Hillard, and D.J. Morafka. 2002. Selective spring foraging by juvenile desert tortoises (*Gopherus agassizii*) in the Mojave Desert: evidence for adaptive nutritional strategy. Chelonian Conservation and Biology 4: 341–352.

Oldemeyer, J.L. 1994. Livestock grazing and the desert tortoise in the Mojave Desert. In: *Biology of North American Tortoises* (R.B. Bury and D.J. Germano, eds). Fish and Wildlife Research 13, National Biological Survey, Washington, DC, pp. 95–103.

Pianka, E.R. 1967. On lizard species diversity: North American flatland deserts. Ecology 48: 333–351.

Pianka, E.R. 1986. *Ecology and Natural History of Desert Lizards: Analysis of the Ecological Niche and Community Structure.* Princeton University Press, Princeton, NJ.

Russell, F.E. and C.M. Bogert. 1981. Gila monster: its biology, venom, and bite – a review. Toxicon 19: 341–359.

Turner, F.B., P.A. Medica, and C.L. Lyons. 1984. Reproduction and survival of the desert tortoise in Ivanpah Valley, California. Copeia 1984: 811–820.

Van Devender, T.R. (ed.) 2002. *The Sonoran Desert Tortoise: Natural History, Biology, and Conservation.* University of Arizona Press, Tucson, AZ.

Walde, A.D., D.K. Delaney, M.L. Harlen, and L.L. Parker. 2007. Osteophagy by the desert tortoise (*Gopherus agassizii*). Southwestern Naturalist 52: 147–149.

Woodbury, A.M. and R. Hardy. 1948. Studies of the desert tortoise, *Gopherus agassizii*. Ecological Monographs 18: 145–200.

Zimmerman, L.C. and C.R. Tracy. 1989. Interactions between the environment and ectothermy and herbivory in reptiles. Physiological Zoology 62: 374–409.

Of soils and mice

Benson, S.B. 1936. Concealing coloration among some desert rodents of the southwestern United States. University of California Publications in Zoology 40(1): 1–69.

Bradt, G.W. 1932. The mammals of the Malpais, an area of black lava in the Tularosa Basin, New Mexico. Journal of Mammalogy 13: 321–328.

Dice, L.R. 1947. Effectiveness of selection by owls of deermice (*Peromyscus maniculatus*) which contrast in color with their background. University of Michigan Contributions to Vertebrate Biology 34: 1–20.

Dice, L.R. and P.M. Blossom. 1937. *Studies of Mammalian Ecology in Southwestern North America with Special Attention to the Colors of Desert Mammals.* Publication 485, Carnegie Institute, Washington, DC.

Hoekstra, H.E. 2006. Genetics, development, and evolution of adaptive pigmentation in vertebrates. Heredity 97: 222–234.

Lewis, T.H. 1949. Dark coloration in reptiles of the Tularosa Malpais, New Mexico. Copeia 1949: 181–184.

Norris, K.S. and C.H. Lowe. 1964. An analysis of background color-matching in amphibians and reptiles. Ecology 45: 565–580.

Deserts and predators

Beck, B.B., C.W. Engen, and P.W. Gelfand. 1973. Behavior and activity cycles of Gambel's quail and raptorial birds at a Sonoran Desert waterhole. Condor 75: 466–470.

Tomoff, C.S. 1974. Avian species diversity in desert scrub. Ecology 55: 396–403. (5–86% of desert passerines nested in vegetation with thorns or spines.)

Wagner, F.H. and R.D. Graetz 1981. Animal-animal interactions. In: *Arid Land Ecosystems: Structure, Functioning, and Management* (D.W. Goodall and R.A. Perry, eds). Cambridge University Press, New York, NY, pp. 51–84.

Pygmies of the sagebrush steppe

Elias, B.A., L.A. Shipley, S. McCusker, et al. 2014. Effects of genetic management on reproduction, growth, and survival in captive endangered pygmy rabbits (*Brachylagus idahoensis*). Journal of Mammalogy 94: 1282–1292.

Green, J.S. and J.T. Flinders. 1980. Brachylagus idahoensis. Mammalian Species 125: 1–4.

Green, J.S. and J.T. Flinders. 1980. Habitat and dietary relationships of the pygmy rabbit. Journal of Range Management 33: 136–142.

Larrucea, E.S. and P.F. Brussard. 2008. Habitat selection and current distribution of the pygmy rabbit in Nevada and California. Journal of Mammalogy 89: 691.

Lyman, R.L. 1991. Late quarternary biogeography of the pygmy rabbit (*Brachylagus idahoensis*) in eastern Washington. Journal of Mammalogy 72: 110–117.

Shipley, L.A. 2008. Pygmy rabbits in peril in the U.S.A. Available at http://www.actionbioscience.org/biodiversity/shipley.html (accessed 29 December 2014).

Shipley, L.A., E.M. Davis, L.A. Felicetti, et al. 2012. Mechanisms for eliminating monoterpenes of sagebrush by specialist and generalist rabbits. Journal of Chemical Ecology 38: 1178–1189.

Thines, N.J., L.A. Shipley, and R.D. Sayler. 2004. Effects of cattle grazing on ecology and habitat of Columbia Basin pygmy rabbits (*Brachylagus idahoensis*). Biological Conservation 119: 525–534.

Wilson, T.L., F.P. Howe, and T.C. Edwards, Jr. 2011. Effects of sagebrush treatments on multi-scale resource selection by pygmy rabbits. Journal of Wildlife Management 75: 393–398.

Woods, B.A., J.A. Rachlow, S.C. Bunting, et al. 2013. Managing high-elevation sagebrush steppe: do conifer encroachment and prescribed fire affect habitat for pygmy rabbits. Rangeland Ecology and Management 66: 462–471.

Desert quail, rainfall, and vitamin A

Gullion, G.W. 1960. The ecology of Gambel's quail in Nevada and the arid southwest. Ecology 41: 518–536.

Heffelinger, J., F. Guthery, R. Olding, et al. 1999. Influence of the precipitation, timing, and summer temperatures on reproduction of Gamble's quail. Journal of Wildlife Management 63: 154–161.

Hungerford, C.R. 1964. Vitamin A and productivity in Gambel's quail. Journal of Wildlife Management 28: 141–147.

Sowls, L.K. 1960. Results of a banding study of Gambel's quail in southern Arizona. Journal of Wildlife Management 24: 185–190.

Swank, W.G. and S. Gallizioli. 1954. The influence of hunting and rainfall upon Gambel's quail populations. Transactions of the North American Wildlife Conference 19: 283–296.

Sailing stones

Kletetschka, G., R.LeB. Hooke, A. Ryan, et al. 2013. Sliding stones of Racetrack Playa, Death Valley, USA: the roles of rock thermal conductivity and fluctuating water levels. Geomorphology 195: 110–117.

Lorenz, R.D., B.K. Jackson, J.W. Barnes, et al. 2011. Ice rafts not sails: floating rocks at Racetrack Playa. American Journal of Physics 79: 37–42.

Messina, P. and P. Stoffer. 2000. Terrain analysis of the Racetrack Basin and the sliding rocks of Death Valley. Geomorphology 35: 253–265.

Norris, R.D., J.M. Norris, R.D. Lorenz, et al. 2014. Sliding rocks on Racetrack Playa, Death Valley National Park: first observation of rocks in motion. PLoS ONE 9(8): e105948.

Wheeled threats to deserts

Belnap, J. 2002. Impacts of off-road vehicles on nitrogen cycles in biological soil crusts: resistance in different US deserts. Journal of Arid Environments 52: 155–165.

Belnap, J. and D.A. Gillette. 1997. Disturbance of biological soil crusts: impacts on potential wind erodibility of sandy desert soils in SE Utah. Land Degradation and Development 8: 355–362.

Bury, R.B. and R.A. Luckenbach. 2002. Comparison of desert tortoise (*Gopherus agassizii*) populations in an unused and off-road vehicle area in the Mojave Desert. Chelonian Conservation and Biology 2: 457–463.

Davidson, E. and M. Fox. 1974. Effects of off-road motorcycle activity on Mojave Desert vegetation and soil. Madroño 22: 381–390.

Eckert, R.E., Jr, M.K. Wood, W.H. Blackburn, and F.F. Peterson. 1979. Impacts of off-road vehicles on infiltration and sediment production of two desert soils. Journal of Range Management 32: 394–397.

Lovich, J.E. and D. Bainbridge. 1999. Anthroprogenic degradation of the Southern California desert ecosystem and prospects for natural recovery and restoration. Environmental Management 24: 309–326.

Vollmer, A.T., B.G. Maza, P.H. Medica, et al. 1976. The impact of off-road vehicles on a desert ecosystem. Environmental Management 1: 115–129.

Webb, R.H. and H.G. Wilshire. (eds) 2011. *Environmental Effects of Off-Road Vehicles: Impacts and Management in Arid Regions.* Springer, New York, NY. (Reprint of the original 1983 edition.)

Infobox 7.1. Agave, margaritas, and bats

Arita, H.Y. and D.E. Wilson. 1987. Long-nosed bats and agaves: the tequila connection. Bats 5: 3–5.

Dalton, R. 2005. Alcohol and science: saving the agave. Nature 438: 1070–1071.

Hensley, A.P. and K.T. Wilkins. 1988. Leptonycteris nivialis. Mammalian Species 307: 1–4.

Moreno-Valdez, A., R.L. Honeycutt, and W.E. Grant. 2004. Colony dynamics of *Leptonycteris nivialis* (Mexican long-nosed bat) related to flowering *Agave* in northern Mexico. Journal of Mammalogy 85: 453–459.

Nobel, P.S. 2003. *Environmental Biology of Agaves and Cacti.* Cambridge University Press, New York, NY.

Sanchez, R. and R.A. Medellín. 2007. Food habits of the threatened bat, *Leptonycteris nivialis* (Chiroptera: Phyllostomatidae) in a mating roost in Mexico. Journal of Natural History 41: 1753–1764.

Weir, J. 2009. *Tequila: A Guide to Types, Flights, Cocktails and Bites.* Ten Speed Press, Berkeley, CA.

Infobox 7.2. Desert seeds and an endangered whale

Foster, K.E. and N.G. Wright 1980. Jojoba: an alternative to the conflict between agricultural and municipal ground-water requirements in the Tucson area, Arizona. Ground Water 18: 31–36.

Gentry, H.S. 1958. The natural history of jojoba (*Simmondsia chinensis*) and its cultural aspects. Economic Botany 12: 261–295.

Gentry, H.S. 1972. Plant a seed and save a whale. Saguaroland Bulletin 26: 44–47.

Goodin, J.R., and D.K. Northington. (eds) 1985. *Plant Resources of Arid and Semiarid Lands.* Academic Press, New York, NY.

McGraw, L. 2000. Simmondsin from jojoba checked for appetite suppression. Agricultural Research 48: 21.

Sherbrooke, W.C. 1989. Seedling survival and growth of a Sonoran Desert shrub, jojoba (*Simmondsia chinensis*), during the first ten years. Southwestern Naturalist 34: 412–424.

Sherbrooke, W.C. and E.F. Haase. 1974. *Jojoba: A Wax-Producing Shrub of the Sonoran Desert.* Arid Lands Resource Paper No. 5, Office of Arid Land Studies, University of Arizona, Tucson, AZ.

Infobox 7.3. El tigre! El gran gato de Las Americas

Audubon, J.J. and J. Bachman. 1989. *Audubon's Quadrupeds of North America.* Wellfleet Press, Secaucus, NJ. (An unabridged reprint of the original 1846 work, entitled *The Viviparous Quadrupeds of North America.*)

Brown, D.E. and C.A. Lopez Gonzalez. 2000. Notes on the occurrences of jaguars in Arizona and New Mexico. Southwestern Naturalist 45: 537–546.

Larson, S.E. 1997. Taxonomic re-evaluation of the jaguar. Zoo Biology 16: 107–120.

McCain, E.B. and J.L. Childs. 2008. Evidence for resident jaguars (*Panthera onca*) in the southwestern United States and the implications for conservation. Journal of Mammalogy 89: 1–10.

Rabinowitz, A.R. 1999. The present status of jaguars (*Panthera onca*) in the southwestern United States. Southwestern Naturalist 44: 96–100.

Seymour, K.L. 1989. Panthera onca. Mammalian Species 340: 1–9.

Schmidly, D.J. 2004. *The Mammals of Texas,* revised edition. University of Texas Press, Austin, TX.

Sowls, L.K. 1997. *Javelinas and Other Peccaries: Their Biology, Management, and Use.* Texas A&M Press, College Station, TX.

Taylor, W.P. 1947. Recent record of the jaguar in Texas. Journal of Mammalogy 28: 66. (Cites early records and the last verified specimen for Texas.)

CHAPTER 8

Chaparral and Pinyon-Juniper Woodlands

This (pinyon) is undoubtedly the most important food-tree in the Sierra, and furnishes the Mono, Carson, and Walker River Indians with more and better nuts than all the other species taken together...

John Muir (1894)

Chaparral is a name popularly applied to various shrub communities in western North America. However, from a strict ecological view, chaparral is represented primarily by the shrubby vegetation extending from the foothills of the Sierra Nevada to the Pacific coastline. A second and separate region of chaparral occurs in parts of Arizona and Mexico.

In its physiognomy, mature chaparral is dense, dwarfed, and one-layered; from a distance, the shrubby tangle resembles a velvety blanket extending across the slopes (Fig. 8.1). Chaparral is derived from *chaparro*, a term for the dense shrublands in the Mediterranean region of Spain. The suffix *–al* means "place of" in Spanish; hence chaparral emerged when the Spanish explored and settled in California. The term is the origin of the word *chaps*, the protective leggings ranch hands wear when working cattle in this vegetation.

From a central location in California chaparral reaches northward into southern Oregon and southward into Baja California (Fig. 8.2). Among other features, chaparral develops: (a) in a Mediterranean climate (i.e., hot dry summers and cool, wet winters); (b) in accordance with a regime of fire; and (c) as a closed canopy of shrubs with leathery, evergreen leaves (i.e., sclerophyllic leaves). The shrubs in chaparral communities vary in height from 1 to 3 m (3–10 feet), but they are characteristically uniform at any one location.

Chaparral communities extend from sea level to 2000 m (6562 feet) and, within this range, generally favor steep slopes with nutrient-deficient and poorly developed soils. In southern California, Santa Ana winds at times intensify the dry conditions in summer. These strong winds drive hot, dry air from interior desert areas westward across southern California, producing low humidity and other conditions that favor devastating wildfires. Chaparral often forms a mosaic with other communities – grasslands or oak woodlands – depending on fire frequency at each location.

Chaparral communities yield no products of immediate economic value, nor are they especially productive as grazing areas. However, because they form dense cover on erosion-prone hillsides, chaparral offers important protection for watersheds. In particular, winter rains may initiate devastating mudslides after fires temporarily remove chaparral.

Woodlands characterized by pinyon (both *pinon* and *pinyon* appear in the literature with equal regularity, the latter an attempt to reflect the Spanish pronunciation of *piñon*) pine and juniper occur across a large region of western North America. These communities extend from Colorado and Utah into New Mexico, Arizona, and northern Mexico, with some representatives in southern Idaho, Oregon and parts of California and Nevada (Fig. 8.2). Pinyon-juniper is the characteristic vegetation of the mesa region of the Colorado Plateau, the mountains of the Great Basin, and the foothills of the southern Rocky Mountains.

Pinyon-juniper woodlands occupy the lowest and warmest forested zone in the United States in climatic conditions generally characterized as semiarid. These woodlands typically occur between 1370 and 2440 m (4500–8000 feet). However, individual pinyons sometimes ascend to 3200 m (10,500 feet) on south- and west-facing slopes in Arizona and scattered junipers may descend to 910 m (3000 feet). Variations in elevation, topography, and geography account for differences in amounts and distribution of annual precipitation in the regions occupied by pinyon-juniper woodlands. Seasonal precipitation, for example, varies along a northwest to southeast gradient. Rainfall from Pacific winter and spring storms provides most of the moisture in the woodlands of Nevada and northern Utah, but about 75% of the annual precipitation in New Mexico occurs during the warm season (i.e., summer and early fall).

Ecology of North America, Second Edition. Brian R. Chapman and Eric G. Bolen.
© 2015 John Wiley & Sons, Ltd. Published 2015 by John Wiley & Sons, Ltd.

Figure 8.1 White ceanothus, often in association with chamise, characterizes chaparral communities along the coasts of California and southern Oregon. Whereas ceanothus rapidly regenerates from seedlings after fires, recent studies have indicated that too-frequent burns (less than 20 years apart) can eliminate ceanothus and other fire-cued plant species from the chaparral community. Photo courtesy of Richard W. Halsey of the California Chaparral Institute (http://california chaparral.com).

Features and adaptations of chaparral

Sclerophytic leaves are the key feature of chaparral vegetation. Their tough, leathery texture counters wilting during the dry summers and frequent droughts typical of the region. The current foliage does not fall until the next set of leaves emerges, so chaparral is evergreen and retains its foliage during droughty periods. The evergreen physiognomy distinguishes chaparral from the periodically deciduous but otherwise similar shrubby vegetation occurring elsewhere (e.g., montane shrublands, sometimes identified as petran chaparral). The evergreen nature of chaparral also allows the shrubs to resume photosynthetic activity rapidly after a chance summer rainstorm; dormancy ends quickly when soil moisture is available. Evergreen sclerophylls do not wilt, even in severe drought, and the leaves therefore remain ready for maximum light absorptions and the resumption of photosynthesis.

As well as coping with drought, chaparral communities also deal in various ways with fire. Many of the dominant species regenerate from sprouts which appear soon after fires destroy the aboveground vegetation. These sprouts originate from numerous buds in the root crown just below ground level. This area – a burl – enlarges as the cycle of fire and regrowth continues, eventually developing a woody platform up to 4 m (13 feet) in width. Chaparral burls in some cases may be 250 or more years old. At times, the centers of burls rot away and the sprouts emerge from the remaining root crown in a circular pattern. Post-fire sprouting is independent of rainfall; water reserves in the root system sustain the

new growth at any time of year. Because the sprouts arise from a common root crown, the resulting plants are multi-stemmed, which increases the shrubby physiognomy of chaparral. The evergreen nature of chaparral shrubs also ensures a ready supply of aboveground fuel; fires therefore spread across the landscape when optimal conditions – low humidity, low fuel moisture, and high winds – coincide with an ignition source (Infobox 8.1).

Sprouting offers chaparral shrubs several advantages: (a) they retain their already established position; (b) they initiate regrowth after fires with nourishment from a fully developed root system; (c) their first-year growth is rapid compared to seedlings; and (d) their regrowth suppresses the seedlings of other plants. Herbaceous annual plants may be abundant for a year or two following a fire, but the developing overstory of chaparral sprouts soon blocks sunlight from reaching other plants in the community.

Other species of chaparral-forming shrubs regenerate from seeds **scarified** by heat, that is, the seeds do not germinate unless first exposed to high temperatures which crack their hard coverings and let in water and nutrients. Fires do not stimulate germination *per se* but instead weaken the seed coat and allow water to reach the dehydrated embryo. These seeds are not adapted for dispersal, but instead persist in the **seed bank** near the parent plant. With this strategy, the seeds remain in place awaiting the eventual arrival of fire rather than dispersing to sites elsewhere.

All told, these features enable chaparral to maintain itself as a climax community. Some ecologists nonetheless speculate that, if fires were curtailed for long enough,

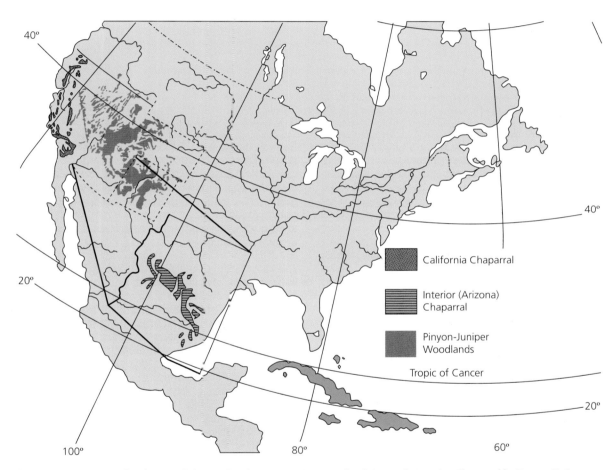

Figure 8.2 Approximate distribution of chaparral and pinyon-juniper woodlands in North America. Illustrated by Tamara R. Sayre.

Infobox 8.1 Fire-loving beetles

Most mobile animals move away from the impending threat of forest fires. This behavior is not universal, however, for many insects do just the opposite: they fly directly to a fire. Indeed, entomologists in North America have identified dozens of pyrophilic (i.e., fire-loving) species of flies, wasps, and beetles that rapidly converge on forests devastated by fires. In particular, large numbers of black fire beetles often descend on burns where embers still glow in a smoldering, tree-blackened landscape. Although seemingly suicidal, this behavior is actually key for the survival of this insect, which occur in pinyon and other pine forests in North America.

These small wood-boring beetles possess remarkable abilities to disperse, often appearing in environments (e.g., seashores) far removed from forests or forest fires. However, the behavior of black fire beetles came under closer scrutiny when the beetles notoriously alighted on crowds at football stadiums awash with cigarette smoke or swarmed at sawmills, oil fires, smelters, campfires, forest fires, and other places producing clouds of smoke. Olfactory sensors located in their antennae allow the beetles to detect certain chemicals contained in the smoke. Once smoke is detected, black fire beetles orient toward the source in ways similar to the searching behavior used by bloodhounds. To locate their quarry, they zig-zag in a series of movements to determine the strongest source of scent, constantly repeating the zig-zags to stay on course.

Eventually, swarms of black fire beetles congregate in a highly unstable zone just behind the advancing inferno. Here, conditions change capriciously – wind direction and wind speed shift constantly, firebrands scatter in all directions, and flare-ups suddenly burst from smoldering logs and other flammable debris. The beetles, which typically fly just 1 m or so (about 3 feet) above ground (i.e., below burning tree branches), dodge these deadly situations using infrared heat sensors (pit organs) located on their mesothorax at the base of the anterior pair of legs. Each pit organ contains many tiny spherical receptors, each smaller than the diameter of the finest human hair. The receptors are filled with water and efficiently absorb infrared radiation; when they sense infrared heat, the receptors instantly swell, thereby applying pressure on highly responsive mechano-sensitive nerve cells. Thus equipped, the beetles can detect sudden changes in radiant flux and avoid lethal hazards within the active fire zone.

Rather uniquely, heat and smoke apparently stimulate the flights, mating, and ovipositing in black fire beetles. Because they arrive at still-smoking forest fires, the beetles gain a significant advantage for reproducing in the absence of either predators or competitive species of wood-boring insects. Some evidence also suggests that their offspring develop more quickly when the adults lay their eggs in bark warmed by the fire's infrared radiation. Beetles originating from eggs deposited in trees recently killed by some means other than fire (i.e., at ambient temperatures, as trees blown over by wind) require about a year to develop into pupae, whereas those arising from eggs laid in fire-warmed bark do so in two months. Furthermore, by laying their eggs in recently killed trees, the beetles subvert the defenses of living trees; that is, the eggs or pupae will not be damaged by the trees' own cellular growth or drowned in sticky resin. In sum, the behavioral and physical adaptations of these fire-loving beetles permit exploitation of a dangerous habitat necessarily avoided by most other insects. Photo courtesy of Stephen P. L. Luk.

oak woodlands would eventually replace chaparral, but supporting evidence is lacking. Chaparral fires are unusually hot and consume virtually all of the above-ground biomass, but the community quickly replaces itself without competition from other vegetation. These events differ from those in longleaf-pine communities, where periodic fires maintain a stable subclimax of pines by eliminating the final stages of succession (e.g., oak and other hardwoods). The fires in these communities are also less intense, killing the understory of young hardwoods but rarely harming the mature pines in the overstory (*Longleaf pine forests*, Chapter 4).

The stems of some non-sprouting species of chaparral shrubs exhibit ribbon-like strips of exposed wood; these develop when strips of bark die. The same occurs in other species, except that new growth may surround the dead strips creating a stem wrinkled with vertical fissures. Such reductions in vascular tissue seem to be associated with droughty conditions and may increase the longevity of non-sprouting species.

Coastal (California) chaparral

Chaparral, like the term forest or mangrove, is a type of vegetation represented by several taxa and communities, of which three associations in California are paramount.

Chamise chaparral

Nearly pure stands of chamise (a vigorous sprouter) cloak the slopes in southern California and northern Baja California (Fig. 8.2). When fully grown, chamise forms virtually impenetrable stands with dense canopies of interlocking branches. Such tight canopy cover precludes the development of herbaceous understory except in the years following fires. Chamise grows on nutrient-poor soils and, partly for this reason, it recovers from fire more slowly in comparison to other types of chaparral.

Chamise produces narrow leaves somewhat resembling those of spruce; the plants therefore have a large surface area relative to their volume. Thin stems further increase the high surface:volume ratio, which enhances ignition of the vegetation. Like most kinds of shrubs in chaparral, the leaves of chamise are resinous and highly flammable. Many of the branches in otherwise healthy chamise begin to die at a relatively early age, thereby adding to the volume of fuel within the stands (i.e., leaf litter). Most ecologists agree that such features actually increase the probability of fire.

Manzanita chaparral

This community is somewhat more limited in extent than other chaparral associations. About 12 species of manzanita are included in the flora of this community and, depending on the species, sprout after fires or regenerate from heat-scarified seeds. Sprouting species increase in the community if fires occur often enough to prevent other species from reaching maturity; that is, the other non-sprouting species burn before producing seed. Manzanita chaparral generally develops at higher elevations, near the lower edges of the montane coniferous forests, where the soils are moister and somewhat deeper than on the lower slopes. The cooler temperatures at these sites often produce precipitation in the form of snow, fog, and freezing rain, hence manzanita communities are

sometimes known as "cold chaparral." Fires usually occur less frequently (at 50- to 100-year intervals) in manzanita chaparral, perhaps because of the cool, moist regime where these communities develop.

Manzanita chaparral commonly includes other shrubs, including chamise, although some stands are monotypic. After the first rain in the wake of a fire, the blackened surface may be obscured with a herbaceous carpet of colorful flowers and other short-lived vegetation.

Ceanothus chaparral

Where conditions along the coast are somewhat wetter, this community may develop as a monotypic stand (i.e., any one of several species of ceanothus) or in association with chamise and other shrubs. Unlike chamise, the branches of ceanothus plants do not intertwine. Ceanothus regenerates from seedlings that grow rapidly after fires. Nodules on the roots of these shrubs bear nitrogen-fixing bacteria.

Ceanothus recovers rapidly after fires, but the plants are relatively short-lived and, at least in southern California, other species invade when the stands age and the canopy opens. Because of these circumstances, Ceanothus chaparral in southern California represents a seral community rather than climax vegetation. However, in northern California and southern Oregon, ceanothus achieves climax status and replaces chamise as a dominant species.

Other chaparral communities

Communities dominated by California scrub oak develop at sites where conditions are somewhat more mesic (e.g., north-facing slopes). Many other shrubs are associated with scrub oak, making this one of the richer types of chaparral. California scrub oak and most of the associated species readily sprout after fires. Vines contribute to the tangle of cover in the canopy and, as usual for all kinds of chaparral, lack herbaceous understories except for ferns. This community is sometimes known as broad-sclerophyll chaparral because the oaks bear wide leaves in comparison to chamise. Leaf litter is deeper than in other types of chaparral, reaching depths of 20 cm (8 inches). In fact, it may take 30 years or more before enough leaf litter accumulates for the successful germination of scrub oak seeds. When the frequency of fires precludes deep accumulations of litter (e.g., intervals of 15 years or less), scrub oaks and other shrubby vegetation may be replaced by non-native weedy, grassland communities.

Other types of chaparral form from red shanks, mountain mahogany, and other sclerophyllic species. Red shanks bear rust-colored, shaggy bark and chartreuse, feather-like foliage. Because of these features, red shanks chaparral is widely regarded as the most attractive of the several chaparral communities.

Interior (Arizona) chaparral

Chaparral also occurs in Arizona and nearby Mexico. In Arizona, these communities generally develop between 1050 and 1850 m (3440–6070 feet), whereas those in Mexico occur at somewhat higher elevations. Above these elevations, interior chaparral gives way to a zone of woodland vegetation (e.g., ponderosa or pinyon pine). Desert vegetation, including desert grasslands, forms the lower border.

Interior chaparral develops in a climatic regime unlike the Mediterranean pattern that favors coastal chaparral. Precipitation in this case peaks twice a year – from December to March and during a period of thunderstorms in July and August – instead of the winter rainfall typical of the Pacific coastline. The growing season for each varies accordingly, with interior chaparral developing new growth during spring and summer.

Several of the principal species regenerate from sprouts after fires, but others produce immense seed crops. The latter persist for decades in seed banks until fires stimulate their germination. Shrub live oak is a common dominant of interior chaparral in Arizona. Other species are the counterparts of those found in coastal chaparral (e.g., ceanothus and manzanita), but chamise is absent. Similarly, interior chaparral develops with a closed canopy of a relatively uniform height. At drier sites with rocky soils, desert plants – particularly thorny shrubs – join the flora of interior chaparral communities. Fire frequency is low compared to most types of coastal chaparral; where oak is abundant, interior chaparral burns at intervals of 80–100 years.

In northern Mexico, what is known as Coahuilan chaparral includes representatives of the same genera that characterize chaparral elsewhere. However, although desert ceanothus and a few other species are shared with the communities in Arizona, Coahuila scrub oak replaces shrub live oak as a dominant species.

The fauna of interior chaparral lacks distinction. Instead of producing some endemic species, for example, the fauna includes representatives of more widely distributed mammals such as eastern cottontails and white-footed mice. Likewise, the avifauna is represented by scrub jays and other species occurring elsewhere in vegetation with similar physiognomy.

Pinyon-juniper woodlands

Distribution and ecology

Pinyon-juniper woodlands generally do not extend in unbroken blankets across vast areas but instead form a a highly irregular patchwork across foothills, slopes, and canyon walls (Fig. 8.3a). These discontinuities often interrupt the lowermost belts or zones of forest communities in mountainous regions, but the physiognomy of pinyon-juniper remains essentially unchanged (i.e., open woodlands of low, round-crowned trees, which are often bushy and contorted). In places, however, pinyon-juniper occupies a distinctive zone in the profile of montane vegetation (Chapter 9). Unlike the physiognomy of forests, the canopy of woodland vegetation is not closed and the crowns of individual trees seldom touch. Woodlands also differ from savannas, where the trees are even farther

Figure 8.3 (a) Pinyon-juniper woodlands in south-central New Mexico likely represent the appearance of this community prior to the arrival of the Euro-Americans. Open grasslands cover the valley floor (lower right), with pinyon pines and junipers scattered across the intervening hills and ridges. Fires historically maintained the mosaic of this vegetation relative to its topographic position. (b) Mature stands of pinyon-juniper woodlands typically develop in relatively dense clusters of both species. Grassy areas in these communities are often heavily grazed. Photos courtesy of Carleton M. Britton.

apart and where the herbaceous understory vegetation – often grasses – represents the dominant stratum.

Pinyon-juniper woodlands typically develop at elevations lying between sagebrush-grassland and an upper zone dominated by ponderosa pine. Stands of juniper and, to a lesser extent pinyon, often expand downward on slopes where grazing has reduced competition from grasses. In consequence, the total area of pinyon-juniper has increased since Euro-Americans settled the Southwest. Birds and other wildlife readily disperse the seeds of both pinyon and juniper trees, thereby aiding expansion of the community into adjacent areas.

Pinyon pines occur on steep slopes of canyons, mesas, and other sites with dry, rocky soils. The trees grow slowly (those with diameters of 15–25 cm or 6–10 inches may well be more than 100 years old) and mature in 250–350 years with a few living for 800 years. The root systems of pinyon and juniper extend beyond the crowns of individual trees and fully occupy the open areas in between. In many locations pinyon pines develop a shrub-like physiognomy, but on better sites they acquire straight stems topped by rounded crowns (Fig. 8.3b). Some ecologists refer to pinyon-juniper woodlands as a forest of "pygmy conifers."

Four species comprise a group commonly known as pinyon or nut pines, including two with relatively restricted distributions: Parry pinyon and Mexican pinyon. We deal primarily with the species known simply as pinyon pine, which is widely distributed throughout western North America. Junipers in these communities also vary by region and include several species, of which one-seeded juniper and Rocky Mountain juniper are representative. Shrubs include mountain mahogany, but this and other understory vegetation is not limited to pinyon-juniper communities. Heavy grazing has removed much of the original cover of grasses.

The influence of altitude is evident when the communities on the north and south rims of the Grand Canyon are compared. Pinyon-juniper woodlands occupy much of the Coconino Plateau on the south rim, where the elevation is about 2134 m (7000 feet). In contrast, a forest of ponderosa pine covers most of the Kaibab Plateau on the higher north rim (see Chapter 9 for more information).

Human uses

Pinyon pines produce large crops of mast every 2–4 years, but only a few of these successfully produce seedlings. Only seeds that find suitable conditions soon after falling actually germinate. Rodents harvest large amounts of pinyon mast, but humans have also gathered pinyon "nuts" for centuries. Shoshone and other tribes in the Great Basin sustained themselves in winter on pinyon nuts. Native Americans, and later Euro-Americans, also obtained firewood from pinyon-juniper woodlands, which eventually produced ecological changes at some locations. Firewood is still harvested in some places in New Mexico, often in excess of the growth rates for replacement. Native Americans also valued juniper as structural framework for their pueblos. All told, the history of the Southwest was greatly entwined with products obtained from pinyon-juniper woodlands (e.g., fenceposts and mine timbers). Today, place names in the Southwest commonly reflect the inspiration of pinyon pine (e.g., Pine Nut Mountains, Nevada) or juniper vegetation, although juniper is usually designated as "cedar" as in Cedar City, Utah.

Chaparral and fire

Water-repellant soils

Fires can produce a water-repellent layer within the soil profile of chaparral communities. In between fires, hydrophobic substances (organic compounds) accumulating in the ground litter briefly slow infiltration and absorption of water at the surface, but moisture eventually reaches the underlying soil. During fires, however, the hydrophobic materials are vaporized and the gasses move downward, thereby forming the water-repellent layer deeper in the soil profile. The depth of the water-repellent layer, as well as its thickness, varies with the intensity of the fire. However, the temperature gradients developing across the upper soil profile are of even greater significance; that is, the gradients facilitate the downward movement of the vaporized substances which then condense on mineral soil particles. Sandy and other coarse-textured soils are more hydrophobic than finer soils, which have more surface area per particle. Coarse-textured soils therefore require smaller amounts of hydrophobic materials to effect a full coating.

A layer of water-repellent soil therefore lies below and parallel to the soil surface after fires sweep across chaparral. Hence, rainfall water infiltrates the upper horizon but penetrates downward only to the layer of water-repellent soil. When the upper layers are saturated the water flows laterally, eroding the surface soils and carrying off nutrients. Moreover, because the water-repellent layer reduces the depth to which water flows downward, storms often produce more water than can be absorbed, leading to floods. These conditions, amplified by the relatively steep slopes on which chaparral develops, contribute to serious post-fire mudslides.

The ecological significance, if any, of water-repellent soils is not clear. Although they retard infiltration, water-repellent soils also reduce evaporation, thereby providing a key mechanism for water conservation in a semiarid region. Another idea suggests that water-repellency complements the effects of **allelopathy** and further inhibits the growth of herbaceous plants beneath the shrub canopy. In any case, the water repellent layer is not distributed uniformly across entire watershed but occurs more as a mosaic whose pattern seems related to pre-fire conditions such as canopy cover and litter accumulations.

Post-fire vegetation

Coastal chaparral develops as a continuous community covering large areas, which is why wildfires typically sweep across large areas of terrain. Lightning is the natural cause of fires in chaparral, although humans have undoubtedly increased the frequency of burns in historical times. Little biomass remains aboveground when chaparral burns. Surface temperatures may reach hundreds of degrees for several minutes, making chaparral fires among the hottest occurring in natural vegetation. Santa Ana winds increase the intensity and extent of fires in coastal chaparral.

A temporary community of herbaceous plants develops in the first year after fire removes the climax vegetation. Most of these plants are annuals, which are sometimes known as "fire annuals" because they bloom and then disappear until the next fire. The seeds of such plants necessarily remain viable for many years. Most of these produce bright flowers; freshly burned sites are often colorful gardens in the spring after a fire. Curiously, these seeds do not germinate in response to heat but instead are stimulated by organic chemicals leached from the litter of charred wood.

The diversity of the post-fire herbaceous community usually peaks in the first years after a fire. Diversity diminishes, however, as herbaceous perennials replace the annual vegetation. Meanwhile, chaparral quickly begins to regenerate burned areas – sprouts may appear within weeks – and by the fourth year, new growth generally dominates these sites. The regrowth of chaparral produces even-aged stands, which accounts for the rather uniform height of the mature vegetation.

Wildlife and chaparral fires

Questions naturally arise about the impact of fires on the welfare of wildlife. In chaparral, fires not only occur with considerable frequency (especially in the wake of human settlement) but fires in these communities are typically very hot. Flame temperatures may exceed 1090°C (2000°F) and those at the soil surface reach a maximum of about 760°C (1400°F). Nonetheless, even a thin layer of dry soil offers a remarkable amount of insulation for burrowing animals. Subsurface temperatures are far less than those at the surface. At soil depths of about 8 cm (3 inches) the temperature seldom exceeds 65°C (150°F), and the differential between surface and subsurface temperatures increases at a greater rate thereafter. Based on field tests, rodents succumb only when subsurface temperatures reach 59–63°C (138–145°F). Animals in burrows therefore normally find adequate protection from the temperatures generated during chaparral fires. Even temperature-sensitive animals such as reptiles may

survive in rocky crevices, whereas deer and other large animals easily escape most fires. Singed animals are seldom noticed by observers at controlled burns, nor do animals always flee hysterically from fires as might be supposed. On the other hand, because they live in houses constructed of dried sticks, woodrats are vulnerable to fires in chaparral communities.

Overall, habitat changes after fires influence the composition and abundance of the mammalian fauna far more than fire-related mortality. Birds respond similarly, with mourning doves and other seed-eating species moving into burned areas to take advantage of the now-exposed food sources on the soil surface. Insects are more exposed to predation after fires, which increases the percentage of insectivorous birds in the post-burn avifauna for a time. The loss of overstory cover also attracts both mammalian and avian predators to newly burned chaparral (e.g., coyotes and hawks). Predation, at least temporarily, likely represents a greater ecological influence on chaparral wildlife than direct mortality from fire. Thereafter however, prey populations recover as chaparral shrubs reclaim the burned areas. Fires therefore produce significant changes in the composition of wildlife populations, changes that are the direct result of the vegetation's radically altered physiognomy and not because of large-scale deaths.

Highlights

Allelopathy in chaparral

The phenomenon known as allelopathy was introduced earlier as a possible factor in the spacing of desert shrubs (Chapter 7). Allelopathy is better illustrated in chaparral, where it represents another dimension of physiological ecology.

Because fire plays a crucial role in the development and maintenance of chaparral, particularly where chamise is the dominant species, some ecologists refer to a "fire cycle" when describing the ecology of chamise and other types of chaparral communities. Integrated into this cycle, however, is a toxic substance produced from the leaves of mature chamise and the subsequent suppression of understory vegetation. The toxin results from the normal metabolic activities of chamise and, because the toxin is water soluble, rains transport it into the soil beneath the plants. More toxin is produced in the intervals between rains, so it steadily accumulates in the upper 2–3 cm (1 inch) of the soil profile. When sufficient, the same rainfall would normally trigger the seed bank – representing a rich herbaceous flora – but the toxin is at its peak at this time and inhibits germination. The seeds of a few resistant species do germinate, but the root systems of the seedlings cannot compete for soil moisture and thereafter perish.

Although not all ecologists fully agree, the phenomenon of allelopathy restrains the development of herbaceous vegetation in mature stands of chamise until fire consumes the aboveground biomass and initiates regeneration of the climax community. Fire itself apparently destroys any residual toxin in the soil, and the young regrowth of chamise produces only limited amounts of toxin. As a result, the soil is temporarily free of toxin, and herbaceous vegetation flourishes until the chamise matures and again produces enough toxin to suppress the understory. Chamise accordingly has evolved a means – allelopathy – of controlling other vegetation. Conversely, understory herbs have evolved their own means of coping with allelopathy, namely propagules that persist in a seed bank. Additionally, by responding adaptively to the toxin, these plants delay germination until site conditions favor their chances of reaching reproductive maturity. Similarly, the seeds of chamise itself seem inhibited by the toxin produced by mature plants as they too delay germination until a period following a fire when seedlings have a better chance of surviving.

Animal associates in coastal chaparral

Mountain quail favor slopes covered with chaparral. In summer, some coveys of these secretive birds extend into montane communities as high as 3050 m (10,000 feet), but in fall they descend into chaparral at lower altitudes. These yearly movements – altitudinal migration – are made on foot and may cover distances of up to 32 km (20 miles). Mountain quail are distinctively marked: their chestnut-colored flanks are prominently highlighted by white bars, and they bear a crest of two long, thin feathers, which at a distance often appear as a single plume. These birds are the largest quail in temperate North America, and they are locally abundant. Nonetheless, hunters bag relatively few mountain quail largely because of their secretive habits, their tendency to run instead of fly when pursued, and the difficulty of hunting in dense chaparral. Mountain quail also occur elsewhere (e.g., southern Idaho), but these are introduced populations and lie outside the species' original range on the Pacific slope of North America.

In places, the distribution of mountain quail and California quail overlap; interspecific competition is avoided, however, because of dietary and other differences. Mountain quail feed on acorn mast, fruits, and seeds in the fall, flowers and leafy foods in winter, and bulbs in the spring and summer, whereas California quail are heavily dependent on the seeds and greens of annual plants.

Chaparral-covered foothills also provide the primary habitat for California thrashers. The distribution of a smaller bird, the wrentit, also coincides with chaparral, although some are found at higher elevations in an adjacent zone of shrubby conifers. Within this range, the plumage of wrentits varies from reddish-brown in northern birds to gray in southern populations.

Three mammals are closely affiliated with chaparral: the brush rabbit, the California mouse, and the dusky-footed woodrat. Brush rabbits, despite their abundance, are not widely hunted, likely because the dense vegetation offers hunters limited access. This species does not use burrows as regularly as rabbits elsewhere, perhaps reflecting the excellent cover afforded by chaparral. The range of California mice extends southward from San Francisco Bay, whereas dusky-footed woodrats lives in chaparral and other habitats from Oregon to Baja California. However, where these two species overlap in distribution, they commonly dwell together in the stick dens constructed by dusky-footed woodrats.

Whereas the overall vertebrate fauna is diverse – rodents and reptiles are among the well-represented taxa – few other species are closely tied to chaparral communities. Sizeable populations of mule deer forage in chaparral for example, but they also occur in montane, desert, and other communities in much of western North America. Nonetheless, the widespread availability of chaparral sprouts after fires significantly increases the forage available to mule deer, and their number increase proportionally. Reproductive output, for example, jumps from 60–85 fawns per 100 does in heavy, unburned chaparral to 115–140 fawns at sites where "spot burns" produce small openings of 2–4 ha (5–10 acres). However, as the regrowth of chaparral matures, the deer population may exceed the supply of forage. Increased hunting remains about the only means to avoid a "crash" in these temporarily inflated deer populations.

Lizards and burned chaparral

Western fence lizards, a darkly colored species in the group known as spiny lizards, perch on the blackened stems of recently burned chaparral. On these occasions the lizards seldom sun on light-colored rocks, although rocky sites are readily available in most chaparral communities. The lizards and the burned stalks are closely color-matched (at least for a time), and provide another example of adaptive coloration. Additionally, the initial regrowth of sprouts at the base of the burned shrubs forms a small thicket of protective cover. These conditions develop in chaparral consisting of shrub oaks, ceanothus, and/or chamise, but not in manzanita. In the latter species, fire sloughs off the bark and exposes the nearly white wood beneath.

Thereafter, however, progressively fewer spiny lizards perch on the burned stalks as the chaparral recovers from the fire. The char on the surface of the stems gradually wears away, leaving the stalks lighter in color and no longer offering a protective background. By the third year after a fire, western fence lizards perch less often on the stalks and instead favor rocky locations (sites with crevices and little fuel), where the lizards might better survive fires.

The temporary nature of the lizards' behavior, which is keyed to the occurrence of fires, is noteworthy: the lizards select protectively colored backgrounds precisely at times when cover is reduced and the abundance of predators increases. Hawks are among the predators of lizards responding to habitats opened by fire.

Western fence lizards play a unique role in the **epizootiology** of Lyme disease: they provide blood meals for many of the immature ticks occurring in chaparral communities. In doing so, however, black-legged ticks – the primary vector for Lyme disease in coastal California – acquire a protein that kills the pathogenic bacteria. Consequently, bites from these "cleansed" ticks can no longer transmit Lyme disease to new hosts. On test plots where lizards were experimentally relocated, the tick population declined significantly; only a small percentage survived by switching to dusky-footed woodrats as hosts. Apparently, the risk of humans acquiring Lyme disease might decrease wherever lizards remain numerous, whereas those ticks that feed exclusively on rodents remain carriers of the infectious bacteria that threaten human health.

Pinyon jays

Pinyon-juniper woodlands, sometimes complemented by ponderosa pine, provide key habitat for pinyon jays, one of the more unusual species in the crow and jay family (Corvidae). Pinyon jays begin nesting as early as February, even at high elevations. Flocks of pinyon jays at times include thousands of birds that may become nomadic and move long distances in years when the cone crop fails. The flocks remain highly integrated all year round, including during the breeding season. Large crops of cones apparently serve as cues for breeding. Under experimental conditions, the presence of green cones in the cages of captive pinyon jays stimulated development of their reproductive organs.

Pinyon jays often cache pine seeds on the south sides of trees, which are the first snow-free locations after winter storms. The birds feed as a unit, lacking aggressive encounters between individuals, and deploy sentinels to protect the flock. When breeding begins, pairs court in relative isolation from the flock, but respond to a special "dinner call" as a cue to reassemble. Pinyon jays typically nest in the same area for several years, thereby establishing traditional grounds for their breeding activities. Hatchlings are fed by the parents, often assisted by "helpers" in the form of non-breeding birds of the same species. After fledging, flocks consisting of both adults and young from nearby nests feed together in tightly knit aggregates until late summer, when they begin to collect and cache pinyon seeds. The birds hammer open and extract the seeds from green cones, transporting two dozen or more seeds at a time in their throats. The seed caches serve as sources of winter food but may be visited throughout the year. Courting birds feed each other before nesting, as do pairs maintaining bonds after breeding ends. In sum, the food supplies and well-developed social behavior of pinyon jays seem closely linked.

Human influences

Chaparral fires sometimes destroy human habitations and harm the integrity of watersheds. The number and frequency of fires increases with the growth of human populations, especially in coastal California. In response, firebreaks were cut across large areas of chaparral in the 1920s. Later, much wider lanes – known as fuelbreaks – were cut and maintained in herbaceous vegetation for both fire control and improved wildlife habitat. At selected locations in some fuelbreaks, islands of chaparral were left intact as escape cover and travel stops for wildlife.

Except for sprouts, chaparral provides little nourishment for livestock or wildlife. To improve grazing, ranchers sometimes remove the brush with herbicides or mechanical means, then seed the cleared areas with exotic perennial grasses. This form of management requires large infusions of energy and money for both the initial treatment and continued maintenance. Similar modifications reduce fire hazards where urbanization has spread into chaparral-dominated landscapes. Despite recurring fires, however, most ecologists believe chaparral represents the best protective cover for watersheds in this region of North America. Meanwhile, the public steadily gains greater appreciation for chaparral.

Human activities, especially overgrazing and fire suppression, have imposed major changes in the original structure, composition, and distribution of pinyon-juniper woodlands. Intensive grazing, in particular, expanded pinyon and juniper communities into areas formerly dominated by grasses and shrubs. Seedlings of the woody plants, especially juniper, gain an advantage when the palatable grasses disappear. Additionally, the grass cover once supplied ample fuel for fires, which periodically arrested fuller development of the woody vegetation. Soil erosion also increased as herbaceous ground cover disappeared. In comparison to the previous 300 years, erosion in the last 100 years at one location in Colorado jumped 400% with the introduction of livestock at the site. Unfortunately, the understory did not recover after grazing and other anthropogenic impacts were reduced, presumably because the seeds of native plants were no longer available (no parent plants remained to supply seeds, and the seed bank was lost with topsoil erosion). In ironic consequence, sizeable areas of pinyon-juniper woodlands have been completely cleared and seeded with grasses to restore grazing.

Readings and references

Introduction

Brown, D.E. (ed.) 1982. Biotic communities of the American Southwest – United States and Mexico. Desert Plants 4: 1–342. (See pp. 91–99 for descriptions of coastal and interior Chaparral.)

Cole, K.L., J. Fisher, S.T. Arundel, et al. 2007. Geographical and climatic limits of needle types of one- and two-needled pinyon pines. Journal of Biogeography 35: 257–269.

Detling, L.E. 1961. The chaparral formation of southeastern Oregon with consideration of its post-glacial history. Ecology 42: 348–357.

Keeley, J.E. and S.C. Keeley. 1988. Chaparral. In: *North American Terrestrial Vegetation* (M.G. Barbour and W.D. Billings, eds). Cambridge University Press, New York, NY, pp. 166–207.

Miller, R.F. and P.E. Wigland. 1994. Holocene changes in semi-arid pinyon-juniper woodlands: response to climate, fire, and human activities in the Great Basin. BioScience 44: 465–474.

Features and adaptations of chaparral

Biswell, H.H. 1974. Effects of fire on chaparral. In: *Fire and Ecosystems* (T.T. Kozlowski and C.E. Ahlgren, eds). Academic Press, New York, NY, pp. 321–364.

Countryman, C.M. and C.W. Philpot. 1970. Physical characteristics of chamise as a wildland fuel. Research Paper PSW-66, Pacific Southwest Forest and Range Experiment Station, USDA Forest Service, Berkeley, CA.

Davis, C.B. 1973. "Bark stripping" in *Arctostaphylos* (Ericaceae). Madroño 22: 145–149.

Hanes, T.L. 1965. Ecological studies on two closely related chaparral shrubs in southern California. Ecological Monographs 35: 213–235.

James, S.M. 1984. Lignotubers and burls – their structure, function and ecological significance in Mediterranean ecosystems. Botanical Review 50: 225–266.

Jones, M.D. and H.M. Laude. 1960. Relationships between sprouting in chamise and physiological condition of the plant. Journal of Range Management 13: 210–214.

Keeley, J.E. 1975. The longevity of nonsprouting *Ceanothus*. American Midland Naturalist 93: 504–507.

Keeley, J.E. 1977. Seed production, seed populations in soil, and seedling production after fire for two congeneric pairs of sprouting and non-sprouting chaparral shrubs. Ecology 58: 820–829.

Keeley, J.E. 1987. Role of fire in seed germination of woody taxa in California chaparral. Ecology 68: 434–443.

Keeley, J.E. and P.H. Zedler. 1978. Reproduction of chaparral shrubs after fire: a comparison of sprouting and seedling strategies. American Midland Naturalist 99: 142–161.

Kummerow, J., J.V. Alexander, J.W. Neel, and K. Fishbeck. 1978. Symbiotic nitrogen fixation in *Ceanothus* roots. American Journal of Botany 65: 63–69.

Parsons, D.J., P.W. Rundel, R. Hedlund, and G.A. Baker. 1981. Survival of severe drought by a non-sprouting chaparral shrub. American Journal of Botany 68: 215–220.

Rundel, P.W. and D.J. Parsons. 1979. Structural changes in chamise (*Adenostoma fasciculatum*) along a fire-induced age gradient. Journal of Range Management 32: 452–466.

Coastal (California) chaparral

Haidinger, T.L. and J.E. Keeley. 1993. Role of high fire frequency in destruction of mixed chaparral. Madroño 40: 141–147.

Halsey, R.W. 2008. *Fire, Chaparral, and Survival in Southern California*, second edition. Sunbelt Publications, San Diego, CA.

Keeley, J.E. 1995. Future of California floristics and systematics: wildfire threats to the California flora. Madroño 42: 175–179.

Moritz, M.A., J.E. Keeley, E.A. Johnson, and A.A. Schaffner. 2004. Testing a basic assumption of shrubland fire management: how important is fuel age? Frontiers in Ecology and the Environment 2: 67–72.

Quinn, R.D. and S.C. Keeley. 2006. *Introduction to California Chaparral*. University of California Press, Berkeley, CA.

Vogl, R.J. and P.K. Schorr. 1972. Fire and manzanita chaparral in the San Jacinto Mountains, California. Ecology 53: 1179–1188.

Wilson, R.C. and R.J. Vogl. 1965. Manzanita chaparral in the Santa Ana Mountains, California. Madroño 18: 47–62.

Interior (Arizona) chaparral

Bolander, D.H. 1982. Chaparral in Arizona. General Technical Report PSW-58, Pacific Southwest Forest and Range Experiment Station, USDA Forest Service, Berkeley, CA.

Charmichael, R.S., O.D. Knipe, C.P. Pase, and W.W. Brady. 1978. Arizona chaparral: plant associations and ecology. USDA Forest Service Research Report Paper RM-202. Rocky Mountain Forest and Range Experiment Station, Fort Collins, CO.

Dick-Peddie, W.A. 1999. New Mexico Vegetation: Past, Present and Future. University of New Mexico Press, Albuquerque, NM.

Huebner, C.D., J.L. Vankat, and W.H. Renwick. 1999. Change in the vegetation mosaic of central Arizona USA between 1940 and 1989. Plant Ecology 144: 83–91.

Pase, C. and D. Brown. 1994. Interior chaparral. In: *Biotic Communities: Southwestern United States and Northwestern Mexico*, second edition. (D.E. Brown, ed.). University of Utah Press, Salt Lake City, UT, pp. 95–99.

Pinyon-juniper woodlands

Barney, M.A. and N.C. Frischknecht. 1974. Vegetation changes following fire in the pinyon-juniper type in west-central Utah. Journal of Range Management 27: 91–96.

Floyd, L.M., D.D. Hanna, W.H. Romme, and M. Colyer. (eds) 2003. *Ancient Pinyon-Juniper Woodlands: A Natural History of Mesa Verde*. University Press of Colorado, Boulder, CO.

Howell, J. 1941. Pinon and juniper woodlands of the southwest. Journal of Forestry 39: 542–545.

Jameson, D.A. and E.H. Reid. 1965. The pinyon-juniper type of Arizona. Journal of Range Management 18: 152–153.

Lanner, R.M. 1975. Pinon pines and junipers of the southwestern woodlands. In: *The Pinyon-Juniper Ecosystem, A Symposium*. Agricultural Experiment Station, Utah State University, Logan UT, pp. 1–7.

Merkle, J. 1952. An analysis of a pinyon-juniper community at Grand Canyon, Arizona. Ecology 33: 375–384.

Miller, R.F. and J.A. Rose. 1999. Fire history and western juniper encroachment in sagebrush steppe. Journal of Range Management 52: 550–559.

St. Andre, G., H.A. Mooney, and R.D. Wright. 1965. The pinyon woodland zone in the White Mountains of California. American Midland Naturalist 73: 225–239.

Tausch, R.J. and N.E. West. 1988. Differential establishment of pinyon and juniper following fire. American Midland Naturalist 119: 174–184.

Woodbury, A.M. 1947. Distribution of pygmy conifers in Utah and northeastern Arizona. Ecology 28: 113–126.

Zlotin, R.I. and R.R. Parmenter. 2008. Patterns of mast production in pinyon and juniper woodlands along a precipitation gradient in central New Mexico (Sevilleta National Wildlife Refuge). Journal of Arid Environments 72: 1562–1572.

Chaparral and fire

Water-repellant soils

DeBano, L.F. 1974. Chaparral soils. In: *Symposium on Living with the Chaparral: Proceedings*. (M. Rosenthal, ed.). Sierra Club Special Publication, San Francisco, CA, pp. 19–26.

DeBlano, L.F. 1981. Water repellant soils: a state-of-the-art. General Technical Report PSW-46, Pacific Southwest Forest and Range Experiment Station, USDA Forest Service, Berkeley, CA.

DeBlano, LF. 2000. The role of fire and soil heating on water repellency in wildland environments: a review. Journal of Hydrology 231–232: 195–206.

Hubbert, K.R. and V. Oriol. 2005. Temporal fluctuations in soil water repellency following wildfire in chaparral steeplands, southern California. International Journal of Wildland Fire 14: 439–447.

Post-fire vegetation

Hanes, T.L. 1971. Succession after fire in the chaparral of southern California. Ecological Monographs 41: 27–52.

Horton, J.S. and C.J. Kraebel. 1955. Development of vegetation after fire in the chamise chaparral of southern California. Ecology 36: 244–262.

Keeley, J.E. and S.C. Keeley. 1981. Postfire regeneration of California chaparral. American Journal of Botany 68: 524–530.

Keeley, S.C. and M. Pizzorno. 1986. Charred wood stimulated germination of two fire-following herbs of the California chaparral and the role of hemicellulose. American Journal of Botany 73: 1289–1297.

Keeley, S.C., J.E. Keeley, S.M. Hutchinson, and A.W. Johnson. 1981. Postfire succession of the herbaceous flora in southern California chaparral. Ecology 62: 1608–1621.

Sweeney, J.R. 1956. Responses of vegetation to fire. A study of the herbaceous vegetation following chaparral fires. University of California Publications in Botany 28: 143–249.

Zedler, P.H., C.R. Gautier, and G.S. McMaster. 1983. Vegetation change in response to extreme events. The effect of a short interval between fires in California chaparral and coastal scrub. Ecology 64: 809–818.

Wildlife and chaparral fires

Biswell, H.H. 1969. Prescribed burning for wildlife in California brushlands. Transactions of the North American Wildlife and Natural Resources Conference 34: 438–444.

Countryman, C.M. 1964. Mass fires and fire behavior. Research Paper PSW-19, Pacific Southwest Forest and Range Experiment Station, USDA Forest Service, Berkeley, CA. (Cites data for fire temperatures in chaparral communities.)

Howard, W.E., R.L. Fenner, and H.E. Childs, Jr. 1959. Wildlife survival on brush burns. Journal of Range Management 12: 230–234.

Highlights

Allelopathy in chaparral

Christensen, N.L. and C.H. Muller. 1975. Relative importance of factors controlling germination and seedling survival in *Adenostoma* chaparral. American Midland Naturalist 93: 71–78.

Halsey, R.W. 2004. In search of allelopathy: an eco-histoical review of the investigation of chemical inhibition in California coastal sage scrub and chamise chaparral. Bulletin of the Torrey Botanical Club 131: 343–367.

McPherson, J.K. and C.H. Muller. 1969. Allelopathic effects of *Adenostoma fascicularum* "chamise," in the California chaparral. Ecological Monographs 39: 177–198.

McPherson, J.K., C.H. Chou, and C.H. Muller. 1971. Allelopathic constituents of the chaparral shrub *Adenostoma fasciculatum*. Phytochemistry 10: 2925–2933.

Muller, C.H. 1966. The role of chemical inhibition (allelopathy) in vegetational composition. Bulletin of the Torrey Botanical Club 93: 332–351.

Muller, C.H., R.B. Hanawalt, and J.K. McPherson. 1968. Allelopathic control of herb growth in the fire cycle of California chaparral. Bulletin of the Torrey Botanical Club 95: 225–231.

Animal associates in Coastal chaparral

Biswell, H.H., R.D. Taber, D.W. Hedrick, and A.M. Schultz. 1952. Management of chamise brushlands for game in the north coast region of California. California Fish and Game 38: 453–484. (Source of data for deer populations.)

Connell, J.H. 1954. Home range and mobility of brush rabbits in California chaparral. Journal of Mammalogy 35: 392–405.

Davis, J. 1967. Some effects of deer browsing on chamise sprouts after fire. American Midland Naturalist 77: 234–238.

Gutierrez, R.J. 1980. Comparative ecology of the mountain and California quail in the Carmel Valley, California. The Living Bird 18: 71–93.

Hedricks, J.H. 1968. Control burning for deer management in California. Proceedings of the Tall Timbers Fire Ecology Conference 8: 219–233.

Kundaeli, J.N. and H.C. Reynolds. 1972. Desert cottontail use of natural and modified pinyon-juniper woodland. Journal of Range Management 25: 116–118.

Merritt, J.F. 1974. Factors influencing the local distribution of *Peromyscus californicus* in northern California. Journal of Mammalogy 55: 102–114.

Mossman, A.S. 1955. Reproduction of the brush rabbit in California. Journal of Wildlife Management 19: 177–184.

Swank, W.G. 1958. *The Mule Deer in Arizona Chaparral and an Analysis of Other Important Deer Herds*. Arizona Game and Fish Department Wildlife Bulletin 3, Phoenix, AZ.

Taber, R.D. and R.F. Dasmann. 1958. *The Black-Tailed Deer of the Chaparral*. Game Bulletin No. 8, California Department of Fish and Game, Sacramento, CA.

Vogl, R.J. 1967. Wood rat densities in southern California manzanita chaparral. Southwestern Naturalist 12: 176–179.

Lizards and burned chaparral

Kahn, W.C. 1960. Observations on the effect of a burn on a population of *Sceloporus occidentalis*. Ecology 41: 358–359.

Lane, R.S. 1990. Susceptibility of the western fence lizard *Sceloporus occidentalis* to the Lyme borreliosis spirochete *Borrelia burgdorferi*. American Journal of Tropical Medicine and Hygiene 42: 75–82.

Lane, R.S. and G.B. Quistad. 1998. Borreliacidal factor in the blood of the western fence lizard (*Sceloporus occidentalis*). Journal of Parasitology 84: 29–43.

Lillywhite, H.B. and F. North. 1974. Perching behavior of *Scelophorus occidentalis* in recently burned chaparral. Copeia 1974: 256–257.

Lillywhite, H.B., G. Friedman, and N. Ford. 1977. Color matching and perch selection by lizards in recently burned chaparral. Copeia 1977: 115–121.

Swei, A., R.S. Ostfeld, R.S. Lane, and C.J. Briggs. 2011. Impact of experimental removal of lizards on Lyme disease risk. Proceedings of the Royal Society B 278(1720): 2970–2978.

Pinyon jays

Balda, R.P. and G.C. Bateman. 1971. Flocking and annual cycle of the pinon jay, *Gymnorhinus cyanocephalus*. Condor 73: 287–302.

Ligon, J.D. 1974. Green cones of the pinon pine stimulate late summer breeding in the pinon jay. Nature 250: 80–82.

Marzluff, J.M. and R.P. Balda. 2010. *The Pinyon Jay: Behavioral Ecology of a Colonial and Cooperative Corvid*. T&AD Poyser, London, UK.

Human influences

Arno, R.S. 1971. Evaluation of pinyon-juniper conversion to grass. Journal of Range Management 24: 188–197.

Baker, W.L. and D.J. Shinneman. 2004. Fire and restoration of pinyon-juniper woodlands in the western United States: a review. Forest Ecology and Management 189: 1–21.

Blackburn, W.H. and P.T. Tueller. 1970. Pinyon and juniper invasion in black sagebrush communities in east-central Nevada. Ecology 51: 841–848.

Burkhardt, J.W. and E.W. Tisdale. 1976. Causes of juniper invasion in southwestern Idaho. Ecology 57: 472–484.

Evans, R.A. 1988. Management of pinyon-juniper woodlands. General Technical Report INT-249, Intermountain Research Station, USDA Forest Service, Ogden, UT

Fogg, G.G. 1966. The pinyon pines and man. Economic Botany 20: 103–105.

Gray, J.R., J.F. Fowler, and M.A. Bray. 1982. Free-use fuelwood in New Mexico: inventory, exhaustion, and energy equations. Journal of Forestry 80: 23–26.

Hurst, W.D. 1976. Management strategies within the pinyon-juniper ecosystem. Rangeman's Journal 3: 5–7

Jacobs, B.F. 2011. Spatial patterns and ecological drivers of historic piñon-juniper woodland expansion in the American southwest. Ecography 34: 1085–1095.

Koniak, S. and R.L. Everett. 1982. Seed reserves in soils of successional stages of pinyon woodlands. American Midland Naturalist 108: 295–303.

Lanner, R.M. 1981. *The Pinon Pine: A Natural and Cultural History*. University of Nevada Press, Reno, NV.

Miller, R.F. and P.E. Wigand. 1994. Holocene changes in semiarid pinyon-juniper woodlands: response to climate, fire, and human activities in the US Great Basin. BioScience 44: 465–474.

Rosenthal, M. (ed.) 1974. *Symposium on Living with Chaparral: Proceedings*. Sierra Club Special Publication, San Francisco, CA.

Samuels, M.L. and J.L. Betancourt. 1982. Modeling the long-term effects of fuel wood harvest on pinyon-juniper woodlands. Environmental Management 6: 505–515.

Infobox 8.1. Fire-loving beetles

Evans, W.G. 1964. Infrared receptors in *Melanophila acuminata* De Geer. Nature 202: 211.

Evans, W.G. 1966. Perception of infrared radiation from forest fires by *Melanophila acuminata* De Geer (Buprestidae, Coleoptera). Ecology 47: 1061–1065.

Evans, W.G. 2010. Reproductive role of infrared radiation sensors of *Melanophila acuminata* (Coleotera: Buprestidae) at forest fires. Annals of the Entomological Society of America 103: 823–826.

Linsley, E.G. 1943. Attraction of *Melanophila* beetles by fire and smoke. Journal of Economic Entomology 36: 341–342.

Linsley, E.G. and P.D. Hurd, Jr. 1957. *Melanophila* beetles at cement plants in southern California (Coleoptera, Burpestidae). Coleopterists' Bulletin 11: 9–11.

Schmitz, H. and H. Bleckmann. 1997. Fine structure and physiology of the infrared receptor of beetles of the genus *Melanophila* (Coleoptera: Buprestidae). International Journal of Insect Morphology and Embryology 26: 205–215.

Schmitz, H. and H. Bousack. 2011. Modelling a historic oil-tank fire allows an estimation of the sensitivity of the infrared receptors in pyrophilous *Melanophila* beetles. PLoS ONE 7(5): e37627.

Sowards, L.A., H. Schmitz, D.W. Tomlin, et al. 2001. Characterization of beetle *Melanophila acuminata* (Coleoptera: Buprestidae) infrared pit organs by high-performance liquid chromatography/mass spectrometry, scanning electron microscope, and Fourier transform-infrared spectroscopy. Annals of the Entomological Society of America 94: 686–697.

CHAPTER 9
Montane Forests

Climb the mountains and get their good tidings. Nature's peace will flow into you as sunshine flows into trees. The winds will blow their own freshness into you, and the storms their energy, while cares will drop off like autumn leaves.

John Muir

Two large mountain ranges – the Sierra Nevada and the Rocky Mountains – dominate western North America. The Wasatch, Cascades, and a few other less extensive ranges of western mountains are also of ecological interest. The communities in each of these differ somewhat, but for the most part they share a common feature: altitudinal zonation. Alpine tundra tops the taller peaks (Chapter 2) and, in the foothills, the montane vegetation generally gives way to pinyon-juniper woodlands (Chapter 8). In between are zones of subalpine, upper montane, and lower montane vegetation, collectively the Montane Forest.

The relative position of each zone remains the same throughout North America, but the altitude where each occurs changes with latitude, that is, a given zone develops at progressively higher altitudes toward the south. Eventually, the upper zones disappear as the montane environment becomes too warm for these to develop. In Arizona and Mexico, for example, subalpine forests only develop atop the tallest mountains, and alpine tundra disappears altogether. Conversely, to the north, the progressively cooler environment eventually eliminates all but the upper zones. Pinyon-juniper woodlands and forests characterizing the lower montane zone disappear from the zonal sequence in Idaho, and the upper zones develop at progressively lower altitudes. Farther to the north, the subalpine zone in the Montane Forest merges with the Boreal Forest (Chapter 3). A broad ecotone occurs at the end of the Montane Forest where the dominance of Engelmann spruce and subalpine fir gives way to the boreal species of white spruce. On a map, this transition lies at approximately 57°N. In places, however, white spruce extends southward into the Rocky Mountains (e.g., northern Wyoming). This and other southward extensions of northern biota into western mountains are ecologically akin to the Appalachian Extension described in Chapter 4.

Features and adaptations

Significant differences arise between the vegetation developing on slopes with northern and southern exposures: the band representing each zone extends farther upward on south-facing slopes, whereas the same band dips downward on north-facing slopes where conditions are relatively cooler. Evaporation is reduced on north-facing slopes, and snowdrifts often persist into early summer. Soil moisture in such sites is therefore greater when compared to locations at the same elevation on south-facing slopes. Exposure-related environmental differences in temperature and moisture therefore exert major influences on community development at corresponding elevations. In some places, for example, Douglas fir covers a north-facing slope, but ponderosa pine occupies the opposite slope at the same elevation. Similarly, one slope may be covered with conifers, whereas shrubby vegetation may occupy the opposite side.

Coniferous trees dominate the major zones (described in the following sections) and, for the most part, these feature the same adaptations shown by vegetation in the Boreal Forest (Chapter 3). Heavy snowfall, particularly in the Sierra Nevada, exerts a strong influence on the biota each winter, as do long periods of cold. Some animals, among them elk and mule deer, move down the slopes for greater protection in winter and then return to higher elevations for the warmer months. These movements are known as **altitudinal migration,** and they are as regular for many species as the more familiar north–south **latitudinal migration** typical of waterfowl and other birds.

Rocky outcrops are a prominent feature of most western mountain ranges. Crustose lichens colonize these barren surfaces, but generally do not play a major role in the establishment of other vegetation. In the Sierra Nevada, for example, weathering slowly produces cracks and

Ecology of North America, Second Edition. Brian R. Chapman and Eric G. Bolen.
© 2015 John Wiley & Sons, Ltd. Published 2015 by John Wiley & Sons, Ltd.

fissures in which fine particles and organic matter collect. Woody plants, including ponderosa and Jeffery pines, eventually gain a foothold in these sites without intervening successional stages.

Montane Forest zones

Lower montane zone

A community dominated by ponderosa pine marks the lowest zone in the Montane Forest (Fig. 9.1). Ponderosa pine, also called "western yellow pine," ranges across much of the mountainous backbone of North America. As well as reaching from northern Mexico to southern Canada, ponderosa pine also occurs on "montane islands" within the Great Basin and the Black Hills (*Black Hills*, this chapter). Small groves of ponderosa pine also interrupt the mixed prairie on ridges and eroded buttes in western North Dakota. The development of ponderosa pine communities varies considerably throughout this extensive distribution as a result of differences in elevation, soil types, and exposure. At lower elevations in the Rockies the stands are relatively open and sometimes include oaks, whereas denser, essentially monotypic, forest develops near the upper elevational limits of ponderosa pines. Ponderosa pine is well represented in the Cascades but a closely related species, Jeffery pine, either shares this zone with ponderosa pine or forms pure stands in the Sierra Nevada. Elsewhere in the Montane Forest, ponderosa pine characterizes a seral community rather than climax vegetation.

Fire is the recurring ecological force in the ecology of ponderosa pine, as shown by the incidence of fire scars among the annual rings on large stumps or in cores bored from living trees (Fig. 9.2). For example, in climax communities in the Sierra Nevada prior to 1875, fires occurred with an average frequency of 6–9 years. Ponderosa pine forests therefore developed in a regime of frequent burning, which reduced the fuels necessary for extremely hot, cataclysmic wildfires. In other words, frequent "cool" fires were normal events, from which mature trees remain protected by their thick, corky bark. The regular occurrence of fires represents a natural way of thinning overstocked stands, thereby releasing the survivors for renewed growth. As a result, ponderosa pine communities generally develop with a grassy understory beneath a canopy of well-spaced trees. The openness of these forests, in which the trees sometimes comprise no more than 25% of the ground cover, is very unlike the thick growth in the upper montane zone (following section).

Anthropogenic factors, specifically fire suppression and the introduction of livestock, produce significant changes in the community structure described above. Grazing steadily reduces the grassy ground cover, presumably lessening competition and increasing the survival of more pine seedlings. Without periodic "cool" fires, the open forest of ponderosa pine shifts to much denser stands where growth stagnates (e.g., 45-year-old trees with arm-thick trunks). Eventually, immensely destructive fires rage through the fuel-rich thickets and annihilate what once was a robust forest.

Upper montane zone

Douglas fir represents the climax vegetation in the upper part of the montane zone. Subordinate species are uncommon, but blue spruce and white fir are included on moist sites in the central and southern

Figure 9.1 Ponderosa pine dominates the lower montane zone. When burned periodically, the stands mature into large, well-spaced trees with grassy understories, whereas aspens and other species indicate earlier stages of succession. Reproduced by permission of Brian R. Chapman.

Figure 9.2 Fire scars, such as those shown on this slab of ponderosa pine, provide ecologists with estimates of fire frequencies in various types of forests. Ponderosa pine is among the species normally experiencing frequen burns, as indicated here by scars from fires in 1856, 1863, 1868, and 1879 on a tree established about 1835. Reproduced by permission of Carleton M. Britton.

Rockies. Northward, grand fir and white spruce share a measure of dominance with Douglas fir. More-or-less pure forests of Douglas fir are common however, and the closed canopy within these dense stands generally limits development of a herbaceous understory. Red fir – the largest species of fir in North America – is prominent in this zone in both the Sierra Nevada and Cascade ranges.

Fires typically establish monotypic stands of quaking aspen or lodgepole pine, either of which represents a lower successional stage in this zone. Some ecologists nonetheless believe lodgepole pine, at least in some eco-logical settings, may represent a climax community lying between Douglas fir and the subalpine zone. In any event, lodgepole pine is uncommon in the southern Rockies but aspen occurs widely throughout the western mountains and elsewhere in North America. Because it can regenerate from sprouts, aspen has a particular advantage of not only surviving but also expanding after moderate fires.

Subalpine zone

Deep snows regularly fall in this zone, especially on the west-facing slopes of the Sierra Nevada where the atmosphere is still richly laden with moisture (i.e., before the rain shadow takes hold). Winds are also severe and, together with the extreme cold and porous rocky soils, form a desiccating environment where only the hardiest species can survive. Ground litter is minimal; fuel accumulations are therefore low and fires are not

usually a major influence. Many of the trees necessarily anchor their roots among the cracks in boulders and rocky ledges.

Two trees – Engelmann spruce and subalpine fir – are indicator species of the subalpine zone in much of the Rocky Mountains. Both are relatively slender in profile, presumably to shed the heavy snowfall. In the northern Rockies, where Montane Forest grades into Boreal Forest, white spruce gradually becomes more prominent. Whitebark pine, mountain hemlock, and foxtail pine are representative species in the subalpine zone of the Sierra Nevada. Whitebark pine occurs regularly at the timber line where some individuals are markedly stunted and, in highly exposed areas along ridges or windswept slopes, the trees may form low mats of about 1 m (40 inches) in height. In sheltered areas however, whitebark pine develops erect stems up to 15 m (50 feet) tall with tough limbs capable of withstanding heavy snowfall and strong winds. Foxtail pine is similar in appearance to bristlecone pine, but it does not live nearly as long.

In favorable locations, trees in the subalpine zone may grow upright in more-or-less typical stands or, where conditions are particularly severe, develop into gnarled and stunted individuals. This distinctive growth-form known as krummholz, described in Chapter 2, marks the most rigorous, wind-driven sites where subalpine trees persist. Subalpine fir is among the species developing as krummholz at the timber line, but its spiral-like crowns also reach heights of up to 30

Figure 9.3 The timber line, as shown here in Glacier National Park, indicates the upper limit for tree growth on mountainsides. Pikas (inset), related to rabbits, find year-round shelter in crevices on talus slopes and other rocky sites near the timber line. Reproduced by permission of Brian R. Chapman.

m (100 feet) at the edges of subalpine meadows and other favorable locations.

Douglas fir represents a subclimax community in the subalpine zone, as do quaking aspen and lodgepole pine. Succession toward climax proceeds very slowly, as might be expected in the harsh conditions at high elevations. Near the upper limit of the zone (the timber line), the climax vegetation replaces itself after fires without passing through seral stages. Following fires, the exposed mineral soils become favorable habitat for seedlings of both Engelmann spruce and subalpine fir, but Engelmann spruce does not achieve maximum seed production until the trees reach 200–250 years of age. With the prolonged absence of fire, however, a thick accumulation of needles inhibits seedling development, and the two climax species then regenerate by somewhat different means. Subalpine fir reproduces asexually by layering, whereas the seeds of Engelmann spruce germinate primarily on the surfaces of fallen logs (i.e., a nutrient-rich seedbed atop the infertile needle layer).

The understory in the subalpine zone is richest on west-facing slopes where mosses, lichens, and herbs find favorable moisture. Conversely, herbaceous growth is diminished on the drier, east-facing slopes within the rain shadow.

Subalpine zones in the Sierra Nevada and the northern Rockies are the home of the American pika, a peculiar relative of rabbits. American pikas are kitten-sized and, unlike rabbits, have short rounded ears and no visible evidence of a tail. They live in colonies in rockslides and talus slopes near the timber line (Fig. 9.3), but extend to much lower elevations at the northern edge of their range (e.g., coastal British Columbia). Because of their isolation on mountaintops, American pikas may include 35 subspecies. Another species of pika occurs in Alaska and adjacent parts of Canada, separated from those in the northern Rockies by about 800 km (500 miles).

American pikas do not hibernate; instead, they "farm" vegetation in the alpine meadows near their rocky habitat (Fig. 9.3 inset). They selectively cut grasses and other herbaceous plants which are high in protein, lipids, calories, and water content. The harvest is piled in small "haystacks" for curing and, weeks and months later, is consumed as winter food. The haystacks – often a bushel or more in volume – are well sheltered by rocks but are always placed where exposed to sunlight. Despite their farming activities, American pikas can survive most winters without their hay supplies. When ecologists experimentally removed the haystacks in autumn, survivorship the following spring did not differ from those pikas whose haystacks remained available throughout the winter.

Associated habitats

Mountain parks and meadows

Large, naturally treeless sites known as mountain parks are scattered throughout the forests of the Rocky Mountains. Estes Park, North Park, and South Park, all in Colorado, are among the best known of these areas. Mountain parks are locations shaped by ecological phenomena and should not be confused with public lands

Figure 9.4 Openings known as mountain parks occasionally dot subalpine zones in the Montane Forest. South Park in the Front Range of the Rocky Mountains, shown here, is the largest of three similar mountain parks in Colorado. In winter, deep drifts fill the snow glade behind the thin line of trees (center) at the edge of the park. Reproduced by permission of Brian R. Chapman.

set aside for recreational and scenic values (i.e., state or national parks). Somewhat similar openings at higher altitudes are known as mountain meadows, and result from disturbances such as beaver occupation, avalanches, frost pockets, or fires. Mountain parks interrupt continuous stands of mature trees with other kinds of vegetation, but the composition of these communities varies somewhat from site to site (Fig. 9.4). In general, sagebrush often dominates west-facing slopes, whereas grasslands cover east-facing slopes. Sites within the subalpine zone typically support various combinations of willow and alder thickets along with a rich carpet of grasses and sedges. Those in the Sierra Nevada consist of similar plants (sedges, grasses, and willows) as well as some representatives of the heath family.

In their physiognomy, mountain parks seem analogous to mountain balds in the southern Appalachian Mountains (Chapter 4). The ecological forces responsible for shaping mountain parks in the Rocky Mountains are not clearly understood, similar to the situation for mountain balds in eastern North America. Mountain parks have abrupt edges, and trees from the surrounding forest seem unable to invade. Seedlings do not survive their first summer, apparently because their roots die from insufficient moisture. Some ecologists suggest that mountain parks develop and persist because of soil-related phenomena (e.g., soil structure that does not favor moist root zones). Conversely, the distribution of meadows in the Sierra Nevada seems correlated with shallow water tables where the moisture remains high all year round.

Another idea proposes that park vegetation – Cinnabar Peak in this case, a 73-ha (180-acre) area of forbs, shrubs, and grasses at 2758 m (9050 feet) in the Medicine Bow Mountains of Wyoming – moves slowly across forested sites in response to snowdrifts caused by the prevailing winds. As the snow sweeps across the open park, it drops in the forest about 75 m (245 feet) beyond the edge, much as a drift forms on the lee side of a snow fence. The trees in this zone die from the weight of the drift, creating a treeless "snow glade" running parallel to the current edge of the forest. Snow in the glade often exceeds depths of 7 m (23 feet) during the winter, and the deep snowbank persists well into the summer. Even trees at the edge of the glade are severely deformed by the heavy accumulation of snow. Eventually, the remaining row of trees, known as a ribbon forest, dies from wind damage and herbaceous vegetation takes over (i.e., the snow glade is no longer isolated and becomes part of the park). The process then begins anew, with another snow glade forming deeper into the forest on the windward side of the park. As the cycle repeats itself, the park creeps slowly along in one direction in a leapfrog manner. Meanwhile, young trees encroach on the leeward side, slowly establishing a forest. The overall result is the gradual movement of the park community.

An alternative analysis of Cinnabar Park indicates that the vegetation is not shifting, but that the grassy area appears stabilized because of a surface layer of fine soils. According to this analysis fire originally cleared the site, after which erosion exposed an underlying layer of stones. New deposits of fine, windblown soil then blanketed the stony stratum, whereas surface soils in the adjacent forests are coarse-textured. Because water evaporates rapidly from the fine-textured soils within the park, percolation is reduced downward beyond the layer of stones. These conditions, which exclude trees because their deep root systems cannot gain adequate soil moisture, favor plants tolerant of drier soils (e.g., grasses).

Explanations concerning the origin and persistence of mountain parks, such as those for mountain balds in the southern Appalachians, are varied and, as illustrated at Cinnabar Park, sometimes conflicting. Soil texture and

Figure 9.5 The Black Hills stand alone in the midst of the northern Great Plains, but geologists regard the formation as part of the Rocky Mountains. Forests of ponderosa pine dominate the rugged terrain but, overall, the vegetation is a heterogeneous mixture of plants from other regions. Reproduced by permission of John B. Pascarella.

its effect on soil moisture seem paramount, but these conditions may be linked to historical events such as fire, whereas features such as snow depth seemingly exert greater influence at other locations. All told, the ecological reasons for these tree-free areas in otherwise forested regions undoubtedly differ from site to site and therefore preclude a unifying concept.

Black Hills

For the Sioux, *Paha Sapa* or Black Hills describes the dark mountains rising above the plains in what is today western South Dakota and adjacent Wyoming. Forests of ponderosa pine account for the dark appearance of the Sioux's sacred mountains. The Black Hills cover about 10,360 km² (4000 square miles) and rise up to 1220 m (4000 feet) above the plains. Harney Peak, the highest point, extends 2207 m (7242 feet) above sea level. The montane zone prevails throughout the Black Hills, and species and communities associated with higher elevations are rare or altogether absent: Douglas fir is absent as a well-defined subalpine zone.

The Black Hills are a mountainous "blister" arising in isolation on the northern Great Plains, but geologists nonetheless consider the formation as part of the Rocky Mountains (Fig. 9.5). Structurally, the hills are a flat-topped dome of Precambrian rocks (granite, for the most part) encircled by progressively younger strata. They formed from uplifts whose hilly Central Basin of crystalline rock is surrounded by a limestone plateau. Along the outer edge of the plateau lies the Red Valley, a grassy trough encircling the Central Basin. A distinctive sandstone ridge – the "Dakota Hogback" – with a

steep-sided inner edge marks the border separating the Black Hills from the Great Plains.

A prominent gap in the ridge system served for millennia as a route for bison traveling between the grasslands within the Red Valley and those of the Great Plains. Early explorers and mountain men gained access to the Black Hills through the same gap. Spurred by rumors of gold, a troop led by George Armstrong Custer (1839–1876) marched into the Black Hills in 1874 in violation of a treaty with the Sioux. Custer's expedition fortunately included naturalist George Bird Grinnell (1849–1938), who consequently provided detailed biological observations of the Black Hills. As for Custer, another venture into Sioux lands northwest of the Black Hills in 1876 ensured him of a lasting place in history.

Because of their unique geography, the Black Hills support a heterogeneous biota largely derived from elements occurring elsewhere. Bur oak, for example, reaches its western limit in the Black Hills, and lodgepole pine extends no farther east. White spruce, a principal species of the Boreal Forest, also thrives at cool, moist sites in the Black Hills. Biodiversity is somewhat richer than might be expected for a region of similar size at the same latitude elsewhere in North America. Based on a flora of more than 1260 species of plants, the vegetation in the Black Hills consists of species from the Rocky Mountains (30%), plains (17%), deciduous forest (9%), and boreal forest (6%), as well as those from arid areas in southwestern North America (4.5%). Only a small part of the flora may be endemic (0.05%), with the remainder representing introduced or widespread species.

Similarly, the mammalian fauna includes many widespread species (e.g., raccoon), but also those of grasslands (e.g., black-tailed prairie dog), Rocky Mountains (e.g., mountain sheep), boreal forest (e.g., lynx), deciduous forest (e.g., fox squirrel) and a few associated with deserts and plains (e.g., black-tailed jackrabbit). In terms of percentages, species representing grasslands, mountains, and boreal forest form equal amounts (14% each) of the overall mix of mammals, followed closely by mammals associated with deserts (10%) and deciduous forest (9%).

The avifauna is also heterogeneous, with only the white-winged junco showing strong biogeographic affinities with the Black Hills. White-winged juncos are year-round residents in the Black Hills, the only place where this distinctively marked subspecies of the dark-eyed junco regularly breeds. Their habitat includes pine and spruce forests, as well as stands of deciduous trees, where they nest under the shelter of rock ledges, sod banks, logs, or exposed roots. White-winged juncos apparently evolved when glaciers forced their ancestors southward into the Black Hills (which remained ice free during the Pleistocene). When the glaciers retreated, the ancestral junco population returned north; some remained in the Black Hills however, where they continued breeding in isolation and evolved their distinctive plumage.

Coyotes, which today are widespread in North America including the Black Hills, warranted a special note in Grinnell's 1874 observations. Although coyotes were singularly abundant on the tablelands bordering the Black Hills, Grinnell did not observe "a single specimen (after) penetrating into the hills proper." Instead, Grinnell noted, coyotes were replaced by gray wolves, which the naturalist found much in evidence. "I found the gray wolf to be one of the common animals in the Black Hills, and hardly a day passed without my seeing several individuals of this species."

With settlement by Euro-Americans, however, wolves soon came under heavy pressure in the Black Hills area. Reports of the decimation include some 500 killed on a large ranch between 1895 and 1897, as well as 925 wolf carcasses obtained by the US Forest Service in 1907. Because of such ruthlessness the wolf population steadily declined, although a few individuals remained (also to be shot) in the Black Hills as late as 1928. Meanwhile, as competition with wolves diminished, coyotes moved into the Black Hills, replacing the wolves and increasing in numbers after the turn of the century.

Another member of Custer's expeditionary force, surveyor William Ludlow, commented on the absence of trout in the Black Hills, despite the abundance of streams "which seemed as though made expressly for that fish which requires an unfailing flow of cold, pure water. There could be no finer streams in the world than these (but) we found nothing but some small chub and a species of sucker....' Ludlow had observed a significant biogeographical feature of the Black Hills: their island-like isolation had precluded the colonization of trout. Specifically, the waters of intervening drainages on the plains were too warm, too turbid, and too devoid of oxygen to serve as corridors for trout that otherwise might have dispersed eastward from stream systems in the Rocky Mountains. This apparent oversight of nature, however, was remedied in 1893 when a shipment of trout traveling by stagecoach in milk cans arrived in the Black Hills. By 1900, the region had its own hatchery, and trout thereafter became an established element in the Black Hills fauna.

Redwoods and sequoias

The western mountains of North America are home to two of the world's largest trees. Each species – the coast redwood and giant sequoia – is a colossus of nature and, late in the 19th century, emerged as icons of conservation (Fig. 9.6). For John Muir (1838–1914), the majestic stands of these immense trees served as centerpieces for protecting wilderness (Infobox 9.1).

Like a forest of stately columns, mature redwoods reach heights of more than 91 m (300 feet) with their lowest branches as much as 46 m (150 feet) above ground. Redwood forests grow in the coastal "fog belt" from central California to southern Oregon, where they are sometimes included as components of the Temperate Rain Forest (Chapter 10). They occupy a narrow zone, seldom more than 15 km (10 miles) wide, which runs parallel to the coastline. However, because redwoods do not tolerate salt spray, this zone often lies some distance inland from the coastline. Some of the finest redwood stands develop on flatlands bordering rivers, where flooding deposits layers of new sediments. At favorable sites, redwoods dominate the forest almost to the exclusion of other trees. Western hemlock and Douglas fir are among the associated species at other locations.

Redwoods produce immense numbers of seed, but few of these germinate unless fire first removes the thick layer of ground litter accumulating beneath the canopy. Combined with the fact that many of the seeds are sterile, seedlings are rarely encountered. Redwoods, unlike most other conifers, also regenerate from the vigorous and rapid growth of stump sprouts, which are sometimes called "cathedral rings." Heavy bark, which may be 15–30 cm (6–12 inches) thick, protects older trees from fire, but even moderately light ground fires readily kill younger trees. In addition to scars from fires, older trees often possess complex crowns. Storms frequently snap the tops and high limbs of lofty redwoods, thereby creating temporary gaps in their crowns. Redwoods respond by sprouting new trunks – a process known as reiterating – developing structural complexity in an ever-expanding crown. Several recently studied large redwoods possessed crowns composed of more

(a)   (b)

Figure 9.6 Majestic stands of redwoods in the forests of California deeply moved the Scottish immigrant John Muir, whose inspired and relentless fight for their protection saved many of these trees for the enjoyment of future generations. Reproduced by permission of (a) Brian R. Chapman and (b) Photograph used with permission of Carl L. Chapman.

than 100 reiterated trunks, some of which reached 2.6 m (8.5 feet) in diameter.

The wide crowns of tall redwoods support distinctive communities that were recently discovered when daring biologists explored the tree tops. More than 200 species of bryophytes and lichens and 13 species of vascular plants occur on the thick branches and reiterated trunks in the crowns of massive redwoods. As the epiphytic plants decay, moisture-holding soil accumulates on the branches and in the crotches between limbs and trunks. Wandering salamanders act as top predators in the canopy community, where they feed on a wide variety of invertebrates inhabiting the epiphytic plants and soils. Marbled murrelets nest in epiphytes on large branches (*A Seabird in the Forest*, Chapter 10).

In one sample more than 40 redwoods reached heights exceeding 110 m (360 feet), which is near the theoretical maximum height of 122–130 m (400–427 feet). Beyond this limit, the upward hydraulic conductivity produced by transpiration in the crown cannot overcome the pull of gravity plus the friction of water running through the xylem. For many years, biologists assumed that redwoods surmounted these limitations by their leaves absorbing water from fog. Now, however, it appears that

fog plays a greater role in suppressing water loss from leaves instead of providing supplemental water. Although there may be limits to how tall a redwood may grow, the trees continually add more wood to their trunks and branches.

Redwood lumber is famous for its durable qualities, and large areas within the species' range are privately managed as commercial forests. By 2000, logging had claimed 95% of the old-growth redwood forests, and most of the stands logged before 1930 had been cut a second time. After logging mature trees, the community structure of soil arthropods changes in redwood stands, which in turn may negatively affect decomposition processes and nutrient cycling in the regenerating forests. Younger redwood forests also lack the structural and biological diversity characteristic of the old-growth forests that now remain only in parks and preserves.

A related species – giant sequoia – is limited to the west-facing slopes of the Sierra Nevada of California at elevations of 1500–2380 m (5000–7800 feet). With an average height of about 80 m (265 feet), giant sequoias are not as tall as redwoods but their trunks are about three times thicker; their diameters approach thicknesses of 9 m (30 feet) near ground level. Because of lightning

Infobox 9.1 John Muir (1838–1914), patron saint of wilderness

For John Muir, the majestic stands of redwood and sequoia became the centerpiece for protecting wilderness. Muir, a Scottish immigrant at the age of 11 years, worked on his father's farm then attended the University of Wisconsin where he studied chemistry, geology, and botany. Muir never graduated and emigrated to Canada to avoid fighting in the Civil War. He returned to the United States in 1866, eventually finding work in Indiana in a wagon-wheel factory. A factory accident nearly cost Muir his sight. During his recovery in a darkened room, Muir decided to leave the industrial world to spend his life exploring wild areas and studying plants. He began his quest by *walking* from Indiana to Florida, taking the "wildest, leafiest, and least trodden way" possible. Once there, he visited Cuba where he studied plants as well as shells, then left for New York to begin a voyage to California.

In California, Muir immediately resumed his treks, notably those in the Sierra Nevada with which he developed a life-long spiritual attachment. His observations led him to postulate that Yosemite and nearby valleys had been shaped by glaciers, and not by earthquakes as held by the prominent geologists of the day. His views were proved correct, bolstered by an alpine glacier he later discovered below Merced Peak. But it was in the Sierra's giant trees that he found his *cause célèbre*. Muir was a passionate writer who wrote incessantly about protecting the trees from any threat, including "hooved locusts" (as he called sheep). His articles and books inspired layman and Congress alike and ultimately led to the establishment of Yosemite and Sequoia national parks. Other stands, particularly groves of sequoias, were later added to the National Forest System.

Muir's unyielding fight to save wildness attracted notable figures to the Sierras, among them essayist-philosopher Ralph Waldo Emerson (1803–1882) and President Theodore Roosevelt (1858–1919). He initially befriended Gifford Pinchot (1865–1946), first chief of the US Forest Service, but the two later fell out over the protection of wilderness. Indeed, the views of the deeply religious Muir were rooted in the philosophies of Emerson and Henry David Thoreau (1817–1862), namely that communion with nature brings humans closer to their divine creator. Nature, he felt, is a temple too often sullied by humans and their economic activities. For Muir, this meant complete protection of wilderness and the biota therein, whereas Pinchot adhered to a pragmatic view of nature: one based on the economic values of natural resources.

Hetch Hechy, a valley whose splendor rivaled that of Yosemite, became Muir's final great challenge and the breaking point in his relationship with Pinchot. The city of San Francisco wanted to dam the valley to create a reservoir which, in Muir's eyes, was an unimaginable sin against a "temple of nature." Muir fought vigorously against the project, delaying it for several years, but with the support of Pinchot it was eventually authorized in 1913. Muir died the following year, some saying from a broken heart over the loss of Hetch Hechy.

Muir is remembered as the "patron saint of the American wilderness." He founded the Sierra Club in 1892, which still upholds the protection of wilderness areas, and served as its first president until his death 22 years later. A long list of sites bear his name, including the glacier he discovered in Alaska, a wilderness area, trails, groves of redwoods, a mineral (muirite), and several species and subspecies of plants and animals. Two commemorative stamps have borne his likeness as well as California's quarter in the state series issued by the US Mint. In addition to hundreds of articles and essays, including a paper on giant sequoias published by the American Association for the Advancement of Science, he wrote several books including *The Mountains of California*, *Our National Parks*, *A Thousand-mile Walk to the Gulf*, and *The Yosemite*, which stressed the ecological importance of natural areas, now an accepted principle of conservation biology. In the words of a colleague, his work "sung the glory of nature like another Psalmist."

strikes, the crowns of giant sequoias are often rounded. Fire is a crucial component in the ecology of these trees and, without periodic burning, litter accumulates on the forest floor and limits reproduction. Groves of giant sequoias represent fire climax communities, which will yield eventually to white fir if fires are suppressed over the long term. While extremely durable, the wood is brittle and is no longer cut commercially.

Bristlecone pine forest

Ecologists and other scientists for years believed that sequoias were the oldest living organisms on Earth. Many of these giants indeed are about 2000 years old, a feat complementing the grandeur of their immense physical presence. However, bristlecone pines more than 4000 years old were discovered in the 1950s. Unlike sequoias, bristlecone pines seldom exceed a height of about 11 m (35 feet) and, depending on site conditions, their weather-sculpted trunks may be twisted and gnarled (Fig. 9.7). The contorted appearance of the older trees emanates an engaging, almost mythical, quality suggesting timeless endurance.

Bristlecone pines persist in scattered stands in the subalpine zone near the timber line. On sites with favorable soil and moisture conditions, the trees develop fully and often mix with limber pine and Engelmann spruce. Bristlecone pines dominate some locations, but they are intolerant to shading and therefore compete poorly with other subalpine species. Where the environment is harsher, however, the stress sometimes produces twisted, ill-formed trees to the exclusion of other species. Such sites are most often at higher elevations where the rocky

Figure 9.7 Bristlecone pines characterize the subalpine forest zone in some regions of North America. With favorable conditions the trees grow in a characteristic form, but stresses produce twisted and gnarled trees at higher elevations, some of which attain ages of more than 6000 years. Reproduced by permission of R. Scott Beasley.

soils are poorly developed and limited in moisture. Areas near outcrops of bedrock represent typical locations for ancient bristlecone pines. These trees typically lose most of their bark, which exposes and erodes the underlying wood and limits the foliage development; crown size is closely related to the amount of live bark. The dieback of bark invariably occurs in bristlecone pines at least 1500 years old and, in one case, only 8% of the bark remained on the trunk of a tree at least 4900 years old. Such trees apparently cope with reduced foliage and bark by adding new wood slowly to only a small part of the stem, striking a balance between growth and the limited amount of living tissue. Interestingly, bristlecone pines attain their great ages only where environmental conditions are stressful, but the reasons for this ecological anomaly remain unclear.

In the White Mountains of California, where 4000-year-old trees were initially discovered, bristlecone pines are the principal subalpine species. Few of the trees in this area are small – seedlings are rare or absent – whereas others have diameters of much larger size (e.g., 90–150 cm or 36–60 inches). A size distribution of this type suggests that the White Mountain stand originated at a time when climatic or other conditions were more suitable; the absence of seedlings and small trees suggest poor reproduction in historical times. If so, the exceptionally long life span of bristlecone pines may represent an adaptation by which the species persists through long periods of unfavorable conditions.

The seeds of bristlecone pines are small and winged and therefore adapted for wind dispersal. This feature seems of greatest consequence in the milder areas at lower elevations where bristlecone pines face competition from other species. Conversely, in the rigorous subalpine environments at the timber line, many seeds are dispersed by Clark's nutcrackers. These birds store the seeds in shallow caches, some of which later germinate and produce a clump of stems. Because these clumps occur with far greater frequency at higher elevations, Clark's nutcrackers seem more effective than wind as dispersal agents at sites where bristlecone pines persist for centuries. Clark's nutcrackers also disperse the seeds of other conifers, including limber pine, whitebark pine, and pinyon pine. The **coadaptation** between these trees and the birds represents an example of mutualism.

Because of their exceptional longevity, bristlecone pines represent an important means of determining long-term climatic patterns in North America. Analyses of their growth rings – dendrochronology – reveal remarkable fluctuations in climate since the Ice Age ended. The samples may be collected from living trees as well as fallen logs. While still living, the old pines produce proportionally more resin-bearing tissues and experience less decay. Similarly, when old trees eventually die, the logs remain unusually resistant to rot for many centuries afterward. Logs collectively representing a time span as long as 9000 years have been sampled to determine weather patterns for that period.

Figure 9.8 Huge fires swept across Yellowstone National Park in 1988 (inset), burning large areas of forest but killing relatively few large animals. A decade later, fallen snags burned by the fire contribute organic nutrients to the soil, enhancing primary production and improving foraging conditions of many animals such as the elk shown here. Reproduced by permission of Brian R. Chapman and (inset) Jeff Henry, the National Park Service.

A site designated as the Ancient Bristlecone Pine Forest specifically protects old stands in the White Mountains of California. One magnificent tree within this forest – the "Patriarch" – is the largest of its species. However, at an age of just 1500 years it is scarcely the oldest.

Fire in montane forests

As noted repeatedly in previous chapters, fire represents a significant ecological force in a variety of communities (e.g., longleaf pine). Nonetheless, because fire can also threaten humans and their property, justifications for tolerating natural fires on lands managed by state and federal governments are not always appreciated by the public at large. This issue reached its zenith in 1988 when wildfires spread across a huge part of Yellowstone National Park and adjacent lands (Fig. 9.8). This massive fire subsequently became an important ecological laboratory for understanding the dynamics of fire in Montane Forest ecosystems and the focus of our discussion.

The grandeur of Yellowstone, although long known to "mountain men," gained little credibility until the vibrant landscapes of artist Thomas Moran (1837–1926) were displayed in the eastern United States and to members of Congress. Then-President Ulysses S. Grant signed the legislation proclaiming Yellowstone a national park in 1872, the first in the United States as well as in the world. A formal policy of fire suppression, which began when the US Army initiated administration of the park in 1886, continued with the creation of the National Park Service in 1916. However, stimulated by a review of

issues concerning wildlife management in Yellowstone, the National Park Service modified the fire-suppression policy in 1972: natural fires (those ignited by lightning) could burn unchecked as long as human lives and property were not threatened. Such fires therefore became part of land management on federal lands in the Yellowstone area, including the park and adjacent national forests.

In 1988, nearly ideal conditions prevailed for a rash of almost 250 fires in the Yellowstone area: prolonged drought, strong and persistent winds, adequate fuel, and numerous lightning storms (not forgetting human carelessness) contributed to the ignitions. Precipitation in 1988, for example, was the lowest on record, only 36% of normal for June, July, and August. Once ignited, the fires expanded rapidly and, with their growing size and intensity, spread across creeks and other landforms that usually act as natural barriers and deterrents (e.g., the fire leapt across the Grand Canyon of the Yellowstone, which is more than 0.4 km or 0.25 mile wide at its narrowest point). Some fires advanced at the exceptional rate of 16 km (10 miles) per day. In part, the flames spread from tree to tree through the forest canopy, producing a "crown fire" which can spread without requiring heavy fuel accumulations at ground level. The strong winds also carried live embers (firebrands) as far as 1.5 km (2.5 miles) beyond the fire front, creating spot fires elsewhere. Human efforts to control the fires proved fruitless, and about 36% (321,000 ha or 793,000 acres) of Yellowstone National Park eventually burned as did areas in adjacent forests.

Because of variations in wind velocity, fuel supplies, and other factors, the Yellowstone fires burned unevenly and produced a checkerboard of differential patterns in plant

Infobox 9.2 The underwater bird

The avifauna along the mountain streams of western North America includes a bird with what seems to be the remarkable ability of "walking" underwater. The dipper – water ouzel is another name – is a slate-colored, stubby-tailed bird resembling a large wren. John Muir (see Infobox 9.1) fondly likened dippers to hummingbirds of mountain waterfalls in which "the bird and stream are inseparable, songful and wild, gentle and strong – the bird ever in danger in the midst of the stream's mad whirlpools, yet seemingly immortal." When dippers blink, a silvery membrane flashes conspicuously across their eyes. This special structure acts in the same way as a contact lens for vision and protection underwater.

Dippers are solitary and bob about on the rocks of rushing streams before suddenly slipping underwater to forage on **benthic** invertebrates and small fishes. They may significantly reduce the densities of some kinds of insect larvae, perhaps in some instances competing with trout for certain foods. While moving about underwater, dippers propel themselves with their wings, including movements on the stream bottom where they use their unwebbed feet to grasp stones as they "walk" in search of food.

Dippers seldom remain submerged for more than 10 seconds and rarely forage at depths of more than 0.6 m (2 feet). Like most aquatic birds, they maintain their plumage using secretions from well-developed oil glands, although those in dippers are usually large in relation to their body size. Dippers build large ball-like nests constructed of fresh moss, located on rocky walls behind waterfalls or along streamsides. Their nests are so tightly woven that the mosses, nourished by the mist, continue growing. The dipper family is represented worldwide by only five species; just one of these occurs in North America.

mortality and ash deposition across the landscape. Such a mosaic, in turn, influenced the distribution and behavior of forest animals. Similarly, the aquatic biota in many of Yellowstone's streams and lakes was subject to post-fire influences such as accelerated erosion and sedimentation, as well as altered water chemistry and nutrient availability.

These conditions may last for several years but eventually return to their pre-fire state as the forest regrows. Nonetheless, despite some episodes of heavy sediment loads and nitrogen enrichment, the 1988 Yellowstone fires did not significantly affect trout populations, suggesting that their invertebrate foods were unharmed (Infobox 9.2).

Millions of lodgepole pines were killed in the 1988 Yellowstone fires. Approximately 25 million metric tons of standing and downed dead wood, known as coarse woody debris, were left in the aftermath of the firestorms. Immediately after the intense fire, about 24% of the forest floor was covered by downed logs and charred stumps, and this percentage increased for two decades as more snags fell. The tangle of coarse woody debris modified the forest floor environment; it sometimes impeded the movements of large animals, but the debris became a primary source of organic nutrients enriching the generally infertile forest soils (Fig. 9.8). Contrary to popular concepts at the time, the fires consumed only about 8% of the coarse woody debris already on the forest floor. An equal amount of woody debris was converted to charcoal, providing a pool of organic material to enrich the forest soil for many years.

Secondary succession is partly controlled by seed availability; lodgepole pine, a fine example of a fire-adapted species, therefore establishes thick stands soon after fires (Fig. 9.9). Lodgepole pine produces serotinous

Figure 9.9 Fires stimulate the regrowth of lodgepole pine, which thereafter develops in thick monotypic stands. Reproduced by permission of Bob Stevenson and the National Park Service.

cones which open when heated, releasing large quantities of seed for germination on the mineral soils exposed by fires. Moreover, lodgepole pines mature rapidly, producing cones as early as 10 years of age and heavy crops every 2–3 years after 25 years. Many of the cones remain attached to the trees for 20 years or more, so an immense reservoir of seeds builds up in the intervals between fires. Curiously, however, some cones are not serotinous, and the incidence of these apparently depends on the nature of the previous disturbance. Lodgepole pines produce more serotinous cones after fires initiate secondary succession but, when the disturbance is caused by storms or insect attacks, the regrowth includes a larger percentage of non-serotinous cones. With long intervals between fires (i.e., more than one generation of lodgepole pine), fewer serotinous cones are produced and the reserve of seeds is reduced when a fire eventually occurs.

An ecological surprise from the 1988 Yellowstone fires arose from the post-fire dynamics of aspen. As expected, sprouts appeared from the roots of burned aspen clones, but many small aspen seedlings also appeared as far as 15 km (9 miles) away from the nearest clone. Conventional wisdom at the time assumed that aspen clones in the northern Rocky Mountains do not reproduce by seed (a belief undermined by the appearance of the seedlings). Aspen, which had been relatively rare in Yellowstone prior to the fire, consequently became widespread in areas once dominated by conifers. However, aspen is again becoming scarce as the overstory of lodgepole pine continues to expand.

The understory recovered rapidly. Some herbaceous plants, such as fireweed and silvery lupine, sprouted just days after the fires were extinguished. These and other perennial plants with deep roots and rhizomes survived because the soil was charred only to a mean depth of 14 mm (0.5 inch) even in areas affected by intense surface and crown fires. Perennial grasses, herbs, and shrubs with protected roots and rhizomes also flowered profusely the following year. Within four years, native vegetation again carpeted the understory in burned areas. Such rapid regeneration likely prevented the invasion of exotic species, which did not increase in abundance or extent after the fires.

Forage production increased on burned areas after the fire, with new growth even more palatable and rich in protein than unburned vegetation. Elk heavily browsed aspen seedlings, for example, sometimes resulting in shrub-like regrowth. Weight and other measures of body condition thereafter improved, leading to normal or even above-average survival of elk calves. Moreover, the fires created additional summer ranges, which remain suitable for elk for as long as 30 years in severely burned stands of lodgepole pine. The 1988 fires ultimately improved the carrying capacity, supporting even larger herds than before, and by 1993 the elk population had recovered from its post-fire dip.

Few large mammals died in the Yellowstone fires, despite a summer herd of about 31,000 elk and lesser numbers of bison, deer, and moose. Surveys conducted in the park after the fires subsided revealed 246 dead elk, or about 1% of the summer population. Most of the dead elk succumbed to smoke inhalation, as determined by the presence of soot caked in their tracheas. Fewer than 20 bison, deer, and moose and only one grizzly bear were killed. The dead animals were found only at sites where high winds fanned the fires. Although few large animals died the immediate supply of carrion attracted large numbers of ravens, coyotes, and other scavengers. The incidence of elk and other ungulates in the diet of grizzly bears jumped more than three-fold in the months immediately after the fires. In fact, many grizzly bears moved into still-smoldering areas in search of food.

Although more than 25 years have passed since the Yellowstone fires of 1988 captured public attention, the foregoing must be considered short-term results. Nonetheless, some observations can be made about the ecological aftermath of large-scale episodic disturbances. For example, despite their intensity and extensive nature (and the headlines), the 1988 fires did not threaten the welfare of any species. Wildlife losses were minimal, at least among the larger grazing animals, and some animals died only because they were trapped by backfires started by firefighters. Often, elk and other ungulates continued grazing in meadows, undisturbed by the firestorms raging in nearby forests, quite unlike the general public's perception of behavior (e.g., panic, as depicted in the movie *Bambi*). However, habitats changed – often drastically – benefiting some species but harming others and altering the relative abundance of many organisms in and near the burned areas. Foraging habitat increased for grazing and browsing animals (e.g., elk), but habitat diminished for red squirrels and other species requiring mature forests. For a while, predation of small animals increased because of temporary reductions in cover, a result similar to the post-fire responses in chaparral (Chapter 8). To summarize, the 1988 Yellowstone fires produced a mosaic of burned, partially burned, and unburned habitat and again rejuvenated and enriched the Yellowstone area in an age-old ecological process.

Highlights

Western chipmunks and competitive exclusion

A single species of chipmunk occurs east of the Mississippi River. In comparison, the fauna of western North America includes 15 species, 8 of which are found in the Sierra Nevada. Each of the western species is similar in appearance, which suggests they evolved from a common

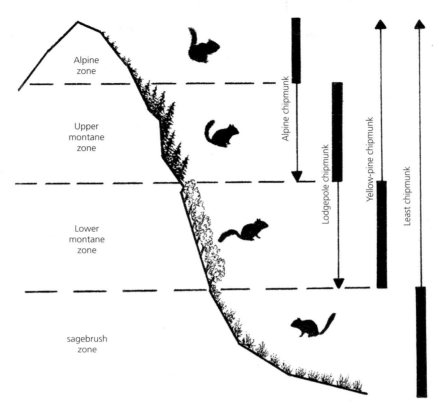

Figure 9.10 The distribution of four species of western chipmunks illustrates the concept of competitive exclusion. In terms of its ecological tolerances, each species could live in at least one other zone (thin arrow) but is kept within a narrower range because of competition from a neighboring species. The total length of each line (arrow plus bar) represents a fundamental niche, whereas the more restricted area is a realized niche (bar only). Illustrated by Tamara R. Sayre, based on Heller (1971).

ancestor in relatively recent times. The interspersion of mountains across a broad landscape – best shown in the **Basin and Range Province** – is among the primary reasons for the geographic isolation of several species. Significant differences in habitat, typically the result of changes in altitude, also account for the genetic isolation and eventual speciation of western chipmunks.

Four species of chipmunks inhabit the eastern slopes of the Sierra Nevada, where their geographic ranges are grossly **sympatric** but where they nonetheless remain ecologically separated according to elevational zones (Fig. 9.10). Alpine chipmunks occupy mountain meadows and pine forests on the upper zones of the Sierra Nevada. Next is the lodgepole chipmunk, which inhabits forests of its namesake lodgepole pine and other conifers, followed by the yellow-pine chipmunk in a zone of ponderosa, Jeffery and pinyon pines. Least chipmunks occupy the lowest elevations in the sequence, living in a sagebrush-dominated community.

Each species in this complex of chipmunks maintains its ecological integrity because of **competitive exclusion**.

Simply put, this concept states that two species with similar needs for resources cannot coexist in the same habitat. One species, because it is better adapted, eventually displaces the other. The adaptation may of course be physical, physiological, or behavioral, or often (as is the case with the four chipmunks) some combination of these.

Lodgepole chipmunks, the most aggressive of the four species, defend their habitat from encroachment by neighboring species above (alpine chipmunks) and below (yellow-pine chipmunks). Their feisty behavior secures lodgepole chipmunks the shade they need to survive (they are the least heat-tolerant of the four species). To avoid sharing the food and other resources in its cooler habitat with the two other species, the lodgepole chipmunk maintains exclusive ownership with highly aggressive behavior.

Another example of competitive exclusion occurs farther downslope, where the habitat of least chipmunks borders that of the slightly larger yellow-pine chipmunk. In this case, the yellow-pine chipmunk is more aggressive

Figure 9.11 A good "cone year" for the whitebark pine in western Montana. In comparison to other species of pines in North America, the lipid-rich seeds of whitebark pine are exceptionally large in relation to cone size (inset); the cone length is 7 cm (2.75 inches). Reproduced by permission of Stephen F. Arno, USDA Forest Service and (inset) Elizabeth D. Bolen.

and prevents its smaller neighbor from encroaching into the pine forests. Least chipmunks, although physiologically able to live anywhere on the slopes of the Sierra Nevada, can tolerate heat far better than any of the other species; they are **eurythermal**. Conversely, the other chipmunks cannot cope with desert-like conditions, enabling least chipmunks to claim their warmer habitat by default. This relationship was demonstrated experimentally when ecologists removed yellow-pine chipmunks, and least chipmunks quickly expanded upslope into the "empty" forest. However, when the experiment was reversed, yellow-pine chipmunks did not invade the much warmer sagebrush community.

The latter relationship also illustrates the difference between a **fundamental niche** and a **realized niche**. Least chipmunks, with their broad tolerance to temperature, can live almost anywhere within their geographic range and therefore have a wide fundamental niche. Because of competitive exclusion by yellow-pine chipmunks, the realized niche of least chipmunks is however relatively narrow and they remain confined to habitat where heat stress keeps out yellow-pine chipmunks.

Squirrels, bears, and pine cones

Red squirrels are closely associated with coniferous forests in much of North America, including the broad expanse of Boreal Forest (Chapter 3). In montane forests, as well as elsewhere in their huge range, red squirrels rely on large caches of cones for their winter food supply (Fig. 3.9). The caches or "middens" also provide a ready source of food for black and grizzly

bears. In the subalpine forests of the northern Rocky Mountains, for example, grizzlies relish the large nuts borne in the cones of whitebark pine (Fig. 9.11). The bears raid the squirrel caches in preparation for hibernation, rapidly acquiring body fat from whitebark pine nuts whose lipid content is almost 80% (a gram of the energy-rich pine nuts, when mature, yields about 7 kilocalories, an amount exceeding the richness of chocolate and nearly as much as butter). Grizzly bears also seek caches as a sure source of high-energy food in the spring after emerging for hibernation. When plundering a cache to uncover the deeply buried cones, a hungry grizzly often plows up the immediate landscape, leaving behind excavations much like those they make in Tundra when hunting for Arctic ground squirrels (Chapter 2). Black bears also forage on pine nuts and sometimes climb trees to secure cones, whereas grizzlies, which cannot climb, rely primarily on the caches of red squirrels for pine nuts.

The caches of lipid-rich nuts of whitebark pine affect grizzly bears in other ways. Cub production, for example, improves in years when whitebark pines are heavily laden with cones. Conversely, when cones are less abundant, sows are poorly nourished and their cubs experience greater mortality. Sows fattened on pine nuts also wean their cubs earlier in years when cones are plentiful. For whitebark pine, exceptional cone production occurs at intervals of 3–5 years, although these so-called "cone years" are not readily predictable.

An abundance of cones from whitebark pines provides at least one other benefit for grizzly bears in Yellowstone

National Park: in years with heavy cone production, the bears forage from August to October at higher elevations where contact with humans is less likely. In years with poor cone crops, however, grizzlies more often visit camping grounds, which requires trapping and moving the bears to remote areas of the park. Unfortunately, some transplanted bears persist in returning and these animals are summarily dispatched; two–three times more bears are killed in years with poor cone crops.

Three pathogens seriously limit whitebark pine in the northern Rocky Mountains and therefore indirectly affect grizzly bears: white pine blister rust, an exotic disease mentioned in Chapter 4; the mountain pine beetle; and dwarf mistletoe. White pine blister rust attacks all species of five-needled pines, of which whitebark pine is the most susceptible. The disease kills the upper cone-bearing branches but can eventually kill the entire tree. Along with insects and other diseases, dwarf mistletoe (a parasitic plant) also attacks trees weakened by white pine blister rust, further reducing their growth and cone production.

Fire suppression likewise diminishes the distribution and abundance of whitebark pine, which then gives way to Engelmann spruce and subalpine fir in the course of succession. Additionally, fire suppression at lower elevations creates older stands of lodgepole pine that are particularly susceptible to mountain pine beetles. The destructive beetles then move upslope and attack whitebark pines in the subalpine zone. A natural fire interval of 50–350 years is sufficient to maintain vigorous stands of whitebark pine in subalpine forests

Because of these limitations, whitebark pine has become less abundant in the last century and, lacking remedy, may be extirpated from the alpine forest in another century or two. In part, the decline of whitebark pine is monitored by examining seed residues in bear droppings and the kinds of cones in red squirrel caches.

Sky islands in Arizona

Isolated populations are often at risk, situations most commonly encountered on oceanic islands disturbed by humans. However, mountaintops also represent islands where unique populations may develop after generations of reproductive isolation (e.g., salamanders in the southern Appalachians as described in Chapter 4). Other "sky islands" are represented by a forest of Engelmann spruce and corkbark fir (a subspecies of subalpine fir) atop the Pinaleño Mountains, which includes Mt Graham, one of the highest peaks in Arizona. This stand of subalpine forest – the largest in southern Arizona – provides much of the habitat for several endemic taxa including the white-bellied vole, the Mt Graham pocket gopher, Rusby's mountain fleabane, and the southernmost subspecies of red squirrel.

Mt Graham red squirrels, which evolved during some 10,000 years of isolation, represent a classic example of allopatric speciation. Unlike the other subspecies of red squirrels, the Mt Graham red squirrel is listed as an endangered species. Years of logging and development have destroyed and fragmented parts of the subalpine forest in the Pinaleño Mountains, leaving about half of the original area as suitable habitat for Mt Graham red squirrels. Other risks to this small population of red squirrels include its complete isolation without the possibility of immigration and large fluctuations in cone crops. Moreover, Abert's squirrels were introduced into the ponderosa pine forests on Mt Graham, thereby becoming a competitive influence at lower elevations of the habitat occupied by the red squirrels. All told, just about any set of bad circumstances (e.g., back-to-back years of poor cone crops coupled with a large fire or disease in either trees or squirrels) could eliminate the population in an **extinction vortex**.

The 3267 m (10,720 feet) height of Mt Graham also made the mountaintop attractive to astronomers who planned to build several observatories on the same "island" occupied by the red squirrels. Although construction was immediately challenged under the mandate of the Endangered Species Act of 1973, a permit was nonetheless issued in 1988 and work began on three telescopes, with allowances for later adding four more. In doing so, the University of Arizona (builder of the observatories), along with state and federal agencies, agreed to several measures including the creation of a refuge with limited human access, the reforestation of potential habitat, and monitoring the impacts of construction on the red squirrel population. The agreement also initiated a 10-year research program on Mt Graham red squirrels and their ecology.

Population estimates for Mt Graham red squirrels are based on the number of occupied caches (piles of cone debris used as sources of winter food). Year-to-year differences in cone production, as expected, caused fluctuations in the number of squirrels. Fires and insect infestations severely damaged the spruce and fir forests in the late 1990s, resulting in a population decline from about 350 squirrels in 2001 to approximately 212 in 2011. A minimum of 300 adults in the spring population each year remains the official goal of the agencies responsible for protecting the squirrels, but this level has not been attained in most years. However, based on initial comparisons with undisturbed locations elsewhere on Mt Graham, construction of the observatories has not harmed the squirrel population.

Mt Graham red squirrels generally establish their caches at sites where the structural features ("old growth") include mature trees, numerous logs, and heavy canopy cover. The squirrels apparently seek these locations expressly because of their survival value. Such sites are not common on Mt Graham however and, despite efforts to reforest disturbed areas, centuries will pass before the replanted areas mature as "old growth." Fires and marginal cone crops remain concerns and, without a full measure of short-term conservation, the small population of Mt Graham red squirrels may not survive long enough to benefit from long-term remedies.

Monarchs in winter

Migratory behavior is widely associated with birds and only rarely with insects. Nonetheless, the yearly travels of monarch butterflies rival the migratory behavior of any bird. Each year multitudes of these colorful insects journey northward, moving only part way along their route before stopping to breed. The adults lay their eggs on milkweed plants, on which the caterpillars later feed and acquire cardiac glycosides. The latter chemicals are bitter, emetic, and toxic to heart muscle, thereby making the monarchs distasteful to their avian predators. The second generation continues migrating northward but again stops *en route* to reproduce. Each time, the adults die after breeding. The cycle repeats itself once more before this or perhaps a fourth generation eventually reaches the northern edges of the species' breeding range in southern Canada. As autumn approaches, the last generation then begins to migrate southward, completing its journey all the way to the wintering grounds – some travel nearly 4000 km or 2486 miles – without further reproduction and only rarely stopping to feed.

This remarkable sequence of events clearly indicates that the behavior culminating in the northward migration of monarch butterflies is transmitted genetically from one generation to the next (i.e., the adults die well before their offspring hatch, so learning is not involved). Moreover, each individual in the late summer population returns unerringly to its ancestral wintering area, a location it has never visited before.

Based on their migration routes and wintering grounds, monarch butterflies separate into two geographically distinct populations. The first is a western population, which overwinters along the California coast where the moderate climate protects the butterflies from severe freezes. Pacific Grove is one of the better-known sites where immense numbers of monarchs cluster on the branches of Monterey pines and other trees native to coastal California. Curiously, the blue gum, an Australian species introduced in the 1800s, is among the trees favored by the population overwintering in California. Presumably, the exotic trees proved attractive to a North American butterfly simply because the native trees steadily disappeared with human development. With the introduction and spread of blue gum, the number of sites where monarchs overwinter in California has increased. Today, the western population clusters each winter at approximately 300 locations from just north of San Francisco to the Mexican border. About 180,000 butterflies overwinter at the largest of these locations.

The second and largest population ranges from the Rocky Mountains eastward across North America. The number of butterflies in this population is immense (annual estimates vary from hundreds of millions to a billion), but their wintering grounds remained a mystery until early 1975 when a massive concentration was discovered in the volcanic mountains of south-central Mexico (Fig. 9.12). Chief among these sites are climax stands of oyamel fir lying between 2800 and 3400 m (9200–11,150 feet) of altitude. At these and other locations in the mountains, most of which lie within an area of just 65 km² (25 square miles), the butterflies find the lush cover and wet conditions favoring their survival. In particular, this "fog belt" protects the masses of butterflies from freezing; the moisture-laden air at these locations protects the monarchs by acting as a thermal blanket just as the forest canopy serves as an umbrella against excessive wetting during winter storms. However, if thinning opens the canopy, heat escapes from the forest and large numbers of butterflies become wet and freeze: the primary cause of winter mortality. The structure of their clusters also lessens wetting, and therefore the risk is reduced for butterflies inside the clusters.

The importance of fully protecting the forests at these special sites is self-evident. Regrettably, clearing for agriculture, uncontrolled timber harvests, fires, and livestock grazing threaten some of these areas. These disturbances profoundly alter the almost pure stands of oyamel fir and their understory of lichens and mosses, producing a subclimax forest in which pines mix with the firs. Oyamel fir is prized lumber and the largest trees are selectively removed, leaving behind forests with greatly reduced basal areas. Even worse, clear-cutting (which has occurred in a few areas) leads to soil erosion, nutrient loss, and damaged watersheds on steep mountainsides where oyamel forests occur.

In 1986 the Mexican government designed 5 key locations, containing 8 of the 14 specific wintering sites, as conservation areas whose express purpose is to protect the winter habitat of monarch butterflies. These locations also were designated *Reserva de la Biosfera Mariposa Monarca* (Monarch Butterfly Biosphere Reserve) in 2000, followed by their declaration as an UNESCO World Heritage Site in 2003. Unfortunately, these titles offer little additional protection because most of the reserve's 13,550 ha (33,482 acre) core zone remains privately owned and subject to small-scale logging, subsistence farming, grazing, and other uses. Current efforts encourage local communities to adopt an ecotourism-based economy centered on butterfly zones and forest conservation.

Bears and moths

Grizzly and black bears are classic examples of omnivores, eating in season everything from berries and grasses to ground squirrels and salmon, along with an occasional deer, elk, or moose. In the northern Rocky Mountains, however, grizzlies and some black bears add moths (Fig. 9.13) to their late summer diet, as recorded

(a)   (b)

Figure 9.12 Millions of monarch butterflies winter in a few groves of oyamel fir in the mountains of central Mexico after migrating across the eastern two-thirds of North America each year to reach this habitat. (a) A single tree may be blanketed by thousands of butterflies and (b) hundreds cluster on a single branch. A smaller western population overwinters at numerous smaller sites along the California coastline. Reproduced by permission of Monarch Watch (www.MonarchWatch.org).

Figure 9.13 In the northern Rocky Mountains, grizzly bears rummage in talus slopes for cutworm moths, a rich source of lipids, and may devour nearly 40,000 daily in late summer. After feeding nocturnally on the nectar of montane wildflowers, moths cluster in the interstices of jumbled rocks (inset) where they are especially accessible to bears in the cool morning and evening hours. Reproduced by permission of Frank van Manen and (inset) Jonathan Coop.

in Montana (including Glacier National Park), Wyoming, British Columbia and Alberta.

The life cycle of army cutworms starts in fall when they lay their eggs in the soil of the Great Plains. The larvae hatch the following spring, feed on a variety of plants (including commercially valuable alfalfa and small grains), then pupate underground. By late May or early June, the adults emerge and migrate *en masse* to the Rocky Mountains, likely to escape the heat of summer at lower elevations. The migration may cover as much as 470 km (300 miles). On arrival they move upward, feeding at night on the nectar of alpine and subalpine flowers which begin blossoming just when the moths arrive. During the day, they seek shelter in spaces between the rocks and stones of **talus slopes,** again apparently as a means to avoid heat. The moths return to the Great Plains in late summer or early fall, mate, lay eggs, and then die.

For bears in the northern Rockies, particularly grizzlies, the aggregations of moths provide an abundant and surprisingly rich source of food just prior to the time when the bears cease feeding and den for the winter (*Bears, salmon, and the forest*, Chapter 10). Because the moths are themselves storing lipids and energy in preparation of their fall migration and breeding activities, they increase their body mass by 70% and lipid content by 40% in just eight weeks. The result is an energy source that exceeds many other bear foods (Table 9.1). Bears search in the rocks for clusters of moths and may eat nearly 40,000 per day (1700 per hour) in years when these insects are abundant. In the cool morning hours the moths congregate within 10 cm (4 inches) of the talus surface, but to avoid heat they move deeper into the rocks as the day warms before again heading to the surface as night approaches. The bears' feeding activities match these movements; they forage for moths twice a day in the morning and evening, when the insects are near the surface. This synchrony likely favors a net gain in energy (i.e., times when the reward exceeds the energy expended in digging, which would not occur during midday).

Table 9.1 Energy content of army cutworm moths in comparison to some other foods of grizzly and black bears. Source of data: Pritchard and Robbins (1990) (except for moths, which is reported by French et al. 1994).

Food	kcal g^{-1}
Army cutworm moths	7.91
Deer	7.32
Pinyon pine nuts	6.48
Cutthroat trout	5.71
Ground squirrels	5.28
White clover	4.83
Blueberries	4.47

To summarize, army cutworm moths represent an abundant source of energy that is concentrated and readily obtained during a critical period in the life history of bears. Moreover, foraging for moths takes place in the more remote areas of the Rockies and reduces the potential for encounters between humans and bears during the tourist season.

Readings and references

Introduction

Peet, R.K. 1978. Latitudinal variation in southern Rocky Mountain forests. Journal of Biogeography 5: 275–289.

Sawyer, D.A. and T.E. Kinraide. 1980. The forest vegetation at higher altitudes in the Chiricahua Mountains, Arizona. American Midland Naturalist 104: 224–241.

Storer, T.I. and R.L. Usinger. 1963. *Sierra Nevada Natural History*. University of California Press, Berkeley, CA.

Taber, R.D. (ed.) 1969. *Coniferous Forests of the Northern Rocky Mountains. Proceedings of the 1968 Symposium*. Center for Natural Resources, University of Montana, Missoula, MT.

Features and adaptations

Rundel, P.W. 1975. Primary succession on granite outcrops in the montane southern Sierra Nevada. Madroño 23: 209–220.

Thomas, J.W. (ed.) 1979. *Wildlife Habitats in Managed Forests, the Blue Mountains of Oregon and Washington*. Agriculture Handbook No. 553, USDA Forest Service, Portland, OR. 512 pp.

Montane Forest zones

Daubenmire, R.F. 1943. Vegetational zonation in the Rocky Mountains. Botanical Review 9: 325–393.

Peet, R.K. 1981. Forest vegetation of the Colorado Front Range: composition and dynamics. Vegetation 45: 3–75.

Lower montane zone

Dieterich, J.H. and T.W. Swetnam. 1984. Dendrochronology of a fire-scarred ponderosa pine. Forest Science 30: 238–247.

Dodge, M. 1972. Forest fuel accumulation – a growing problem. Science 177: 139–142.

Haller, J.R. 1959. Factors affecting the distribution of ponderosa and Jeffery pines in California. Madroño 15: 65–96.

Kilgore, B.M. and D. Taylor. 1979. Fire history of a sequoia-mixed conifer forest. Ecology 60: 129–142. (Reports on fire scars and fire frequencies for ponderosa pine.)

Potter, L.D. and D.L. Green. 1964. Ecology of ponderosa pine in western North Dakota. Ecology 45: 10–23.

Wells, P.V. 1983. Paleobiogeography of montane islands in the Great Basin since the last glacio-pluvial period. Ecological Monographs 53: 341–382.

White, A.S. 1985. Presettlement regeneration patterns in a southwestern ponderosa pine stand. Ecology 66: 589–594.

Upper montane zone

Dye, A.J. and W.H. Mor. 1977. Spruce-fir forest at its southern distribution in the Rocky Mountains, New Mexico. American Midland Naturalist 97: 133–146.

Fahey, T.J. and D.H. Knight. 1986. Lodgepole pine ecosystems. BioScience 36: 610–617.

Moir, W.H. 1969. The lodgepole pine zone in Colorado. American Midland Naturalist 31: 87–98.

Morgan, M.D. 1969. Ecology of aspen in Gunnison County, Colorado. American Midland Naturalist 82: 204–228.

Stohlgren, T.J. and R.R. Bachand. 1997. Lodgepole pine (*Pinus contorta*) ecotones in Rocky Mountain National Park, Colorado, USA. Ecology 78: 632–641.

Subalpine zone

Billings, W.D. 1969. Vegetational pattern near alpine timberline as affected by fire–snowdrift interactions. Vegetatio 19: 192–207.

Broadbrooks, H.E. 1965. Ecology and distribution of the pikas of Washington and Alaska. American Midland Naturalist 73: 299–335.

Conner, D.A. 1983. Seasonal changes in activity patterns and the adaptive values of haying in pikas (*Ochotona princeps*). Canadian Journal of Zoology 61: 411–416.

Huntly, N.J., A.T. Smith, and B.L. Ivins. 1986. Foraging behavior of the Pika (*Ochotona princeps*), with comparisons of grazing versus haying. Journal of Mammalogy 67: 139–148.

Millar, J.S. and F.C. Zwickel. 1972. Characteristics and ecological significance of hay piles of pikas. Mammalia 36: 657–667.

Romme, W.H. 1982. Fire and landscape diversity in subalpine forests of Yellowstone National Park. Ecological Monographs 52: 199–221.

Smith, A.T. 1974. The distribution and dispersal of pikas: consequences of insular population structure. Ecology 55: 1112–1119.

Wardle, P. 1968. Engelmann spruce (*Picea engelmannii* Engel.) at its upper limits on the Front Range, Colorado. Ecology 49: 483–495.

Whipple, S.A. and R.L. Dix. 1979. Age structure and successional dynamics of a Colorado subalpine forest. American Midland Naturalist 101: 142–158.

Associated habitats

Mountain parks and meadows

Doering, W.R. and R.G. Reider. 1992. Soils of Cinnabar Park, Medicine Bow Mountains, Wyoming, USA: indicators of park origin and persistence. Arctic and Alpine Research 24: 27–39.

Knight, D.H. 1994. *Mountains and Plains: The Ecology of Wyoming Landscapes*. Yale University Press, New Haven, CT. (Discussion of mountain parks on pp. 193–200.)

Koterba, W.D. and J.R. Habeck. 1971. Grasslands of the North Fork Valley, Glacier National Park, Montana. Canadian Journal of Botany 49: 1627–1636.

Miles, S.R. and P.C. Singleton. 1975. Vegetative history of Cinnabar Park in Medicine Bow National Forest, Wyoming. Soil Science Society of America Proceedings 39: 1204–1208.

Root, R.A. and J.R. Habeck. 1972. A study of high elevational grassland communities in western Montana. American Midland Naturalist 87: 109–121.

Vale, T.R. 1978. Tree invasion of Cinnabar Park in Wyoming. American Midland Naturalist 100: 277–284.

Black Hills

Bailey, R.M. and M.O. Allum. 1962. *Fishes of South Dakota*. Miscellaneous Publication 119, Museum of Zoology, University of Michigan, Ann Arbor, MI.

Brown, P.M. and B. Cook. 2006. Early settlement forest structure in Black Hills ponderosa pine forests. Forest Ecology and Management 223: 284–290.

Froiland, S.G. 1990. *Natural History of the Black Hills and Badlands*. Revised edition. Center for Western Studies, Augustana College, Sioux Falls, SD.

Jackson, D.D. 1972. *Custer's Gold: The United States Cavalry Expedition of 1874*. University of Nebraska Press, Lincoln, NE.

Ludlow, W. 1875. *Report of a Reconnaissance of the Black Hills of Dakota made in the Summer of 1874*. Engineering Department, US Army, Washington, DC.

McIntosh, A.C. 1931. A botanical survey of the Black Hills of South Dakota. Black Hills Engineer 19: 159–276.

Miller, A.H. 1941. Speciation in the avian genus *Junco*. University of California Publications in Zoology 44: 173–434.

Over, W.H. and G.M. Clement. 1930. Nesting of the white-winged junco in the Black Hills of South Dakota. Wilson Bulletin 42: 28–31.

Pettingill, O.S., Jr. and N.R. Whitney, Jr. 1965. *Birds of the Black Hills*. Publication 1, Cornell Laboratory of Ornithology, Ithaca, NY.

Progulske, D.R. 1974. *Yellow Ore, Yellow Hair, Yellow Pine: A Photographic Study of a Century of Forest Ecology*. Bulletin 616, South Dakota Agricultural Experiment Station, Brookings, SD.

Raventon, E. 1994. *Island in the Plains: A Black Hills Natural History*. Johnson Books, Boulder, CO.

Rich, F.J. (ed.) 1981. *Geology of the Black Hills of South Dakota and Wyoming*. American Geological Institute, Alexandria, VA.

Shepperd, W.D. and M.A. Battaglia. 2002. *Ecology, Silviculture, and Management of Black Hills, Ponderosa Pine*. General Technical Report RMRS-GTR-97, USDA Forest Service, Rocky Mountains Research Station, Fort Collins, CO.

Turner, R.W. 1974. *Mammals of the Black Hills of South Dakota and Wyoming*. Miscellaneous Publication No. 60, Museum of Natural History, University of Kansas, Lawrence, KS. (Source of Grinnell's quotes regarding wolves and coyotes.)

Weedon, R.R. and P.M. Wolken. 1990. The Black Hills environment. In: *Megafauna and Man, Discovery of America's Heartland* (L.D. Agenbroad, J.L. Mead, and L.W. Nelson, eds). The Mammoth Site, Hot Springs, ND, pp. 123–135.

White, E.M., J.R. Johnson, and J.T. Nichols. 1969. Prairie–forest transition soils of the South Dakota Black Hills. Soil Science Society of America Proceedings 33: 932–936.

Redwoods and sequoias

Azevedo, J. and D.L. Morgan. 1974. Fog precipitation in coastal California forests. Ecology 55: 1135–1141.

Burgess, S.S.O. and T.E. Dawson. 2004. The contribution of fog to the water relations of *Sequoia sempervirens* (D. Don): foliar uptake and prevention of dehydration. Plant, Cell & Environment 29: 229–239.

Dawson, T.E. 1998. Fog in the redwood forest: ecosystem inputs and use by plants. Oecologia 117: 476–485.

Florence, R.G. 1965. Decline of old-growth redwood forests in relation to some soil microbiological processes. Ecology 46: 52–64.

Hoekstra, J.M., R.T. Bell, A.E. Launer, and D.D. Murphy. 1995. Soil arthropod abundance in coast redwood forest: effect of selective timber harvest. Environmental Entomology 24: 246–252.

Koch, G.W., S.C. Sillett, G.M. Jennings, and S.D. Davis. 2004. The limits to tree height. Nature 428: 851–854.

Rundel, P.W. 1971. Community structure and stability in the giant sequoia groves of the Sierra Nevada. American Midland Naturalist 85: 478–492.

Rundel, P.W. 1972. Habitat restriction in giant sequoia: the environmental control of grove boundaries. American Midland Naturalist 87: 81–99.

Sawyer, J.O., J. Gray, G.J. West, D.A. Thornburg, et al. 2000. History of redwood and redwood forests. In: *The Redwood Forest: History, Ecology, and Conservation of the Coast Redwoods* (R.F. Noss, ed.). Island Press, Washington, DC, pp. 7–38.

Sillett, S.C. and R. Van Pelt. 2007. Trunk reiteration promotes epiphytes and water storage in an old-growth redwood forest canopy. Ecological Monographs 77: 335–359.

Stone, E.C. and R.B. Vasey. 1968. Preservation of coast redwood on alluvial flats. Science 159: 157–161.

Zinke, P.J. 1977. The redwood forest and associated north coast forests. In: *Terrestrial Vegetation of California* (M.G. Barbour and J. Major, eds). John Wiley & Sons, New York, NY, pp. 679–698.

Bristlecone pine forest

Bailey, D.K. 1970. Phytogeography and taxonomy of *Pinus* subsection *Balfourianae*. Annals of the Missouri Botanical Garden 57: 210–249.

Baker, W.L. 1992. Structure, disturbance, and change in the bristlecone pine forests of Colorado, USA. Arctic and Alpine Research 24: 17–26.

Beasley, R.S. and J.O. Klemmedson. 1980. Ecological relationships of bristlecone pine. American Midland Naturalist 104: 242–252.

Curry, D.R. 1965. An ancient bristlecone pine in eastern Nevada. Ecology 46: 564–566. (Records almost total loss of bark on a 4900-year-old living tree.)

Cutright, P.R. 1969. *Lewis and Clark: Pioneering Naturalists*. University of Illinois Press, Urbana, IL.

Ferguson, C.W. 1968. Bristlecone pine science and esthetics. Science 159: 839–846.

Ferguson, C.W. 1969. A 7104-year annual tree-ring chronology for bristlecone pine, *Pinus aristata*, from the White Mountains, California. Tree-ring Bulletin 29: 3–29.

Hutchins, H.E. and R.M. Lanner. 1982. The central role of Clark's nutcracker in the dispersal and establishment of whitebark pine. Oecologia 55: 192–201.

LaMarche, V.C., Jr. 1969. Environment in relation to age of bristlecone pine. Ecology 50: 53–59.

LaMarche, V.C., Jr and H.A. Mooney. 1972. Recent climatic change and development of the bristlecone pine (*P. longaeva* Bailey) krummholz zone, Mt Washington, Nevada. Arctic and Alpine Research 4: 61–74. (Dendrochronology determined from tree-rings in logs and those in still living trees.)

Lanner, R.M. 1982. Adaptations of whitebark pine for seed dispersal by Clark's nutcracker. Canadian Journal of Forest Research 12: 391–402.

Lanner, R.M. 1988. Dependence of Great Basin bristlecone pine on Clark's nutcracker for regeneration at high elevations. Arctic and Alpine Research 20: 358–362.

Lanner, R.M. 1996. *Made For Each Other, A Symbiosis of Birds and Pines*. Oxford University Press, New York, NY.

Lanner, R.M. 2007. *The Bristlecone Book: A Natural History of the World's Oldest Trees*. Mountain Press Publishing Company, Missoula, MT.

Vander Wall, S.B. and R.P. Balda. 1977. Coadaptations of the Clark's nutcracker and the pinon pine for efficient seed harvest and dispersal. Ecological Monographs 47: 89–111.

Wright, R.D. and H.A. Mooney. 1965. Substrate-oriented distribution of bristlecone pine in the White Mountains of California. American Midland Naturalist 73: 257–284.

Fire in montane forests

Anderson, J.E. and W.H. Romme. 1991. Initial floristics in lodgepole pine (*Pinus contorta*) forests following the 1988 Yellowstone fires. International Journal of Wildland Fire 1: 119–124.

Aplet, G.H. 2006. Evolution of wilderness fire policy. International Journal of Wilderness 12: 9–13.

Bartlett, R. 1977. From imagination to reality: Thomas Moran and Yellowstone. Prospects: Annual of American Cultural Studies 3: 111–124.

Blanchard, B.M. and R.R. Knight. 1990. Reactions of grizzly bears, *Ursus arctos horribilis*, to wildfire in Yellowstone National Park, Wyoming. Canadian Field-Naturalist 104: 592–594.

Christensen, N.L., J.K. Agee, P.F. Brussard, et al. 1989. Interpreting the Yellowstone fires of 1988, ecosystem responses and management implications. BioScience 39: 678–685.

Houston, D.B. 1973. Wildfires in northern Yellowstone National Park. Ecology 54: 1111–1117.

Jakubas, W.J., R.A. Garrott, P.J. White, and D.R. Mertens. 1994. Fire-induced changes in the nutritional quality of lodgepole pine bark. Journal of Wildlife Management 58: 35–46.

Keiter, R.B. and M.S. Boyce. 1991. *The Greater Yellowstone Ecosystem: Refining America's Wilderness Heritage*. Yale University Press, New Haven, CT

Knight, D.H. and L.L. Wallace. 1989. The Yellowstone fires: issues in landscape ecology. BioScience 39: 700–706.

Leopold, A.S. 1963. Study of wildlife problems in national parks. Transactions of the North American Wildlife and Natural Resource Conference 28: 28–45.

Lotan, J.E. 1976. Cone serotiny–fire relationships in lodgepole pine. Proceedings of the Tall Timbers Fire Ecology Conference 14: 267–278.

Muir, P.S. and J.E. Lotan. 1985. Disturbance history and serotiny of *Pinus contorta* in western Montana. Ecology 66: 1658–1668.

Romme, W.H. and D.G. Despain. 1989. Historical perspective on the Yellowstone fires of 1988. BioScience 39: 695–699.

Romme, W.H. and D.G. Despain. 1989. The long history of fire in the Greater Yellowstone Ecosystem. Western Wildlands 15: 10–17.

Romme, W.H., M.G. Turner, L.L. Wallace, and J.S. Walker. 1995. Aspen, elk, and fire in northern Yellowstone National Park. Ecology 76: 2097–2105.

Singer, F.J. and M.K. Harter. 1996. Comparative effects of elk herbivory and 1988 fires on northern Yellowstone National Park grasslands. Ecological Applications 6: 185–199.

Singer, F.J., W. Schreier, J. Oppenheim, and E.O. Garton. 1989. Drought, fires, and large mammals, estimating the 1988 severe drought and large-scale fires. BioScience 39: 716–722. (Source of data concerning mammals killed in the Yellowstone fires.)

Taylor, D.L. 1973. Some ecological implications of forest fire control in Yellowstone National Park, Wyoming. Ecology 54: 1394–1396.

Turner, M.G., W.H. Romme, and D.B. Tinker. 2003. Surprises and lessons from the 1988 Yellowstone fires. Frontiers in Ecology and the Environment 1: 351–358.

Turner, M.G., W.H. Romme, D.B. Tinker, R.A, Reed, and G.A. Tuskan. 2003. Postfire aspen seedling recruitment across the Yellowstone (USA) landscape. Landscape Ecology 18: 127–140.

Wu, Y., M.G. Turner, L.L. Wallace, and W.H. Romme. 1996. Elk survival following the 1988 Yellowstone fires a simulation experiment. Natural Areas Research Journal 16: 198–207.

Highlights

Western chipmunks and competitive exclusion

Brown, J.H. 1971. Mechanisms of competitive exclusion between two species of chipmunks. Ecology 52: 305–311.

Chappel, M.A. 1978. Behavioral factors in the altitudinal zonation of chipmunks (*Eutamias*) Ecology 59: 565–579.

Hardin, G. 1960. The competitive exclusion principle. Science 131: 1292–1297.

Heller, H.C. 1971. Altitudinal zonation of chipmunks (*Eutamias*): interspecific aggression. Ecology 52: 312–319.

Heller, H.C. 1972. Altitudinal zonation of chipmunks (*Eutamias*): adaptations to aridity and high temperature. American Midland Naturalist 87: 296–313.

Hutchinson, G.E. 1957. Concluding remarks. Cold Spring Harbor Symposia on Quantitative Biology 22: 415–427. (Introduces the concepts of fundamental and realized niches.)

Sheppard, D.H. 1971. Competition between two chipmunk species (*Eutamias*). Ecology 52: 320–329.

Squirrels, bears, and pine cones

Arno, S.F. 1986. Whitebark pine cone crops – a diminishing source of wildlife food? Western Journal of Applied Forestry 1: 92–94.

Finley, R.B., Jr. 1969. Cone caches and middens of *Tamiasciurus* in the Rocky Mountains. In: *Contributions in Mammalogy* (J.K. Jones, Jr, ed.). University of Kansas Museum of Natural History Miscellaneous Publication 51, Lawrence, KS, pp. 233–273.

Kendall, K.C. 1983. Use of pine nuts by grizzly bears and black bears in the Yellowstone area. International Conference on Bear Research and Management 5: 166–173.

Kendall, K.C. and S.F. Arno. 1990. Whitebark pine: an important but endangered wildlife resource. In: *Proceedings of Symposium on Whitebark Pine Ecosystems: Ecology and Management of a High-Mountain Resource* (W.C. Schmidt and K.J. McDonald, eds). USDA Forest Service General Technical Report INT–270, Ogden, UT, pp. 264–273.

Mattson, D.J. and C. Jonkel. 1990. Stone pines and bears. In: *Proceedings of Symposium on Whitebark Pine Ecosystems: Ecology and Management of a High-Mountain Resource* (W.C. Schmidt and K.J. McDonald, eds). USDA Forest Service General Technical Report INT–270, Ogden, UT, pp. 223–236.

Mattson, D.J., B.M. Blanchard, and R.R. Knight. 1992. Yellowstone grizzly bear mortality, human habituation, and whitebark pine seed crops. Journal of Wildlife Management 56: 432–442.

Reinhart, D.P. and D.J. Mattson. 1990. Red squirrels in the whitebark zone. In: *Proceedings of Symposium on Whitebark Pine Ecosystems: Ecology and Management of a High-Mountain Resource* (W.C. Schmidt and K.J. McDonald, eds). USDA Forest Service General Technical Report INT–270, Ogden, UT, pp. 256–263.

Schwartz, C.C., M.A. Haroldson, K.A. Gunther and D. Moody. 2006. Distribution of grizzly bears in the Greater Yellowstone ecosystem in 2004. Ursus 17: 63–66.

Tomback, D.F., S.F. Arno, and R.E. Keane. (eds) 2001. *Whitebark pine Communities: Ecology and Restoration.* Island Press, Washington, DC.

Sky islands in Arizona

Angel, R., B. Walsh, P. Strittmatter, and N. Woolf. 1995. The saga of Mt. Graham. Astronomy 23(7): 16–19.

Blount, S.J. and J.L. Koprowski. 2012. Small mammal response to post–fire conditions: case of the endangered Mount Graham red squirrel. Southwestern Naturalist 57: 8–15.

Hoffmeister, D.F. 1956. Mammals of the Graham (Pinaleno) Mountains, Arizona. American Midland Naturalist 55: 257–288.

Istock, C.A. and R.S. Hoffman. (eds) 1995. *Storm over a Mountain Island, Conservation Biology and the Mt Graham Affair.* University of Arizona Press, Tucson, AZ.

Koprowski, J.L., M.I. Alanen, and A.M. Lynch. 2005. Nowhere to run and nowhere to hide: response of endemic Mt. Graham red squirrels to catastrophic forest damage. Biological Conservation 126: 491–498.

Rhodes, T.C. and P.N. Wilson. 1995. Sky islands, squirrels, and scopes: the political economy of an environmental conflict. Land Economics 71: 106–121.

Sanderson, H.R. and J.L. Koprowski. (eds) 2009. *The Last Refuge of the Mt Graham Red Squirrel: Ecology of Endangerment.* University of Arizona Press, Tucson, AZ.

Smith, A.A. and W.R. Mannan. 1994. Distinguishing characteristics of Mount Graham red squirrel midden sites. Journal of Wildlife Management 58: 437–445.

Stromberg, J.C. and D.T. Patten. 1991. Dynamics of the spruce-fir forests in the Pinaleno Mountains, Graham County, Arizona. Southwestern Naturalist 36: 37–48.

Warshall, P. 1994. The biopolitics of the Mt Graham red squirrel (*Tamiasciuris* (sic) *hudsonicus grahamensis*). Conservation Biology 8: 977–988.

Zugmeyer, C.A. and J.L. Koprowski. 2009. Severely insect-damaged forest: a temporary trap for red squirrels? Forest Ecology and Management 27: 464–470.

Monarchs in winter

Alonso-Mejía, A., E. Rendon-Salinas, E. Montesinos-Patiño, and L.P. Brower. 1997. Use of lipid reserves by monarch butterflies overwintering in Mexico: implications for conservation. Ecological Applications 7: 934–947.

Anderson, J.B. and L.P. Brower. 1996. Freeze-protection of over-wintering monarch butterflies in Mexico: critical role of the forest as a blanket and an umbrella. Ecological Entomology 21: 107–116.

Brower, L.P. 1995. Understanding and misunderstanding the migration of monarch butterfly (Nymphalidae) in North America: 1857–1995. Journal of the Lepidopterists' Society 49: 304–385.

Calvert, W.H. and L.P. Brower. 1986. The location of monarch butterfly *Danaus plexippus* L. overwintering colonies in Mexico in relation to topography and climate. Journal of the Lepidopterists' Society 40: 164–187.

Frey, D., K.L.H. Leong, D. Fredricks, and S. Raskowitz. 1992. Clustering patterns of monarch butterflies (Lepidoptera: Danadiae) at two California central coast overwintering sites. Annals of the Entomological Society of America 85: 148–153.

Lane, J. 1981. The status of monarch butterfly overwintering sites in Alta, California. Atala 9: 17–20.

Malcolm, S.B. 1993. Conservation of monarch butterfly migration in North America: an endangered phenomenon. In: *Biology and Conservation of the Monarch Butterfly* (S.B. Malcolm and M.P. Zalucki, eds). Science Series No. 38, Natural History Museum, Los Angeles County, CA, pp. 357–361.

Marriott, D. 1997. *Where to See the Monarchs in California: Twenty-five Selected Sites*. The Monarch Program, San Diego, CA.

Masters, A.R., S.B. Malcolm, and L.P. Brower. 1988. Monarch butterfly *Danaus plexippus* thermoregulation behavior and adaptations for overwintering in Mexico. Ecology 69: 458–467.

Nunez, J.C.S. and L.V. Garcia. 1993. Vegetation types of monarch butterfly overwintering habitat in Mexico. In: *Biology and Conservation of the Monarch Butterfly* (S.B. Malcolm and M.P. Zalucki, eds). Science Series No. 38, Natural History Museum, Los Angeles County, CA, pp. 287–293.

Ramirez, M.I., J.G. Azcárate, and L. Luna. 2002. Effects of human activities on monarch butterfly habitat in protected mountain forests, Mexico. Forestry Chronicle 79: 242–246.

Snook, L.C. 1993. Conservation of the monarch butterfly reserves in Mexico: focus on the forest. In: *Biology and Conservation of the Monarch Butterfly* (S.B. Malcolm and M.P. Zalucki, eds). Science Series No. 38, Natural History Museum, Los Angeles County, CA, pp. 363–375.

Tucker, C. 2004. Community institutions and forest management in Mexico's Monarch Butterfly Reserve. Society and Natural Resources 17: 569–587.

Urquhart, F.A. and N.R. Urquhart. 1976. The overwintering site of the eastern population of the monarch butterfly (*Danaus plexippus*: Danaidae) in southern Mexico. Journal of the Lepidopterists' Society 30: 153–158.

Bears and moths

French, S.P., M.G. French, and R.R. Knight. 1994. Grizzly bear use of army cutworm moths in the Yellowstone ecosystem. In: *Bears, Their Biology and Management: Proceedings of the 9th International Conference on Bear Research and Management* (J. Claar and P. Schullery, eds). Bear Biology Association, University of Tennessee, Knoxville, pp. 398–399.

Mattson, D.J., C.M. Gillin, S.A. Benson, and R.R. Knight. 1991. Bear feeding activity at alpine insect aggregation sites in the Yellowstone ecosystem. Canadian Journal of Zoology 69: 2430–2435.

Pritchard, G.T. and C.T. Robbins. 1990. Digestive and metabolic efficiencies of grizzly and black bears. Canadian Journal of Zoology 68: 1645–1651.

White, D., Jr, K.C. Kendall, and H.D. Picton. 1998. Grizzly bear feeding activity at alpine army cutworm moth aggregation sites in northwest Montana. Canadian Journal of Zoology 76: 221–227.

White, D., Jr, K.C. Kendall, and H.D. Picton. 1998. Seasonal occurrence, body composition, and migration potential of army cutworm moths in northwest Montana. Canadian Journal of Zoology 76: 835–842.

White, D., Jr, K.C. Kendall, and H.D. Picton. 1999. Potential energetic effects of mountain climbers on foraging grizzly bears. Wildlife Society Bulletin 27: 146–151.

Infobox 9.1. John Muir (1838–1914), patron saint of wilderness

Clarke, J.M. 1979. *The Life and Adventures of John Muir*. Sierra Club Books, San Francisco, CA.

Ehrlich, G. 2000. *John Muir: Nature's Visionary*. National Geographic Society, Washington, DC.

Miller, C. 2001. *Gifford Pinchot and the Making of Modern Environmentalism*. Island Press, Washington, DC.

Turner, F. 1985. *Rediscovering America: John Muir in His Time and Ours*. Viking, New York, NY.

Williams, D.C. 2002. *God's Wilds: John Muir's Vision of Nature*. Texas A&M University Press, College Station, TX.

Wooster, D. 2008. *A Passion for Nature: The Life of John Muir*. Oxford University Press, New York, NY.

Infobox 9.2. The underwater bird

Bakus, G.J. 1959. Observations on the life history of the dipper in Montana. Auk 76: 190–207.

Goodge, W.R. 1959. Locomotion and other behavior in the dipper. Condor 61: 4–17.

Harvey, B.C., and C.D. Marti. 1993. The impact of the dipper, *Cinclus mexicanus*, predation on stream benthos. Oikos 68: 431–436.

Muir, J. 1901. *Our National Parks*. Houghton Mifflin, Boston, MA, and New York, NY.

Muir, J. 1977. *The Hummingbird of the California Waterfalls*. Outbooks, Olympic valley, CA. (Reprint of an article originally published in 1878 in Scribner's.)

Price, F.E., and C.E. Bock. 1983. *Population Ecology of the Dipper (Cinclus mexicanus) in the Front Range of Colorado*. Studies in Avian Biology 7, Cooper Ornithological Society.

Teague, S.A., A.W. Knight, and B.N. Teague. 1985. Stream microhabitat selectivity, resource partitioning, and niche shifts in grazing caddisfly larvae. Hydrobiologia 128: 3–12.

Thut, R.N. 1970. Feeding habitats of the dipper in southwestern Washington. Condor 72: 234–235.

Willson, M.F. and K.M. Hocker. 2008. Natural history of nesting American dippers (*Cinclus mexicanus*) in southeastern Alaska. Northwestern Naturalist 89: 97–106.

CHAPTER 10
Temperate Rain Forest

Destroying rainforest for economic gain is like burning a Renaissance painting to cook a meal.

Edward O. Wilson

A relatively narrow belt of huge coniferous trees – Temperate Rain Forest – extends along the Pacific coastline from northern California to southern Alaska (Fig. 1.3). Whereas ecological complexity is greater in their tropical counterparts, climax communities in Temperate Rain Forest generally contain more biomass. Despite the publicity given to rain forests in the tropics, rain forests are far rarer in the world's temperate zones (e.g., coastal New Zealand).

In the tropics, the trees are generally evergreen and broad-leaved; cone-bearing, needle-leaved evergreens characterize Temperate Rain Forest however, and the few species of broad-leaved trees are deciduous (e.g., bigleaf maple). Tropical Rain Forest features an unusually rich community of canopy-dwelling organisms, a structure well represented by vines and lianas, and trees with immense buttresses (features not matched in temperate zones; see Table 10.1). In both types of forest, however, the trees grow to immense sizes and, left undisturbed, live for centuries. Uncut stands of Temperate Rain Forest in North America are commonly designated as "old growth."

These forests, true to their name, receive large amounts of rainfall, but the amount varies greatly along the coasts of British Columbia and southern Alaska. On the Olympic Peninsula in Washington, however, annual precipitation averages 356 cm (140 inches), more than anywhere else in continental North America. About 70% of the rainfall occurs during the winter in southern areas of the Pacific Northwest (e.g., Washington and Oregon), whereas heavy fogs conserve moisture during the drier summer months. Because of their great heights, many trees penetrate fog banks drifting high above ground and "fog drip" (condensation forming on the canopy's heavy foliage) adds considerable moisture directly to the forest floor. Fogs also preserve moisture by lessening evapotranspiration from the vegetation, the result of reduced sunlight and lower daytime temperatures. The climate is generally mild and, even in winter, temperatures in

Temperate Rain Forest rarely reach the extremes occurring farther inland.

Species composition varies with location but western hemlock, western redcedar, Sitka spruce, and Douglas fir are major components in most areas (Fig. 10.1). Pacific silver fir, grand fir, and Port-Orford cedar also occur in these forests. Douglas fir is a pioneer species but, because of its longevity, it often persists to dominate large areas in old-growth forests. In time, however, western hemlock and other shade-tolerant conifers eventually join the canopy. Bigleaf maple is one of the few deciduous species characteristic of Temperate Rain Forests. In fact, in some areas the biomass of this and other deciduous trees is less than a thousandth that of conifers. The species is aptly named; bigleaf maple leaves resemble other maples, but they reach widths of up to 38 cm (15 inches). A profusion of mosses, lichens, and ferns generally drapes the trunks and branches of bigleaf maples.

Shrubs in the understory vegetation include several species, of which huckleberries and tangles of vine maples are among the key forage species for Roosevelt elk in the Olympic Peninsula. Saplings of the climax species, because of their shade tolerance, occur throughout the understory. A large percentage of the herbaceous species bear white flowers. Even more distinctive, however, is the architectural catalog of leaf shapes among the herbaceous plants, including profiles resembling hearts, shamrocks, and butterfly wings. Ferns, mosses, and clubmosses are richly represented in the ground cover, and these form lush carpets beneath the huge trees.

What is old-growth forest?

Old-growth forests in the Pacific Northwest consist of coniferous trees of great age and size. These long-lived trees reach heights of 50–90 m (165–295 feet) and diameters of 2 m (>6 feet). Stands of old-growth forest include trees of ages 500 years or more; these ancient trees are

Ecology of North America, Second Edition. Brian R. Chapman and Eric G. Bolen.
© 2015 John Wiley & Sons, Ltd. Published 2015 by John Wiley & Sons, Ltd.

Table 10.1 General comparison between Temperate Rain Forest and Tropical Rain Forest. Compiled from various sources including Kirk and Franklin (1992).

Temperate Rain Forest	Tropical Rain Forest
Seasonally cool to moderately warm	Warm to hot all year round
Rainfall mostly during winter; summers drier, but foggy	Seasonal rains, often in heavy bursts
Highest terrestrial biodiversity in North America	Immense biodiversity on a global basis[1]
More biomass per area	Less biomass per area[2]
Trees lack buttresses	Most trees with buttresses
Needle-leaved evergreens	Broad-leaved evergreens
1–2 canopy layers	2–4 canopy layers
Mostly ground-dwelling fauna	Large canopy fauna
No venomous snakes	Several venomous snakes
Epiphytes: mosses, lichens, and ferns	Epiphytes: orchid and pineapple families
Soils relatively rich	Soils heavily leached
Numerous national parks and forests; low human density	Greater vulnerability to human activities

[1] About half of all known species occur in Tropical Rain Forests, which occupy only 2% of Earth's surface area. For example, ecologists recorded 835 species of trees on a single 0.5 km² (0.2 square mile) plot in Malaysia – more than occur in all of North America – and in Panama, a 1 ha (2.5 acre) plot of tropical forest contained about 42,000 species of plants.

[2] Temperate Rain Forests include a higher percentage of large trees; they therefore contain about 40% more biomass per area than Tropical Rain Forests.

not just scattered individuals, but instead represent major components of an extensive evergreen forest. Minimum size and age criteria for defining an old-growth forest vary among authorities but the following are suitable standards: (a) the dominant trees have diameters of at least 1 m (39 inches); and (b) an overstory dominated by trees more than 200 years old.

Several structural features also distinguish old-growth forests from younger stands: (a) the form of the living trees themselves; (b) a multilayered canopy of several species; (c) numerous **snags** and logs; and (d) unaltered by logging. At 175–250 years of age, the trees start showing old-growth characteristics which include not only their large trunks but also large, irregular branch systems, deep crowns, and often broken tops. Snags provide habitat for various kinds of wildlife. Vaux's swifts, for example, depend heavily on large, hollow snags in old-growth forests for their nesting and roosting sites. Snags originate when old trees die and, when the snags fall, these litter the forest floor as logs of huge size and great biomass. In western Oregon, the logs in some stands of old-growth forest average 190 metric tons per ha (85

tons per acre), with three times that amount littering slopes next to streams. Because of their high moisture content logs provide important habitat for amphibians, and they may also serve as shelters and corridors for small mammals venturing into cleared areas within the forest matrix. Some logs also fall into forest streams, where they become important to the structure and nutrient cycling of aquatic communities (*More about logs*, this chapter). Primarily because of their enormous accumulations of huge logs, old-growth forests are highly retentive of nutrients.

The huge logs on the floor of old-growth forests have another function. Seedlings (particularly those of Sitka spruce and western hemlock) often gain their initial footholds on logs, which provide a rich source of nutrients. These are known as **nurse logs**, on which young trees develop in rows called **colonnades**. This linear configuration persists as the seedlings grow; colonnades of mature trees are therefore a distinctive feature of the Temperate Rain Forest. In time, however, the nurse logs decay and disappear. Because the roots of the young trees grow downward and around nurse logs into the mineral soil below, trees in the colonnades are left standing on what seem like stilts – their roots – after the nurse logs vanish (Fig. 10.2). Similarly, when a seedling grows atop a stump, a single tree on stilts is left behind when the stump eventually disappears.

Given their great age, old-growth forests eventually experience random events such as insect attacks, lightning strikes, and wind damage, thereby adding patchiness to their structure. All told, old-growth forests exhibit a large degree of complexity in both their vertical and horizontal planes. Moreover, the fauna in these areas included specialists whose evolutionary development was shaped by the unique structure of old-growth forests. Northern spotted owls are the best-known specialists in the ancient forests of the Pacific Northwest, where they come close to being an obligate species.

Features and adaptations

Valleys of rain forest

Exceptional conditions favoring Temperate Rain Forest occur in four valleys on the western side of the Olympic Peninsula. Each of these valleys – the Hoh Valley is representative – is sculpted with a classic U-shaped cross-section, the tell-tale sign of glacial history. The soils on the broad floors of these valleys are deep, well-drained deposits of materials transported during four periods of glacial activity. Along the valley bottoms, Sitka spruce, western hemlock, and lesser numbers of western red cedar reach their zenith as a rainforest community (Infobox 10.1).

The moisture regime in these valleys, coupled with mild regional temperatures, provides an ideal climate for

(a) **(b)**

Figure 10.1 (a) Lush undergrowth of ferns is evident in old-growth forests in the Olympic Peninsula . (b) In addition to Sitka spruce and western hemlock, the climax community in old-growth forests includes western red cedar. Photos courtesy of Brian R. Chapman.

Figure 10.2 Nurse logs provide sites where seedlings become established in old-growth forests. This tree is midway in its development on a nurse log, which eventual decays leaving a mature tree seemingly growing on stilts. A colonnade forms when a row of such trees is established on a single nurse log. Photo courtesy of Brian R. Chapman.

Temperate Rain Forests. Because of their orientation, the valleys receive cool moist air draining down from the mountains and into the forest, and fogs and rains drift in from the ocean. Precipitation at the upper ends of the valleys, where the crest of the Olympic Mountains forms a headwall 48 km (30 miles) inland from the ocean, may reach about 508 cm (200 inches). The heavy clouds act as a thermal blanket, keeping temperatures warmer in winter and cooler in summer. To summarize, these valleys experience unending saturation and produce a marvel of vegetation: the wet realm of Temperate Rain Forest.

Infobox 10.1 The elk that saved a forest

Elk, or wapiti, once ranged across much of North America, extending as far east as present-day Pennsylvania and including large herds on the prairies. Today, elk are largely restricted to the mountainous areas of western North America. The range of one subspecies, heavier and darker than other taxa of elk, extends from northern California northward along the Pacific Coast into Washington where substantial numbers thrive in the rain forests of the Olympic Peninsula (e.g., Queets Valley). Roosevelt elk also occur on Vancouver Island, British Columbia, and introduced populations were established on Afognak Island, Alaska in 1928 and other Alaskan islands in 1987.

The elk in this coastal population were named in honor of Theodore Roosevelt (1858–1919) who, as president, set aside part of the Olympic Peninsula as Mount Olympus National Monument (later known as Olympic National Park). Roosevelt's action preserved a large area of magnificent rain forest, but his primary goal was protection for the dwindling elk herd. Curiously, ruthless hunting was diminishing the elk herds not so much for sport or meat but because elk teeth – the two large and somewhat unusual canines on their upper jaws – were in fashion as decorations of gentlemen's watch fobs! These teeth, with their ivory-like texture, sold for the then-princely sums of $10–25 each, which indeed encouraged commercial exploitation. Those from older animals were stained with shades of rich brown and commanded even higher prices. The intervention of Roosevelt, a dedicated big-game hunter, curtailed the senseless trade. The elk population soon recovered and their habitat, a magnificent rain forest, was protected. Photo courtesy of Brian R. Chapman.

Epiphytes, canopy roots, and "scuzz"

Epiphytes are plants that grow on other plants, from which they gain physical support but no nutritional benefits (epiphytes are not parasites). A few epiphytes occur regularly in other communities, including certain areas of Eastern Deciduous Forest where Spanish moss often drapes trees in the southeastern coastal plain (Chapter 4). In Temperate Rain Forest, however, epiphytes reach their apex of development in North America. More than 125 species of mosses, lichens, clubmosses, and other species flourish in an unusually rich epiphytic community and thereby become another hallmark of old-growth forests.

Virtually all trees in old-growth forests bear epiphytic vegetation. Shaggy cushions of epiphytes are particularly evident on bigleaf maple, but the underlying reasons for this association are not known. When dry, the biomass of epiphytes on a single bigleaf maple averages almost 36 kg (78 pounds). Slightly lesser amounts grow on large Douglas fir or other conifers in an old-growth forest. Epiphytes also festoon the tangled branches of vine maple. The thick carpet of epiphytic mosses and lichens itself forms the strata on which other plants become anchored, including those otherwise typical of ground cover (e.g., ferns, but also flowering plants). The decomposition of this vegetation, together with other litter and dust, produce fertile patches of "perched" soils on top of large branches.

Epiphytes increase the nutrient-gathering capabilities of forest vegetation by intercepting far more mist, rainfall, and particulate matter than a bare branch or trunk alone. Moreover, in addition to encountering air and water-borne nutrients, epiphytes also hold on to nutrient surpluses that might otherwise be exported from the community. All told, epiphytes are efficient at collecting and retaining nutrients from atmospheric sources, thereby enriching the luxuriance of old-growth forests.

Figure 10.3 Canopy roots capture nutrients from mats of epiphytes and other organic matter accumulating in the crowns of some trees in old-growth forests. Shown here is a network of canopy roots about 18 m (59 feet) aboveground on the trunk of a mature bigleaf maple, one of the species commonly developing this feature. A thick blanket of epiphytes on the trunk has been removed to expose the underlying system of canopy roots and bark. Photo courtesy of Nalini M. Nadkarni.

Trees in both temperate and tropical rain forests gain access to the nutrient resources produced by epiphytic communities partly by putting forth canopy roots (Fig. 10.3). These structures arise directly from the trunks of bigleaf maples and vine maples but have not been discovered on any of the conifers in Temperate Rain Forest. Canopy roots form extensive networks beneath the mats of epiphytes, where they absorb moisture and nutrients from the wealth of organic matter accumulating high above the forest floor. Some canopy roots branch downward and these may penetrate into the ground where they merge with the belowground root system. Young trees which have not yet acquired thick growths of epiphytes lack canopy root systems. On a mature bigleaf maple, however, canopy roots originate along the trunk from a dozen or more sites – junctions with branches – ranging from 2 to 20 m (6.5–66 feet) above ground level and extending as far as 10 m (33 feet) outward on branches. Because of their canopy root systems, trees in rain forests obtain nutrients produced and retained aboveground by epiphytes, an adaptation of importance in ecosystems where heavy rainfall rapidly leaches nutrients from the soil.

Although several species of epiphytes may adorn a single tree, one species of foliose lichen – *Lobaria oregana* – may account for about half of the epiphytic biomass in some areas of old-growth forest. This species and some others exhibit site-specific preferences for habitat. For example, some epiphytes develop primarily on the moist sides of trunks, whereas others appear mainly on the drier sides. Mosses generally occur only on the bottom half of the trees. *Lobaria oregana* also requires the cool, moist conditions found under the canopy and, if exposed to drier conditions, becomes dormant. This lichen fixes nitrogen; its presence therefore seems important to the nutrient economy of old-growth forests. Large amounts of the lettuce-like *Lobaria* fall to the forest floor, where it provides an accessible and nutritious food for deer, elk, and other animals.

At a much finer scale are the algae, bacteria, and both single-celled and multicellular fungi living on the surfaces of countless needles in old-growth forests. These microorganisms are technically known as **epiphylls**, literally meaning "on leaves." Others live inside the needles and are therefore **endophylls**. This complex community of microorganisms – profuse beyond measure or imagination – undoubtedly interacts with larger organisms in the forest, although precise knowledge of these relationships remains meager. Popularly known as "**scuzz**," epiphylls and endophylls perhaps enrich the forest with nutrients by supplying the system with nitrogen. Scuzz may also produce toxins that protect the needles from harmful fungi and herbivorous insects, but this relationship is among the many that awaits confirmation. Similarly, scuzz perhaps forms important links in the food webs of mites and other tiny creatures, which in turn are prey for spiders and predaceous insects. If so, such an abundant year-round food base maintains the predator population at a level high enough to curtail the seasonal buildup of herbaceous insects. In other words, perhaps because of scuzz, old-growth trees may lose little of their foliage to leaf-eating insects.

More about logs

Large logs fall into and across the numerous streams in old-growth forests, where they influence both the physical and biological characteristics of the aquatic systems. Large amounts of needles, cones, and other litter also enter these streams, but these materials do not physically alter small streams or remain in place; they wash downstream. Logs, in contrast, persist in place for decades, adding nutrients and structure to streams on a long-term basis. Moreover, some of the litter remains trapped behind log dams where it undergoes additional processing by aquatic organisms before being exported downstream into other environments. Indeed, of all of the organic matter entering small streams in old-growth forests, 60–70% is retained long enough to be incorporated into local food webs, with the balance carried downstream to other communities.

In old-growth forests, a large part of the food available in **first-** and **second-order streams** is derived from wood, but the food materials moving downstream into third-, fourth-, and higher-order streams becomes progressively smaller. In other words, the influence of the forest steadily diminishes as the stream increases in size. This relationship is reflected in the composition of invertebrates living in these streams. In first-order streams, insects and other invertebrates capable of shredding and gouging coarse woody foods are dominant, but these give way to grazing organisms as algae gains importance in the food webs of higher-order streams. Algae become more prevalent in food webs because larger streams have greater exposure to sunlight, whereas first- and second-order streams deep in the forest are heavily shaded. As a result, the fauna increases in its biomass and species diversity in the progression from first- to higher-order streams as fish and predators enter the community structure.

Logs also obstruct and alter the flow of small streams. These natural dams persist for many decades because, in comparison to logs on land, wood decays slowly underwater where there is less oxygen and diminished bacterial activity. Quiet backwaters collect behind the logs, and plunge pools develop in front. Logs and other woody debris falling into spawning streams trap salmon carcasses (*Bears, salmon, and forest enrichment*, this chapter), thereby retaining these sources of nutrients within the old-growth forest community. As many as 22 species of birds and mammals feed on the dead fish, which become entangled in proportion to the amount of woody debris littering the streambeds. However, the capacity to catch and retain salmon carcasses is reduced when the logs are removed to improve navigation and stream flow, which perhaps influences nutrient cycling and food webs in ways not yet fully appreciated.

Logs also fall into larger streams and rivers, and these wash downstream to the coastline where many accumulate at high-tide lines. The dead trees form a tangle of driftwood unmatched elsewhere on the coasts of North America; large areas of beach in the Pacific Northwest may be covered by driftwood more than 3 m (10 feet) deep. The accumulation stabilizes the beaches and offers protected areas where pioneering plants gain footholds along the shoreline. Logs remaining afloat are colonized by barnacles and other marine life, and logs sinking to the sea floor fuel an energy-poor environment. Biodiversity accordingly increases at sites where ocean currents accumulate this debris. Various kinds of wood-boring invertebrates establish a food chain which includes organisms feeding on their wastes, as well as predators. Some of these ecological associations may be impaired, however, as lumbering encroaches on old-growth forests and fewer trees are flushed to the sea.

Succession on glacial till

Glaciers are common features along the rugged coastline of northwestern North America, where mountains intercept moisture-rich air masses moving inland from the Pacific Ocean. At lower elevations this results in heavy rainfall, whereas equally heavy snowfall occurs in the mountains. Glaciers develop where snowfall regularly exceeds snowmelt, and ice forms from compaction and recrystallization of the accumulated snow (*Rivers of ice*, Chapter 12). In places, the thick ice sheets meet the sea, calving icebergs as the glaciers march forward. Elsewhere along the coast the ice is retreating, revealing fjords gouged in the underlying rock millennia earlier.

Retreating glaciers offer sites where primary succession can be monitored. The newly exposed surface strata are known as **glacial till**, which are unsorted deposits of rock fragments mixed with finely ground materials dropped from the melting ice. These sites lack a defined soil profile and contain essentially no organic matter or nitrogen. Succession on glacial till is therefore closely linked to soil development. Moreover, a reasonably precise time line may be delineated for succession on glacial till; that is, because the rate of glacial retreat can be determined, ages can be estimated for communities at sites once covered by ice.

In coastal Alaska, the successional sequence begins with a pioneer community of fireweed, mosses, and other Arctic herbs, accompanied by dwarf willows. Dryas (a type of rose) is an important pioneer because it forms mats, whereas the other vegetation is thin and provides little ground cover. A community of shrubby willows appears next in the sequence. About 35–40 years after becoming ice free, the sites are dominated by tall thickets of alder and, at this point, ground cover is nearly complete. Sitka spruce slowly invades the alders and marks the transition into forest when the sites are about 100 years old. During the next 100 years, two species of hemlock – western and mountain – steadily mix with the spruce, resulting in the spruce-hemlock climax typical of this region of Temperate Rain Forest after 200 years.

The glacial till itself changes significantly during the 200 years in the successional sequence. Leaves and roots steadily add organic matter, and nitrogen and other nutrients gradually enrich the developing soil. A forest floor consisting of 13–20 cm (5–8 inches) of organic matter accumulates after 200 years. Alders represent a particularly important stage in the sequence, partly because soil fertility is improved by nitrogen-fixing bacteria associated with their root systems. A net accumulation of about 62 kg ha^{-1} (55 pound per acre) of nitrogen occurs each year during the period when alder thickets dominate the sites. Alder thickets which grow vigorously on the nitrogen-deficient till therefore play a major role in establishing conditions favorable for the final stages of succession.

Infobox 10.2 Spirit bears, the other white bear

A maze of islands and channels borders the deeply sculptured coastline of British Columbia. The islands, like the adjacent mainland, are cloaked in Temperature Rain Forest, although in places logging has cleared large tracts of old-growth timber. What is different, however, are the 400–500 white bears that roam parts of the north-central coast, particularly on Gribbell, Princess Royal, and Roderick islands; a few also occur on the adjacent mainland. These are the Spirit bears, a genetic variant of the more familiar black bear. They are not albinos (as their eyes and skin are normally pigmented) or polar bears (an entirely different and far-off species). Instead, Spirit bears are a white-phase "morph" occurring in one of the 16 subspecies of black bear (i.e., white bears do not represent a separate taxon; other bears of the same subspecies have normal coloration). Also known as Kermode bears or simply as Kermodes, the white bears were named in honor of Francis Kermode, a former director of the Royal British Columbia Museum.

Other than their unique whiteness (some are cream-colored), Spirit bears' physical appearance and biology are much like others of their subspecies (e.g., size and reproduction, but see below). Their white coat is caused by a difference in a single recessive **allele** on a gene associated with melanism. A substitution of a single amino acid for another produces the obvious difference, but only when a cub receives this recessive allele from both parents. Thus, two black-colored bears can produce a white cub if each parent carries the recessive allele. The trait may have resulted from genetic isolation of a relatively small population located in an insular landscape, perhaps an offshore glacial refugium harboring bears and other wildlife during the Pleistocene. Gribbell Island, where more than one-third of the bears are white, may be where **genetic drift** established the trait sometime before the Wisconsin glaciation (i.e., the last glacial event, ending about 10,000 years ago).

Why the trait persists may be related to an advantage white bears have when fishing for salmon during daylight hours: they are twice as efficient at catching fish as normally colored bears (by night, the fishing success of bears of both colors is essentially the same). Spirit bears are less visible to salmon, which suggests an adaptation favored by natural selection. Spirit bears typically feed closer to the shoreline than their black counterparts so, even when they each feed on similar foods (e.g., berries), the white bears acquire plant foods enriched by marine nutrients. They move even closer to the beaches and forage on barnacles, mussels, and shoreline vegetation in summer, whereas the black-coated bears generally remain farther inland. By fall, bears of both colors seek salmon but fishing success is greater for the white bears.

Spirit bears hold a prominent place in the mythology of Native Peoples in the region. According to legend, white bears originated when the raven – the creator of the world – arrived from heaven and turned the land green, as it is today. Nonetheless, the raven wished to leave a reminder of the time when the land was covered all year round with snow, and so turned every tenth bear white. The raven decreed that the white bears would live forever in peace.

The raven's hopes were challenged when, much to the dismay of conservationists, loggers targeted Princess Royal Island and other prime sites in the world's largest remaining area of Temperate Rain Forest. Spirit bears emerged as the regional icon for saving this treasure of biodiversity just as pandas have long served as the world-wide symbol for conservation. The issue simmered for years, but in 2006 conservationists and loggers, along with government officials and Native Peoples, produced the Great Bear Rainforest Agreements that established a core conservancy of 1.78 million ha (4.4 million acres), which includes nearly 202,000 ha (500,000 acres) of Spirit bear habitat, all closed to logging, mining, and hunting. Another 4.04 million ha (10 million acres) will be sustainably managed with practices that, while timbered, will leave intact 50% of the natural level of old-growth forest. All told, the raven should be pleased as his white bears will continue living in peace.

Highlights

Bears, salmon, and forest enrichment

The distribution of huge bears coincides with the northern part of the Temperate Rain Forest region. These bears – the world's largest – are found on islands along the coastline as well as on the adjacent mainland in southern Alaska (Infobox 10.2). Variously known as Alaskan brown, big brown, Kodiak, or simply brown bears, they attain weights of up to 680 kg (1500 pounds) and, when standing erect, commonly reach heights of 2.4 m (8 feet) or more. Because they are so large, brown bears were once considered a distinct species; they are now regarded as a subspecies of grizzly bear, however.

Like smaller-sized grizzlies, brown bears have "dish-shaped" faces (i.e., a concave facial profile) and noticeable humps above their shoulders. Their claws are huge but, in contrast to black bears, neither brown bears nor other forms of grizzlies can climb trees. Brown bears do not live in the interior of old-growth forests, instead favoring more exposed habitats within their geographical range (e.g., thickets, grassy meadows, and open areas near streams on valley floors or above the timber line).

Brown bears are solitary, or asocial, for most of their biological year. However, to exploit dependable sources of high-energy foods, brown bears at times gather in aggregations, the most notable of which occur during salmon "runs" (i.e., upstream movements to spawning

Figure 10.4 Brown bears, a large subspecies of grizzly bear, congregate each year at traditional fishing areas (usually waterfalls or rapids) to catch salmon "running" upstream to spawn. The bears establish a social hierarchy to settle disputes for the better fishing sites in these areas. Photo courtesy of Barrie K. Gilbert.

areas). The aggregations of bears represent a long-standing seasonal tradition at sites (often rapids or water-falls) sometimes termed "ecocenters," which attract not only bears but also other carnivores and scavengers (e.g., gulls and ravens). Thanks to their greater experience older bears sometimes catch salmon mid-leap, but bears of every age grab at fish swimming in shallow water. Cubs continue the fishing tradition when they come to these locations with their mothers, although they initially seek fish scraps as these wash downstream. Brown bears interact socially on these occasions, but only for about 45 days each year.

Brown bears establish a behavioral organization based on social dominance, which begins anew each year after the bears arrive at their fishing grounds (Fig. 10.4). At first their social behavior is no different than that which occurs at other times of the year, namely wariness of each other and long-range avoidance, but it also includes aggressive encounters followed by brief chases. Aggressive behavior declines rapidly during the first 10 days, and the bears begin to tolerate each other as long as adjacent bears do not violate their neighbor's **individual distance**, a separation of about 2 m (6.5 feet). Bears regularly display low-intensity threats as a means of reinforcing the order of social dominance. However, when a skirmish occurs, the loser sometimes meets a third bear and picks a fight, displaying what is known as **redirected behavior**.

The overriding consequence of social dominance is its effect on when and where each bear fishes, which generally affects an individual's fishing success (i.e., socially

dominant bears usually fish at the better locations in the streambed). In terms of natural selection, bears of higher social rank presumably gain an advantage, that is, better nutrition and weight gains which subsequently enhance their reproductive success and winter survival. Social organization also minimizes fighting and maximizes the time for fishing.

Along the coastline of western North America, five species of salmon are included in a single genus collectively known as Pacific salmon. Although they share a common characteristic – after reaching maturity in the Pacific Ocean, they return to the same freshwater river where they hatched years earlier, spawn, then die *en masse* after just one breeding effort per individual – their sizes and life histories vary by species. Millions of salmon are involved in the yearly spawning runs which occur in coastal rivers and their tributaries from California to Alaska. The males of some species are known for their distinctively hooked jaws, humped back, and bright red coloration, all of which are spectacularly visible in shallow water (e.g., sockeye salmon; Fig. 10.5).

Juvenile salmon leave freshwater weighing up to 20 g (0.7 ounces) and return from the ocean weighing up to 10 kg (22 pounds) or more. In other words, Pacific salmon gain at least 90% of their adult weight while at sea. While maturing in the ocean (the exact locations are not well known) salmon therefore acquire considerable body mass well-laden with nutrients and energy destined to be transported – and incorporated – into two far-off systems: freshwater spawning areas and the surrounding riparian zones in the Temperate Rain Forest. Such an

Figure 10.5 Sockeye salmon represent one of five salmon species that spawn in river systems from California to Alaska. The young fish mature in the ocean, then return years later to breed just once before dying at the same freshwater locations where they hatched. In the ocean these are silvery with dark backs but, as shown here, they turn dark red and develop hooked jaws when returning upstream. Photograph courtesy of Alaska Department of Fish and Game.

"uphill" flow is contrary to the usual workings of ecology in which energy and nutrients travel "downhill" (i.e., from watershed to ocean).

Three or more years after the juvenile salmon journey downstream to mature in the Pacific Ocean, they return as adults to the same streams and areas where they hatched several years before. Pacific salmon locate their spawning areas in freshwater streams using a keen sense of smell. This remarkable ability (a type of homing behavior) occurs because juvenile salmon, after hatching, imprint on the distinctive odors peculiar to their natal streams. Further, after years of wandering in the ocean, the salmon recognize and unerringly follow these odors when they move upstream to spawn. Salmon also detect and employ the odors of other streams as navigational cues during their upstream migration, and may additionally differentiate between the odors from their own population and those from other populations.

The predictable and stream-specific nature of homing behavior in Pacific salmon also founded the seasonal movements of two predators – brown bears and salmon sharks – to traditional salmon streams. When the freshwater salmon run begins, brown bears await upstream to capture returning fish that survive the gauntlet of salmon sharks, which greet returning salmon at the mouths of their natal rivers and streams. Reaching lengths exceeding 2.6 m (8.6 feet), salmon sharks time their movements in the northern Pacific Ocean to meet salmon at the saltwater–freshwater interface.

In spawning areas, typically gravel-covered stream-beds in shallow water, the decomposing carcasses of dead salmon add an abundance of marine-derived nutrients,

thereby directly and indirectly enriching the food chain in the nursery area for their offspring. On land, the distribution of salmon-derived nutrients is facilitated by both brown (or grizzly) and black bears. The salmon are a staple source of lipids that bears must accumulate to sustain their metabolic needs when they cease foraging and den for the winter. Bears are not true hibernators as their body temperature does not drop to ambient levels; during winter they must therefore continue to fuel metabolic activities for as long as 7 months. Lipid reserves are especially critical for females, which give birth and begin lactation during the denning period. In fact, lipids from salmon represent about 80% of the body mass gained by bears with access to salmon streams; the skin, brain, and, if a female, unspawned eggs from the carcasses provide the main sources of lipids.

The fishing behavior of bears is a popular scene appearing on nature programs, calendars, and other media. What is not often shown, however, is a bear carrying a fish into the nearby forest to avoid the aggressive behavior of other bears. Bears typically leave behind more than half of the biomass of each fish. On average, bears capture at least 30% of all salmon moving upstream, but the predation rate may exceed 90% in small streams little wider than the bears themselves. Of these bears transport 43–68%, depending on water depth at the capture site, into the surrounding forest. The result is a significant source of enrichment available on the ground well into the forest; the effected zone often extends 500 m (1640 feet) or more from the river bank. Similar results, but to a lesser extent, stem from other species that prey or scavenge on salmon (e.g., gray wolves, mink, ravens, bald eagles, and gulls). At one location in Alaska, estimates indicated that bears transported an average of nearly 3000 kg (6614 pounds) of carcasses per year into the forest adjacent to a 2 km (1.2 mile) stretch of stream. This in turn represented the deposition of almost 90 kg (198 pounds) of nitrogen and 12 kg (26 pounds) of phosphorous at a rate per unit of area approaching commercial fertilization applications for managed forests.

The impact of salmon-derived nitrogen increases growth rates of climax species (e.g., western hemlock, Sitka spruce, and Douglas fir) along with understory vegetation. During the course of decomposition, about two-thirds of the nitrogen in a salmon carcass enters the pool of soil nitrogen available to these plants. The effects of nitrogen loading are somewhat localized because the bears do not randomly deposit the carcasses on the forest floor and, in fact, may favor certain areas. Other factors, however, further spread nitrogen derived from salmon; these include scavengers that carry off all or parts of the carcasses, defecation and urination, and the life cycles of flies and other insects (e.g., blowflies hatch from larvae feeding on the carcasses and thereafter enter food chains in other parts of the forest).

Those salmon that escape predators eventually spawn then die when their bodies are exhausted, an extreme case of starvation technically known in this case as **senescent death**. Pacific salmon do not feed during their spawning runs (which in some cases may be distances measuring hundreds of kilometers) and therefore rely entirely on their reserves to sustain their energetic requirements for the journey; these needs are heightened by swimming against the current and, often, surmounting strong rapids and low waterfalls. After dying, their carcasses move downstream until they are trapped and decompose behind logs, gravel bars, or other debris at or relatively near the spawning site, thereby enriching the immediate area. Others settle in pools formed by obstructions in the steam bed. **Hyporheic flow** likewise may transport carcass-borne nutrients from the stream into riparian areas; this occurs where stream water mixes with groundwater below and on the sides of the stream bed, thereby allowing the exchange of nutrients and other water-borne materials. Trees and other riparian vegetation therefore gain access to this enrichment when it seeps into their root zones. Flood waters may also carry salmon carcasses from the stream well into the surrounding riparian zones. The remainder of the carcasses washes downstream and into estuaries where it becomes food or enrichment for even more organisms (e.g., algae).

Additional nutrients originate from salmon eggs, which are comparatively higher in energy and nutrients than the content of carcasses. Moreover, unless disturbed, they remain wedged between pebbles and gravel on the stream bottom where they were originally laid and hence become available to the immediate aquatic fauna. Mortality rates estimated at 87–93% for the eggs of pink and chum salmon indicate the impact of predators and scavengers, including juvenile salmon. Unhatched eggs may sometimes be retained on site even longer than carcasses.

Indirectly, enrichment from salmon-derived nutrients influences a variety of other species in addition to those that directly feed on salmon (e.g., bears), their carcasses (e.g., gulls and other scavengers), or obtain nutrients from the soil (e.g., vegetation). For example, the streamside densities of songbirds (including golden-crowned kinglets, varied thrushes, and winter wrens) may be higher below than above a waterfall (i.e., a barrier to further upstream movement of salmon, hence providing natural, but sharp separation between sites with and without salmon). Some of these birds may be responding to increased food availability in the form of aquatic and terrestrial invertebrates that were nourished by salmon carcasses. Winter wrens seem representative of this relationship; when analyzed, isotopes in their tissues show the presence of salmon-derived nutrients. Other songbirds, particularly those feeding on fruits, may be responding to the increased productivity of plants growing near salmon streams.

The impact of salmon-derived nutrients therefore extends to a wide range of organisms at various trophic levels in the Temperate Rain Forest, including both aquatic and terrestrial environments. Unfortunately, the ecological significance of these relationships emerged long after commercial fishing, dams, and other anthropogenic forces began curtailing the spawning runs of salmon. Some estimates suggest historic runs of 228–351 million salmon are now reduced to 142–287 million fish, or a 20–40% loss in their numbers. Moreover, because average weights of salmon have decreased, the loss in biomass is even greater with today's runs diminished by 47–61% from the past. The implications are clear: continued reductions in salmon populations will further reduce the wealth of nutrients once transported from open ocean to inland forest as part of an age-old cycle of life and death.

A seabird in the forest

A slow but steady shower of needles falls from the immense canopy in these ancient forests. Most of this debris reaches the forest floor, but some also lodges on large limbs well above ground level. This material accumulates and develops into platforms of organic matter known as "perched soil." Lichens and mosses colonize these microhabitats where birds such as rufous hummingbirds build their nests, thereby adding even more material.

Marbled murrelets are somewhat unexpected visitors to these locations. Many seabirds, including other kinds of murrelets and their relatives (22 species, all in the family Alcidae) characteristically nest on cliffs along coastlines (*Waterbird colonies*, Chapter 11). In contrast, marbled murrelets nest high above ground on platforms of organic debris in old-growth forests. The platforms require about 150 years to reach a stage of development – a sort of "microsuccession" – in which they become suitable as nesting sites. In British Columbia, Sitka spruce tends to contain more platforms than other trees, but nests often occur in Douglas fir elsewhere in the Pacific Northwest. However, the availability of suitable platforms is of greater importance than the species of tree. Young trees may be utilized for nesting if they are infected with mistletoe, which forms broom-like structures equivalent in area to the platforms on the large branches of older trees. Some marbled murrelets also nest on the ground, especially along the upper coastline of Alaska. Overall, marbled murrelets occur from the Pacific rim of North America from California to Alaska and in Asia from Russia to Japan.

For their nests, the birds usually select habitat within 60 km (37 miles) of the coastline, but at times the birds have been found as far inland as 88 km (55 miles). Nest trees in the Pacific Northwest are usually 300 to more than 500 years of age and range in height from 56 to 61 m (184–200 feet). The nests themselves are situated on branches about 45 m (148 feet) above ground on average. Given these circumstances, the nests of marbled murrelets prove difficult to find and, in fact, the first nest was not discovered until 1974. Marbled murrelets lay a single egg which the parents take turns to incubate during 24-hour shifts, but many other aspects of their nesting ecology are still uncertain or are documented with meager field data.

Nestling marbled murrelets are fed herrings, anchovies, and other small fishes that their parents often carry one at a time from the sea several times per day. The energy expended to feed their young therefore limits, in part, their inland breeding range. Based on current evidence, however, the birds do not necessarily select their breeding areas based on the availability of food along the coastline; that is, after breeding ends and the young birds have fledged, marbled murrelets may move elsewhere along the coast, presumably seeking areas where food might be more plentiful.

Marbled murrelets are among several species of seabirds that become entangled and drown in gill nets while foraging underwater. Oil spills additionally kill large numbers of murrelets (e.g., about 8400 died in the *Exxon Valdez* spill in 1989). Nonetheless, the permanent loss of nesting habitat poses the greatest long-term threat to marbled murrelets, and their ecological dependence on old-growth forests seems comparable to the more publicized status of northern spotted owls (see *Ecological controversy*, this chapter). Marbled murrelets are now listed as a threatened species in the southern part of their range (i.e., British Columbia to California). Moreover, the phenomenon of seabirds nesting on lofty platforms in ancient trees provides additional evidence that old-growth communities cannot survive with rotations designed for commercial timber production (i.e., 80–100 years).

Some small mammals and their ecology

Voles are stocky, mouse-like rodents found in a wide range of habitats (e.g., grasslands, woodlands, and marshy areas). However, red tree voles may spend their entire lives high in the branches of a single tree. Their diet consists of the needles of Douglas fir, western hemlock, and other dominant species in Temperate Rain Forest, where the rodents build bulky nests on branches near the main trunk of these same species. The nests grow larger with age, probably because each successive generation of voles adds more materials. The nests are constructed at all levels of the canopy, although most are found in the lower one-third. Some trees contain two or more nests, suggesting colonial behavior.

Red tree voles travel slowly through the branches and, when moving to an adjacent tree, cross on interdigitating branches; they do not leap as tree squirrels do. Their long tails serve as counterbalances and, at times, may grasp a twig to avoid falling. When descending a tree, they move downward head-first and sometimes jump from heights of 20 m (66 feet) or more, but they land with a bounce,

uninjured. Red tree voles also move about on the ground, where their cryptic coloration helps conceal them from predators. In addition to their arboreal habits, red tree voles are of interest because they are prey for an endangered species – northern spotted owls – and live in forests subject to heavy lumbering and fragmentation.

Northern flying squirrels, Douglas squirrels (also known as chickarees), and Townsend chipmunks are among the other small mammals whose distributions include Temperate Rain Forest. In the same way as red tree voles, they form important links in food chains within the forest. However, because of their own diets, some species of small mammals are involved in a subtle yet crucial process in forest ecology: the distribution of fungi. Root-inhabiting organisms, known as mycorrhizal fungi, are symbiotic partners with coniferous trees. The trees depend on the fungi for the absorption of nutrients from the soil, and the fungi feed on the sugars manufactured by the trees. Some mycorrhizal fungi also produce antibiotics, thereby protecting the trees from diseases, and others release enzymes that stimulate root growth. In most cases, neither the trees nor their mycorrhizal fungi can survive without each other. However, because they live underground where their thread-like mycelia entwine and fuse with tree roots, mycorrhizal fungi cannot disperse their spores in the same way as fungi living above ground (e.g., mushrooms, whose spores are dispersed by air currents). By locating (the fungi have species-specific odors), excavating, and eating the fleshy parts of mycorrhizal fungi, small mammals therefore serve as the primary vectors for the dispersal of spores vital to the conifers in the forest. The spores of mycorrhizal fungi pass through digestive tracts with no loss of viability and are dispensed in the abundance of droppings produced daily by small mammals. Just one dropping may contain a half-million spores. By inoculating soils and tree roots throughout the forest with their spore-rich droppings, small mammals may more than offset their more widely publicized role as seed eaters, for which they are sometimes blamed for harming, rather than helping, forest regeneration.

Banana slugs

Temperate Rain Forest from northern California to southern Alaska is home for the largest terrestrial mollusk in North America. Banana slugs, so named because of their yellow coloration, attain lengths of 30.5 cm (12 inches). At minimum, their density in Douglas fir and western hemlock communities is estimated at 2500 slugs per ha (1012 per acre), with a biomass of 35 kg ha^{-1} (14.2 pounds per acre). With favorable conditions – cool, moist sites with a substrate of mosses and a thin understory of herbaceous plants – these parameters may double for local populations of banana slugs (Fig. 10.6).

Banana slugs are more than curiosities in old-growth forests. They perform a fundamental role in nutrient

Figure 10.6 Banana slugs are distinctive natives of moist coniferous forests and play a significant role in decomposition and nutrient cycling. Photo courtesy of Brian R. Chapman.

cycling. Specifically, banana slugs consume fresh and senescent plant materials instead of dead plants, thereby speeding the process of cycling nutrients within the community. Banana slugs feed on the fresh foliage of at least 19 species of plants in addition to mosses and fungi, but even greater preference is shown for the senescent leaves of alder and other plants. Their diet also includes decaying ground litter, but these materials are probably consumed more for their fungal content than for the nutrients in the dead leaves. All told, the varied diet of banana slugs produces about 25 kg ha^{-1} (225 pounds per acre) of fecal deposits each year, which represents a significant turnover of nutrients in the understory of old-growth forests.

Pacific yew

The importance of preserving biodiversity is often justified on the basis of aesthetics, ecological functions, and even morality. More commonly, practical reasons are cited, particularly in the context of medical history (e.g., quinine, the original preventative for malaria, was derived from the bark of the cinchona tree).

Native Americans once valued Pacific yew because its wood was ideal for bows, harpoons, and other tools, but they also concocted potions of varying efficacy from the bark, needles, and berries of yew trees. More recently, however, persons concerned with timber and other forest products had little regard for yews, which grow slowly and lack straight trunks. Pacific yews prosper

in the understory of old-growth forests, where they tolerate the shade of the closed canopy. In fact, yews may be damaged by sunlight when the canopy cover is less than 50%. The trees are not numerous under the best of circumstances and, as the old-growth forests are cut, yews also decline. Logging machinery often kills yews in the process of harvesting western hemlock and other commercially valuable trees. Fire is another limiting factor and, because yews require centuries to recover from this or other disturbances, Pacific yews only survive in forests managed with exceptionally long **rotations**.

A relatively new discovery – taxol – adds emphasis to the practical reasons for maintaining biodiversity. Taxol, a complex compound present in the bark of yews, offers an effective treatment for some types of cancer; it inhibits cell division, thereby curtailing further growth of the malignancy. Yews throughout the world have endophytic fungi in their bark that produce taxol and, because of the limited accessibility to Pacific yews in old-growth forests in North America, other species later gained greater importance as sources of raw material. Nonetheless, taxol derived from Pacific yews initiated the discovery of its modern-day medicinal values. Originally, about 27 kg (60 pounds) of bark were needed to produce a single treatment, but the compound has now been successfully synthesized in the laboratory and approved for the treatment of ovarian, breast, and lung cancers, as well as certain cancers of the head and neck.

Giant salamanders and other amphibians

Old-growth forests provide habitat for the coastal giant salamander and the California giant salamander, each of which reaches lengths of 30.5 cm (12 inches). These unusual salamanders, heavy bodied and clumsy in appearance, climb well and may be found in bushes and on the sloping trunks of trees. Even more unusual, however, is their ability to produce vocal sounds sometimes described as suggestive of a barking dog.

The dorsal surface of giant salamanders is coarsely marbled with black markings on a brown, gray, or purplish background. These markings create a light and dark patchwork, which perhaps mimic the dappling produced by sunlight reaching through the canopy of the forest floor. Both giant salamanders are accordingly sometimes active diurnally. Their diet includes snails, worms, and insects – quarry typical of most salamanders – but also larger prey such as small snakes, shrews, mice, and even other salamanders.

Old-growth forests also house other amphibians. Among these are the Olympic torrent salamander, an aquatic species strongly linked to rapidly flowing streams in old-growth forests and the Oregon slender salamander, which occurs only in the Cascade Range in Oregon. For habitat, Oregon slender salamanders rely heavily on the interior of large, well-decayed logs, structures abundant in ancient stands but rarely available in forests managed with short rotations (i.e., those producing wood products). The aptly named coastal tailed frog also occurs only in the Pacific Northwest, where it prefers swift, well-shaded streams in forests mature enough to have closed canopies. The stumpy "tail" appears only on adult males and is used during mating to transfer sperm internally to the tail-less females (the organ is not a true tail). Tailed frogs are voiceless, and their tadpoles have sucker-like mouths with which they cling to rocks in the swift current.

Giant salamanders, Olympic torrent salamanders, and coastal tailed frogs – all stream-dwelling amphibians – are susceptible to the fine sediments produced during and after logging. These sediments fill the spaces between rocks in the streambed, thereby eliminating the habitat where the larval forms of these amphibians find refuge. Some of these species also have low tolerances for high temperatures, which may result if logging removes the shade-producing canopy. Once extirpated by logging or wildfires, species such as the Olympic torrent salamander may not re-colonize a stream for 40–60 years afterward.

Mount St Helens

A breath-taking event occurred in the Cascade Range when Mount St Helens erupted on 18 May 1980 (Fig. 10.7). The eruption spewed no lava, but hurled huge amounts of ash 19 km (12 miles) into the sky, across a large area of Washington, and onto lands in adjacent states. Volcanic ash, along with cinders and pumice, is a component of **tephra**, a term for the airborne materials deposited from a volcanic eruption. The blanket of tephra, which in places was almost 1 m (39 inches) thick, created a moonscape of desolation. This feature – burial – distinguishes volcanic eruptions from wildfires and most kinds of ecological disturbances.

(a)

(b)

Figure 10.7 Mount St. Helens (a) before and (b) after it exploded in 1980. The huge blast and blanket of ash devastated a large area of forest in Washington, as shown by the absence of vegetation. Photos courtesy R. P. Hoblett and Harry Glicken of the US Geological Survey.

The blast, which produced an explosion equivalent to more than 2700 atomic bombs, leveled almost 650 km^2 (250 square miles) of forest, including 250- to 650-year-old stands of Douglas fir, Pacific silver fir, and western hemlock. Viewed from above, the devastated forests resembled fields of scattered straw. The explosion knocked down trees 27 km (17 miles) from the crater, which itself lost enough height (460 m or about 1500 feet) to drop in rank from the 5th-highest peak in Washington to the 13th. Although not considered one of the major volcanic disturbances in Earth's geologic history, Mount St Helens quickly became one of the best-studied biogeological events in scientific history.

Estimates suggest that as many as 5000 deer, 1500 elk, and 200 black bears were among the wildlife killed by the eruption, but the number of birds killed is beyond calculation. Most insects likewise perished, often because the ash scarified the cuticles of their exoskeletons, thereby causing desiccation and death. The biota in 26 lakes also suffered; some 11 million fish likely died as a direct result of the eruption. The clear, cold waters of Spirit Lake, just north of the volcano, were heated to almost 39°C (100°F) and became clouded with tephra, which quickly caked the gills of aquatic organisms. All fish, including their food chains, were destroyed.

Virtually all aboveground vegetation, from mature trees to herbaceous understory, simply disappeared in the blast zone on the north side of the crater. Farther removed from the volcano, the blast knocked down trees in immense numbers. Despite the immense destruction, careful examination revealed a surprising array of survivors. At least seven species of ants emerged from their underground galleries or from logs, and various kinds of aquatic life survived in some places under a protective cover of ice. In places, pocket gophers also survived in their underground burrows, especially where snow cover provided additional protection, and their activities initially helped restore these sites. Snow cover also effectively shielded some plants from both heat stress and the force of the blast. The aptly name perennial, fireweed, and another herbaceous plant, pearly everlasting, are representative of the plants established soon after the eruption. These pioneer species and others provided cover for other immigrants as succession advanced in the mid-1990s.

Contrary to expectations, colonists – as opposed to survivors – played a secondary role in the ecological recovery of Mount St Helens. Colonization had long been regarded as the primary process for restoring damaged landscapes, but evidence at Mount St Helens proved otherwise. Nonetheless, Roosevelt elk wandered across the blast zone shortly after the eruption, disturbing the tephra crusts and leaving hoof prints in which windborne seeds might lodge (i.e., the depressions served as microhabitats for colonizing plants). On the other hand, the elk severely overgrazed much of the vegetation as it emerged from the blanket of tephra. Large numbers of insects (including their eggs), other invertebrates, and seeds rained into the blast area. Some 75 species of spiders, for example, sailed on their webs into the tephra-covered landscape. Ladybird beetles also arrived within a year of the eruption and preyed on aphids that infested the fireweed colonizers, thereby reestablishing one of the first post-blast food chains. The first fish – a rainbow trout – mysteriously turned up in 1993 as did another in 1994. Most likely someone illegally stocked the trout, but the fish may somehow have colonized Spirit Lake from a nearby lake where ice protected the water during the eruption.

However, dispersal is only part of the equation. Newcomers may have to endure rigorous conditions once arrived, and relatively few species are adapted for both dispersal and hardship. Successful colonists display well-developed **ecesis**, the ability to disperse and survive in harsh environments. Lupine successfully colonized the tephra deposits at Mount St Helens, in part because of its nitrogen-fixing capabilities. Once established, lupine colonies sometimes became centers where other organisms gain footholds (i.e., lupine created microhabitat for other plants as well as insects).

One long-term result, yet to be determined, concerns the patchy distribution of organisms surviving the eruption. This is known as the **founder effect**, in which the genetic composition of the few survivors (and their offspring, of course) may differ from a larger population of the same species. If the survivors have less genetic variation, then some characteristics of these organisms may differ enough to distinguish them from the original population (i.e., a different genetic pool is established at the patch site). The differences in genetic variation may be amplified over time, creating a new and distinctive endemic population (i.e., similar to the way a small number of "founders" colonize oceanic islands and develop unique characteristics). Interestingly, at least one endemic taxon occurs on each of several other relatively inactive volcanoes in the Cascades although none have so far evolved on Mount St Helens (presumably because its repeated eruptions erase the initial stages of endemic development).

All told, nature proved remarkably resilient to the huge blast and the recovery pattern relied more on the abundance of survivors and less on colonists. Close study revealed the presence of 90% (or more than 230 species) of the original vegetation three years after the eruption although low successional species, as might be expected, were heavily represented in the flora. Douglas fir, western hemlock, and other characteristic species of old-growth forests are recovering slowly; indeed, the forest vegetation may not reach full maturity before Mount St Helens erupts again. Even before the 1980 eruption, the forests on Mount St Helens were well below the tree line on other peaks in the Cascades, evidence that the volcano erupts often enough to preclude reforestation of the mountain's upper slopes.

(a)   (b)

Figure 10.8 (a) Loggers prepare to cut a giant fir late in the 19th century in western Washington. (b) The northern spotted owl remains a focal point in debates concerning the management of old-growth forests. Continuing losses of habitat from logging and other disturbances have imperiled the owl across significant areas of its range. The future extent of lumbering in the remaining old-growth ecosystems remains uncertain. Photos courtesy of (a) the Forest History Society and (b) the National Park Service.

Ecological controversy

Timber production remains a significant economic force in much of the Pacific Northwest, where some aspects of forest management have generated a heated controversy at the national level. In general terms, the controversy concerns elimination of old-growth forests, which are being cut for two economic reasons. First, they contain immense amounts of valuable timber. These trees, because of their great size, each yield thousands of board feet (the standard unit of measurement for raw timber). Second, clearing the land of old-growth forests makes way for replacement stands of young, more rapidly growing trees; rapid growth shortens the **rotation**, the interval between commercial harvests. Once eliminated, these forests will be cut every 80–100 years and will never again return to old-growth (Fig. 10.8a).

In more specific terms, however, northern spotted owls became the flash point in the controversy (Fig. 10.8b). Spotted owls depend heavily on old-growth forests, and logging (typically, clear-cutting) reduced or eliminated some of their populations. In response, state agencies provided the northern subspecies of spotted owl with various degrees of legal protection, which was bolstered when the US Fish and Wildlife Service listed the birds as "threatened" and therefore subject to full protection under the Endangered Species Act of 1973. Additionally, regulations pursuant to enactment of the National Forest Management Act of 1976 charge the USDA Forest

Service to maintain viable populations of all native species on lands managed by that agency. That mandate includes spotted owls, which the Forest Service regards as both a "sensitive species" and an "indicator species" of old-growth forest ecosystems. The welfare of northern spotted owls accordingly came to bear on the management of federal, state, and privately owned old-growth forests. For some, the issue was reduced simply to "jobs or owls," whereas ecologists wished to preserve not just an owl population but the structure and function of an entire ecosystem.

Spotted owls are not shy around humans, who can often approach the birds quite closely before they fly. The owls feed primarily on small mammals, especially arboreal or semiarboreal species such as flying squirrels, but they also prey on insects and birds. Pairs occupy and defend territories year after year as long as the habitat remains suitable. When nesting, which does not occur every year, spotted owls usually lay two eggs. The female alone incubates the eggs and, later, broods the nestlings while her mate provides most of the food for both the female and the owlets. Spotted owls seem intolerant of high temperatures and, to reduce heat stress in summer, they roost in sites with cool microclimates.

Populations of northern spotted owls continue to decline – the subspecies is nearly extirpated in British Columbia – primarily from lost habitat and competition with a related species, the barred owl. Logging, conversion to younger forests, fire, wind throws, insect outbreaks, tree diseases, and both urban and suburban developments all claim old-growth habitat. The eruption of Mount St Helens additionally destroyed large areas of old-growth habitat for spotted owls. However, the greatest population declines seem coincident with areas invaded by barred owls. Barred owls, which have recently expanded their range from eastern North America into northern California and the Pacific Northwest, are habitat generalists with greater dispersal abilities and higher reproductive rates than spotted owls. Larger and more aggressive, barred owls disrupt the nesting efforts of spotted owls and compete for their food; at times, they kill spotted owls. Accordingly, the US Fish and Wildlife Service considered removing barred owls from certain areas of spotted owl habitat, but the experiment was opposed by animal rights and conservation groups and remains in limbo and untested.

The "jobs or owls" controversy continues, albeit moderated somewhat after enactment of the National Forest Plan of 1994 which allows some timber allotments while safeguarding other areas of old-growth forest. Nonetheless, spotted owl populations still diminish by an annual average of 7.3%, presumably because of competition from barred owls. The proposal to remove barred owls also generated new arguments, of which the justification for "killing one species to save another" is a poignant example. Some loggers believe barred owls should be allowed to displace spotted owls – seemingly a natural process – thereby opening tracts of once-protected forest to commercial cutting. Regardless of whatever proposals come forth, the controversy seems sure to fester for years to come.

Readings and references

Introduction

Alaback, P.B. 1982. Dynamics of understory biomass in Sitka spruce–western hemlock forests of southeast Alaska. Ecology 63: 1932–1948.

Norse, E.E. 1994. *Ancient Forests of the Pacific Northwest*. Island Press, Washington, DC.

Orians, G., and J. Schoen. (eds) 2013. *North Pacific Temperate Rainforests: Ecology and Conservation*. University of Washington Press, Seattle, WA.

Waring, R.H. and J.F. Franklin. 1979. Evergreen coniferous forests of the Pacific Northwest. Science 204: 1380–1386.

What is old-growth forest?

Agee, J.K. 1993. *Fire Ecology of Pacific Northwest Forests*. Island Press, Washington, DC.

Bolsinger, C.L. and K.L. Wedell. 1993. Area of old-growth forests in California, Oregon, and Washington. Resource Bulletin PNW-RB 197. USDA Forest Service, Portland, OR.

Booth, D.E. 1994. *Valuing Nature: The Decline and Preservation of Old-Growth Forests*. Rowman & Littlefield, Lanham, MA.

Cline, S.P., A.B. Berg, and H.M. Wight. 1980. Snag characteristics and dynamics in Douglas-fir forests, western Oregon. Journal of Wildlife Management 44: 773–786.

Franklin, J.F., K. Cromack, Jr, W. Denison, et al. 1981. *Ecological Characteristics of Old-Growth Douglas-Fir Forests*. General Technical Report PNW-118, USDA Forest Service, Portland, OR.

Mannan, R.W. and E.C. Meslow. 1984. Bird populations and vegetation characteristics in managed and old-growth forests, northeastern Oregon. Journal of Wildlife Management 48: 1219–1238

Mannan, R.W., E.C. Meslow, and H.M. Wight. 1980. Use of snags by birds in Douglas-fir forests, western Oregon. Journal of Wildlife Management 44: 787–797.

Meslow, E.C., C. Maser, and J. Verner. 1981. Old-growth forests as wildlife habitat. Transactions of the North American Wildlife and Natural Resources Conference 46: 329–335.

Spies, T.A. and J.F. Franklin. 1988. Old growth and forest dynamics in the Douglas-fir region of western Oregon and Washington. Natural Areas Journal 8: 190–201.

Features and adaptations

Valleys of rain forest

Kirk, R. and J. Franklin. 1992. *The Olympic Rain Forest, An Ecological Web*. University of Washington Press, Seattle, WA.

Matthews, D. 1992. *Cascade-Olympic Natural History: A Trailside Reference*. Raven Editions, Portland, OR.

Whitney, S.R. and R. Sandelin. 2003. *Field Guide to the Cascades and Olympics*, second edition. Mountaineers Books, Seattle, WA. (Source for four valleys of rain forest.)

Epiphytes, canopy roots, and "scuzz"

Bernstein, M.E. and G.C. Carroll. 1977. Internal fungi in old-growth Douglas-fir foliage. Canadian Journal of Botany 55: 644–653.

Carroll, G.C. 1995. Forest endophytes: pattern and process. Canadian Journal of Botany 73 (Supplement 1): S1316–S1324.

Nadkarni, N.M. 1981. Canopy roots: convergent evolution in rainforest nutrient cycles. Science 214: 1023–1024.

Nadkarni, N.M. 1984. Biomass and mineral capital of epiphytes in an *Acer macrophyllum* community of a temperate moist coniferous forest, Olympic Peninsula, Washington state. Canadian Journal of Botany 62: 2223–2228.

Pike, L.H. 1978. The importance of epiphytic lichens in mineral cycling. Bryologist 81 247–257.

Pike, L.H., W.C. Denison, D.M. Tracy, et al. 1975. Floristic survey of epiphytic lichens and bryophytes growing on old-growth conifers in western Oregon. Bryologist 78: 389–402.

Pike, L.H., R.A. Rydell, and W.C. Denison. 1977. A 400-year-old Douglas-fir tree and its epiphytes: biomass, surface area, and their distributions. Canadian Journal of Forest Research 7: 680–699.

Vit, D. and M. Ostafichuk. 1973. Folicolous bryophytes and lichens of *Thuja plicata* in western British Columbia. Canadian Journal of Botany 51: 561–580.

More about logs

Cederholm, C.J. and N.F. Peterson. 1985. The retention of Coho salmon (*Oncorhynchus kisutch*) carcasses by organic debris in small streams. Canadian Journal of Fisheries and Aquatic Science 42: 1222–1225.

Grier, C.C. and R.S. Logan. 1977. Old-growth *Pseudotsuga menziesii* communities of a western Oregon watershed: biomass distribution and production budgets. Ecological Monographs 47: 373–400.

Maser, C. and J.R. Sedel. 1994. *From the Forest to the Sea, the Ecology of Wood in Streams, Rivers, Estuaries, and Oceans*. St Lucie Press, Delray Beach, FL.

Maser, C., R. Anderson, K. Cromack, Jr, et al. 1979. Dead and down woody material. In: *Wildlife Habitats in Managed Forests: The Blue Mountains of Oregon and Washington* (J.W. Thomas, ed.). USDA Forest Service, Agricultural Handbook 553, Portland, OR, pp. 78–95.

Strahler, A.N. 1952. Dynamic basis of geomorphology. Geological Society of America Bulletin 63: 923–938. (Seminal reference for classifying streams as first-order, second-order, etc.)

Succession on glacial till

Bormann, B.T. and R.C. Sidle. 1990. Changes in productivity and distribution of nutrients in a chronosequence at Glacier Bay National Park, Alaska. Journal of Ecology 78: 561–578.

Chapin, F.S, III, L.R. Walker, C.L. Fastie, and L.C. Sherman. 1994. Mechanisms of primary succession following deglaciation at Glacier Bay, Alaska. Ecological Monographs 64: 149–175.

Crocker, R.L. and J. Major. 1955. Soil development in relation to vegetation and surface age at Glacier Bay, Alaska. Journal of Ecology 43: 427–448.

Highlights

Bears, salmon, and forest enrichment

Bilby, R.E., B.R. Frensen, and P.A. Bisson. 1996. Incorporation of nitrogen and carbon from spawning coho salmon into the trophic system of small streams: evidence from stable isotopes. Canadian Journal of Fisheries and Aquatic Sciences 53: 164–173.

Bilby, R E., B.R. Frensen, P.A. Bisson, and J.K Walter. 1998. Response of juvenile Coho salmon (*Oncorhynchus kisutch*) and steelhead (*O. mykiss*) to the addition of salmon carcasses in two streams in southwestern Washington, USA. Canadian Journal of Fisheries and Aquatic Sciences 55: 1908–1918.

Cederholm, C.J., D.B. Houston, D.L. Cole, and W.J. Scarlett. 1989. Fate of Coho salmon (*Oncorhynchus kisutch*) carcasses in spawning streams. Canadian Journal of Fisheries and Aquatic Sciences 46: 1347–1355.

Christie, K.S. and T.E. Reimchen. 2005. Post-reproductive Pacific salmon, *Oncorhynchus* spp., as a major nutrient source for large aggregations of gulls, *Larus* spp. Canadian Field-Naturalist 119: 202–207.

Christie, K.S. and T.E. Reimchen. 2008. Presence of salmon increases passerine density on Pacific Northwest streams. Auk 125: 51–59.

Darimont, C.T., T.E. Reimchen, and P.C. Paquet. 2003. Foraging behavior by gray wolves on salmon streams in coastal British Columbia. Canadian Journal of Zoology 81: 349–353.

Egbert, A.L. and A.W. Stokes. 1976. The social behavior of brown bears on an Alaskan salmon stream. In: *Bears: Their Ecology and Management* (M.R. Pelton, J.W. Lentfer, and G.E. Folks, eds). IUCN New Series 40, Morges, Switzerland, pp. 41–56.

Gard, R. 1971. Brown bear predation on sockeye salmon at Karluk Lake, Alaska. Journal of Wildlife Management 35: 193–204.

Gende, S.M., R.T. Edwards, M.F. Willson, and M.S. Wipfli. 2002. Pacific salmon in aquatic and terrestrial ecosystems. BioScience 52: 917–928.

Gende, S.M., A.E. Miller, and E. Hood. 2007. The effects of salmon carcasses on soil nitrogen pools in a riparian forest in southeastern Alaska. Canadian Journal of Forest Research 37: 1194–1202.

Gende, S.M., T.P. Quinn, M.F. Willson, et al. 2004. Magnitude and fate of salmon-derived nutrients and energy in a coastal stream ecosystem. Journal of Freshwater Ecology 19: 149–160.

Gresh, T., J. Lichatowich, and P. Schoonmaker. 2000. An estimation of historic and current levels of salmon production in the northeast Pacific ecosystem: Evidence of a nutrient deficit in the freshwater systems of the Pacific Northwest. Fisheries 25(1): 15–21.

Groot, C., T.P. Quinn, and T.J. Hara. 1986. Responses of migrating adult sockeye salmon (*Oncorhynchus nerka*) to population-specific odors. Canadian Journal of Zoology 64: 926–932.

Hasler, A.D. 1966. *Underwater Guide-Posts, Homing of Salmon*. University of Wisconsin Press, Madison, WI.

Hasler, A.D. and A.T. Scholz. 1983. Olfactory imprinting and homing in salmon: investigations into the mechanism of the imprinting process. Zoophysiologie 14, Springer-Verlag, New York, NY.

Helfield, J.M. and R J. Naiman. 2001. Effects of salmon-derived nitrogen on riparian forest growth and implications for stream productivity. Ecology 82: 2403–2409.

Hilderbrand, G.V., T.A. Hanley, C.T. Robbins, and C.C. Schwartz. 1999. Role of brown bears (*Ursus arctos*) in the flow of marine nitrogen in a terrestrial ecosystem. Oecologia 121: 546–550.

Hocking, M.D. and T.E. Reimchen. 2006. Consumption and distribution of salmon (*Oncorhynchus* spp.) nutrients and energy by terrestrial flies. Canadian Journal of Fisheries and Aquatic Sciences 63: 2076–2086.

Hulbert, L.B., A.M. Aires da Silva, V.F. Callucci, and J.S. Rice. 2005. Seasonal foraging movements and migratory patterns of female *Lamna ditropis* tagged in Prince William Sound, Alaska. Journal of Fish Biology 67: 490–509.

Larkin, G.A. and P.A. Slaney. 1997. Implications of trends in marine-derived nutrients flow to south coastal British Columbia salmonid production. Fisheries 22(11): 16–24.

Luque, M.H. and A.W. Stokes. 1976. Fishing behavior of Alaskan brown bears. In: *Bears: Their Biology and Management* (M.R. Pelton, J.W. Lentfer, and G.E. Folk, eds). IUCN New Series 40, Morges, Switzerland, pp. 71–78.

Nagasawa, K. 1998. Predation by salmon sharks (*Lamna ditropis*) on Pacific salmon (*Oncorhynchus* spp.) in the North Pacific Ocean. Bulletin of the North Pacific Anadromous Fish Commission 1: 419–432.

Reimchen, T.E. 2000. Some ecological and evolutionary aspects of bear–salmon interactions in coastal British Columbia. Canadian Journal of Zoology 78: 448–457.

Quinn, T.P., S.M. Carlson, S.M. Gende, and H. Rich, Jr. 2009. Transportation of Pacific salmon carcasses from streams to riparian forests by bears. Canadian Journal of Zoology 87: 195–203.

Stokes, D. 2014. *The Fish in the Forest: Salmon and the Web of Life*. University of California Press, Berkeley, CA.

Stonorov, D. 1972. Protocol at the annual brown bear fish feast. Natural History 81: 66–73, 90–94.

Stonorov, D. and A.W. Stokes. 1972. Social behavior of the Alaskan brown bear. In: *Bears: Their Ecology and Management* (S. Herrero, ed.). IUCN New Series 23, Morges, Switzerland, pp. 232–242.

Stouder, D.J., P.A. Bisson, and R.J. Naiman. 1997. *Pacific Salmon and their Ecosystems*. Chapman & Hall, New York, NY.

Sugai, S.F. and D.C. Burrell. 1984. Transport of dissolved organic carbon, nutrients and trace metals from the Wilson and Blossom rivers to Smeaton Bay, southeast Alaska. Canadian Journal of Fisheries and Aquatic Sciences 41: 180–190.

Willson, M.F., S.M. Gende, and B.H. Marston. 1998. Fishes and the forest, expanding perspectives on fish–wildlife interactions. BioScience 48: 455–462.

A seabird in the forest

Binford, L.C., B.G. Elliott, and S.W. Singer. 1975. Discovery of a nest and the downy young of the marbled murrelet. Wilson Bulletin 87: 303–319. (This discovery ended a quest that captured the imagination of North American ornithologists for 185 years.)

Carter, H.R. and M.L. Morrison. (eds) 1992. Status and conservation of the marbled murrelet in North America. Proceedings of the Western Foundation for Vertebrate Zoology 5(1): 1–133.

Marshall, D.B. 1988. The marbled murrelet joins the old-growth forest conflict. American Birds 42: 202–212.

Marshall, D.B. 1988. Status of the marbled murrelet in North America, with special emphasis on populations in California, Oregon, and Washington. Biological Report 88(30), US Fish and Wildlife Service, Washington, DC.

Ralph, C.J., G.L. Hunt, Jr, M.G. Raphael, and J.F. Platt. (eds) 1995. Ecology and conservation of the marbled murrelet. General Technical Report PSW-GTR-152. Pacific Southwest Research Station, USDA Forest Service, Albany, CA.

Banana slugs

Richter, K.O. 1979. Aspects of nutrient cycling by *Ariolimax columbianus* (Mollusca: Arionidae) in Pacific Northwest coniferous forests. Pedobiologia 19: 60–74.

Some small mammals and their ecology

Corn, P.S. and R.B. Bury. 1986. Habitat use and terrestrial activity by red tree voles (*Arborimus longicaudus*) in Oregon. Journal of Mammalogy 67: 404–406.

Fogel, R.D. and J.M. Trappe. 1978. Fungus consumption (mycophagy) by small mammals. Northwest Science 52: 1–31.

Gillesburg, A.-M. and A.B. Carey. 1991. Aboreal nests of *Phenacomys longicaudus* in Oregon. Journal of Mammalogy 72: 784–787.

Graham, S.A. and G.W. Mires. 2005. Predation on red tree voles by owls and diurnal raptors. Northwestern Naturalist 86: 38–40.

Hamilton, W.J., III. 1962. Reproductive adaptations of the red tree mouse. Journal of Mammalogy 43: 486–504.

Maser, C., J.M. Trappe, and R.A. Nussbaum. 1978. Fungal–small mammal interrelationships with emphasis on Oregon coniferous forests. Ecology 59: 799–809.

Maser, C., J.M. Trappe, and D.C. Ure. 1978. Implications of small mammal mycophagy to the management of western coniferous forests. Transactions of the North American Wildlife and Natural Resources Conference 43: 78–88.

Trappe, J.M. and C. Maser. 1976. Germination of spores of *Glomus macrocarpus* (Endogonaceae) after passage through a rodent digestive tract. Mycologia 68: 433–436.

Trappe, J.M. and D.L. Luoma. 1992. The ties that bind: fungi in ecosystems. In: *The Fungal Community, Its Organization and Role in the Ecosystem* (G.C. Carroll and D.T. Wicklow, eds). Vol. 9, second edition. Marcel Decker, Inc. New York, NY, pp. 17–27.

Pacific yew

Busing, R.T., C.B. Halpern, and T.A. Spies. 1995. Ecology of Pacific yew (*Taxus brevifolia*) in western Oregon and Washington. Conservation Biology 9: 1199–1207.

Hartzell, H., Jr. 1991. *The Yew Tree: A Thousand Whispers*. Biography of a Species. Hulogosi, Eugene, OR.

Holton, R.A., C. Somoza, H.B. Kim, et al. 1994. First total synthesis of taxol. 1. Functionalization of the B ring. Journal of the American Chemical Society 116: 1597–1598.

Holton, R.A., H.B. Kim, C. Somoza, et al. 1994. First total synthesis of taxol. 2. Completion of the C and D rings. Journal of the American Chemical Society 116: 1599–1600.

Stierle, A., G. Strobel, and D. Stierle. 1993. Taxol and taxane production by *Taxomyces andreanae*, an endophytic fungus of Pacific yew. Science 260 (5105): 214–216.

Whelan, J. 2002. Targeted taxane therapy for cancer. Drug Discovery Today 7: 90–92.

Giant salamanders and other amphibians

Carey, A.B. 1989. Wildlife associated with old-growth forests in the Pacific Northwest. Natural Areas Journal 9: 151–162.

Corn, P.S. and R.B. Bury. 1989. Logging in western Oregon: responses of headwater habitats and stream amphibians. Forest Ecology and Management 29: 39–58.

Jackson, C.R., C.A. Sturm, and J.M. Ward. 2001. Timber harvest impacts on small headwater channels in the Coast Ranges of Washington. Journal of the American Water Resources Association 37: 1553–1549.

Jackson, C.R., D.P. Batzer, S.S. Cross, et al. 2006. Headwater streams and timber harvest: channel, macroinvertebrate, and amphibian response and recovery. Forest Science 53: 356–370.

Parker, M.S. 1991. Relationship between cover availability and larval Pacific giant salamander density. Journal of Herpetology 25: 355–357.

Parker, M.S. 1994. Feeding ecology of stream-dwelling Pacific giant salamander larvae (*Dicamptodon tenebrosus*). Copeia 1994: 705–718.

Stoddard, M.A. and J.F. Hayes. 2005. The influence of forest management on headwater stream amphibians at multiple scales. Ecological Applications 15: 811–823.

Mount St Helens

Andersen, D.C. and J.A. MacMahon. 1985. Plant succession following Mount St Helens volcanic eruption: facilitation by a burrowing rodent, *Thomomys talpoides*. American Midland Naturalist 114: 62–69.

Antos, J.A. and D.B. Zobel. 1982. Snowpack modification of volcanic tephra effects on forest understory near Mount St. Helens. Ecology 63: 1969–1972.

Bilderback, D.E. (ed.) 1987. *Mount St. Helens, 1980: Botanical Consequences of the Explosive Eruptions*. University of California Press, Berkeley, CA.

Carson, R. 1990. *Mount St. Helens: The Eruption and Recovery of a Volcano*. Sasquatch Books, Seattle, WA.

Dale, V.H., F.J. Swanson, and C.M. Crisafulli. (eds) 2005. *Ecological Responses to the 1980 Eruption of Mount St. Helens*. Springer Science, New York, NY.

Del Moral, R. 1983. Initial recovery of subalpine vegetation on Mount St. Helens, Washington. American Midland Naturalist 109: 72–80.

Del Moral, R. 2007. Vegetation dynamics in space and time: an example from Mount St. Helens. Journal of Vegetation Science 18: 479–488.

Del Moral, R. and I.L. Lacher. 2004. Vegetation patterns 25 years after the eruption of Mounts St. Helens, Washington. American Journal of Botany 92: 1948–1956.

Del Moral, R., J.M. Saura, and J.M. Emenegger. 2010. Primary succession trajectories on a barren plain, Mount St. Helens, Washington. Journal of Vegetation Science 21: 857–867.

Edwards, J.S. 1986. Arthropods as pioneers – recolonization of the blast zone on Mt. St. Helens. Northwest Environment Journal 2: 63–73.

Edwards, J.S. and L.M. Schwartz. 1980. Mount St. Helens ash: a natural insecticide. Canadian Journal of Zoology 59: 714–715.

Franklin, J.F., J.A. MacMahon, F.J. Swanson, and J.R. Sedell. 1985. Ecosystem responses to the eruption of Mount St. Helens. National Geographic Research 1: 198–216.

Frenzen, P.M., and C.M. Crisafulli. 1990. Mount St. Helens ten years later: past lessons and future promise. Northwest Science 64: 263–267.

Halpern, C.B., P.M. Frenzen, J.E. Means, and J.F. Franklin. 1990. Plant succession in areas of scorched and blown-down forest after the 1980 eruption of Mount St. Helens, Washington. Journal of Vegetation Science 1: 181–194.

Larson, D. 1993. The recovery of Spirit Lake. American Scientist 81: 166–177.

Lawrence, D.B. 1938. Trees on the march: notes on the recent volcanic and vegetational history of Mount St. Helens. Mazama 20(12): 49–54.

MacMahon, J.A. 1982. Mount St. Helens revisited. Natural History 91(5): 14–22.

MacMahon, J.A., R.R. Parmenter, K.A. Johnson, and C.M. Crisafulli. 1989. Small mammal recolonization of the Mount St. Helens volcano: 1980 – 1987. American Midland Naturalist 122: 365–387.

Mullineaux, D.R. and D.R. Crandell. 1981. The eruptive history of Mount St. Helens. In: *The 1980 Eruptions of Mount St. Helens, Washington* (P.W. Lipman and D.R. Mullineaux, eds). Professional Paper 1250, US Geological Survey, Washington, DC, pp. 3–15.

Wissmar, R.C., A.H. Devol, J.T. Staley, and J.F. Sedell. 1982. Biological responses of lakes in the Mount St. Helens blast zone. Science 216: 178–181.

Zobel, D.B. and J.A. Antos. 1997. A decade of recovery of understory vegetation buried by volcanic tephra from Mount St. Helens. Ecological Monographs 67: 317–344.

Ecological controversy

Barrows, C.W. 1981. Roost selection by spotted owls: an adaptation to heat stress. Condor 83: 302–309.

Carey, A.B., J.A. Reid, and S.P. Horton. 1990. Spotted owl home range and habitat use in southern Oregon Coast Ranges. Journal of Wildlife Management 54: 11–17.

Carey, A.B., S.P. Horton, and B.L. Biswell. 1992. Northern spotted owl: influence of prey base and landscape character. Ecological Monographs 62: 223–250.

Dark, S.J., R.J. Gutiérrez, and G.I. Gould, Jr. 1998. The barred owl (*Strix varia*) invasion in California. Auk 115: 50–56.

Dietrich, W. 2010. *The Final Forest: Big Trees, Forks, and the Pacific Northwest*. University of Washington Press, Seattle, WA.

Dixon, K.R. and T.C. Juelson. 1987. The political economy of the spotted owl. Ecology 68: 772–776.

Doak, D. 1989. Spotted owls and old-growth logging in the Pacific Northwest. Conservation Biology 3: 389–396.

Forsman, E.D., E.C. Meslow, and M.J. Strub. 1977. Spotted owl abundance in young versus old forests. Wildlife Society Bulletin 5: 43–47.

Forsman, E.D., E.C. Meslow, and H.M. Wight. 1984. Distribution and ecology of the spotted owl in Oregon. Wildlife Monographs 87: 1–64.

Forsman, E.D., T.J. Kaminski, J.C. Lewis, et al. 2005. Home range and habitat use of northern spotted owls on the Olympic Peninsula, Washington. Journal of Raptor Research 39: 365–377.

Forsman, E.D., K.M. Dugger, E.M. Glenn, et al. 2011. Population demography of northern spotted owls. Studies in Avian Biology 40: 1–120.

Pearson, R.R. and K.B. Livezey. 2003. Distribution, numbers, and site characteristics of spotted owls and barred owls in the Cascade Mountains in Washington. Journal of Raptor Research 37: 265–275.

Salwasser, H. 1987. Spotted owls; turning a battleground into a blueprint. Ecology 68: 776–779.

Simberloff, D. 1987. The spotted owl fracas: mixing academic, applied, and political ecology. Ecology 68: 766–772.

Singleton, P.H., J.F. Lehmkuhl, W.L. Gaines, and S.A. Graham. 2010. Barred owl space use and habitat selection in the eastern Cascades, Washington. Journal of Wildlife Management 74: 285–294.

Thomas, J.W., E.D. Forsman, J.B. Lint, et al. 1990. *A Conservation Strategy for the Northern Spotted Owl*. USDA Forest Service, Portland, OR.

US Fish and Wildlife Service. 2011. Revised recovery plan for the northern spotted owl (*Strix occidentalis caurina*). US Fish and Wildlife Service, Portland, OR.

US Fish and Wildlife Service. 2012. Draft environmental impact statement for experimental removal of barred owls to benefit threatened northern spotted owls. Federal Register 77: 14036–14039.

Wiens, J.D., R.G. Anthony, and E.D. Forsman. 2011. Barred owl occupancy surveys within the range of the northern spotted owl. Journal of Wildlife Management 75: 531–538.

Infobox 10.1. The elk that saved a forest

Brinkley, D. 2010. *The Wilderness Warrior: Theodore Roosevelt and the Crusade for America*. Harper Perennial, New York, NY. (See pp. 304–307 for the naming of Roosevelt elk and pp. 810–811 for the political history of Olympic National Park.)

Cutright, P.R. 1956. *Theodore Roosevelt, The Naturalist*. Harper & Bros., New York, NY.

Greer, K. 1968. Elk teeth as ornaments. Montana Wildlife (Feb.): 14–17.

Harper, J.A., J.H. Harn, W.W. Bentley, and C.F. Yocom. 1967. The status and ecology of the Roosevelt elk in California. Wildlife Monographs 16: 1–49.

McCabe, R.E. 1982. Elk and Indians: historical values and perspectives. In: *Elk of North America, Ecology and Management* (J.W. Thomas and D.E. Toweill, eds). Stackpole Books, Harrisburg, PA, pp. 61–123.

Paul, T.W. 2009. Game transplants in Alaska. Technical Bulletin 4, second edition. Alaska Department of Fish and Game, Juneau, AK. (See pp. 33–36 for elk.)

Schwartz, J.E., II and G.E. Mitchell. 1945. The Roosevelt elk on the Olympic Peninsula, Washington. Journal of Wildlife Management 9: 295–319.

Troyer, W.A. 1960. The Roosevelt elk on Afognak Island, Alaska. Journal of Wildlife Management 24: 15–21.

Infobox 10.2. Spirit bears, the other white bear

Klinka, D. R. and T. E. Reimchen. 2009. Adaptive coat colour polymorphism in the Kermode bear of coastal British Columbia. Biological Journal of the Linnean Society 98: 479–488.

Marshall, H. D. and K. Ritland. 2002. Genetic diversity and differentiation of Kermode bear populations. Molecular Ecology 11: 685–697.

Ritland, K., C. Newton, and H. D. Marshall. 2001. Inheritance and population structure of the white-phased "Kermode" black bear. Current Biology 11: 1468–1472.

Sachs, J. S. 2010. Icon for an endangered ecosystem. National Wildlife 48(2): 30–35.

CHAPTER 11

Coastal Environments

Before the land rose out of the ocean, and became dry land, chaos reigned; and between the high and low water mark, where she is partially disrobed, a sort of chaos reigns still, which only anomalous creatures can inhabit.

Henry David Thoreau

Coastal environments, with their noisy waterbirds, salt air, and relentless surf, have for eons captivated humans. As anthropologist, philosopher and science writer Loren Eiseley (1907–1977) observed, seashores release an "ancient urge" within us to shed our shoes and scavenge among seaweed and driftwood like "homesick refugees of a long war." Extensive coastlines border North America, each with unique ecological environments, but all offering opportunities for study, reflection, and shoeless enjoyment.

The coast is the frontier between the land and the sea. Most of North America's coastline borders either the Atlantic or Pacific oceans or, in places, the Arctic Ocean and Bering Sea. The Gulf of Mexico, Hudson Bay, and the Gulf of California (Sea of Cortez) are next in scale. Etched into this extensive coastline are somewhat smaller areas such as Chesapeake Bay, Puget Sound, and shallow coastal lagoons. Coastlines often embrace a wealth of islands, including the Florida Keys and the far-flung Aleutian Archipelago, along with Ellesmere and others the size of small countries and many more just mere specks of land. At an even finer scale oyster reefs, seagrass meadows, shifting dunes, tidal pools, and kelp forests gain our attention. The shoreline itself varies from verdant marshes or rocky shores to sandy beaches along a gradient, south to north, ranging from the humid subtropics to the frigid Arctic. In short, these landforms – and their biota – represent a lengthy and varied menu, but here a limited number of selections must suffice to taste the ecological richness of the North American coastline.

Currents and climates

"The coldest winter I ever spent was a summer in San Francisco". This insight of unknown origin (but often associated with the wit of Mark Twain or W. C. Fields) nonetheless rings true for thinly clad travelers visiting San Francisco in July or August. The winds, humidity, and especially the temperature in any coastal zone fall under the heavy influence of the oceanic currents flowing offshore. San Francisco is no exception.

The waters of the world's major oceans remain in constant motion. Waves present the most obvious expression of these movements, but both surface and subsurface currents also flow for great distances in somewhat predictable directions. These currents, generated primarily by a combination of wind and the **Coriolis effect**, form as a product of the Earth's rotation. Responding to the Earth's spin, dominant ocean currents flow westward at the Equator where they create a huge vortex or **gyre**. This gyre moves northward in a clockwise motion along the eastern coastlines of northern continents, then eastward nearer the pole, and finally southward along the western coastlines of northern continents. A similar, but counterclockwise gyre circulates in the oceans of the Southern Hemisphere. In either case, the waters warm as these currents traverse the equatorial zone.

The Gulf Steam – the northward flow of the North Atlantic Gyre – warms the eastern shores of North America, producing a moderate climate. Conversely, after flowing northward along the eastern shores of Asia, the currents forming the North Pacific Gyre move westward near the Arctic Circle where the waters cool, then turn southward along North America. These cold waters form the California Current, which hugs its namesake coastline, but flows farther offshore along the more northern parts of the western coast (e.g., British Columbia). Cold, moist air above the California Current blows inland during the day creating the climate-supporting temperate rain forests (Chapter 10) on west-facing coastal mountains as well as the conditions for which San Francisco is so well known.

Ecology of North America, Second Edition. Brian R. Chapman and Eric G. Bolen.
© 2015 John Wiley & Sons, Ltd. Published 2015 by John Wiley & Sons, Ltd.

Features and adaptations

Rocky seashores and tidal pools

In many coastal areas, the continent exposes a hard shoulder to the sea. The analogy is appropriate; rocky seashores are more extensive along the northernmost third of the Atlantic and Pacific coasts. Anyone familiar with the coastlines of Monterrey in California or Bar Harbor in Maine, for example, retains a clear image of foaming waves crashing onto rocks (Fig. 11.1R). Thousands of waves daily batter rocky seashores, but these sites often provide a biologically rich environment for a biota highly adapted to severe challenges. Here, organisms contend with pounding waves and the swirl of retreating waters but also deal with complete inundation, soon followed by periodic exposure to the dehydrating effects of the sun and air, changes in salinity and temperature, and occasional pollutants.

Many animals of rocky seashores moderate their exposure to harsh conditions by moving with the tides or finding cover in crevices between rocks. Some, such as small fishes, sea lice, and many sea snails, are cryptically colored and, whether underwater or exposed, their camouflages reduce predations. The bodies of small cryptically colored fishes known as blennies lack scales; instead, they have a layer of thick mucus allowing them to hide and slide into small holes or other tiny refugia. Conversely, many crabs and some sea slugs advertise themselves with bright colors and rely on other means of defense: well-developed pincers in one case, toxic tissues in the other.

Not all residents of rocky seashores move rapidly. Plants attached to rocks and fixed or sedentary animals (e.g., barnacles) persist in narrow and distinctive zones, each stacked vertically in relation to their exposure. The uppermost zone, the **supratidal zone** (also known as the splash zone), receives some moisture as spray from breaking waves, but it is more often exposed to the desiccating effects of wind, sun, and high temperatures. Common inhabitants include green algae, cyanobacteria, lichens, limpets, barnacles, isopods, and crabs (Fig. 11.1L). Between the highest and lowest limits of the normal tidal range lies the **intertidal zone** (or littoral zone) which, on closer inspection, separates into three bands depending on differences between exposure and inundation. The supratidal band, submerged only during high tides, is home for barnacles, sea stars, and sea anemones. Next, brown algae, barnacles and mobile animals flourish in the mid-littoral band, where water turbulence and exposure alternate with tidal changes. Finally, and almost always under water, lies the lower littoral band where characteristic species such as sea lettuces, sea urchins, sea cucumbers, tube worms, and shrimps face the intense forces of wave turbulence.

When the tide retreats, pools forming in the holes and other depressions in rocky shorelines temporarily isolate an aquarium-like collection of organisms representing the intertidal zone (Fig. 11.2). With eloquence, Rachael Carson (Infobox 3.2) observed, "Tide pools contain mysterious worlds within their depths, where all the beauty of the seas is subtly suggested and portrayed in miniature." Almost any small marine organism can be trapped in tidal pools, which for many stand alone as a feature of rocky seashores.

Few natural rock outcrops occur along the southern Atlantic and Pacific coasts of North America and, with the exception of some small **serpulid reefs** near Baffin Bay, Texas, coastal outcrops are otherwise absent along

Figure 11.1 Right: waves crash into the hard surfaces of rocky seashores, limiting the types of organisms that can inhabit such harsh environments. Left: organisms that inhabit rocky seashores and other hardened shoreline habitats arrange themselves into vertical zones based on their respective abilities to withstand wave energy, tidal changes, and moisture variation. Photos courtesy of Brian R. Chapman.

(a)   (b)

Figure 11.2 (a) Tide pools form in depressions on rocky seashores when the tide recedes, (b) trapping a diverse assemblage of organisms often representing a colorful cross-section of the biotic community. Photos courtesy of Jack Kushman Koch.

the Gulf of Mexico covered in this book. However, large rocks or chunks of concrete have been widely deployed to maintain navigable channels or to prevent beach erosion. In time, these artificial structures develop the same littoral zones, often involving colonies of the same species as those which occur on natural rocky shorelines.

Like all coastal environments, rocky seashores remain threatened by pollution and other human disturbances. Rivers may wash raw sewage downstream, including toxic chemicals and bacteria, often harming life on rocky seashores. Oil spills, such as the Exxon Valdez spill in Prince William Sound, Alaska, represent the most widely publicized of these events; toxic hydrocarbon compounds kill marine organisms and the black tar-like residues smother both life and habitat. Unfortunately, the richness of many locations may be reduced when humans indiscriminately collect marine organisms. Many a trip to the seashore ends with smelly shellfish or hermit crabs lying forgotten and decaying in the family car.

Sandy seashores

Waves slam against an unyielding fortress on rocky shorelines, but on sandy seashores waves end their long journey on gentle slopes that offer little resistance to the impact. Most waves originate from the frictional drag of wind on the ocean's surface. Sustained wind pushes surface water forward, generating a crest followed by a trough.

Ocean waves, or swells, sometimes travel great distances before reaching a shoreline. In shallow waters near a **high-energy beach**, the bottom of a wave encounters the seafloor which forces the crest upward and increases the wave's steepness and height. Moments later, the base of the wave drags and slows while the crest continues at its original speed; a breaker results when the crest crashes forward. This action lifts bottom sediments, which are re-deposited as a ridge that enlarges into an offshore sand bar. When waves crash on a well-developed bar, their energy continues ashore in smaller slower waves before dissipating into foamy ripples on the sandy beach. In contrast, **low-energy beaches** occur on bays and other sheltered areas where the waves are seldom large or powerful enough to build offshore sand bars.

Waves usually break at an angle to the shoreline, generating a **longshore current** that moves water and sediments parallel to the shoreline. Incoming waves transport sediments toward the beach where they are deposited when the wave dies; the accumulated sand forms a berm at the high-tide line. Slightly higher than the dry **backshore** beyond it, the berm protects the backshore and coastal dunes from waves except under the most extreme conditions. When tides recede, sand

Figure 11.3 Coppice dunes form in backshore habitats where immobile objects such as driftwood capture sand. Railroad vine, beach tea, and other pioneer plants establish forming low dunes that may eventually be stabilized by sea oats and other grasses with dense root systems. Photo courtesy of Brian R. Chapman.

atop the berm dries and blows to the backshore. Where wind flow is impeded by driftwood, plants, or debris on the backshore, the sand settles and collects as **coppice dunes** on the lee side of the obstructions, sites eventually colonized by hardy pioneer species (Fig. 11.3). Railroad vine, beach tea, and shoreline sea purslane stabilize coppice dunes, contribute organic matter, and capture more sand. American beach grass stabilizes dunes along the northern Atlantic coast but is replaced by sea oats from North Carolina to the Gulf of Mexico. Dune formation occurs along the entire length of a beach, eventually producing a foredune ridge facing the sea.

Dune formation on the Pacific Coast follows a similar pattern, but involves different species of plants. Deep-rooted mat-forming plants such as beach strawberry, silver beachweed, and yellow sand verbena represent examples of backshore pioneers. Red fescue is a desirable native species that solidifies larger dunes, but more than a century ago European beach grass was introduced to speed dune formation. The introduced grass spread, rapidly becoming an **invasive species** whose dense stands eliminated native plants. In some areas, the unwanted grass eliminated nesting habitat for western snowy plovers, an endangered species, but attempts to eradicate the invader have been largely unsuccessful.

The **surf zone**, where waves break, initially seems devoid of life. However, the soft sediments comprising the shoreward face of an offshore bar provide nutrient-rich habitats for burrowing organisms such as moon snails, lettered olives, and sand dollars. Lunar tides and ever-changing winds regularly alter the length and position of the **swash zone**, the intertidal zone where waves die. Because swash zones are exposed alternately to sun and currents produced by the constant landward rush – the swash – of breaking waves, followed by the backwash as the water ebbs, swash zones also appear to lack biological diversity. Hundreds of species occur here in fact, most of which are bacteria, fungi, diatoms, and **meiofauna** occupying the interstitial spaces between sand grains. Larger organisms inhabiting the intertidal zone must burrow quickly when exposed by wave currents. More than 90% of the **macrofauna** on ocean beaches are surf clams, mole crabs, and polychaetes, filter feeders that migrate vertically through the sand column in response to wave action. Tiny fish ride the swash to feed opportunistically on particulate matter and exposed meiofauna. Beachgoers are amused by the activities of sanderlings and other short-legged shorebirds that rush seaward following the backwash to prey on exposed macrofauna, then hustle shoreward to escape an oncoming wave. Shorebirds with longer legs and bills (e.g., willets, godwits, and dowitchers) also probe for food in the swash zone, but forage while wading in the swirling waves.

Waves carry debris to the upper limits of the intertidal zone where a **wrack** line forms from deposits of flotsam. Seashells, detached fragments of kelp, seagrasses, and other marine plants along with dead marine animals commonly occur in wrack lines. Nutrients leach out of decomposing wrack and filter through the sand, producing the base of the food chain for the intertidal zone and the nearshore marine ecosystem. Gulls, terns, shorebirds, rodents, and carnivores scavenge along the wrack line where isopods, insects, and other organic matter offer abundant sources of food. Ruddy turnstones flip shells and other debris to expose prey. At night, ghost crabs emerge from burrows located higher on the beach to wet their gills in the surf before foraging at the wrack line.

The foredunes and backshore are occupied by many rodents, including some rare or endangered forms. Subspecies of the oldfield mouse and Ord's kangaroo rat occupy sandy seashores; these have paler pelage than their inland counterparts, an adaptation to the color of beach sands that lessens predation. Lizards, snakes, marsh rabbits, eastern cottontails, raccoons, coyotes, and gray foxes make regular foraging trips to sandy seashores. Many shorebirds, including black skimmers and Wilson's plovers, nest on barren sands or among sparse clumps of vegetation. Gulls and terns also nest in backshore areas, but place their nests in denser vegetation. Six of the seven living species of sea turtles nest on the sandy beaches of North America, but otherwise spend their lives at sea; the faithful return of the nesting females to the same beaches where they hatched years earlier illustrates the phenomenon of **philopatry.**

The profile of sandy seashores remains in constant flux because beach sand moves continually up or down seashores. During the summer months, beaches usually increase in width. Rivers contribute the most sediment to beaches by transporting sands from inland areas to the sea, but shoreline erosion is another source of beach materials. Beach erosion caused by storm-generated waves occurs commonly in winter or during hurricanes (*Natural disturbances*, this chapter).

Chesapeake Bay

The geologic history of Chesapeake Bay, North America's largest tidal **estuary,** began some 35 million years ago when a **bolide** struck the Atlantic coast near present-day Cape Charles, Virginia. The resulting crater initiated a drainage pattern that eventually formed the lower valley of the Susquehanna River. Later still – about 10,000 years ago – waters from melting glaciers raised sea levels, slowly drowning the valley and creating a broad, shallow embayment. Currently extending 322 km (200 miles) with a surface area of 9920 km² (3840 square miles), Chesapeake Bay gained due notice from early colonists including Captain John Smith (1580–1631) who recorded that "Heaven and earth have never agreed to frame a better place for man's habitation." A century ago, hundreds of sailboats called skipjacks plied the bay towing oyster dredges but, as oyster stocks declined, so did the skipjacks. Today, only a few of these remarkable boats survive but nonetheless provide a unique icon for initiatives designed to restore the bay's weakened health.

More than 400 tributaries, draining a multistate watershed of almost 165,760 km² (64,000 square miles), contribute fresh water to Chesapeake Bay; the Susquehanna, James, and Potomac rivers provide about 80% of the total. Much of the estuary is therefore **oligohaline** (essentially fresh) with salinities of 0.5–10 parts per thousand (ppt). Nearer the ocean the waters become **polyhaline**, varying between 19 ppt and 35 ppt (seawater). Between these extremes lies a **mesohaline** zone where salinities fluctuate from 11 to 18 ppt. The bay is shallow; about a quarter of it is less than 2 m (6 feet) deep. In combination, the Chesapeake's water depth, salinity gradient, freshwater inflow, and constant tidal circulation create one of the world's most productive estuarine systems.

Submerged aquatic vegetation (SAV) varies with salinity and largely consists of seagrasses (*Seagrass meadows*, this chapter). Near the mouth of the Susquehanna, wild celery, sago, and other pondweeds represent plants typical of low-saline areas, whereas eelgrass, sea lettuce, and wigeon grass dominate the extensive areas of saltier water. Given adequate sunlight and water quality, the bay's expansive shallows can potentially support almost 243,000 ha (600,000 acres) of SAV, but aerial surveys currently indicate that no more than 10% of this potential is realized. SAV provides food and cover for a host of species, whose populations and welfare remain inexorably entwined with this vegetation. In particular, blue crabs, striped bass, and, collectively, waterfowl deserve mention in any discussion of Chesapeake Bay.

The distribution of blue crabs stretches along the Atlantic coastline from Nova Scotia to Argentina, but only in Chesapeake Bay do they share a role of keystone species and symbol of the region's economy and culture. Blue crab populations, like those of oysters and other resources, suffered as environmental ills steadily stressed the bay's ecosystem late in the 20th century. Nonetheless, blue crabs are resilient and still represent the most valuable fishery in Chesapeake Bay; in 2012, more juveniles entered the population (a process known as **recruitment**) than ever recorded previously.

In part, the resilience of blue crabs stems from their reproductive capabilities. When spawning, a single female can produce 750,000 to 3.2 million eggs protected in an orange-colored "sponge" attached to her abdomen. After hatching, the larvae leave the bay and move offshore, later returning to mature in the cover of sea-grasses. Blue crabs also tolerate **hypoxia** (conditions of low oxygen) for long periods, and thus cope with the degraded waters in many areas of Chesapeake Bay.

Striped bass ("rockfish") occur from the St Lawrence estuary in Canada to the St John's River in Florida, but Chesapeake Bay provides the primary spawning area for 70–90% of the stocks along the Atlantic coast. Adults commonly weigh 15 kg (33 pounds) with some growing much larger. Stripers are **anadromous** but, at times, some (often males) do not return to the ocean after spawning in freshwater streams and rivers. Once fertilized, their eggs drift downstream, hatch as larvae, then develop into post-larvae; both stages prey on microscopic animals. They mature into juveniles in nursery areas in sounds and estuaries where, after 2–4 years, they journey into the Atlantic Ocean.

Like many other estuarine species, striped bass fell on hard times when pollution steadily overwhelmed Chesapeake Bay. Overfishing also contributed significantly to the collapse of the striper fishery. After plunging to just under 9 million fish in 1982, the population averaged 58 million fish during the first decade of the current century. Stripers are no longer overfished but other problems remain, including the emergence of mycobacteriosis, a disease attacking the kidneys and other internal organs. Current evidence links the disease with poor nutritional health, again indicating the importance of restoring the entire ecosystem of Chesapeake Bay.

Chesapeake Bay has a long history as habitat – and hunting – for huge numbers of ducks and other waterfowl, particularly during the winter months. The once-extensive beds of seagrasses attracted, among others, overwintering canvasbacks, American wigeon, and tundra swans, and maintained a year-round population of black ducks. With a population estimated at 1 million ducks and geese, the birds seemed inexhaustible and weathered a level of market hunting unmatched in North America that lasted until outlawed in 1918. Market hunters on Chesapeake Bay (and elsewhere) operated huge shotguns known as "puntguns" from which a single shot easily killed dozens of ducks, usually as they dabbled for food. Concurrently, waterfowl hunting initiated the art form of carving decoys, of which some of the most treasured originated from Chesapeake Bay.

In the 1960s, SAV in the bay faltered as a result of increased water turbidity, excessive nutrient loads, and other types of impacts. Between 1971 and 1978, SAV declined by 65% in Maryland alone and, as the decline continued, so did the numbers of waterfowl wintering on Chesapeake Bay. Canvasbacks, in particular, suffered from the loss of wild celery and sago pondweed, whereas Canada geese and snow geese escalated dramatically, particularly on the Bay's Eastern Shore where they fed on waste corn and weeds on farmlands instead of SAV. Large numbers of tundra swans left the bay when SAV declined and migrated to coastal locations in North Carolina where they still overwinter.

Mute swans, native to parts of Eurasia, made attractive additions to parks and aviaries in North America, where they eventually escaped and established feral populations. Today, more than 14,000 mute swans occur in the Atlantic Flyway, primarily in Chesapeake Bay, where they consume large amounts of SAV and destroy even more when their feeding activities expose roots and other underground structures. Because few mute swans migrate, the damage continues all year round.

Chesapeake Bay faces numerous environmental ills including **dead zones**. These develop in freshwater or marine systems where oxygen has been so depleted that organisms can no longer survive. Dead zones result when excessive loads of nutrient-rich sediments flow into aquatic systems, particularly nitrogen and phosphorus from fertilizers applied on farmlands throughout the watershed. Industrial and municipal wastes add to the mix. The flush of nutrients stimulates dense blooms of algae that, with sediments, block sunlight from reaching SAV. Without sunlight, SAV dies and, eventually, so do the algae. In turn, decomposer bacteria thrive on the rotting mass of vegetation, deoxygenating the water and creating a hypoxic dead zone. In Chesapeake Bay, nutrients carried by the Susquehanna River greatly influence the extent of the dead zone.

Although still troubled, much is underway to restore the environmental health – especially water quality – of Chesapeake Bay. The efforts of federal, state, and municipal governments, often in partnerships with private organizations and industry, seek legal and ecological remedies. For example, a hatchery program restocked more than 6 million oysters in a sanctuary where their filtering capabilities helped to improve water quality. In other cases, upgraded water-treatment facilities in the estuary's watershed reduce the discharge of nitrogen, and new laws restrict the release of harmful materials. In all, the ongoing renewal of Chesapeake Bay coincides with the resurging interest in skipjacks, providing links in a common heritage.

Mother Lagoon

On 27 June 1926, J. J. Carroll bravely navigated the Laguna Madre from Flour Bluff, Texas, to "Bird Island," recording the rich birdlife along the way. Little would be changed if he repeated the trip today. Only navigational channels and islands of dredged material bear witness to human activities in the intervening years. The Laguna Madre remains largely unspoiled and highly productive, a rarity for a coastal bay in an industrial era.

The Laguna Madre, or "Mother Lagoon," includes two shallow lagoons bordering the Gulf of Mexico in southern Texas and northern Mexico. In Texas, the lagoon forms an elongated finger of water pointing southward 185 km (115 miles) from Corpus Christi Bay to the Rio Grande Delta. South of the delta, which is approximately 75 km (47 miles) wide, lies the Laguna Madre de Tamaulipas, an ecologically similar lagoon extending about 185 km (115 miles) to Río Soto la Marina in Tamaulipas, Mexico. Because of their proximity and similar physiographic and ecological conditions, the two sites are together regarded as the largest hypersaline lagoon system in the world.

Both lagoons are protected from the Gulf of Mexico by barrier beaches (*Barrier islands*, this chapter), with each divided into "upper" (i.e., northern) and "lower" (i.e., southern) sections. A land bridge approximately 18 km (11 miles) long separates Laguna Madre of Texas into an upper unit that is 75 km (47 miles) long with an average width of about 6 km (3.7 miles). The lower Laguna Madre is slightly longer and wider. The land bridge, variously known as the Salt Flats, Saltillo Flats, Laguna Madre Flats, or the Land Cut, developed from storm washovers and windblown sands. A navigational channel dredged in the 1940s crosses the land bridge, thereby connecting the upper and lower lagoons. In Mexico, a mudflat partially separates the Laguna Madre de Tamaulipas into northern and southern units.

The Laguna Madre lies in a semiarid region with low rainfall, long, hot summers and short, mild winters; notably, evaporation exceeds precipitation. The area also experiences periodic, unpredictable, and sometimes long droughts. Accordingly, the lagoon's western shoreline in Texas is sometimes called the Wild Horse Desert. No rivers empty into the Laguna Madre of Texas, with runoff from otherwise dry creeks after rare but heavy rains providing the only source of fresh water. Likewise, the Laguna Madre in Mexico receives limited flows of fresh water from creeks and arroyos, whereas the Río San Fernando feeds the lagoon only after fulfilling heavy demands for irrigation. After major storms, a relatively new overflow channel transports fresh water from the Río Grande to the Laguna de Tamaulipas.

The Laguna Madre is one of only six hypersaline lagoons in the world. Ocean waters average 35 ppt, but the upper and lower Laguna Madre of Texas each register normal salinities of 42 ppt. However, salinities occasionally reach 150 ppt after long droughts, with a record high of 295 ppt. Conversely, salinity levels approaching fresh water (>5 ppt) occur after wet-weather cycles, torrential rains, and tropical storms. Fish kills resulting from the extreme salinities in the Laguna Madre of Texas are now rare, thanks to better water circulation created by dredged channels.

Unique areas known as wind-tidal flats border the Laguna Madre in both Texas and Tamaulipas. These are broad, flat expanses of barren sand or mud elevated only slightly above the normal range of a lunar tide. As implied, winds and storms, rather than astronomical pressure, push water over these barren areas. Emergent marsh plants cannot survive in this harsh environment, although mats of blue-green algae (Cyanobacteria) intermix with the sediments at some areas. Still, the primary productivity of the algal mats approaches about 40% of the output typical of a cordgrass marsh. Thus enriched, when inundated, the wind-tidal flats become significant feeding grounds for fish, crabs, and wintering and migrating birds.

Shallow waters averaging about 1 m (3.3 feet) in depth cover much of the Laguna Madre and, coupled with their relative clarity, provide extensive sites for the development of seagrass meadows (*Seagrass meadows*, this chapter). Despite its hypersalinity, the Laguna Madre provides habitat and nursery areas to more species of finfish and shellfish than anywhere else on the Texas coast, as well as harboring large populations of resident and migratory waterbirds. About 80% of the North American population of redhead ducks overwinters in the Laguna Madre of Texas.

Except for the wind-tidal flats, only a few natural islands existed in the lower Laguna Madre of Texas at the time of Carroll's voyage. Today however, islands are numerous, the result of huge amounts of spoil removed from dredged channels (Fig. 11.4). Vegetation eventually reduced erosion, with various stages of plant succession reflecting the relative age of each island. The new land was soon occupied by fishing cabins, oil and gas wells, and often rookeries for colonies of nesting waterbirds (*Waterbird colonies*, this chapter). Farther south about 200 natural islands emerge from the Laguna Madre de Tamaulipas, some which are large enough to shelter small fishing villages. A handful of both natural and spoil islands in the Laguna Madre serve as outposts for nesting white pelicans, which otherwise nest far to the north and west of the Texas coast (e.g., northern Utah); the breeding range of these birds illustrates a **discontinuous distribution**.

The Laguna Madre at times experiences "blooms" of harmful phytoplankton, the toxins of which kill larval fish but do not harm adult fish or certain invertebrates. These blooms, called "Texas Brown Tides," sometimes persist for several years but the trigger mechanisms remain somewhat unclear. Once the algal populations increase, they release toxins that may harm their

Figure 11.4 The Gulf Intracoastal Waterway extends for the full length of the Laguna Madre; the dredged material ("spoil") was dumped to the side of the channel, creating a chain of islands. South Bird Island, a natural island visible to the left of the chain, and many of the spoil islands are important nesting sites for birds. Photo courtesy of Brian R. Chapman.

zooplankton and invertebrate predators and the cloudy water impairs photosynthetic activities to the point where seagrass communities fail. Documented blooms of the better-known "red tide" are rare in the Laguna Madre of Texas despite the presence of the causative agent, a dino-flagellate (Infobox 11.1); in Texas, the last fish kill attributed to red tide occurred in 1996. Harmful red tides have not been recorded in the Laguna Madre of Tamalulipas. However, red-tide blooms may develop more often than realized simply because large areas remain unmonitored.

Despite dredging and oil and gas development, the Laguna Madre remains relatively pristine. In Texas, the lagoon's coastline is protected by Padre Island National Seashore, Laguna Atascosa National Wildlife Refuge, and large, privately owned ranches; most of the region surrounding the lagoon in Mexico remains sparsely populated. As a consequence, the entire Laguna Madre ecosystem has suffered less environmental degradation than many coastal embayments elsewhere. No doubt J. J. Carroll would be impressed.

Submergent communities

Seagrass meadows

Eleven species of seagrasses occur naturally in North America. The distribution of these can be assigned by region, beginning with the seven species found on the Pacific coast from Alaska to Baja California. Some of the seagrasses in this flora grow along the entire length of this extensive coastline thanks to the uniformity provided by the cold water of the California Current. Chief among these is eelgrass, which covers large areas in both Alaska (Izembeck Lagoon) and Baja California (*Laugna Ojo de Liebre*) in Mexico, two of the largest eelgrass ecosystems in the world. Wigeon grass is also widely distributed along the eastern Pacific coastline, and shoal grass colonizes disturbed locations.

Eelgrass and wigeon grass also occur on the northern and mid-Atlantic coastline. Because eelgrass thrives in cold water the species reaches its southern limit in the warm waters off North Carolina, where it is replaced to the south by shoal grass. This zone of separation represents a relatively sharp ecotone – at least for marine vegetation – based on each species' tolerance to water temperatures. Because of its intolerance to warm water, the southern limits of eelgrass differ in latitude on the eastern and western coastlines of North America.

Vegetation collectively known as seagrass consists of flowering, monocot plants that complete their entire life cycle in seawater. Most taxonomists currently recognize five families in the group, which excludes the family of true grasses (described in Infobox 5.1). Seagrasses should not be regarded as "seaweeds," which are usually marine algae. Most species have ribbon-like leaves that grow in dense stands known as meadows.

Ecologically, seagrasses provide several important functions as follows.

1 The meadows generally lack significance as spawning areas, but serve as important nursery areas for many species. In many cases, young fish establish a daily routine of feeding in adjacent habitat (e.g., reefs), but return to the meadows for refuge. Eventually, the fish move permanently to other locations after reaching a critical length that reduces the risk of predation.

2 Seagrass leaves are often swathed with **periphyton**, a felt-like coating of diatoms, algae, bacteria, and other microorganisms, which provide the foundation for many marine food chains.

3 Seagrass foliage generates large amounts of biomass that, after decompositon, enrich both the meadows and adjacent communities. This source of organic matter may be transported over long distances and reach benthic communities at considerable depths. In one case, fragments of turtle grass traveled at least 540 km (335 miles).

4 The vast network of roots and rhizomes in seagrass meadows stabilizes bottom sediments. The leaves also slow water movements, thus trapping and accumulating additional sediments and detritus.

Relatively few species graze directly on seagrasses; of those that do, many more occur in subtropical and tropical regions than in temperate zones. The queen conch is one of the more prominent invertebrates in seagrass meadows; green sea urchins often eat turtle grass. Some fishes feed at least in part on seagrasses, but the diet of the buck-tooth parrotfish is almost exclusively turtle grass. Of the world's seven species of sea turtle, only the green turtle is herbivorous. A layer of green fat lying under the carapace (upper shell) accounts for their name; the hue apparently results from their herbivorous diet. In North America, the species nests primarily on the beaches of Mexico, including the shores of Baja California, and on the Atlantic coast of Florida north to North Carolina. Redheads, unlike other species of pochard ducks, feed heavily on vegetation. This relationship is particularly evident in the Laguna Madre in Texas where redheads feed almost entirely on a single species: shoal grass.

Manatees represent the classic (and only) example of a seagrass-eating mammal in North America. The Florida manatee lives all year round along the coastline of its namesake state and seasonally moves as far north as Chesapeake Bay. Manatees also venture into freshwater environments, but in saltwater turtle grass is among the seagrasses in their diet. Manatees, which weigh up to 500 kg (1102 pounds), may eat 10–20% of their body mass per day in aquatic vegetation. They eat both the leaves and rhizomes of seagrasses, sometimes leaving a trail known as a feeding scar in which more than 90% of the biomass may be removed.

The welfare of other numerous species in large measure depends on the health of the eelgrass itself.

Infobox 11.1 Red tide

Algae are essential components of marine and freshwater ecosystems. Under certain conditions, the populations of some species explode producing "blooms" that disrupt ecosystems. "Red tides" are blooms of *Karenia brevis*, a microscopic **dinoflagellate** identified only by its scientific name, that often tint coastal waters red or reddish-brown. The tell-tale discoloration signals the release of toxic compounds that trigger massive fish kills and the deaths of sea turtles, birds, dolphins, and manatees. In the early months of 2013, national headlines reported the deaths of 174 manatees from red tides on the southwestern coast of Florida.

Red-tide blooms occur when dissolved levels of phosphates and nitrates suddenly jump in marine environments. Some blooms transpire naturally when seasonal upwellings of coastal currents redistribute nitrogen and phosphorous compounds otherwise confined to deeper waters. In other cases, winds transport iron-rich sands from dust storms in the Sahara Desert to the Gulf of Mexico where they nourish nitrogen-fixing cyanobacteria. These organisms convert nitrogen gas dissolved in seawater to nitrates, which in turn creates optimal conditions for blooms of red tide. However, **anthropogenic** factors, particularly agricultural runoff and other types of water pollution, likely explain the increased frequency and distribution of red-tide blooms during the last 100 years (see **eutrophication**). For example, recurring red tides on the Texas coast have been linked with phosphorus from fertilizers carried downstream by rivers from the North American interior to the Gulf of Mexico.

Fish kills have long been associated with red-tide blooms. Some believe that Spanish explorer Cabeza de Vaca, shipwrecked in 1528 on the coast of Texas, was the first to report a fish kill attributed to red tide. The diary of a Franciscan monk written in 1648 provides a more reliable account, however. Fatalities result when blooms of *Karenia brevis* contaminate the water with their metabolic products of highly potent **neurotoxins** known as brevetoxins. Bottlenose dolphins and fish-eating birds sometimes consume lethal doses of brevetoxins when feeding on live **planktivorus** fish whose tissues harbor sub-lethal concentrations of the poison. Accumulations of decomposing fish and other organisms killed by red tides prolong the outbreak by adding new sources of nitrogen and phosphorus to further nourish the bloom.

Although the expression "red tide" seems likely to persist, most scientists prefer "harmful algal blooms" (HABs) as a more accurate descriptor; red tides are neither related to tidal fluctuations nor always redden waters. Furthermore, HABs occur worldwide, not all of which are caused by *K. brevis*. On the northeastern coast of North America, for example, HABs result from another species of dinoflagellate, *Alexandrium fundyense*, whose toxins also kill fish *en masse*. A closely related species likewise produces HABs in the eastern Pacific Ocean.

State health agencies monitor HABs along the Gulf of Mexico coast to reduce the chances of human illnesses. The cells of *K. brevis* are uncommonly fragile and easily fracture when wind or waves cause turbulence; minute particles become airborne and are therefore potentially inhaled by boaters and coastal residents, among others. Not everyone who is exposed is affected, but susceptible individuals experience breathing difficulties, irritation of the nose and throat, and burning eyes. Human fatalities may result after eating oysters and clams harvested from areas impacted by HABs. Shellfish concentrate many toxic chemicals including brevetoxins in their tissues, which potentially exposes humans to neurotoxic shellfish poisoning. The effects of the illness vary from abdominal pain, nausea, diarrhea, and vertigo at low concentrations, to death. State agencies therefore restrict shellfish harvests when brevetoxins appear. In what might be a classic case of "good things sometimes come from bad," research designed to find new ways of detecting the toxins led to a discovery that derivatives of the poison might eventually treat cystic fibrosis, a debilitating lung disorder. Photo courtesy of Jace Tunnell.

For example, about 20 species of commercially valuable fishes feed in eelgrass meadows – but rarely on the plants themselves – at some point in their respective life histories. Similarly, the bay scallop is among the mollusks associated with eelgrass communities; as juveniles, they commonly attach to eelgrass leaves and later drop to the bottom sediments where they mature. Large numbers of a small snail – the grass cerith – also occur in eelgrass meadows where they feed on periphyton and in turn enter the food chain of other organisms (e.g., blue crabs). Overall, the fauna living on eelgrass periphyton numbers more than 120 species and includes sponges, various worms (e.g., nematodes and polychaetes), mollusks, and barnacles.

Disaster struck in the early 1930s when an amoeba-like slime mold produced an epidemic loss of eelgrass on the Atlantic coast of North America. The infection (eelgrass-wasting disease) blackens the leaves, destroys the tissues, and then kills the entire plant. In many places, large areas of eelgrass completely disappeared, leaving survivors only in scattered refugia. The effects of the epidemic lasted for three decades, including a diminished scallop fishery in the mid-Atlantic states. A more immediate result was the crash of the Atlantic brant population, which fell to a fraction of its former size before gradually recovering. Black brant on the Pacific coast, where a similar epidemic occurred, shifted their winter range southward to places where eelgrass remained plentiful. Today, at least 70% of all black brant in the Pacific Flyway overwinter in coastal lagoons of Baja California.

The eelgrass disaster also produced reductions in entire assemblages of invertebrates when the vegetation disappeared. A permanent loss occurred with the extinction of the eelgrass limpet, one of a group of mollusks sometimes called "Chinese hats" because of their conical shape. The disappearance of the eelgrass limpet represents the first recorded extinction of a marine invertebrate in an ocean basin.

Regrettably, the lethal slime mold did not disappear. Another epidemic occurred in Great Bay estuary on the New Hampshire–Maine border between 1981 and 1984. Nothing is known about what conditions prompt new eruptions of the disease, but recent evidence determined that eelgrass can survive wasting disease in areas with low salinities.

Forests in the ocean

In contrast to much of the Atlantic coastline, the Pacific Coast of North America features stands of marine vegetation known as kelp "forests." The rich fauna associated with this vegetation likely served as important food resources that sustained the first Paleo-Indians venturing into North America along what anthropologists now call the "kelp highway."

Kelps are large brown algae growing in cold waters in temperate regions throughout the world. As for algae,

kelps lack vascular tissues and, instead of true roots, are anchored to hard bottoms with structures known as hold-fasts. Growing from their long "stems," which commonly reach 15 m (82 feet) or more, are broad leaf-like blades, each equipped with an air-filled float. The floats support vertical growth, allowing kelps to extend upward through the water column. A kelp forest forms when the blades reach the surface to form a floating canopy. Conversely, without a floating canopy, the vegetation is known as a kelp bed. Kelp beds occur along the northern Atlantic coast (e.g., New England), but kelp forests are typical of the Pacific coastline where they extend from California to Alaska. Lacking vascular systems, kelps absorb nutrients directly from seawater. Upwellings associated with the currents along the Pacific Coast of North America maintain a nutrient-rich environment where kelp communities become particularly luxuriant.

With their three-dimensional structure, kelp forests provide major habitat for a wealth of organisms, some of which occur nowhere else. For example, a kelp community in California harbored 98 species of invertebrates (14 of which were characteristic) and 38 species of fishes. Despite such richness, however, relatively few marine animals graze directly on kelp. Grazing may incorporate as little as 10% of the kelp forest's immense production directly into the food web in these communities, with the balance becoming other sources of energy (e.g., detritus). Sea urchins are among the few grazers that feed directly on kelp and, in turn, become the principle prey for sea otters. Based on a study of captive animals, adult sea otters each day require 20–23% of their body weight in food. To meet this demand, adults (weighing about 23 kg or 50 pounds on average) annually consume immense amounts of sea urchins and other foods, arguing for their role as a keystone species.

Because of their valuable fur which is extraordinarily dense, sea otters were once exploited to the verge of extinction; remnant populations survived only in a few remote areas. Without the controlling effect of sea otters, sea urchins soon overgrazed and decimated the kelp forests. In the resulting **tropic cascade**, the kelp community rapidly changed in its composition of species and ecological interactions. With protection and transplants, however, sea otter populations increased and again occupy much of their former distribution. Their recovery in Alaska coincided with a reduction in sea urchin numbers and rapid increases in the distribution and abundance of kelp. According to some ecologists, by preying on sea urchins, sea otters exert a fundamental influence on community structure and thereby fit the definition of a keystone species. Other ecologists, however, believe that sea otters act as a keystone species only in a limited numbers of locations and instead propose that nutrient levels and diseases are among the factors that exert greater influences on the community structure of kelp forests.

Interestingly, the prehistoric rubbish left by Native Americans lends credence to the ecological importance of sea otters to kelp communities. Some 2500 years ago, Aleuts settled in the western Aleutian Islands where they subsisted on a diet of marine animals. Evidence of these foods, primarily shells and the bones of fishes and marine animals, persist today in **middens**; these piles of historical campsite litter can be dated with radiocarbon techniques. The oldest layers in these middens contain the remains of marine mammals, including sea otters, and fishes, but little evidence of shellfish. In deposits about 1000 years old, however, sea urchins and invertebrates are commonplace but the remains of marine vertebrates are rare. More recently, the Aleut middens again contain the discarded bones of fishes, seals, and sea otters rather than sea urchins and shellfish.

These findings suggest a back-and-forth dominance by predator and prey, presumably including corresponding changes in the abundance of kelp. In such a scenario, the Aleut colonists initially encountered kelp forests and an abundance of sea otters and other vertebrates. When these were hunted out, sea urchin and shellfish populations increased, becoming a secondary source of food for the Aleuts but also diminishing the kelp forest. Meanwhile, the Aleuts began to hunt for meat elsewhere. This allowed recovery of the local sea otter population, which again preyed heavily on sea urchins, and the kelp forest recovered. This chain of events supports the idea that sea otters represent a keystone species.

Oyster reefs

Middens of discarded oyster shells scattered along the coasts of North America confirm the prehistoric importance of oysters as food; these occur along the shorelines of essentially every bay and lagoon from Maine to Tamaulipas. Some middens are quite large and reflect human diets for hundreds or even thousands of years. For example, shells of the Olympia oyster, a species native to the Pacific Coast of North America, formed a mound more than 9 m (30 feet) deep at a site occupied for more than 10,000 years near Namu, British Columbia. Likewise, Native Americans in Maine harvested eastern oysters on the Damariscotta River for more than 1000 years, eventually accumulating a midden of enormous proportions. Named for its shape, the "Whaleback Shell Midden" was originally more than 9 m (30 feet) deep and 503 m (1650 feet) wide before it was mined for lime (Fig. 11.5).

Once valued solely as food, oysters are now considered a keystone species in estuarine ecosystems. Aggregations of living oysters, variously referred to as oyster bars, beds, bottoms, banks, or reefs, sometimes cover large areas (Fig. 11.6) and often provide the only hard substrate in bays and lagoons with soft bottoms. Oysters modify the contours of their shells to conform to the topography of adjacent objects, including other oysters, and produce reefs of dense, three-dimensional clusters of living oysters and empty shells.

Parts of an oyster reef typically remain submerged, whereas others parts are periodically exposed in an intertidal zone. Oysters tolerate a wide range of salinities and temperatures, but sites where salinities range from 5 to 35 ppt offer optimum habitat. Oysters attain sexual maturity as males but may later transform into females. Individual females produce more than 5 million eggs and males more than 2.5 billion sperm. Once fertilized, the zygote develops through three larval stages, the last of which descends to the bottom in search of **cultch** (a hard surface such as a rock, piling, or oyster shell which is free of mud or debris). At this point, each larva "sets" by cementing itself to the cultch where it begins a sessile life in a form known as a spat. Spats grow rapidly, reaching about the size of a dime in three months and becoming sexually mature oysters in about three years. About one of every million fertilized eggs survives to become a spat.

THE SHELL HEAPS AT DAMARISCOTTA, ME.

Figure 11.5 Historic photo (postcard) of the Whaleback Shell Midden. Photo courtesy of the Marjorie and Calvin Dodge Collection.

Figure 11.6 Accretions of oyster shells build reefs along the southern coasts of North America at sites in estuaries, tidal creeks, bays, and lagoons. As shown here (at low tide), living oysters form the upper layers of the reefs and overlay accumulations of shells from previous generations; the reefs are inundated at other times in the tidal cycle. The structural irregularities and interstices of oyster reefs create microhabitats for diverse and viable communities of other organisms. Photo courtesy of Troy D. Alphin, Center for Marine Science, University of North Carolina Wilmington.

The dense, irregular surfaces of oyster reefs provide microhabitats for many other animals. This architecture creates a surface area about 50 times greater than a mud bottom, thereby increasing the harborage for macro-invertebrates by one or two orders of magnitude. Clams, scallops, sponges, flatworms, amphipods, isopods, shrimp, crabs, and small fishes are among the various species finding protection, structural support, and foraging grounds in oyster reefs. Whelks, oyster drills, fishes, and other organisms also frequent oyster reefs as secondary consumers (i.e., predators of oysters or reef organisms). The planktonic larva of a **kleptoparasite**, the oyster pea crab, enters the shell of an eastern oyster where it grows to the size of a pea, then develops a thin, flexible carapace and spindly legs, and picks food from the gills of its host.

Oysters filter microalgae and suspended organic particles from water passing over their gills. When feeding, water is drawn into the shell cavity where cilia trap suspended particles on palps surrounding the mouth. A mature oyster filters about 189 L (50 gallons) of water a day and removes phytoplankton, sediments, and organic nutrients from the water column, thus improving water clarity and minimizing the effects of eutrophication. These changes increase light penetration and in turn enhance primary productivity in adjacent seagrass communities. In the process, however, oysters often concentrate heavy metals, toxic compounds, and harmful bacteria, which pose threats to humans who eat oysters from contaminated sites.

Populations of eastern and Olympic oysters have declined throughout their North American range. Although oysters still remain in most locations, many reefs have disappeared or have been reduced to the point that they no longer function as ecosystems, largely because of over harvesting or dredging. Oyster reefs were dredged as sources of lime for fertilizer, mortar, or concrete, and for roadway surfaces. Over many years, oysters were harvested at rates that exceeded annual **recruitment,** and the removal of shells eliminated substrates essential for spat settlement.

Oyster reefs on the Pacific Coast are facing a new threat: ocean acidification. Seawater has become about 30% more acidic since the beginning of the industrial revolution. Water near the ocean bottom is about three times more acidic than surface waters because dead organisms sink to the ocean floor. Carbon released during decay combines with water to form carbonic acid, which normally remains in deeper waters. Carbonic acid is also formed at the surface when carbon dioxide emitted from burning coal, oil, and other natural fuels is absorbed in ocean waters. Strong winds blowing outward from the steep coasts of the Pacific create upwellings that bring corrosive waters from the depths, adding to surface concentrations. As a result, the increased acidity in shallow waters dissolves the delicate shells of oyster larvae and spat.

Efforts to restore oyster reefs on the coasts of North America include translocations of non-native oysters to new areas. Artificial reefs, constructed for the recovery of native species, are initiated by dumping clutch-forming materials (e.g., shells from restaurants) at suitable locations along the coast. Thereafter, these sites are stocked with hatchery-reared larvae or spat that mature and flourish as harvest-free reefs. At some places in Chesapeake Bay, oysters produced on artificial reefs are now legally harvested at locations where natural reefs disappeared long ago. Other artificial reefs intend to improve recreational fishing along with water quality, biodiversity, and other ecological services.

Emergent communities

Atlantic tidal marshes

With few exceptions, tidal marshes edge the coastline of eastern North America. A rocky shoreline characterizes much of Nova Scotia and Maine but thereafter, from southern New England to Texas and into Mexico, a belt of tidal marshes generally marks the transition between maritime and terrestrial environments.

Two grasses dominate the Atlantic tidal marshes: the related species of marsh hay and smooth cordgrass. Marsh hay, the shorter of the two, grows to 1.5 m (4 feet), whereas smooth cordgrass reaches heights of 2.5 m (8 feet). Ecologically, marsh hay generally develops in meadows above the tidal zone on the landward side of the salt marsh (i.e., "high marsh"). Cowlicks – swirled mats of marsh hay – are distinctive features of these meadows, which farmers in the mid-Atlantic states once regularly cut for their rich yield of forage (Fig. 11.7). Smooth cordgrass favors the tidal zone (i.e., "low marsh") and therefore experiences alternating periods of inundation and exposure each day. Both species typically form monotypic stands, covering large areas in which other species are largely excluded. Deep layers of peat often form under both high and low marshes.

Zonation – primarily the result of differing tolerances to salinity and tidal flow – is one of the more obvious features of tidal marshes. Gentle depressions and elevational changes in salt marshes can produce rather large differences in salinity. Soils in the center of depressions known as salt pannes contain far more salt than the higher, better drained locations at the edges. The result is a series of concentric rings of markedly different vegetation, each responding to its own tolerance for soil salinity.

In some cases the center of the panne lacks vegetation altogether; alternatively, halophytes such as glasswort can flourish on waterlogged, salt-encrusted soils. The stems and leaves of several species of saltmarsh plants have salt glands that secrete the excessive salt loads absorbed from soil water. Other plants also develop in distinctive zones or well-delineated clumps within salt marshes (Fig. 11.7).

Salinity decreases toward the landward edge of the high marsh, and other vegetation marks the transition to terrestrial communities. Common cane – often known by its Latin name, Phragmites – is among the plants encountered in these transition zones. Cattail is also among the species marking the transition from salt- to freshwater environments. Woody plants at the edges of the high marsh include silverling and marsh elder.

The Atlantic tidal marsh, including its extension along the Gulf Coast, provides habitat for a wealth of birds and other animals. Few of these represent species whose entire distribution coincides exclusively with this relatively narrow strip of coastal habitat. Diamondback terrapins, however, are a notable exception and provide another fine example of an endemic species (Fig. 11.8). Because of their tasty meat, these turtles faced heavy exploitation for many years. Early in the 20th century, for example, diamondback terrapins fresh from the marsh commanded "$1 an inch" in the marketplace, roughly equivalent to $23 an inch today. The species was later legally protected; populations in several areas are again declining however, in part because of degraded habitat but also because many enter and drown inside crab traps.

Another endemic is the namesake saltmarsh sparrow; the species breeds along the coast from Maine to Virginia and winters farther south on the Atlantic coastline.

Figure 11.7 Vegetation in salt marshes is often sharply delineated forming strands of black needle rush and marsh hay (foreground) on the southern coast of North Carolina. These and other stands of saltmarsh vegetation seldom include other species of plants. Photo courtesy of Elizabeth D. Bolen.

Figure 11.8 Diamondback terrapins are endemic to tidal marshes from Cape Cod to Texas. Females are noticeably larger than males. Note the distinguishing feature of concentric markings and grooves on each of the large scutes. Flecking on the legs, necks, and heads of diamondback terrapins is another characteristic of the species, and individuals have prominent "moustaches" (inset). Photos courtesy of Gilbert S. Grant.

Females nest near the ground and seek vegetation above normal high tides, but unexpected flooding remains a threat to breeding success. Because of the limited distribution of saltmarsh sparrows and their obligate dependence on habitat subject to disturbance, various conservation agencies list the birds as a species of concern.

Research conducted in the 1960s indicated that cordgrass marshes were the most productive of all communities, including tropical rain forests, but later work suggested somewhat lesser production than originally believed. Nonetheless, salt marshes produce large amounts of **detritus** that decomposes in place or washes out of the marsh system, in either case providing a vital source of nutrients. Decomposition increases the food value of many saltmarsh plants that otherwise might contribute little to food chains in coastal communities. Salt marshes also serve as major nursery areas for the larvae of marine organisms of many kinds (e.g., crustaceans).

Large areas of salt marsh were degraded in the 1930s, when government-sponsored projects included ditching coastal wetlands to control mosquitoes. When drained, however, marsh elder and other noxious vegetation invaded the marshes and significantly reduced their value as habitat for waterfowl and other wildlife. This damage included the elimination of ponds containing wigeongrass and, importantly, caused severe reductions in invertebrate populations; as many as 95% of the mollusks and crustaceans disappeared from the smooth cordgrass communities.

Marshes of the Gulf Coast

Nowhere along the border of the Gulf of Mexico are marshes as vast and complex as those developed under the influence of the Mississippi River. These wetlands extend along the coastline both west (barely into eastern Texas) and east (into Alabama) from the river's mouth, but here we focus on those in Louisiana which comprise 96% of the region's wetland area and about 21% of all coastal marshes in the continental United States, excluding Alaska.

Two plains make up this area; the largest (about 75% of the total) is the Deltaic Plain that extends eastward from Vermillion Bay across the southeastern coast of Louisiana. The plain formed when several lobe-like deltas coalesced at the mouth of the Mississippi River, further augmented by the sediment-bearing outflow from distributaries (i.e., lesser rivers such as the Atchafalaya that split off from the main channel as the Mississippi nears the Gulf). The coastline is irregular, including many embayments and chains of barrier islands formed during the river's history. Sand and silt form the bulk of the delta's soils. Unfortunately, the deltaic plain is subject to degradation from both natural and man-made causes, and currently suffers a net loss in area.

The smaller Chenier Plain extends westward from Vermillion Bay into Texas near the mouth of the Sabine River. Like the marshes of the deltaic plain, these marshes also developed from sediments carried by the Mississippi River. However, in this case, the materials (primarily clays) flowed into the Gulf of Mexico and settled along the shoreline. After marsh vegetation developed on this substrate, a beach formed on the outer edge of the marsh. When sediment deposition resumed a new marsh formed, isolating the former beach landward. In time, a series of stranded beach ridges developed with marshes on either side, together forming the Chenier Plain. These old beachfronts or cheniers (French for the live oaks that characterize these landforms) largely curtail tidal exchange, thereby rendering the Chenier Plain relatively invulnerable to erosional damage in comparison to its deltaic counterpart.

Louisiana's marshlands extend inland from the Gulf for 25–85 km (15–53 miles). Several communities develop in response to changes in salinity. Historically, four zones are widely recognized, although some may split these into finer categories. First in the series are saline marshes (27% of the area); the dominant species include those typical of Atlantic tidal marshes, namely smooth cordgrass, marsh hay, and black needle rush. Salinity limits the diversity of species and submerged vegetation is absent.

Brackish marsh (30%) is dominated by marsh hay, joined by salt grass, and Olney's three-square, a bulrush whose leafless stem is uniquely triangular in cross-section. Submerged vegetation appears in this zone, dominated by wigeongrass.

The third zone is intermediate marsh (15%), still dominated by marsh hay but with several additional species including common reed and two forbs, duck potato and waterhyssop. Diversity of the submerged vegetation expands to include several kinds of pondweeds.

Freshwater wetlands, simply known as 'fresh marsh' (28%), complete the series. Maidencane dominates this zone with lesser amounts of common reed and marsh hay, but more duck potato than that which occurs in intermediate marsh. In some locations, however, the dominance of maidencane is replaced by spikerushes, but causes for this transformation remain unclear. Alligatorweed, an invasive species from South America, thrives in this zone. Complexity in the submerged vegetation is increased (e.g., water milfoil, coontail, and bladderworts), floating-leafed plants make an appearance (e.g., white water lily), and an invasive exotic, water hyacinth,

often forms dense mats that clog surface water at the expense of other vegetation. As expected, the greater diversity in the vegetation in this zone influences the richness of higher trophic levels. The inner edge of fresh marsh ends with the beginnings of forest vegetation.

With the exception of the salt marsh, these wetland types represent important winter habitat for migratory waterfowl in the Mississippi Flyway including mallards, northern pintails, gadwalls, and lesser scaup. Historically, snow geese numbering in the hundreds of thousands preferred the brackish marshes as winter habitat, but the conversion of nearby inland areas to agricultural uses, especially rice production, greatly expanded their winter range. Because of their importance to waterfowl and other wildlife, large areas of Gulf marshes are protected as state or federal refuges.

Muskrats also prefer brackish marsh and were once trapped for their pelts. At times, intensive foraging by either muskrats or snow geese produces "eatouts," denuded patches of vegetation. Dense populations of muskrats are particularly damaging because they also uproot vegetation to construct their lodges (Fig. 11.9). Eatouts may take several years to recover. Nutrias – large South American rodents once heralded as a source of valuable fur – were introduced into southern Louisiana in the 1930s, but soon escaped from their pens. Like muskrats, nutria also consume large amounts of marsh vegetation which can delay the recovery of marshes damaged by other factors (they feed on newly emerging shoots). Their burrows may significantly weaken natural levees or those constructed for rice production and other forms of water management. Overall, nutria remain unwanted pests that eventually spread northward along the Atlantic coastline to Maryland.

Controlled burning of brackish marshes was once a widespread management practice for both geese and

Figure 11.9 In addition to creating openings in marshes, the feeding activities of muskrats also may alter the composition of the vegetation. Muskrats construct "houses," as shown here, and small feeding platforms known as "push-ups." Two layers of fur – outer guard hairs and a dense undercoat – protect muskrats (inset) from winter weather and chilly water. Photos courtesy of William R. Clark.

muskrats; fire removed the dense thatch, thereby exposing the rhizomes of Olney's three-square that these species utilize as food. Fires also prevent woody plants from encroaching into fresh marsh. However, fires also may destroy the organic matter that helps bind the small particles of mineral sediments in marsh soils, thus fostering erosion and wetland destruction, particularly in zones of salt marsh.

Gulf marshes face a number of threats, both natural and anthropogenic in nature. In 2000, patches up to 2 ha (5 acres) of marsh – mostly smooth cordgrass – rapidly died from a cause that still remains unknown. In all, well over 40,500 ha (120,000 acres) of coastal wetland were affected. Most of these areas recovered, but "sudden marsh dieback" (also known as "brown marsh phenomenon") appeared again in 2009 on a more limited scale. Strong hurricanes represent a natural threat, and may cause considerable damage when openings in the marsh are greatly enlarged by storm surges that wash away chunks of exposed sod. Hurricanes also may severely reduce some species of wildlife (e.g., muskrats). Nonetheless, such damage bears witness to the value of the marshes as buffers that help retard the inland movement of storm surges.

Of even greater concern, however, is the effect of subsidence on the coastal ecosystem. Some 65–91 km² (25–35 square miles) of wetlands disappear each year, a loss that currently exceeds the rate at which sedimentation forms new marsh. Subsidence – the gradual sinking of land surface – results when the accumulation of deltaic sediments compresses and becomes heavy enough to depress the underlying crust. The marshlands sink as a result of this process, to be replaced by unvegetated expanses of salt water. Moreover, when sea levels rise even more salt water invades. Together, both processes are known as "apparent sea-level rise," and either or both produce the same effect. Human activities have unfortunately accelerated these losses, especially because of saltwater incursion but also from subsidence resulting from the extraction of petroleum. In particular, canals cut through the marshes to facilitate energy production (e.g., access, drilling, and pipelines) allow salt water from the Gulf to reach areas where the vegetation is adapted to lesser salinities. Moreover, various types of flood-control structures reroute fresh water from the Mississippi River and therefore diminish the normal flow of sediments needed to build (and rebuild) wetland habitat in the delta of the river.

Mangrove islands and thickets

North America includes only a few tropical communities; among these are the mangrove thickets in southern Florida and the coastlines of Mexico. Mangroves belong to 20 genera in 16 families of plants, but only a few of these occur in North America (e.g., three species in Florida). Mangrove thickets are also known as mangrove forests.

Mangroves are among the few woody plants tolerant of seawater. Mangroves seem capable of growing in fresh water, but they persist in saltwater environments where there is little competition with other woody plants. Prop or "stilt" roots, a familiar feature of mangroves (Fig. 11.10), arch downward from the trunks and branches into the substrate below where they counter wave action and carry oxygen downward into the anaerobic soils beneath the water. Some mangroves also produce roots growing straight upward from below the water. These pneumatophores form a bed of finger-like projections around the mangroves and aid the plants with gaseous exchange.

Some mangroves reproduce by skipping the usual blueprint for seed production. Instead of producing seeds that remain scattered and dormant until conditions favor their germination, the propagules of red and other mangroves germinate and develop while still attached to the parent plants. The seedlings elongate as they grow, becoming torpedo-shaped as their free ends become steadily heavier (Fig. 11.10, inset). After dropping from the parent plants, these seedlings float off until they reach shallow water. When they strike bottom, the seedlings lodge, grow roots, and begin their growth as trees. In some cases, the seedlings drop into mud beneath the parent trees and begin their growth at the site, thereby adding to the celebrated tangle of mangrove thickets.

Mangrove thickets develop along protected tropical shores where there is little wave action and deep accumulations of mud. The impenetrable tangle of prop roots further slows water movement and causes even more deposition of muddy sediments. As a result, mangrove communities sometimes produce new land on coasts where they occur. Mangrove communities often develop in zones running parallel to the shoreline, and evidence of zonation appears in the three species found in Florida. Red mangrove is the most seaward of these followed by black mangrove, with white mangrove forming the innermost, landward zone. Mangroves also grow on coral substrates in shallow waters offshore where they form small islands of woody vegetation.

As habitat for animals, mangrove thickets are rather unusual. Land animals move throughout the canopy, whereas marine organisms live at the feet of the same trees. The marine fauna is divided into two types: those that live on hard surfaces (the stems and roots) and those living in muddy sediments. Underwater, an isopod often burrows into the prop roots of red mangroves, which apparently stimulates further root production and branching. Herons and egrets are among the birds that nest and roost in the dense canopy of a mangrove forest.

Mangrove communities provide nurseries for shrimp and other crustaceans as well as for various fishes. Ironically, mangrove thickets in some parts of the world have been cleared to construct ponds for **mariculture**,

Figure 11.10 Thickets of mangroves border creeks and other areas in the coastal zones of southern Florida. Mangroves worldwide include a large number of taxa, but red mangrove (shown here) is the most common species in North America. Note prop roots in water. When mature, seedlings (inset) drop, take root near parent plants, and add to the tangled growth or drift off to establish new stands elsewhere. Photo courtesy of Paul N. Gray of the National Audubon Society; (inset) seedlings illustrated by Tamara R. Sayre.

including shrimp production. Hurricanes sometimes cause extensive damage to mangrove communities, and large areas of mangrove thickets are often permanently destroyed by coastal development. Left in their natural state, however, mangrove communities protect the shoreline and their structure creates important centers of biodiversity.

Some associated communities

Barrier islands

Barrier islands parallel coastlines as narrow ribbons of sand separated from the mainland by lagoons, estuaries, marshes, or bays. Inlets known as **passes** often bisect these landforms, cutting larger islands into smaller units or at times forming an island from a peninsula. Storms cut these waterways, which vary in depth, width, and above all their permanence; some remain for decades or longer while others fill in rapidly, returning smaller islands to their former size. For these and other reasons, barrier islands are dynamic locations awash with the ceaseless movements of sand and water.

Barrier islands and beaches develop dune ridges that shield lagoons and the mainland from most waves and storm surges (*Sandy seashores*, this chapter). Barrier islands, while prominent on the Atlantic and Gulf coasts of the United States, exist on only about 15% of the world's marine coastlines. The northernmost barrier

islands on the Atlantic Coast of North America occur along the Massachusetts coast. Larger islands and barrier beaches flank Long Island, New York from Southampton to Rockaway Point, just south of Coney Island. Chains of barrier islands extend from Sandy Hook, New Jersey to Jupiter, Florida on the Atlantic Coast, and from Appalachicola Bay, Florida to Tuxpan, Mexico on the Gulf Coast. Although less well-developed, small barrier islands also occur on the northern coast of Alaska. The Pacific Coast lacks such features, but low, sandy barrier spits studded with dunes extend laterally from many rocky headlands.

Barrier islands result from interactions among sea level, sediment availability, longshore currents, and waves. Most barrier islands are between 3000 and 7000 years old (young by geological standards). Beginning about 18,000 years ago, the ice sheets covering much of Europe and North America began to melt, causing sea levels to increase. When sea levels stabilized, the combination of currents and wave actions accumulated enough sand to form beaches, spits, peninsulas, and barrier islands.

The geological processes leading to the formation of barrier islands have long been a subject of conjecture. No unified hypothesis explains the development of all barrier islands, but the major concepts can be assembled into three main proposals. The oldest idea, the offshore bar theory, suggests that longshore currents add sediments to offshore bars until the bars emerge as islands. Some barrier islands along the Gulf Coast of Florida developed in this manner within a few decades. A competing idea, the spit accretion theory, is based on evidence that sand spits connected to the mainland gradually elongate as longshore currents deposit additional sediments. When severe storms cut passes between low-lying segments of the spit, islands are created. The dune submergence theory, the third hypothesis, suggests that dunes along mainland coasts were isolated as emergent islands when sea levels rose. No evidence exists to demonstrate that barrier islands were formed by coastal drowning, but recent subsidence disconnected Isles Dernieres, Louisiana from the Mississippi River delta.

Embryonic barrier islands enlarge when longshore currents deliver sediments to the swash zone where they are transported landward by waves and winds. The accumulations of sand soon provide plants with opportunities for colonization (*Sandy seashores*). Foredune ridges grow in height until winds can no longer transport sand to the dune's crest. A shallow trough, known as a swale or **slack**, forms between crests of the low dunes to the windward side and the larger dunes of the foredune ridge. Protected from direct exposure to wind and wind-blown sand, slacks generally develop dense associations of grasses and forbs.

On the lee side of the foredune ridge, sand deposits form expanses known as "flats." Halophytic plants characteristic of the backshore are the pioneers, which eventually provide enough cover and organic matter to support dense meadows of little bluestem, other grasses, sedges, prickly pear, beach primrose, wildflowers, and shrubs. Continued enrichment of the soil by nitrogen-fixing plants and the decay of organic debris eventually allow development of forests characterized by evergreen oaks such as live or laurel oaks (*Maritime forests*, this chapter). At Buxton Woods, the largest maritime forest on the Outer Banks of North Carolina, mature stands of evergreen oaks were heavily logged for ship timbers in past centuries. Lacking recurrence of further **anthropogenic** damage, Buxton Woods will gradually return to a climax community similar to that of the precolonial era.

The vegetated flats of Padre Island and other barrier islands and peninsulas in Texas and Mexico experience periodic droughts that may last for years. These conditions produce a grass-dominated prairie of great diversity, but the harsh and unpredictable environment on these flats prevents development of a climax community. Isolated mottes of stunted live oaks dot the higher elevations of northern Padre Island, but few oaks survive farther south on the island.

Isolated fresh- and brackish-water ponds, sloughs, and marshes occur in swales and depressions in island meadows. These ephemeral aquatic habitats tap a shallow aquifer lying just below meadow surface. Rainwater maintains the aquifer, feeding a freshwater lens perched above a denser, underlying table of salt water. Ponds and sloughs attract wading birds and thirsty island wildlife. Cattail, sedges, and rooted plants with floating leaves provide habitat for fish, frogs, toads, and aquatic insects.

Plants clearly help anchor the foredune ridge, but those dunes only partially vegetated or scarred by fire, grazing, or other disturbances remain vulnerable to wind-generated "blowouts." Additionally, winds transport sand from barren areas on the windward side over the ridge, where the deposits form "marching" dunes on the lee side. As they migrate, these transient dunes may bury parts of meadow communities, fill ponds, and occasionally smother areas of maritime forest. The "ghost forest" on Shackleford Banks, North Carolina, consists of cedar stumps, the relics of trees entombed by dunes nearly a hundred years ago. The landward shoreline of barrier islands slopes gently into a lagoon. On the Atlantic Coast, a wide marsh develops on the lee side of most barrier islands (*Atlantic tidal marshes*, this chapter).

Large waves or storm surges associated with tropical cyclones fashion abrupt changes in the topography of barrier islands. Strong currents push large quantities of sand inland over low areas or gaps in the dunes, sometimes cutting sizable channels ("washover passes") completely across the island (Fig. 11.11). Washover sediments deposited on the back side of barrier islands form fan-shaped blankets of moist sand that retard decomposition of the buried marsh vegetation and initiates a layer of peat.

The barrier islands along Alaska's North Slope develop in the same manner as those on the eastern and southern coasts of North America. Longshore currents during a two-month, ice-free summer period deposit sand and gravel eroded along the mainland or carried to the coastline by the Yukon and other rivers. Strong winds constantly winnow the accumulated sediments, leaving behind barren beaches of pebbles. Sand dunes form on the backside of some islands where low halophytes gained a foothold and trapped eolian sands. Despite their harsh conditions, these islands provide refugia for molting waterfowl and shorebirds.

Barrier islands likely face an uncertain future. Accelerated erosion may result from rising sea levels and increasing frequencies of storms and hurricanes. Moreover, dredging and flood-control structures in rivers have reduced the supply of sediments required to rebuild barrier islands. Consequently, most barrier islands along the Atlantic and Gulf coasts are gradually diminishing in size and retreating landward. Despite their vulnerability, barrier islands remain popular locations for human activities. The average density of human populations on barrier islands is three times greater than those of coastal states, and increased by 14% from 1990 to 2000. Fortunately, state and federal governments now protect all or parts of some barrier islands from further development. These include National Seashores, notably

Figure 11.11 Large waves and high water pushed ahead of hurricanes often erode sandy seashores and breach the foredune ridge, resulting in washover passes. Salt water and sand cover the habitats behind the dunes. Photo courtesy of Brian R. Chapman.

those at Cape Cod (Massachusetts), Cape Hatteras (North Carolina), Assateague Island (Virginia), and Padre Island (Texas) and state parks such as Island Beach State Park (New Jersey) and Delaware Seashore (Delaware), among others.

Coral reefs

Any mention of coral reefs typically invokes images of clear azure waters, warm breezes, and colorful sculptures swarming with schools of even more colorful fishes. Coral reefs, sometimes described as "the rainforests of the sea" because of their tropical settings, complex structure, and biological richness, indeed beg comparisons with most diverse ecosystems on Earth (Table 10.1). Only one region in North America, the Florida Reef Tract, offers easy access to the wonders of a coral reef ecosystem. Two other coral reef systems, the East and West Flower Garden Banks, sit atop salt domes in the northwestern Gulf of Mexico approximately 170 km (105 miles) south of Sabine Pass, Texas.

The Florida Reef Tract, the third-largest barrier reef system in the world, is a series of small coral reef patches in the Atlantic Ocean extending 576 km (358 miles) from Biscayne National Park to the southern boundary of Dry Tortugas National Park. The system includes more than 6000 individual reefs in a 6–7 km (4 mile) wide belt that follows a contour 20 m (66 feet) deep a few miles seaward of the Florida Keys. The seaward margins of the outer bank reefs provide barriers that dampen wave action before sloping downward into deeper water. Their leeward sides protect seagrass meadows, expanses of barren sand, or collections of coralline algae growing on jagged fragments of coral skeletons presumably deposited by storms. Gaps between the bank reefs form channels with sandy bottoms. In the shallow waters between

the bank reefs and the Keys lie hundreds of small, rounded patch reefs, most of which consist of star and brain corals attached to hard-bottom substrates. The densest and more diverse reefs of the Florida Reef Tract are found seaward of Key Largo and Elliott Key.

Coral reefs associated with the North American coasts develop at sites with specific environmental conditions: clear, shallow waters; tropical or subtropical water temperatures; solid substrates for attachment; and enough circulation from either currents or wave action to provide oxygen and waste removal. Another group, known as deep-water or cold-water corals, live in far deeper (and much darker) parts of the ocean where they survive without zooanthellae.

Coral structures form from the calcium carbonate secretions of tiny marine organisms (polyps) living in compact colonies. Each polyp is a miniature version of a more familiar animal, the sea anemone. Like the anemone, a ring of tentacles armed with stinging cells called nematocysts surrounds the mouth and captures planktonic organisms, including larval fish. Although coral polyps are carnivorous, most corals depend on unicellular algae for most of their energy. The algae, known as zooanthellae, live symbiotically within a polyp's tissues where they photosynthetically convert sunlight into energy, adding color to their hosts in the process. In exchange for their photosynthetic products, the zooanthellae gain access to sunlight and the nutrients derived from the coral's feeding activities.

The polyp colonies covering the surface of a coral head constantly adds mass to the skeletal structure with their secretions of calcium carbonate. Over many generations, an architecture typical of each coral species slowly forms (e.g., the distinctively branched staghorn coral). The hard skeleton serves as the colony substrate, but also

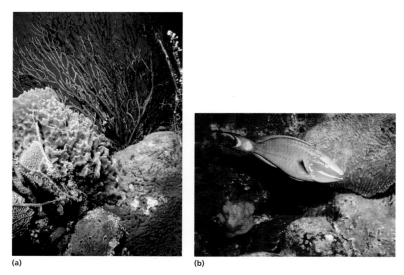

(a)   **(b)**

Figure 11.12 (a) Reefs along the coasts of Florida are composed of stony and soft corals. Soft corals are represented by the branched, tree-like structure at the top; an example of a stony coral is located at middle left. Many of the other structures are sponges. (b) Located atop salt domes in the Gulf of Mexico, coral reefs of the Flower Garden Banks do not contain the species diversity found elsewhere but the reefs remain relatively undisturbed. Note the parrotfish, which grazes on coral. Photos courtesy of (a) Florida Keys National Marine Sanctuary (National Oceanic and Atmospheric Administration) and (b) Quenton Dokken (digitized by Mike Smith).

affords the polyps protection from predators. When disturbed or threatened, the polyps withdraw into the surrounding coral structure.

Biologists place corals into two subclasses, commonly identified as stony corals and soft corals (Fig. 11.12a). Stony corals, which build calcareous reefs, have six or fewer lines of symmetry and contain zooanthellae. The most common stony corals on the Florida reefs include brain corals, elkhorn corals, star corals, and pillar corals. The soft corals, such as sea fans, sea pens, and sea whips, have eight tentacles and lack zooanthellae; their flexible skeletal structures respond to the motion of waves. The skeletons of soft corals consist of proteinaceous compounds perforated with interconnecting pores that allow the polyps to share nutrients within the colony.

Although the Florida reefs lie close to the northernmost limits for tropical corals, species diversity is similar to the reef systems in the Caribbean Sea. The complex structure of a coral reef community and its associated habitats provides sanctuaries and feeding areas for myriad species. More than 50 species of hard corals and 37 species of soft corals occur in the Florida reef community, supplemented by nearly 1400 species of marine plants and animals of which 500 species are fish.

Patch reefs feature a "halo" of exposed sand that extends 2–10 m (6.5–33 feet) outward from the base of the coral structure into the surrounding beds of seagrasses. The inner parts of the halo, which are completely barren, are maintained by herbivorous fishes and sea urchins seeking cover in the reef from predators during the day, but venturing out at night to forage on the vegetation. When predators are removed halos become less obvious because the herbivores, with less risk, can move farther from the reef for food.

Fishermen discovered the Flower Garden Banks in the Gulf of Mexico in 1936 (Fig. 11.12b). These are the northernmost coral reefs on North America's continental shelf and named for their bright colors that are clearly visible from the surface. Each reef (there are two) caps an uplifted salt dome at the margin of the continental shelf. Although smaller and less biologically diverse than the reefs in Florida their remote location means that they have been virtually undisturbed by human impact. The Flower Garden Banks boast 21 species of corals, 27 sponges, more than 300 invertebrates, and over 170 fishes, including whale sharks and manta rays.

Florida's coral reefs are troves of biological diversity but disruptions to the food web may alter their structure. For example, after decades of overharvesting (solely for their fins) of silvertip and other sharks at reefs in tropical oceans, their prey populations – snappers and other midsize fishes collectively known as **mesopredators** – steadily increased. In turn, the expanded population of mesopredators depleted the herbivorous species (e.g., most notably, algae-eating parrotfish) and thereby imperiled the reef itself; this top-down process is known as a **trophic cascade**. Freed from grazing pressure by parrotfish, algae flourished and smothered the young corals needed to maintain the coral communities. Because of these relationships, it seems likely that apex predators (sharks in this case) play a role in the recovery of coral reefs damaged by storms or bleaching (i.e., destructive bottom-up processes).

Coral reefs are vulnerable ecosystems easily injured by human activities but seem to recover relatively quickly from hurricane damage and localized, short-term disturbances. Despite their rather sturdy appearance, corals are however ill adapted to recover from long-term environmental changes. A change in water temperature of more than 1–2°C (1.8–3.6°F) or a slight alteration of salinity can stress corals to the point where they expel their zooanthellae and die. Lacking their sources of color, dead coral appears white, a condition described as coral bleaching. Some diseases and pollutants also cause bleaching, and reefs near coastal towns and cities more often show greater signs of stress than those that are more isolated.

Efforts to protect coral reefs in Florida began in the 1960s. Today, all parts of the Florida Reef Tract are protected either by the Florida Keys National Marine Sanctuary or by the Florida Department of Environmental Protection. Likewise, the Flower Garden Banks National Marine Sanctuary ensures that these reefs maintain their pristine condition in the Gulf of Mexico. The mandate of the Marine Sanctuary Program wisely considers reef systems as part of a larger watershed where runoff, pollutants, and human disturbances are minimized as part of the protection program.

Maritime forests

A distinctive type of forest hugs the coastlines of the southeastern United States and Gulf states. This community, aptly known as maritime forest, extends from the southern border of Virginia to Texas. Stands of maritime forest are typically associated with the vegetation on barrier beaches (e.g., the Outer Banks of North Carolina), but they also develop along the shorelines of estuaries. However, because of beachfront homes and other coastal developments, few blocks of undisturbed maritime forest still remain. These fragments embody the bleak reality of an endangered ecosystem.

The characteristic species of maritime forests are by no means unique to coastal locations, but together they form a community shaped by a powerful agent: salt spray. The airborne spray from ocean waves: (a) distinctively sculpts the forest canopy; and (b) imports calcium, potassium, and other metallic cations into the nutrient-poor sandy soils (e.g., quartz sands) at these sites. Maritime forests usually originate in low spots on the beach where enough shelter exists to establish thickets of wax myrtle and yaupon, species whose waxy foliage is resistant to salt burns. Openings in these thickets provide protected sites where live oaks and pines gain footholds and continue to develop the forest.

Live oak is the dominant species in most maritime forest communities, but laurel oak is a common associate in the maritime forests of North Carolina. The durable strength of its wood enables live oak to withstand the punishing winds of coastal storms and hurricanes. Such strength, together with the desirable curvature of their limbs, brought live oak to the notice of 19th century shipwrights. Crews of "live oakers" were dispatched from the shipyards of New England to roam through maritime forests in search of structural timber. The commercial value of live oak eventually diminished with the development of steel-hulled ships, but by then large areas of maritime forests had been exploited beyond recovery. Maritime forests today are threatened by coastal development, but a few of the remaining exemplars fall under the vigorous protection of conservation groups (e.g., sites at Cumberland Island, Georgia and in North Carolina, Bald Head Island and Buxton Woods on Hatteras Island).

The typical overstory of a mature maritime forest on the Atlantic coast also includes loblolly pine and red cedar. Red bay, dogwood, dwarf palmetto, ironwood, and yaupon characterize the shrub layer, often accompanied by tangles of greenbriar, grape, poison ivy, and other vines. The tight canopy greatly reduces light penetration, hence limiting herbaceous ground cover.

A sloping canopy is a defining feature of the maritime forest (Fig. 11.13). Once ascribed to the effects of strong offshore winds, the wedge-shaped contours in fact result from salt spray repeatedly killing successive generations of terminal buds on the seaward side of the canopy. Conversely, terminal buds on the downwind side of the canopy survive and continue growing. Live oak, with its thickly cuticled sclerophyllic leaves, is particularly adapted for coping with the salty aerosol. Plants beneath the live oak canopy are sheltered and thrive in a largely salt-free environment, but openings in the canopy produce a severely damaged understory. When undisturbed, maritime forests shelter a diverse biota that could not otherwise endure the harsh conditions of the beachfront environment.

Highlights

Synchrony at Delaware Bay

Each spring, an ancient spawning ritual unfolds on the sandy beaches of Delaware Bay. During May and June the "living fossil" in the form of the horseshoe crab, a species unchanged for millions of years, crawls ashore to spawn in the intertidal zone, particularly along the bay's coastline in southern New Jersey (Fig. 11.14). They come by the tens of thousands to stage one of the more important ecological events occurring in any coastal environment in North America. Lesser numbers of horseshoe crabs also breed on sheltered, low-energy

Figure 11.13 Salt spray from nearby surf sculpts the profile of maritime forests, as shown here in a canopy of live oak near Fort Fisher, North Carolina. Photo courtesy of Elizabeth D. Bolen.

Figure 11.14 Masses of horseshoe crabs breed on some Atlantic beaches. The eggs they deposit in the sand provide essential food for shorebirds migrating to their northern breeding grounds in the spring. Photo courtesy of Gregory Breese and the US Fish and Wildlife Service.

beaches elsewhere along the Atlantic and Gulf coasts, but Delaware Bay is also the largest staging area in the Atlantic Flyway for migrant shorebirds; therein unfolds an ecological relationship hard to match.

Horseshoe crabs are one of several animals dubbed with inaccurate common names, in this case because they are not crabs or even crustaceans. Instead, just four rather similar species comprise an entire class, Mesostomata, within the phylum Arthropoda (a huge group that includes insects, crustaceans, and other jointed-legged creatures). Their closest relatives are in fact spiders and scorpions. The three other species of horseshoe crabs occur in the Indo-Pacific region, and the fossil history of the group spans at least 240 million years.

Females encounter a throng of males at the water's edge, one of which grasps a female with a pair of claspers, the first of five pairs of leg-like appendages hidden under a dome-shaped shell. Thus attached, the male is dragged ashore behind the female (which are larger than males); one or two additional males sometimes grasp the first, forming a chain. Near the seaward edge of the high-tide line, each female excavates a nest where up to 5 thousand pearly green eggs no bigger than a pinhead are laid, then fertilized by the male. Additional clutches are laid on successive high tides until each female has produced about 20 clutches totaling 80,000 to 100,000 eggs each breeding season. Given a breeding population of 1–2 million horseshoe crabs, the density of their eggs at the waterline may exceed 100,000 m^{-2} (9300 per square feet). In all, the eastern shore of Delaware Bay becomes the incubator for billions of eggs.

Just at this time, thousands of migrating shorebirds of at least six species, including ruddy turnstones, sanderlings, and semipalmated sandpipers, arrive at Delaware

Bay. Red knots are particularly prominent; they overwinter on the coasts of Patagonia and Tierra del Fuego at the southern tip of South America. The first part of their northward migration includes some stopovers and covers about 6700 km (4200 miles) to northern Brazil, followed by a non-stop flight of 6400 km (4000 miles) to Delaware Bay. The birds arrive thin and in need of food to regain their body mass and for energy to make the remainder of their journey (another 2900 km or 1800 miles, again non-stop) *en route* to their nesting grounds in the High Arctic of North America. They also need energy reserves to increase the chances of successful reproduction. Horseshoe crab eggs, laden with energy-rich nutrients and seemingly in limitless supply, fulfill the birds' needs for fuel at exactly the right time.

The birds spend the next 2–3 weeks in what is essentially a feeding frenzy, thereby increasing their body mass by 70–80% in about two weeks. Overall, the shorebirds (all species) consume about 488,972 kg (539 tons) of eggs each spring, a staggering amount, yet safely below the threshold needed to maintain the population of horseshoe crabs. The loss of eggs is partly offset by the fact that the horseshoe crabs continue spawning well after the birds have resumed their northward migration.

Horseshoe crabs are no less curious in their physiology than they are valuable to human health. The blood of these ancient invertebrates is blue, thanks to a base of copper instead of iron. Even more unusual is the response of certain blood cells – known as amebocytes – to bacterial, viral, and other types of contamination. This relationship, discovered in the 1960s, was perfected to ensure the purity and safe production of vaccines and medical devices. To obtain the reagent – it has never been synthesized – thousands of horseshoe crabs are collected each year and bled under sterile and humane conditions. Up to 30% of the blood is removed, after which the animals are returned to their habitat (most rebound within a week). Some features of their compound eyes also prove useful for understanding human visual disorders.

These remarkable circumstances are not without peril, however, as human activities threaten the population of horseshoe crabs spawning at Delaware Bay, in turn threatening red knots and other shorebirds migrating along the Atlantic coast. In the past, horseshoe crabs were used to manufacture fertilizer, and large numbers are still caught to use as bait in eel traps. By the 1990s, some estimates indicated that the availability of their eggs at some locations on Delaware Bay had diminished by as much as 70%.

Likewise, recent surveys of red knots on their wintering areas in South America indicate significant population declines; numbers at a major site dropped by nearly 50% in two years. The obvious conclusion is that, lacking adequate energy to fuel their needs, fewer birds survived the rigors of migration.

Because of its obvious importance to red knots and other shorebirds, Delaware Bay was named the first site in the Western Hemisphere Reserve Network, an international effort to protect wetlands of importance to shorebirds and their management. Horseshoe crabs also gained protection. In 2001, the National Marine Fisheries Service halted the harvest of horseshoe crabs on a 3885 km² (1500 sqaure miles) offshore sanctuary at the mouth of Delaware Bay.

Waterbird colonies

Few places are as vibrant – at once busy, noisy, and smelly – as the nesting areas of colonial waterbirds. Some colonies are small, sometimes with fewer than a dozen nests, but most are larger with many thousands of breeding birds. Off the tip of Vancouver Island, British Columbia, huge numbers of breeding Cassin's auklets (500,000 pairs, about half of the world's population), rhinoceros auklets (40,000 pairs), tufted puffins (30,000 pairs), and common murres (4000 pairs) crowd a seabird colony on Triangle Island. Such colonies develop where food, particularly fish, is both abundant and available, along with the appropriate physical structure (often cliffs, but also vegetation or bare ground; Fig. 11.15). Some colonies are associated with freshwater environments, but many more rely on marine resources.

By definition, a colony (or rookery) develops where nests are closely packed, often with the intervening distance based on the pecking reach of the nearest neighbor. In single-species colonies, this behavior establishes a circular and often vigorously defended territory around each nest. Many colonies consist of several species, each nesting at slightly different locations within the colony's area. Many seabird colonies remain active year after year, the result of strong site fidelity ("homing" behavior) of both adults and many of their offspring. However, other birds may be attracted to suitable sites by birds already in residence, which may help wildlife managers re-establish colonies at abandoned locations (Infobox 11.2).

Ornithologists once believed colonies evolved in response to the limited availability of nesting sites. Colonies may also act as information centers, where birds watch the arrival and departure of others to determine sources of food. Protection from terrestrial predators offers another explanation for colonial nesting, especially for colonies established on secluded islands and cliffs. The concept of "safety in numbers" also helps to detect and repel predators, and invaders – human or otherwise – may experience energetic attacks, including mobbing and aerial defecation. To protect their nests, cormorants and some herons regurgitate their crop contents on unwanted visitors. Finally, birds nesting in colonies usually lay eggs at about the same time, hence the chicks hatch in synchrony. This means a predator can consume only a small percentage of the prey available at

(a)   (b)

Figure 11.15 Waterbird colonies take many forms. Royal terns, gulls and skimmers tend to nest in close aggregations on shell or sand substrates where they dig shallow "scrape" nests located just beyond "pecking distance" to their nearest neighbor. (a) Heron, egret, and cormorant colonies are usually situated in trees, but in northern waters (b) narrow ledges on cliffs offer protected nest sites for many seabirds such as these black-legged kittiwakes. Photos courtesy of (a) Brian R. Chapman and (b) John Maniscalco.

Infobox 11.2 The puffin project

Only one species of puffin, the Atlantic puffin, nests on the northern Atlantic Coast of North America. Egg-collecting and unregulated hunting in the 19th century severely reduced the populations nesting on islands off the coast of Maine. During the 20th century, predatory gulls caused further population declines resulting in **extirpation** of the Atlantic puffin from many islands. By the early 1970s, small colonies remained only on Mantinicus Rock and Machias Seal Island.

Atlantic puffins are small, stout seabirds, known for their distinctive black-above and white-below plumage and brightly colored bills. During the breeding season, the large, triangular bill develops bright orange plates with a patch of blue and a yellow border at the base. Billing, a behavior in which the partners rub bills, plays an important role in courtship, pair formation, and mating, all of which occur at sea. Billing continues on shore as part of their monogamous pair-bonding behavior. Each pair selects a nest site and excavates a shallow burrow in a grassy cliff or conceals a nest among rocks. After the single egg is laid, the pair shares incubation duties and cooperate to feed the chick. Puffins fly up to 100 km (62 miles) offshore to feed by swimming underwater using their wings to propel themselves. Puffins use their tongues to hold several small fish – up to 30 – against a series of spines in the roof of their mouth while still capturing other prey. After fledging, young Atlantic puffins require 4–5 years to mature sexually before returning to nest on the same island where they hatched (i.e., **philopatry**).

Atlantic puffins are abundant in other parts of their range, especially in Iceland and Newfoundland, but the two small colonies remaining on the Maine coast were vulnerable to predation, oil spills, and other ecological disturbances. In 1973, after ignoring the advice of senior scientists, Stephen W. Kress decided to restore Atlantic puffins to islands where they formerly nested. Kress began the "Puffin Project" by transplanting two-week-old nestlings from Great Island, Newfoundland, to artificial sod burrows he made on Eastern Egg Rock, a 2.8 ha (7 acre) island located in Muscongus Bay, Maine. Kress ensured that each nesting was fed a handful of vitamin-enriched fish daily until they fledged. From 1973 to 1986 he relocated 954 nestlings, of which 914 successfully fledged. To entice the return of these birds, Kress then placed wooden puffin decoys – life-size and accurately painted – atop large rocks on the island. The strategy worked! Young puffins attracted to Eastern Egg Rock established a population that steadily grew to 104 nesting pairs by 2012.

Unfortunately, the new population of Atlantic puffins attracted predatory gulls. When gulls began killing puffin chicks, Kress again resorted to decoys, in this case those of common terns (small but aggressive seabirds that vigorously defend their territories against gulls). He arranged the decoys to resemble a thriving colony. Moreover, to attract passing terns, Kress broadcast their recorded calls through loudspeakers. As before, this illusion was successful and the terns that re-colonized the island served as bouncers protecting the puffins from excessive gull predation. In 2012, the colony of common terns consisted of 1400 pairs that reared 1600 fledglings, and 144 pairs of roseate terns (an endangered species) successfully fledged 121 young.

The "social attraction" technique developed by Kress is now used worldwide to restore abandoned seabird colonies. The combination of well-designed decoys and recorded calls has re-established populations of 49 species of seabirds in 14 countries. The biologists who once discouraged Kress now use his techniques. Dr. Kress, today widely known as "The Puffin Man," continues his work as the Vice-President for Bird Conservation and Director of the Seabird Restoration Program for the National Audubon Society.

one time, leaving the remainder intact. Moreover, a concentrated nesting period does not provide a long-lasting food supply to support predator populations.

Many herons, egrets, and pelicans employ a harsh strategy when food is in short supply. In these species adult birds begin incubation when the first egg is laid; embryos in these eggs begin developing hours or days earlier than those in later eggs. The result is asynchronous hatching, with the older nestlings gaining an advantage over their younger siblings for a larger share of the food delivered by their parents. All goes well when food is abundant, but in times of want the younger nestlings starve, ensuring survival for the older birds. Under extreme food stress, older nestlings may even kill their smaller, weaker siblings. Asynchronous hatching and siblicide emerge as adaptive strategies that match survival against the availability of adequate food.

Despite their benefits, these strategies incur some costs. Large concentrations of nesting birds may actually attract predators and, in some cases, the toxic properties of their nitrogen-rich droppings kill the very trees supporting their nests, thereby forcing relocation of the colony. Colonial nesting also increases the likelihood of **epizootics.**

Waterbird colonies remain vulnerable to human disturbances. Late in the 19th century, feathers became fashionable in women's wear, particularly hats, and triggered a profitable trade in the plumes of egrets and other birds, whose value ounce-for-ounce equaled the then-current price of gold. Nesting adults were easily killed, leaving their nestlings to starve, and the colonies soon faded away. This outrage produced the Lacey Act (1900), the first federal law designed to restrict trade in wildlife, including feathers and other body parts. Florida, where plume hunting was particularly rampant, protected waterbird colonies with laws (in 1901) and wardens. Significantly, in 1903 President Theodore Roosevelt declared Pelican Island, Florida a federal sanctuary, thus beginning what today is known as the National Wildlife Refuge System. These and other laws, along with changing fashion styles, enabled the gradual recovery of egrets and other colonial waterbirds. Threats remain however, including oil spills and disturbances from boats, jet skis, and other traffic that may cause nesting birds to abandon their colonies. Current and future monitoring, along with vigilance, continue as priorities for the conservation of waterbird colonies.

A whale of a success

Gray whales again flourish in the coastal waters of North America from Alaska to Mexico. Indeed, their current numbers (25,000–26,000) in the eastern North Pacific may equal the estimated pre-settlement population. In the 19th century, however, gray whales were heavily harvested, partly because they stay near the coastline and were therefore easy targets for whalers who could operate from land without long voyages at sea. They also followed a predictable migration schedule – southward in autumn and northward in spring – which added to their vulnerability; at 5790–7725 km (3600–4800 miles) each way, their migration is the longest of any mammal. Given these circumstances, gray whales faced extinction by the early years of the 20th century and did not begin recovering until gaining protection in 1947, first by action of the International Whaling Commission and later by both the Marine Mammal Protection Act (1972) and Endangered Species Act (1973).

Fortunately, gray whales traveling along the North American coastline seldom venture into international waters and thereby remain under the jurisdiction of sovereign nations (Canada, Mexico, and the United States) that could enforce legal restrictions. With such protection, gray whales recovered and, in 1994, were formally "delisted" as an endangered species. Today, the population is numerous enough to attract boat-loads of whale-watching tourists. Scammon's Lagoon in Baja California, Mexico (a primary calving area and once a favored hunting ground for whalers) and other sites on Baja's western shore are now government sanctuaries for gray whales.

Nonetheless, another population on the Asian coast of the North Pacific remains critically endangered, numbering about 130 of which less than 35 are reproductively active females. A third population once found on the eastern and western coastlines of the North Atlantic seems lost forever. Earlier in history, the unrelenting exploitation of another marine mammal in the North Pacific – the immense Steller's sea cow – drove it into the abyss of extinction just 27 years after its discovery, a fate that later seemed imminent for gray whales had legal protection not been enacted and enforced.

Curiously, the recovered population of gray whales may have become too numerous; evidence gathered since 2000 reveals record numbers of dead whales, including those showing signs of starvation. This suggests that their current numbers may exceed **carrying capacity**. One explanation concerns the diminished availability of food because of changing climatic conditions in the eastern North Pacific, but the matter is far from clear.

When gray whales visit Arctic waters in summer, invertebrates known as amphipods provide the food resources necessary for replenishing the body weight the whales lose during the remainder of the year. Little or no feeding occurs during migration, calving, or breeding, hence the whales only acquire energy-rich blubber when feeding in Arctic seas. Amphipods abound in the fine sands on the floor of the Bering Sea in which the whales gouge deep pits when foraging; no other whales feed in this manner. These activities disturb large areas of the seafloor and inject huge volumes of sediment into the water column, which recycles nutrients otherwise trapped in the sediments. In doing so, the disturbed sites become habitat favorable for establishing new colonies of amphipods, creating a feedback mechanism benefiting both predator and prey and perhaps enriching other marine communities in the Bering Sea.

In 2002, two new species of marine worm were discovered feeding on the bones of a gray whale carcass on the seafloor of Monterrey Canyon, California. *Osedax* (meaning "bone devourer") and a genus later discovered to include many additional species lacks either a mouth or stomach. Instead, the worms tunnel into the lipid-rich whale bones with branching, root-like structures that extract fats and oils. These are then broken down by bacteria living *inside* the roots, a unique type of **symbiosis.** Worldwide, estimates suggest that nearly 70,000 whales of all species die each year, with most carcasses supporting a community of *Osedax* and other scavengers–decomposers (e.g., hagfish) for as long as 80 years. So far, a fauna of more than 400 species has been discovered feeding on whale carcasses. Each site, known as a "whale fall," functions as a nutrient-rich reef that contributes to the biodiversity in what otherwise might be a rather barren and food-short benthic environment.

Ecological challenges

Natural disturbances

Each year between early June and late fall conditions between 10 and 20° north latitude, or about 805 km (500 miles) north of the equator, favor the development of storms. Here, the atmosphere becomes unstable as solar heating warms surface waters and increases evaporation rates. Upward convections of warm, moist air create low-pressure areas known as **tropical depressions**, the first stage in **tropical cyclogenesis.**

Within a tropical depression, warm, moist air rises, then cools and condenses into clouds. Condensation releases heat energy, of which a small fraction converts into mechanical energy that pushes clouds even higher and generates thunderstorms. As it ascends, the warm air draws cooler surface air from the storm's periphery to its core, creating intensive updrafts and more condensation. These conditions release even more energy and drive the storm clouds even higher. High in the troposphere, the Coriolis effect begins rotating the clouds in a counterclockwise motion.

Tropical depressions possess maximum sustained surface winds of less than 63 km hr^{-1} (39 mph), and many soon dissipate. However, if the storm intensifies and its winds sustain speeds of 63–127 km hr^{-1} (74 mph) that circulate in a cyclonic pattern, it becomes a **tropical storm** and receives a name. Storms officially become **tropical cyclones** when their surface winds sustain speeds of 127 km hr^{-1} and circulate around a cloud-free "eye" (Fig. 11.16). In the strongest of these – known as **hurricanes** in North America – sustained wind speeds may reach 314 km hr^{-1} (195 mph).

Many climatic forces influence the path of tropical storms and hurricanes in the western hemisphere, but east–west "trade winds" steer most westward. *En route*, their paths often arc northward, a deflection caused by

Figure 11.16 Satellite view of Hurricane Irene on 24 August 2011, when the eye made landfall in North Carolina. Although only a Category 1 storm at landfall, its outer bands reached far to the north, impacting New England and parts of eastern Canada. National Aeronautics and Space Administration.

(a)   **(b)**

Figure 11.17 (a) After Hurricane Hugo destroyed most trees with red-cockaded woodpecker nests in Francis Marion National Forest, wildlife biologists inserted nest boxes into young pines that survived the storm. (b) The box face resembles chipped-away bark surrounding a natural red-cockaded nest entrance, and the white paint simulates resin weep on the bole. The technique proved successful and has been used elsewhere to increase populations of the endangered woodpecker. Photo courtesy of Brian R. Chapman.

the strong influence of the Coriolis effect on the poleward side of the circulation. The same force creates stronger winds on the upper right quadrant of the storm as it swirls forward. Almost inevitably, the path of a tropical cyclone in the Atlantic Ocean, Caribbean Sea, or Gulf of Mexico encounters a coastline. Hurricanes on North America's Pacific coast occur well to the south and travel far out into the Pacific Ocean or turn toward Mexico south of Cabo San Lucas. Landfall officially occurs when a hurricane's center of circulation crosses the shoreline, but the storm's far-reaching outer bands impact coastal areas much sooner.

The Saffir–Simpson scale, which recognizes five categories based on sustained wind speeds, provides a useful indication of a hurricane's strength. Category 1, the weakest hurricanes, sustains wind speeds of 119–153 km hr^{-1} (74–95 mph), with Category 5 producing winds of speeds >252 km hr^{-1} (157 mph). Tropical storms or hurricanes of any intensity may cause significant wind damage and extensive flooding at the coastline and far inland.

Whereas storm damage to human structures is well publicized, impacts on coastal ecosystems receive little or no attention. Storm-driven winds often carry salt spray well inland, destroying vegetation and changing the composition of plant communities. In Rhode Island, for example, salt spray from Hurricane Gloria (1985) affected the leaf physiology of trembling aspens, thereby limiting their survival. Winds and some of the 3000 tornadoes generated by Hurricane Hugo (1989) snapped

or toppled thousands of longleaf pines in Francis Marion National Forest near Charleston, South Carolina. Before the Category 4 storm, 477 breeding groups of endangered red-cockaded woodpeckers nested in the forest; afterward, 65% of the birds were dead or missing and 87% of the 100-year-old trees with their nesting cavities were destroyed. In response, biologists devised and inserted artificial cavities (Fig. 11.17) into younger trees that survived the storm. By 1994, the woodpecker population had nearly returned to its pre-Hugo level.

Storm-driven waves also stir sediments, increase turbidity, uproot seagrass beds, break up mangrove clumps, and shatter reefs of oysters and branched corals. Rocks and boulders dislodged by strong waves crush or dislocate sessile organisms on rocky seashores. The eggs laid in late summer nests of shorebirds or sea turtles often drown under the waves and surge of strong storms. Despite its intensity (Category 5), Hurricane Andrew (1992) did little damage to coral reefs in southern Florida, but at times other storms in the same area have severely damaged both hard and soft corals.

The seaward return of water from exhausted waves – **rip currents** – intensify during a hurricane, eroding beaches and carrying plants, infauna, and sediments out to sea. Unfortunately, attempts to restore eroded beaches with sand imported from other locations may create new environmental issues. Changes in beach profile or greater compaction of the new surface materials can produce conditions unsuitable for sea turtle nests, resulting in significantly reduced nesting success for a year or more following renourishment. Beach renourishment may also smother infaunal organisms of importance to local food chains. Most renourished beaches eventually return to their previous condition in time, although subsequent storms may result in more sand having to be added, setting back the recovery process.

Fish communities in estuaries generally fare well in hurricanes but they may experience temporary changes in composition, principally because of decreasing salinities resulting from heavy inflows of fresh water. Nonetheless, fish communities seem well adapted to hurricane disturbances and may in fact represent an example **resilience theory** in which a community's structure (e.g., predator–prey relationships) remains intact even with an exchange of species.

Superstorm Sandy pounded the northeast coastline of the United States in October 2012. An unusually powerful storm surge devastated thousands of homes, commercial buildings, and infrastructure, but also washed away uncountable tons of sand from beaches where red knots and other migrant shorebirds stop each spring to feast on horseshoe crab eggs (*Synchrony at Delaware Bay*). With the sand gone, however, the horseshoe crabs spawning in 2013 faced the prospect of an unusable beach of exposed sod banks, mud flats, and a jumble of debris that would subsequently trigger a disastrous outcome for the shorebirds. With only a few months remaining before the birds were due to arrive, and funded by several agencies, conservationists removed the rubble and hauled in 35,380 metric tonnes (39,000 tons) of sand to rebuild five critical spawning beaches. The work was completed just before the first horseshoe crabs reached the restored beaches, soon followed right on schedule by hordes of hungry shorebirds. With help, the age-old cycle continued for another year.

Since 1995, maximum sustained wind speeds and duration of storms have increased in the western hemisphere. For example, the average wind speed for all Category 4 and 5 tropical cyclones in 1981 was 230 km hr^{-1} (140 mph), whereas the average wind speed for hurricanes of the same intensities in 2006 was 251 km hr^{-1} (156 mph). Although hurricanes seem to appear in "cycles," changes in global climate may increase both their frequency and intensity in the decades ahead.

Tropical cyclones rarely impact the Pacific Coast of North America, but **tsunamis** pose a constant threat to the continent's western coastline. Most waves of most tsunamis originate from seismic activity, particularly earthquakes or volcanic eruptions occurring in or near the ocean. The western coastline of North America forms a segment in the "Ring of Fire" that rims the Pacific Ocean and produces about 90% of the world's earthquakes, all potential sources of tsunamis. The Ring marks zones of contact between adjacent plates in the Earth's crust; the waves develop when plates on the seafloor shift abruptly and displace immense volumes of water.

Once initiated, a tsunami might cross the entire ocean almost unnoticed, gaining height and destructive power only when it reaches shallow coastal waters. On shore, the wall of water rapidly moves inland and, along with subsequent seaward retreat of the floodwaters, destroys everything in its path.

Although far less common, tsunamis also originate from meteor impacts, landslides, and glacier calvings. In July 1958, an earthquake initiated a massive landslide on a hillside above small Alaskan fjord, Lituya Bay. About 31 million m^3 (40 million cubic yards) of rock splashed into Gilbert Inlet at the upper end of the bay and thereby generated the largest tsunami, a metatsunami, recorded in modern times. As the wave traveled down Lituya Bay toward the ocean, it removed the entire spruce forest, along with the top soil, on either side of the fjord. The tsunami reached a maximum height of 524 m (1720 feet), as indicated by the scars on the mountainsides flanking the bay. In effect, the huge wave created its own timber line that will remain until succession renews the forest on the exposed substrate.

Sea-level rise

The enormity of geological time is difficult to comprehend, as are the physical changes wrought over the millennia. Indeed, Earth has undergone countless transformations, among them drifting continents and shifting shorelines. Climatic and geological phenomena also produced numerous global, or **eustatic**, changes in sea levels in the last 2.5 million years. During the Pleistocene Epoch, Earth's orbit and axial tilt shifted repeatedly, causing alternate periods of cooling and warming. A dip in average global temperature of only 3°C (1.8°F) was sufficient to carpet large areas of the northern hemisphere with immense glaciers. These ice sheets continued to expand as more water turned to ice, which caused eustatic decreases in sea levels. Conversely, when the ice sheets melted during interglacial periods, the release of water produced eustatic increases in sea levels. Sea levels during the Pleistocene rose and fell by approximately 130 m (427 feet), with maximum rates of change reaching about 2 cm (0.8 inch) per year.

Today, we face new interests in changing sea levels. These concerns intensified when eustatic sea levels rose about 3.4 mm (0.014 inch) per year during the 1990s, about twice the average global rate observed during the earlier decades of the 20th century. Some scientists once regarded the increase as a temporary anomaly, but recent assessments indicate that eustatic sea levels are not only rising, but advancing at ever-greater rates because of anthropogenic activities that produce "greenhouse gases." The latter, primarily carbon dioxide and methane (both released by combustion of coal and other fossil fuels) form a blanket that traps and holds solar radiation on Earth's surface. Global temperatures increase as these byproducts steadily accumulate in the atmosphere.

Continued increases in sea levels seem sure to result in extensive beach erosion and wetland loss. With the possible exception of mangroves, such changes will affect the habitats and organisms described in this chapter – from barrier islands to oyster reefs – and likely not for the better. Today, managers of coastal resources deal with this issue as yet another dimension in the ongoing struggle to protect the coastal environments of North America.

Readings and references

The North American coasts

Manor, K.H. 2000. *Ecology of Coastal Waters: With Implications for Management*, second edition. Wiley-Blackwell, New York, NY.

Masselink, G., M. Hughes, and J. Knight. 2011. *Introduction to Coastal Processes and Geomorphology*, second edition. Routledge, New York, NY.

Currents and climates

Barry, R.G. and R.J. Chorley. 2009. *Atmosphere, Weather, and Climate*, ninth edition. Routledge, New York, NY.

Marshall, J. and R.A. Plumb. 2007. *Atmosphere, Ocean, and Climate Dynamics: An Introductory Text*. Academic Press, New York, NY.

Wells, N. C. 2011. *The Atmosphere and Ocean: A Physical Introduction*. Wiley, New York, NY.

Features and adaptations

Rocky seashores and tidal pools

Ackerman, J.D., L.R. Walker, F.N. Scatena, and J. Wunderle. 1991. Ecological effects of hurricanes. Bulletin of the Ecological Society of America 72: 178–180. (Describes the effects of wave action on rocky seashore intertidal organisms.)

Carson, R. 1955. *The Edge of the Sea*. The New American Library, New York, NY. (The quote appears on page 98 in a reprint of the original book, published by Houghton Mifflin, Boston, in the same year.)

Connell, J.H. 1972. Community interactions on marine rocky intertidal shores. Annual Review of Ecology and Systematics 31: 169–192.

Connell, J.H. and M.J. Keough. 1983. Disturbance and patch dynamics of subtidal marine animals on hard substrata. In: *The Ecology of Natural Disturbance and Patch Dynamics* (S.T.A. Pickett and P.S. White, eds). Academic Press, New York. NY, pp. 125–147.

Kapraun, D.F. 1980. Summer aspect of algal zonation on a Texas jetty in relation to wave exposure. Contributions in Marine Science 23: 101–109.

Koehl, M. and A.W. Rosenfeld. 2006. *Wave-Swept Shore: The Rigors of Life on a Rocky Coast*. University of California Press, Berkeley, CA.

Metaxes, A. and R.E. Scheibling. 1993. Community structure and organization of tidepools. Marine Ecology Progress Series 98: 187–198.

Paine, R.T. 1974. Intertidal community structure. Oecologia 15: 93–129.

Paine, R.T. and S.A. Levin. 1981. Intertidal landscape: disturbance and the dynamics of a pattern. Ecological Monographs 51: 145–198.

Raffaelli, D.G. and S.J. Hawkins. 1996. *Intertidal Ecology*. Kluwer Academic Publishers, Netherlands.

Sandy seashores

Britton, J.C. and B. Morton. 1989. *Shore Ecology of the Gulf of Mexico*. University of Texas Press. Austin.

Brown, A.C. and A. McLachlan. 1990. *Ecology of Sandy Shores*. Elsevier, Amsterdam.

Buell, A.C., A.J. Pickart, and J.D. Stuart. 1995. Introduction history and invasion patterns of *Ammophila arenaria* on the north coast of California. Conservation Biology 9: 1587–1593.

Eschmeyer, W.N., E.S. Herald, and H. Hammann. 1983. *A Field Guide to Pacific Coast Fishes of North America*. Houghton Mifflin Company, Boston, MA. (Main source of information about California grunion.)

Lonard, R.I. and F.W. Judd. 1980. Phytogeography of South Padre Island, Texas. Southwestern Naturalist 23: 497–510.

Nelson, W.G. 1993. Beach restoration in the southeastern US: environmental effects and biological monitoring. Ocean and Coastal Management 19: 157–182.

Oertel, G.F. and M. Lassen. 1976. Developmental sequences in Georgia coastal dunes and distribution of dune plants. Bulletin of the Georgia Academy of Science 34: 35–48.

Pearse, A.S., H.J. Humm, and G.W. Wharton. 1942. Ecology of sand beaches at Beaufort, N.C. Ecological Monographs 12: 136–190.

Seabloom, E.W. and A.M. Wiedmann. 1994. Distribution and effects of *Ammophila brevigulata* Fern (American beachgrass) on the foredunes of the Washington coast. Journal of Coastal Research 10: 178–188.

Wagner, R.H. 1964. The ecology of *Uniola paniculata* L. in the dune-strand habitat of North Carolina. Ecological Monographs 34: 79–96.

Chesapeake Bay

Baird, D. and R.E. Ulanowicz. 1989. The seasonal dynamics of the Chesapeake Bay ecosystem. Ecological Monographs 59: 329–364.

Baldassarre, G.A. and E.G. Bolen. 2006. *Waterfowl Ecology and Management*, second edition. Krieger Publishing Co. Malabar, FL.

Berry, D. 2008. *Maryland Skipjacks*. Arcadia Publishing, Chicago, IL.

Brietburg, D.L. 1992. Episodic hypoxia in Chesapeake Bay: Interacting effects of recruitment, behavior, and physical disturbance. Ecological Monographs 62: 525–546.

Cooper, S. and G. Brush. 1993. A 2500-year history of anoxia and eutrophication in Chesapeake Bay. Estuaries 16: 617–626.

Dennison, W., R. Orth, K. Moore, et al. 1993. Assessing water quality with submerged aquatic vegetation. BioScience 43: 86–94.

Ernst, H.R. 2003. *Chesapeake Bay Blues, Science, Politics, and the Struggle to Save the Bay*. Rowman and Littlefield Publishing Group, Lanham, MD.

Horton, J. 2003. *Turning the Tide: Saving the Chesapeake Bay*. Island Press, Washington, DC. (Source for filtering capabilities of oysters.)

Jacobs, J.M. 2007. *Mycobacteriosis in Chesapeake Bay Striped Bass: The Interaction of Nutrition and Disease*. University of Maryland, College Park, MD.

Kennedy, V.S. and L.E. Cronin. (eds) 2007. *The Blue Crab: Callinectes sapidus*. Maryland Sea Grant College, University of Maryland, College Park, MD.

Lippson, A.J. and R.L. Lippson. *Life in Chesapeake Bay*, third edition. 2006. Johns Hopkins University Press, New York, NY.

Orth, R.J. and K.A. Moore. 1983. Chesapeake Bay: An unprecedented decline in submerged aquatic vegetation. Science 222: 51–53.

Poag, W., C. Koberl, and W.U. Reimold. 2004. *The Chesapeake Bay Crater: Geology and Geophysics of a Late Eocene Submarine Impact Structure*. Springer-Verlag, Berlin.

Richardson, R.H. 1973. *Chesapeake Bay Decoys, the Men Who Made Them and Used Them*. Tidewater Publishers, Centerville, MD.

Warner, W.W. 1994. *Beautiful Swimmers: Watermen, Crabs, and the Chesapeake Bay*, revised edition. Back Bay Books, Little, Brown, and Co., New York, NY. (An informative and elegant study of a way of life and the creature at its epicenter.)

Wennersten, J.R. 2001. *The Chesapeake, an Environmental History*. Maryland Historical Society, Baltimore, MD.

Mother Lagoon

Buskey, E.J., B. Wysor, and C. Hyatt. 1998. The role of hypersalinity in the persistence of the Texas "brown tide" in the Laguna Madre. Journal of Plankton Research. 20: 1553–1563.

Carroll, J.J. 1927. Down Bird Island way. Wilson Bulletin 39: 195–208.

Chaney, A. H., B.R. Chapman, J.P. Karges, et al. 1978. *Use of Dredged Material Islands by Colonial Seabirds and Wading Birds in Texas*. US Army Corps of Engineers Dredged Material Research Program Technical Report D-78-8. Vicksburg, MS.

Chapman, B.R. 1988. History of the white pelican colonies in south Texas and northern Tamaulipas. Colonial Waterbirds 11: 275–283.

Copeland, B.J. 1967. Environmental characteristics of hypersaline lagoons. Contributions in Marine Science 12: 207–218.

Copeland, B.J., J.H. Thompson, Jr, and W. Ogletree. 1968. Effects of wind on tidal levels in the Texas Laguna Madre. Texas Journal of Science 20: 196–199.

Hildebrand, H. 1969. Laguna Madre de Tamaulipas: Observations on its hydrography and fisheries. In: *Coastal Lagoons: A Symposium* (A. Ayala Castañares and F. B. Phleger, eds). Universidad Nacional Autónoma de Mexico. México, DF, pp. 679–686.

Lankford, R.R. 1977. Coastal lagoons of Mexico, their origin and classification. In: *Estuarine Processes: Circulation, Sediments, and Transfer of Materials in the Estuary*, Volume II, (M. Wiley, ed.). Academic Press, London, pp. 182–215.

McMahan, C.A. 1968. Biomass and salinity tolerance of shoalgrass and manatee-grass in the lower Laguna Madre. Journal of Wildlife Management 32: 501–506.

Onuff, C.P. 1996. Seagrass responses to long-term light reduction by brown tide in upper Laguna Madre, Texas: distribution and biomass patterns. Marine Ecology Progress Series 138: 219–231.

Pulich, W., Jr. 1980. Ecology of a hypersaline lagoon: The Laguna Madre. In: *Proceedings of the Gulf of Mexico Coastal Ecosystem Workshop* (P. L. Fore and R. D. Peterson, eds). FWS/OBS 80/30. US Fish and Wildlife Service, Albuquerque, NM, pp. 103–122.

Pulich, W., Jr. 1986. Primary production potential of blue–green algal mats on southern Texas tidal flats. Southwestern Naturalist 31: 39–47.

Robertson, B. 1985. *Wild Horse Desert: The Heritage of South Texas*. New Santander Press, Edinburg, TX.

Tunnell, J.W., Jr and F.W. Judd. (eds) 2002. *The Laguna Madre of Texas and Tamaulipas*. Texas A&M University Press, College Station, TX. (Contains chapters written by experts who describe origin and environmental and biological aspects of the Laguna Madre.)

Weller, M.W. 1964. Distribution and migration of the redhead. Journal of Wildlife Management 28: 69–103.

Submergent communities

Seagrass meadows

Audubon, J.J. 1926. The Turtlers. In: *Delineations of American Scenery and Character*. 1970 reprint by Arno Press, New York, pp. 194–202. (The quote appears on page 199.)

Ball, I.J., R.D. Bauer, K. Vermeer, and M.J. Rabenberg. 1989. Northwest riverine and Pacific Coast. In: *Habitat Management for Migrating and Wintering Waterfowl in North America* (L.M. Smith, R.L. Pederson, and R.M. Kaminski, eds). Texas Tech University Press, Lubbock, TX, pp. 429–449.

Bjorndal, K.A. 1980. Nutrition and grazing behavior of the green turtle, *Chelonia mydas*. Marine Biology 56: 147–154.

Camp, D.K., S.P. Cobb, and J.F. Van Breedveld. 1973. Overgrazing of seagrasses by a regular urchin, *Lytechinus variegatus*. BioScience 23: 37–33.

Carlton, J.T., G.J. Vermeij, D.R. Lindberg, D.A. Carleton, and E.C. Dudley. 1991. The first historical extinction of a marine invertebrate in an ocean basin: the demise of the eelgrass limpet *Lottia alveus*. Biological Bulletin 180: 72–80.

Cornelius, S.E. 1977. Food resource utilization by wintering redheads on the lower Laguna Madre. Journal of Wildlife Management 41: 374–385.

Green, E.P. and F.T. Short. (eds) 2003. *World Atlas of Seagrasses*. University of California Press, Berkeley, CA.

Harrison, P.G. and R.E. Bigley. 1982. The recent introduction of the seagrass *Zostera japonica* Aschers. and Graebn. to the Pacific Coast of North America. Canadian Journal of Fisheries and Aquatic Science 39: 1642–1648.

Hemminga, M.A. and C.M. Duarte. 2000. *Seagrass Ecology*. Cambridge University Press, New York, NY.

Kirby, R.E. and H.H. Obricht, III. 1982. Recent changes in the North American distribution of wintering Atlantic brant. Journal of Field Ornithology 53: 333–341.

Kramer, G.W. and R. Migoya. 1989. The Pacific Coast of Mexico. In: *Habitat Management for Migrating and Wintering Waterfowl in North America* (L.M. Smith, R.L. Pederson, and R.M. Kaminski, eds) Texas Tech University Press, Lubbock, TX, pp. 507–528.

Lehman, R.H. 2013. *Marine Plants of the Texas Coast*. Texas A&M University Press, College Station, TX.

Lobel, P.S. and J.C. Ogden. 1981. Foraging by the herbivorous parrotfish, *Sparisoma radians*. Marine Biology 64: 173–183.

McKone, K.L. and C.E. Tanner. 2009. Role of salinity in the susceptibility of eelgrass *Zostera marina* to the wasting disease pathogen *Labyrinthula zostera*. Marine Ecology Progress Series 377: 123–130.

Menzies, R.J., J.S. Zaneveld, and R.M. Pratt. 1967. Transported turtle grass as a source of organic enrichment of abyssal sediments off North Carolina. Deep-Sea Research 14: 111–112.

Mitchell, C.A., T.W. Custer, and P.J. Zwank. 1994. Herbivory on shoalgrass by wintering redheads in Texas. Journal of Wildlife Management 58: 131–141.

Moffitt, J. and C. Cottam. 1941. Eelgrass depletion of the Pacific Coast and its effect upon black brant. Wildlife Leaflet 204, US Fish and Wildlife Service, Washington, DC.

Moore, D. 1963. Distribution of the sea grass *Thalassia* in the United States. Bulletin of Marine Science of the Gulf and Caribbean 13: 329–342.

Orth, R. J., K.L. Heck, Jr, and J. van Montfrans. 1984. Faunal communities in seagrass beds: a review of the influence of plant structure and prey characteristics on predator–prey relationships. Estuaries 7: 339–350.

Pohle, D.G., V.M. Bricelj, and Z. Garcia-Esquivel. 1991. The eelgrass canopy: an above bottom refuge from benthic predators for juvenile bay scallops *Argopecten irradians*. Marine Ecology Progress Series 74: 47–59.

Provancha, J.A. and C.R. Hall. 1991. Observations of associations between seagrass beds and manatees in east central Florida. Florida Scientist 54: 87–98.

Randall, J.E. 1965. Grazing effect on seagrasses by herbivorous reef fish in the West Indies. Ecology 46: 255–260.

Reep, R.L. and R.K. Bonde. 2006. *The Florida Manatee, Biology and Conservation*. University of Florida Press. Gainesville, FL.

Short, F.T., A.C. Mathieson, and J.I. Nelson. 1986. Recurrence of the eelgrass wasting disease at the border of New Hampshire and Maine, USA. Marine Ecology Progress Series 29: 89–92.

Short, F.T., B.W. Ibelings, and C.D. Hartog. 1988. Comparison of a current eelgrass disease to the wasting disease of the 1930s. Aquatic Biology 30: 295–304.

Short, F.T., L.A. Muehlstein, and D. Porter. 1987. Eelgrass wasting disease: cause and recurrence of a marine epidemic. Biological Bulletin 173: 557–562.

Stauffer, R.C. 1937. Changes in the invertebrate community of a lagoon after disappearance of the eelgrass. Ecology 18: 427–431.

Thayer, G.E., W.J. Kenworthy, and M.S. Fonseca. 1984. The ecology of eelgrass meadows of the Atlantic coast: a community profile. US Fish and Wildlife Service, FWS/OBS-3/02.

Van Montfrans, J., R.J. Orth, and S.A. Vay. 1982. Preliminary studies of grazing by *Bittium varium* on eelgrass periphyton. Aquatic Biology 14: 75–89.

Ward, D.H., C.J. Markon, and D.C. Douglas. 1997. Distribution and stability of eelgrass at Izembek Lagoon, Alaska. Aquatic Botany 58: 229–240.

Weller, M.W. 1964. Distribution and migration of the redhead. Journal of Wildlife Management 28: 64–103.

Zieman, J.C. 1982. The ecology of the seagrasses of south Florida: a community profile. US Fish and Wildlife Service, Office of Biological Services, Washington, DC. FWS/OBS-82/25.

Forests in the ocean

Dayton, P.K. 1985. Ecology of kelp communities. Annual Review of Ecology and Systematics 16: 215–245.

Duggins, D.O. 1980. Kelp beds and sea otters: an experimental approach. Ecology 61: 447–453.

Duggins, D. O., C. A. Simenstad, and J. A. Estes. 1989. Magnification of secondary production by kelp detritus in coastal marine ecosystems. Science 245: 170–173.

Erlandson, J.M., H.H. Graham, B.J. Bourque, et al. 2007. The kelp highway hypothesis: marine ecology, and the peopling of the Americas. Journal of Island and Coastal Archaeology 2: 161–174.

Estes, J.A. and J.F. Palmisano. 1974. Sea otters, their role in structuring near-shore Communities. Science 185: 1058–1060. (Proposes sea otters are a keystone species in kelp forests.)

Foster, M.S. and D.R. Schiel. 1985. The ecology of giant kelp forests in California: a community profile. US Fish and Wildlife Service Biological Report 85(7.2).

Foster, M.S. and D.R. Schiel. 1988. Kelp communities and sea otters: keystone species or just another brick in the wall? In: *The Community Ecology of Sea Otters* (G.R. VanBlaricom and J.A. Estes, eds) Ecological Studies, Volume 65. Springer–Verlag, New York, NY, pp. 92–115.

Kvitek, R.G., D. Shull, D. Canestro, E.C. Bowlby, and B.L. Troutman. 1989. Sea otters and benthic prey communities in Washington state. Marine Mammal Science 5: 266–280.

Love, J.A. 1992. *Sea Otters*. Fulcrum Publications, Golden, CO.

Mills, L.S., M.E. Soule, and D.F. Doak. 1993. The keystone species concept in ecology and Conservation. BioScience 43: 219–224.

Paine, R.T. 1969. A note on trophic complexity and community stability. American Naturalist 103: 91–93. (Suggests "keystone species" may govern community structure.)

Pringle, H. 2011. The 1st Americans. Scientific American 305(5): 36–41, 44–45.

Rosenthal, R.J., W.D. Clarke, and P.K. Dayton. 1974. Ecology and natural history of a stand of giant kelp, *Macrocystis pyrifera*, off Del Mar, California. Fisheries Bulletin 72: 670–684.

Simenstad, C.A., J.A. Estes, and K.W. Kenyon. 1978. Aleuts, sea otters, and alternate stable-state communities. Science 200: 403–411.

Oyster reefs

Burrell, V.G. 1986. Species profiles; life histories and environmental requirements of coastal fishes and invertebrates (south Atlantic): American oyster. US Fish and Wildlife Service Biological Report 82(11.570).

Christenson, A.L. 1985. The identification and study of Indian shell middens in eastern North America: 1643–1861. North American Archeologist 6: 227–243.

Cressman, K.A., M.H. Posey, M.A. Mallin, et al. 2003. Effects of oyster reefs on water quality in a tidal creek estuary. Journal of Shellfish Research 22: 753–762.

Doran, E., Jr. 1965. Shell roads in Texas. Geographical Review 55: 223–240.

Galtsoff, P. 1964. The American oyster, *Crassostrea virginica* Gmelin. US Fish and Wildlife Service Bulletin 64: 1–480.

Goldthwait, R.P. 1935. The Damarsicotta shell heaps and coastal stability. American Journal of Science 230: 1–13.

Grabowski, J.H. and C. H. Peterson. 2007. Restoring oyster reefs to recover ecosystem services. In: *Ecosystem Engineers; Plants to Protists* (K. Cuddington, J. Beyers, W. Wilson, and A. Hastings, eds). Academic Press, Burlington, MA, pp. 281–298.

Gutiérrez, J.L., C.G. Jones, D.L. Strayer, and O.O. Irbarne. 2003. Mollusks as ecosystem engineers: the role of shell production in aquatic habitats. Oikos 101: 79–90.

Meyer, D.L. and E.C. Townsend. 2000. Faunal utilization of created intertidal eastern oyster (*Crassostrea virginica*) reefs in the southeastern United States. Estuaries 23: 33–45.

National Research Council. 2004. Non-native oysters in the Chesapeake Bay. National Academy Press, Washington, DC. (Pages 60–72 contain a detailed description of the biology and ecology of the eastern oyster.)

Newell, R.I.E. 1988. Ecological changes in Chesapeake Bay: are they the result of overharvesting the American oyster, *Crassostrea virginica*? In: *Understanding the Estuary: Advances in Chesapeake Bay Research* (M.P. Lynch and E.C. Krome, eds). Chesapeake Research Consortium Publication 129. Baltimore, MD, pp. 536–546.

Plinket, J.T. and M. La Peyre. 2005. Oyster reefs as fish and macroinvertebrate habitat in Barataria Bay, Louisiana. Bulletin of Marine Science 77: 155–164.

Rothschild, B.J., J.S. Ault, P. Goulletquer, and M. Héral. 1994. Decline of the Chesapeake Bay oyster population: a century of habitat destruction and overfishing. Marine Ecology Progress Series 111: 29–39.

Thompson, V.D. and J.E. Worth. 2011. Dwellers by the sea: Native American adaptations along the southern coast of eastern North America. Journal of Archaeological Research 19: 51–101.

Emergent communities

Atlantic tidal marshes

Bayard, T.S. and C.S. Elphick. 2011. Planning for sea level rise: quantifying patterns of saltmarsh sparrow (*Ammodramus caudacutus*) nest flooding under current sea level conditions. Auk 128: 393–403.

Bourn, W.S. and C. Cottam. 1950. Some biological effects of ditching tidewater marshes. Research Report 19, US Fish and Wildlife Service, Washington, DC.

Clarke, J.A., B.A. Harrington, T. Hruby, and F.E. Wasserman. 1984. The effect of ditching for mosquito control on salt marsh use by birds in Rowley, Massachusetts. Journal of Field Ornithology. 55: 160–180.

Coker, R.E. 1920. The diamondback terrapin: past, present, future. Science Monthly 11: 171–186.

Dorcas, M.E., J.D. Wilson, and J.W. Gibbons. 2007. Crab trapping causes population decline and demographic changes in diamondback terrapin over two decades. Biological Conservation 137: 334–340.

Gibbons, J.W., J.E. Lovich, A.D. Tucker, et al. 2001. Demographic and ecological factors affecting conservation and management of diamondback terrapins (*Malaclemys terrapin*) in South Carolina. Chelonian Conservation and Biology 4: 66–74.

Grosse, A.M., J.C. Maerz, J. Hepinstall-Cymerman, and M.E. Dorcas. 2011. Effects of roads and crabbing pressures on diamondback terrapin populations in coastal Georgia. Journal of Wildlife Management 75: 762–770.

Haines, E.B. 1977. The origins of detritus in Georgia salt marsh estuaries. Oikos 29: 254–260.

Miller, W.R. and F.E. Egler. 1950. Vegetation of the Wequetwquock-Pawatuck tidal marshes. Ecological Monographs 20: 141–172.

Teal, J. 1962. Energy flow in the salt marsh ecosystem of Georgia. Ecology 43: 614–624.

Teal, J. and M. Teal. 1969. *Life and Death in the Salt Marsh*. Little, Brown, Boston, MA.

Walsh, J., I. Kovach, K.J. Babbit, and K.M. O'Brien. 2012. Fine-scale population structure and symmetrical dispersal in an obligate salt-marsh passerine, the saltmarsh sparrow (*Ammodramus caudacutus*). Auk 129: 247–258.

Marshes of the Gulf Coast

Carter, J. and B.P. Leonard. 2002. A review of the literature on the worldwide distribution, spread of, and efforts to eradicate the coypu (*Myocaster coypus*). Wildlife Society Bulletin 30: 162–175.

Chabreck, R.H. 1988. *Coastal Marshes, Ecology and Wildlife Management*. University of Minnesota Press, Minneapolis.

Chabreck, R.H., T. Joanen, and S.L. Paulus. 1989. Southern coastal marshes and lakes. In: *Habitat Management for Migrating and Wintering Waterfowl in North America* (L.M. Smith, R.L. Pederson, and R.M. Kaminski, eds) Texas Tech University Press, Lubbock, pp. 249–277.

Chabreck, R.H. and A.W. Palmisano. 1973. The effects of Hurricane Camille on the marshes of the Mississippi River Delta. Ecology 54: 1118–1123.

Coleman, J.M., H.H. Roberts, and G.W. Stone. 1998. Mississippi River Delta: an overview. Journal of Coastal Research 14: 698–716.

Gosselink, J.G. 1984. The ecology of delta marshes of coastal Louisiana: a community profile. US Fish and Wildlife Service FWS/OBS–84/09.

Lynch, J.J. 1941. The place of burning in the management of Gulf Coast wildlife refuges. Journal of Wildlife Management 5: 454–457.

Lynch, J.J., T. O'Neil, and D.W. Lay. 1947. Management significance of damage by geese and muskrats to Gulf Coast marshes. Journal of Wildlife Management 11: 50–76.

Morton, R.A., J.C. Bernier, J.A. Barras, and N.F. Ferina. 2005. Historical subsidence and wetland loss in the Mississippi Delta Plain. Gulf Coast Association of Geological Societies Transactions 55: 555–571.

O'Neil, T. 1949. *The Muskrat in the Louisiana Coastal Marshes, A Study of the Ecological, Geological, Biological, Tidal, and Climatic Factors Governing the Production and Management of the Muskrat Industry in Louisiana*. Louisiana Department of Wild Life and Fisheries, New Orleans, LA.

Otvos, E.G. 2005. Cheniers. In: *Encyclopedia of Coastal Science* (M.L. Swartz, ed.) Springer, Netherlands, pp. 233–235.

Penfound, W.T. and E.S. Hathaway. 1938. Plant communities in the marshlands of southeastern Louisiana. Ecological Monographs 8: 1–56

Russell, J.R. and H.V. Howe. 1935. Cheniers of southwestern Louisiana. Geographical Review 25: 449–461.

Visser, J.M., C.E. Sasser, R.H. Chabreck, and R.G. Linscombe. 1998. Marsh vegetation types of the Mississippi Deltaic Plain, Louisiana, USA. Estuaries 21 (4B): 818–828.

Visser, J.M., C.E. Sasser, R.H. Chabreck, and R.G. Linscombe. 1999. Long-term vegetation change in Louisiana tidal marshes, 1968–1992. Wetlands 19: 168–175.

Visser, J.M., C.E. Sasser, R.H. Chabreck, and R.G. Linscombe. 2000. Marsh vegetation types of the Chenier Plain, Louisiana, USA. Estuaries 23: 318–327.

Mangrove islands and thickets

Craighead, F.C. and V.C. Gilbert. 1972. The effects of Hurricane Donna on the vegetation of southern Florida. Quarterly Journal of the Florida Academy of Science 25: 1–28.

Lugo, A.E. 1980. Mangrove ecosystems: successional or steady state? Journal of Tropical Ecology 2: 287–288.

Lugo, A.E. 1997. Old-growth mangrove forests in the United States. Conservation Biology 11: 11–20.

Lugo, A.E. and S.C. Snedaker. 1974. The ecology of mangroves. Annual Review of Ecology and Systematics 5: 39–64.

Odem, W.E. and C.C. McIvor. 1990 mangroves. In: *Ecosystems of Florida* (R.L. Myers and J.J. Ewel, eds) University of Central Florida Press, Orlando, FL, pp. 517–548.

Odem, W.E., C.C. McIvor, and T.J. Smith, III. 1982. The ecology of mangroves of south Florida: a community profile. US Fish and Wildlife Service FWS/OBS-81-24.

Rabinowitz, D. 1978. Dispersal properties of mangrove propagules. Biotropica 10: 47–57.

Simberloff, D.S. and E.O. Wilson. 1969. Experimental zoogeography of islands: the colonization of empty islands. Ecology 50: 278–296.

Simberloff, D.S., B.J. Brown, and S. Lowrie. 1978. Isopod and insect root borers may benefit Florida mangroves. Science 201: 630–632. (For an alternative view, see Rehm, A. and H.J. Humm. 1973. *Sphaeroma terebrans*: a threat to the mangroves of southwestern Florida. Science 182: 173–174.)

Tomlinson, P.B. 1986. *The Botany of Mangroves*. Cambridge University Press, New York, NY.

Some associated communities

Barrier islands

Alexander, J. and J. Lazell. 1992. *Ribbon of Sand, the Amazing Convergence of the Ocean and the Outer Banks*. University of North Carolina Press, Chapel Hill, NC.

Bourdeau, P.F. and H.J. Oosting. 1959. The maritime live oak forest in North Carolina. Ecology 40: 148–152.

Bratton, S.P. and K. Davison. 1987. Disturbance and succession in Buxton Woods, Cape Hatteras, North Carolina. Castanea 52: 166–179.

Britton, J.C. and B. Morton. 1989. *Shore Ecology of the Gulf of Mexico*. University of Texas Press, Austin, TX.

Bullard, F.M. 1942. Source of beach and river sands on the Gulf Coast of Texas. Geological Society of America Bulletin 53: 1021–1044.

Dolan, R. and H. Lins. 2000. *The Outer Banks of North Carolina*. US Geological Survey Professional Paper 1177-B. (Fourth printing of a report published in 1966.)

Drawe, D.L. and K.R. Katner. 1978. Effect of burning and mowing on vegetation of Padre Island. Southwestern Naturalist 23: 273–278.

Drawe, D.L., K.R. Katner, W.H. McFarland, and D.D. Neher. 1981. Vegetation and soil properties of five habitat types on North Padre Island. Texas Journal of Science 33: 145–157.

Fearnley, S.M., M.D. Miner, M. Kulp, C. Bohling, and S. Penland. 2009. Hurricane impact and recovery shoreline change analysis of the Chandeleur Islands, Louisiana, USA: 1855 to 2005. Geo-Marine Letters 29: 455–466.

Fish, N.H. 1959. Padre Island and the Laguna Madre flats, coastal south Texas. Coastal Geography Conference, Louisiana State University, 2: 103–151.

Fisher, J.J. 1968. Barrier island formation: discussion. Geological Society of America Bulletin 79: 1412–1426.

Hopkins, D.M. and R.W. Hartz. 1978. Coastal morphology, coastal erosion, and barrier islands of the Beaufort Sea, Alaska. US Department of Interior, Geological Survey, Open File Report 78-1063. (Describes environment on Alaskan barrier islands.)

Hoyt, J.H. 1966. Air and sand movements to the lee of dunes. Sedimentology 7: 137–143.

Hoyt, J.H. 1967. Barrier island formation. Geological Society of America Bulletin 78: 1125–1136.

Lonard, R.I. and F.W. Judd. 1980. Phytogeography of South Padre Island, Texas. Southwestern Naturalist 23: 497–510.

Morton, R.A. 2008. Historical changes in the Mississippi-Alabama barrier-island chain and the roles of extreme storms, sea level, and human activities. Journal of Coastal Research 24: 1587–1600.

Otvos, E.G., Jr. 1970. Development and migration of barrier islands, northern Gulf of Mexico. Geological Society of America Bulletin 81: 241–246.

Pilkey, O.H., W.J. Neal, S.R. Riggs, et al. 1998. *The North Carolina Shore and its Barrier Islands*. Duke University Press, Durham, NC.

Reimnitz, E. and D.K. Maurer. 1979. Eolian sand deflation – a cause for gravel barrier islands in arctic Alaska? Geology 7: 507–510.

Riggs, S.R., D. Ames, S. Culver, and D. Mallison. 2011. *The Battle for North Carolina's Coast: Evolutionary History, Present Crisis, and Vision for the Future*. University of North Carolina Press, Chapel Hill, NC.

Smith, E.H. 2002. Barrier islands. In: *The Laguna Madre of Texas and Tamaulipas* (J.W. Tunnell, Jr and F.W. Judd, eds). Texas A&M University Press, College Station, TX, pp. 127–136.

Tackett, N.W. and C.B. Craft. 2010. Ecosystem development on a coastal barrier island dune chronosequence. Journal of Coastal Research 26: 736–742.

Wagner, R.H. 1964. The ecology of *Uniola paniculata* L. in the dune-strand habitat of North Carolina. Ecological Monographs 34: 79–96.

Coral reefs

Baum, J.K. and B. Worm 2009. Cascading top-down effects of changing oceanic predator abundances. Journal of Animal Ecology 78: 699–714.

Brown, B.E. 1997. Coral bleaching: causes and consequences. Coral Reefs 16: 129–138.

Chabanet, P., H. Ralambondrainy, M. Amaniev, et al. 1997. Relationships between coral reef substrata and fish. Coral Reefs 16: 93–102.

Hoegh-Guldberg, O. 1999. Climate change, coral bleaching and the future of the world's corals. Marine and Freshwater Research 50: 839–866.

Kaplan, E.H., R.T. Peterson, and S.L. Kaplan. 1999. *A Field Guide to Coral Reefs: Caribbean and Florida* (Peterson Field Guide Series). Houghton Mifflin, Boston, MA.

Morrissey, J. and J.L. Sumich. 2008. *Introduction to the Biology of Marine Life*, nineth edition. Jones & Bartlett Publishers, Sudbury, MA.

Ogden, J.C., R.A. Brown, and N. Salesky. 1973. Grazing by the echinoid *Diadema antillaurum phillippi* – formation of halos around West Indian reef patches. Science 182: 715–717.

Prugh, L.R., C.J. Stoner, C.W. Epps, et al. 2009. The rise of the mesopredator. BioScience 59: 779–791.

Rezak, R., T.J. Bright, and D.W. McGrail. 1985. *Reefs and Banks of the Northwestern Gulf of Mexico: Their Geological, Biological, and Physical Dynamics*. Wiley, New York, NY.

Ruppert, J.L.W., M.J. Travers, L.L. Smith, et al. 2013. Caught in the middle: combined impacts of shark removal and coral loss on the fish communities of coral reefs. PLoSONE 8(9): e74648.

Sale, P.F., J.A. Guy, and W.J. Steel. 1994. Ecological structure of assemblages of coral reef fishes on isolated patch reefs. Oecologia 98: 83–99.

Schrag, D.P. and B.K. Linsley. 2002. Corals, chemistry, and climate. Science 296: 277–278.

Whitney, E., D.B. Means, A. Rudloe, and E. Jadazewski. 2004. *Priceless Florida: Natural Ecosystems and Native Species*. Pineapple Press, Sarasota, FL.

Maritime forests

Bellis, V.J. 1995. *Ecology of Maritime Forests of the Southern Atlantic Coast: A Community Profile*. National Biological Service, Biological Report 30. Washington, DC.

Bourdeau, P.F. and H.J. Oosting. 1959. The maritime live oak forest in North Carolina. Ecology 40: 148–152.

Boyce, S.G. 1954. The salt spray community. Ecological Monographs 24: 29–67.

Oosting, H J. 1954. Ecological processes and vegetation of the maritime strand in the southeastern United States. Botanical Review 23: 226–262.

Wells, B.W. 1938. A new forest climax: the salt spray climax of Smith Island, NC Bulletin of the Torrey Botanical Club 66: 629–634.

Wells, B.W. and I.V. Shunk. 1938. Salt spray: an important factor in coastal ecology. Bulletin of the Torrey Botanical Club 65: 485–492.

Wood, V.S. 1995. *Live Oaking, Southern Timber for Tall Ships*. Northeastern University Press, Boston, MA.

Highlights

Synchrony at Delaware Bay

Berkson, J. and C.N. Shuster, Jr. 1999. The horseshoe crab: the battle for a true multiple-use resource. Fisheries Management 24: 6–9.

Botton, M.L. and R.E. Loveland. 2000. The diminishing abundance of horseshoe crabs in Delaware Bay: potential impacts on migrant shorebirds. American Zoologist 40: 950–951.

Botton, M.L., R.E. Loveland, and R. Jacobson. 1994. Site selection by migratory shorebirds in Delaware Bay, and its relationship to beach characteristics and abundance of horseshoe crab (*Limulus polyphemus*) eggs. Auk 111: 605–616.

Castro, G. and J.P. Myers. 1993. Shorebird predation on eggs of horseshoe crabs during spring stopover on Delaware Bay. Auk 110: 927–930.

Karpanty, S.M., J.D. Fraser, J. Berkson, et al. 2006. Horseshoe crab eggs determine red knot distribution in Delaware Bay. Journal of Wildlife Management 70: 1704–1710.

Levin, J., H.D. Hochstein, and T.J. Novitsky. 2003. Clotting cells and *Limulus* amoebocyte lysate: an amazing analytical tool. In: *The American Horseshoe Crab* (C.N. Shuster, R.B. Barlow, H.J. Brockmann, eds). Harvard University Press, Cambridge, MA, pp. 310–340.

Morrison, R.I.G., R.K. Ross, and L.J. Niles. 2004. Declines in wintering populations of Red Knots in southern South America. Condor 106: 60–70.

Niles, L.J., J. Bart, H.P. Sitters, et al. 2009. Effects of horseshoe crab harvest in Delaware Bay on red knots: are harvest regulations working. BioScience 59: 153–164.

Waterbird colonies

Alberico, J.A., J.M. Reed, and L.W. Oring. 1991. Nesting near a Common Tern colony increases and decreases Spotted Sandpiper nest predation. Auk 108: 904–910.

Bengston, S.A. 1984. Breeding ecology and extinction of the Greak Auk (*Pinguinus impennis*): anecdotal evidence and conjectures. Auk: 101: 1–12.

Buckley, F.G. and P.A. Buckley. 1972. Hexagonal packing of Royal Tern nests. Auk 94: 36–43.

Burger, J. and C.G. Beer. 1975. Territoriality in the Laughing Gull (*L. atricilla*). Behaviour 55: 307–320.

Chaney, A.H., B.R. Chapman, J.P. Karges, et al. 1978. Use of dredged material islands by colonial seabirds and wading birds in Texas. US Army Corps of Engineers Dredged Material Research Program Technical Report D-78-8. Vicksburg, MS.

Clode, D. 1993. Colonially breeding seabirds: predators or prey? Trends in Ecology and Evolution 8: 336–338.

Danchin, E. and R.H. Wagner. 1997. The evolution of coloniality: the emergence of new perspectives. Trends in Ecology and Evolution 12: 342–347.

Erlich, P.R., D.S. Dobkin, and D. Wheye. 1988. *The Birder's Handbook: A Field Guide to the Natural History of North American Birds*. Simon and Schuster, New York. (Describes bird artifacts worn as clothing and decoration.)

Gaston, A.J. 2004. *Seabirds: A Natural History*. Yale University Press, New Haven, CN.

Kharitonov, S.P. and D. Siegel-Causey. 1988. Colony formation in seabirds. Current Ornithology 5: 223–272.

McIver, S.B. 2003. *Death in the Everglades: The Murder of Guy Bradley, America's First Martyr to Environmentalism*. University Press of Florida, Gainesville. FL.

Mock, D.W., H. Drummond, and C.H. Stinson. 1990. Avian siblicide. American Scientist 78: 438–449.

Tella, J.L. 2002. The evolutionary transition to coloniality promotes higher blood parasitism in birds. Journal of Evolutionary Biology 15: 32–41.

Times Colonist. 2006. Climate quirk kills thousands of birds off Vancouver Island. Victoria Times Colonist. Available at http://www.canada.com/victoriatimescolonist/news/canada/story.html?id=dce2c24e-7b76-419e-bdde-d742c7337ebb (accessed 8 January 2015).

Ward, P. and A. Zahavi. 1973. The importance of certain assemblages of birds as "Information Centres" for food-finding. Ibis 115: 517–534.

Weimerskirch, K. 2013. Seabirds: Individuals in colonies. Science 341: 35–36.

A whale of a success

Anderson, P.K. 1995. Competition, predation, and the evolution and extinction of Steller's sea cow, *Hydrodamalis gigas*. Marine Mammal Science 11 391–394.

Dedina, S. 2000. *Saving the Gray Whale, People, Politics, and Conservation in Baja California*. University of Arizona Press, Tucson.

Glover, A.G., K.M. Kemp, C.R. Smith, and T.G. Dahlgren. 2008. On the role of bore-eating worms in the degradation of marine vertebrate remains. Proceedings of the Royal Society B 275 (1646): 1959–1961.

Johnson, M.K. and C.H. Nelson. 1984. Side-scan sonar assessment of gray whale feeding in the Bering Sea. Science 225: 1150–1152.

Jones, M.L., S.L. Swartz, and S. Leatherwood. (eds) 1984. *The Gray Whale Eschrichtius robustus*. Academic Press, New York.

Moore, S.E., J. Urban R., W.L. Perryman, et al. 2001. Are gray whales hitting "K" hard? Marine Mammal Science 17: 954–958.

Rice, D.W. and A.A. Wolman. 1971. *The Life History and Ecology of the Gray Whale (Eschrichtius robustus)*. Special Publication No. 3, The American Society of Mammalogists..

Smith, C.R. and A.R. Baco. 2003. Ecology of whale falls at the deep-sea floor. Oceanography and Marine Biology Annual Review 41: 311–354.

Swartz, S.L., B.L. Taylor, and D.J. Rugh. 2006. Gray whale *Eschrichtius robustus* population and stock density. Mammal Review 36: 66–84.

Ecological challenges

Natural disturbances

Ackerman, J.D., L.R. Walker, F.N. Scatena, and J. Wunderle. 1991. Ecological effects of hurricanes. Bulletin of the Ecological Society of America 72: 178–180.

Blair, S.M., T.L. Mcintosh, and B.J. Mostkoff. 1994. Impacts of Hurricane Andrew on the offshore reef systems of central and northern Dade County, Florida. Bulletin of Marine Science 54: 961–973.

Brock, K.A., J.S. Reese, and L.M. Erhart. 2009. Effects of artificial beach renourishment on marine turtles: differences between loggerhead and green turtles. Restoration Ecology 17: 297–307.

Chabreck, R.H. and A.W. Palmisano. 1973. The effects of Hurricane Camille on the marshes of the Mississippi River delta. Ecology 54: 1118–1123.

Connell, J.H. and M.J. Keough. 1983. Disturbance and patch dynamics of subtidal marine animals on hard substrata. In: *The Ecology of Natural Disturbance and Patch Dynamics* (S.T.A. Pickett and P.S. White, eds). Academic Press, New York, NY, pp. 125–147.

Doyle, T.W., T. J. Smith III, and M.B. Bobblee. 1995. Wind damage effects of Hurricane Andrew on mangrove communities along the southwest coast of Florida, USA. Journal of Coastal Research 18: 159–168.

Edmiston, H.L., S.A. Fahrny, M.S. Lamb, et al. 2008. Tropical storm and hurricane impacts on a Gulf Coast estuary: Appalachicola Bay, Florida. Journal of Coastal Research 55: 38–49.

Emanuel, K.A. 1987. The dependence of hurricane intensity on climate. Nature 326: 483–485.

Emanuel, K.A. 2005. Increasing destructiveness of tropical cyclones over the past 30 years. Nature 436: 686–688.

Fourqurean, J.W. and L.M. Rutten. 2004. The impact of Hurricane Georges on soft-bottom back reef communities: site- and stress-specific effects on south Florida seagrass beds. Bulletin of Marine Science 75: 239–257.

Goldenberg, S.B., C.W. Landsea, A.M. Mestas-Nunez, and W.M. Gray. 2001. The recent increase in Atlantic hurricane activity: causes and implications. Science 293: 474–479.

Greening, H., P. Doering, and C. Corbett. 2006. Hurricane impacts on coastal ecosystems. Estuaries and Coasts 29: 877–879.

Greenwood, M.F.D., P.W. Stevens, and R.E. Matheson, Jr. 2006. Effects of the 2004 hurricanes on the fish assemblages in proximate southwest Florida estuaries: change in the context of interannual variability. Estuaries and Coasts 29: 985–996.

Hooper, R.G., W.E. Taylor, and S.C. Loeb. 2005. Long-term efficacy of artificial cavities for red-cockaded woodpeckers: lessons learned from hurricane Hugo. In *Red-Cockaded Woodpecker: Road to Recovery* (R. Costa and S.J Daniels, eds). Hancock House Publishers, Surrey, British Columbia, Canada, pp. 430–438.

Killingbeck, K.T. 1988. Hurricane-induced modification of nitrogen and phosphorous resorption in an aspen clone: an example of diffuse disturbance. Oecologia 75: 213–215.

Knutson, T.R., J.L. McBride, J. Chan, et al. 2010. Tropical cyclones and climate change. Nature Geosciences 3: 157–163.

McCoy, E.D., H.R. Mushinsky, D. Johnson, and W.E. Meshaka. 1996. Mangrove damage caused by hurricane Andrew on the southwestern coast of Florida. Bulletin of Marine Science 59: 1–8.

Morton, R.A., and J.A. Barnes. 2011. Hurricane impacts on coastal wetlands: a half-century of storm-generated features from southern Louisiana. Journal of Coastal Research 27: 27–43.

Parker, B.B. 2010. *The Power of the Sea: Tsumanis, Storm Surges, Rogue Waves and our Quest to Predict Disaster*. Palgrave Macmillan, New York.

Peterson, C.H. and M.J. Bishop. 2005. Assessing the environmental impacts of beach renourishment. BioScience 55: 887–896.

Posey, M. and T. Alphin. 2002. Resilience and stability in an offshore benthic community: responses to sediment borrow activities and hurricane disturbance. Journal of Coastal Research 18: 685–697.

Shanks, A.L. and W.G. Wright. 1986. Adding teeth to wave action: the destructive effects of wave-borne rocks on intertidal organisms. Oecologia 69: 420–428.

Steinitz, M.J., M. Salmon, and J. Wyneken. 1998. Beach renourishment and loggerhead turtle reproduction: a seven-year study at Jupiter Island. Journal of Coastal Research 14: 1000–1013.

Tilmant, J.T., R.W. Curry, R.D. Jones, et al. 1994. Hurricane Andrew's effect on marine resources. BioScience 44: 230–237.

Sea-level rise

Davis, R.A., Jr. 2011. *Sea-Level Change in the Gulf of Mexico*. Texas A&M University Press, College Station, TX.

Haq, B.U., J. Nardenbol, and P.R. Vail. 1987. Chronology of fluctuating sea levels since the Triassic. Science 235: 1156–1157.

Kennedy, V.S., R.R. Twilley, J. Kleypas, et al. 2002. *Coastal and Marine Ecosystems & Global Climate Change: Potential Effects on US Resources*. Pew Center on Global Climate Change. Arlington, VA.

Merrifield, M.A., S.T. Merrifield, and G.T. Mitchum. 2009. An anamolous recent acceleration of global sea level rise. Journal of Climate 22: 5772–5781.

National Research Council. 1990. *Sea Level Change*. National Academy of Science, National Academies Press, Washington, DC.

Infobox 11.1. Red tide

Duce, R.A. and N.W. Tindale. 1991. Atmospheric transport of iron and its deposition in the ocean. Limnology and Oceanography 36: 1715–1726.

Flewelling, L.J., J.P. Naar, J.P. Abbott, D.G. Baden, et al. 2005. Red tides and marine mammal mortalities. Nature 435: 755–756.

Gunter, G. 1951. Mass mortality and dinoflagellate blooms in the Gulf of Mexico. Science 113: 250–251.

Gunter, G. 1952. The importance of catastrophic mass mortalities for marine fisheries along the Texas coast. Journal of Wildlife Management 161: 63–69.

Kirkpatrick, B., L.E. Fleming, D. Squicciarini, L.C. Backer, et al. 2004. Literature review of Florida red tide: implications for human health effects. Harmful Algae 3: 99–115.

Magaña, H.A., C. Contreras, and T.A. Villereal. 2003. A historical assessment of *Karenia brevis* in the western Gulf of Mexico. Harmful Algae 2: 163–171.

Mulvaney, K. 2013. *Red tide slaughtering Florida manatees*. Discovery News. Available at http://news.discovery.com/earth/oceans/red-tide-slaughtering-florida-manatees.htm (accessed 5 January 2015).

Potera, C. 2007. Florida red tide brews up drug lead for cystic fibrosis. Science 316: 1561–1562.

Turner, E. and N. Rabalais. 1991. Changes in Mississippi River water quality this century. BioScience 41: 140–147.

Walsh, J.J. and G.J. Kirkpatrick. 2008. Ecology and oceanography of harmful algal blooms in Florida. Continental Shelf Research 28: 1–214.

Walsh, J.J., J.K. Joliff, B.P. Darrow, J.M. Lenes, et al. 2010. Red tides in the Gulf of Mexico: Where, when, and why? Journal of Geophysical Research 111: 1–46.

Walsh, J.J., R.H. Weisberg, J.M. Lenes, F.R. Chen, et al. 2009. Isotopic evidence for dead fish maintenance of Florida red tides, with implications for coastal fisheries over both source regions of the West Florida shelf and within downstream waters of the South Atlantic Bight. Progress in Oceanography 80: 51–73.

Infobox 11.2. The puffin project

Kress, S.W. 1982. The return of the Atlantic Puffin to Eastern Egg Rock, Maine. The Living Bird Quarterly 1: 11–14.

Kress, S.W. 1983. The use of decoys, sound recordings and gull control for re-establishing a tern colony in Maine. Colonial Waterbirds 6: 185–196.

Kress, S.W. 1992. From puffins to petrel. Living Bird 11: 14–21.

Kress, S.W. 2012. Egg Rock update. Newsletter of Project Puffin. Available at http://www.projectpuffin.org (accessed 5 January 2015).

Kress, S.W. and D.N. Nettleship. 1988. Re-establishments of Atlantic Puffins (*Fratercula arctica*) at a former breeding site in the Gulf of Maine. Journal of Field Ornithology 59: 161–170.

Lowther, P.E., A.W. Diamond, S.W. Kress, G.J. Robertson, and K. Russell. 2002. Atlantic Puffin (*Fratercula arctica*). In: *The Birds of North America Online* (A. Poole, ed.). Cornell Lab of Ornithology, Ithaca, NY. Available at http://bna.birds.cornell.edu/bna/species/709 (accessed 5 January 2015).

CHAPTER 12

A Selection of Special Environments

Look deep into nature, and then you will understand everything better.

Albert Einstein

Various locations and ecological associations do not fit easily into Tundra, Desert, or other of the broad ecological units portrayed in previous chapters. Glaciers, granite outcrops, and caves are among these, as are a few well-known places such as the Everglades and the Grand Canyon. Some of these special environments are described here as a conclusion to the overview of North America and its communities.

The Grand Canyon

The Grand Canyon, a colorful maze of steep-walled canyons, isolated plateaus, and desert landscapes in the Colorado Plateau, is 446 km (277 miles) long, nearly 29 km (18 miles) wide, and over 1800 m (600 feet) deep. The visually stunning complex formed about 17 million years ago when uplifts associated with mountain formation raised the plateau and altered weather conditions. The elevational change enhanced precipitation and produced steeper gradients, thus increasing the speed and intensity of runoff. Over time, the Colorado River and its tributaries carved a massive erosional gully through the plateau. From the canyon rim, it is difficult to see or appreciate the ecological diversity within the depths, but river gorges with white-water rapids and languid pools, riparian vegetation, shaded **slot canyons**, wet-weather waterfalls, seeps, and rocky barrens are among the myriad habitats within the canyon system.

John Wesley Powell (1834–1902), a one-armed veteran of the Civil War, led the first successful expedition of the Colorado River in 1869. Traveling downstream from Utah, Powell continued into Arizona where he encountered an immense chasm he aptly named the Grand Canyon (Fig. 12.1). Major Powell, who later became director of the US Geological Survey, was the first American of European descent to travel through the magnificent canyon and geological marvel. Three centuries earlier, Spanish conquistadors had gazed down

from the canyon's rim, but failing to gain access into the gaping abyss, they continued onward across desert and plain in search of golden cities that never were.

The ecological conditions Powell encountered along the Colorado changed a century later when Glen Canyon Dam was constructed upstream of the Grand Canyon. When the gates closed in 1963 – creating Lake Powell – the historic water regime ended on the lower Colorado River, and the altered flow soon modified plant and animal communities in the Grand Canyon.

At least three ecological changes resulted from the new hydrological regime. First, water coursing downstream through the canyon was much cooler than before, the result of being released through the dam from a zone about 76 m (250 feet) below the surface of Lake Powell. Second, the net sediment loads were greatly reduced in the water flowing through Grand Canyon from about 86 million tons to 11 million tons, or an 87% reduction. Finally, Glen Canyon Dam lessened the seasonal and irregular fluctuations in the downstream flow of water, which had previously varied immensely. Snowmelt from the Rocky Mountains, for example, produced high runoff for long periods in late spring and early summer, and flash floods occurred after thunderstorms. Droughts dropped the flow to a fraction of its peak volume, whereas Glen Canyon Dam "smoothed" out these naturally occurring changes in the flow of water through Grand Canyon.

The cool water greatly reduced the habitat which was once thermally suitable for native fish populations. It also created conditions suitable for trout, which were stocked in the main river. Humpback chub, one of eight endemic species in the Colorado River, is among the native fishes affected by both of these circumstances. The cool water interferes with their reproduction and growth, and the trout prey on juvenile chub as well as other native species. Additionally, sandbars no longer developed along the Colorado's banks, and these sites formed backwater eddies where warmer water provided nursery

Ecology of North America, Second Edition. Brian R. Chapman and Eric G. Bolen.
© 2015 John Wiley & Sons, Ltd. Published 2015 by John Wiley & Sons, Ltd.

Figure 12.1 Standing on the rim of the Grand Canyon, 16th century Spanish conquistadores vastly underestimated the canyon's great depth and thereby judged the Colorado River as a mere trickle of muddy water. The riparian zone at the bottom of the canyon acts as an east–west corridor for desert biota, whereas the chasm itself presents a physical barrier, between its north and south rims, for the dispersal of other organisms. Reproduced by permission of Brian R. Chapman.

areas for young chub – habitat of considerable importance in the new cold-water regime. The chub population in the Grand Canyon now depends on the warmer waters of a large tributary for spawning habitat. Humpback chub were therefore among the endangered species appearing (in 1967) on the original federal list and remain there today. The environmental changes wrought by Glen Canyon Dam also affected other fishes endemic to the Colorado River (e.g., three species have been extirpated from the Grand Canyon).

Humpback chub and other organisms gained some relief when the gates of Glen Canyon Dam were opened in 1996. This experiment was designed to restore sandbars, backwater eddies, and other ecological conditions once characteristic of riverine communities in the Grand Canyon. The amount of water released (enough to fill Chicago's 110-story Sears Tower every 17 minutes) still was less than the downstream flow prior to completing the dam. Initial responses from this trial seemed encouraging. For example, sandbar volumes at 34 test sites increased by an average of 53% and new backwater habitats were created because of changes in the shape sandbars. When the dam's operations returned to normal, however, the new habitats eroded and reversed the gains. Since 1996, the experiment has been repeated several times with mixed results. Nonetheless, by 2008 the total population of humpback chub in the Grand Canyon was estimated at 6000–10,000, a significant increase. A combination of events, such as the occasional experimental flooding of the canyon during periods

when the water was warmer than normal and removal of non-native predatory fish (e.g., rainbow trout) from the confluence of the Little Colorado River and the Colorado River, likely contributed to the chub's population growth.

Beyond creating some extraordinary sport fishing, the trout stocked in the Colorado River and its tributaries produced what may be considered a positive outcome. Bald eagles feeding on the thriving trout population represent one of the largest concentrations of wintering eagles in the American Southwest. Few eagles were observed until the rainbow trout fishery was established. Today, however, eagles concentrate regularly on one tributary in the canyon – Nankoweap Creek – where as many as 26 birds fish for spawning trout by wading into the stream and pouncing on fish moving upstream in shallow water. Golden eagles, whose diets more typically consists of small mammals, also catch trout in this manner at Nankoweap Creek.

Before Glen Canyon Dam closed the river, bursts of high water washed away vegetation from the river's flood plain, thereby initiating secondary succession in the scour zone. These circumstances favored plants adapted to disturbances, but even these seldom gained a lasting foothold before another flood erased the community and succession began anew. After the dam closed, however, the downstream flow stabilized and scouring stopped. Thereafter, along with native plants, the **exotic** shrub saltcedar (also called tamarisk) rapidly expanded in the stabilized riparian zones. Saltcedar forms dense thickets, and its deep roots withdraw large

Infobox 12.1 Eugene P. Odum (1913–2002), founder of ecosystem science

By any measure, Eugene Odum remains one of the foremost figures in ecology; indeed, he is widely regarded as the "father of *modern* ecology" because of his pioneering concepts regarding the functional properties of ecosystems. He was fond of noting that "The ecosystem is greater than the sum of its parts." Partly because of the influence of his father (a noted sociologist), Odum believed ecology could act as a link between the natural and social sciences. Because of his efforts, the principles of ecosystem ecology were put into the service of environmentalism and helped establish ecological awareness as part of the national fabric (for which Earth Day is a well-known celebration).

Odum spent most of his career at the University of Georgia. In 1953, he published his landmark textbook, *Fundamentals of Ecology*, which was translated into 12 languages. A pioneer of its genre, the book stood alone in the field until succeeded by a second edition (1959) coauthored with his brother, Howard T. Odum (1924–2002). A fifth edition appeared in 2004 with coauthor G. W. Barrett.

Odum did not coin the term "ecosystem," which was popularized in 1935 by A. G. Tansely, but he did establish the integrative implications of the term as a foundation for our modern concepts of ecology. In particular, he championed the understanding and unraveling of the dynamics of ecosystems, and not just dissecting their parts. This emphasis materialized in 1951, when Odum seized an opportunity that developed when the Atomic Energy Commission decided to establish a plant to produce plutonium on a 101,175 ha (390 square miles) site on the Savannah River. Odum gained funding from the agency to undertake an environmental inventory of the area before construction started, and the work soon began using radioisotopes for tracking the movement of energy and materials through the ecosystem of the area. The resulting studies changed the course of ecology from a focus on structure to one emphasizing function. The facility, at first a collection of makeshift buildings, became the Savannah River Ecology Laboratory.

The Odum brothers completed a landmark study of energy flow in an ecosystem by measuring the metabolic activity of a coral reef; to do so, they sampled the nutrients in water before and after it passed one way through the reef and recorded the differences. Their results indicated that the nutrient contents of the water itself could not sustain the coral, which instead relied on the photosynthetic activities of algae known as zooanthellae living within the coral's calcium skeleton (see *Coral reefs*). The relationship produced a "steady state" of a self-regulating ecosystem. For their work – and unique "black box" approach – the brothers received the Mercer Award from the Ecological Society of America in 1956.

Odum's influence also established the University of Georgia's on-campus Institute of Ecology (now renamed in his honor) and the university's Marine Institute at Sapelo Island. The latter was an ideal location for studying the salt marshes separating the barrier islands from the mainland. He envisioned that the wetland complex represented an operational unit linked together by common energy sources, including those resulting from tidal influences. He later demonstrated the economic value of Georgia's coastal wetlands, which was instrumental in the enactment of the state's Coastal Marshland Protection Act in 1970.

Odum served as president of the Ecological Society of America (1964–1965) and received the society's Eminent Ecologist Award in 1974. He was elected to the National Academy of Sciences in 1970 and named an honorary member of the British Ecological Society in 1974. With his brother Howard, the Royal Swedish Academy awarded Odum with the Crafoord Prize, the highest in environmental science and a recognition often equated with a Nobel Prize. He was named as an Outstanding Educator of America in 1971 and served as a trustee for both The Nature Conservancy and the Conservation Foundation. The American Institute of Biological Sciences awarded him its Distinguished Service Award. Odum was a link in a chain of eminent ecologists, one serving as an academic mentor for another, having earned his doctorate under the guidance of S. Charles Kendeigh (see Infobox 1.1, re Shelford).

amounts of water directly from the water table (it is a phreatophyte). Saltcedar competes with native vegetation in most locations and provides a good example of an introduced species that thrives in altered ecosystems (Infobox 12.1). Despite its generally undesirable qualities, saltcedar provides some species of birds with nesting habitat in the new riparian communities. Of these, the southwestern willow flycatcher is the most notable because of its precarious status as a federally listed endangered species.

The native flora along the Colorado River reflects its course through three of North America's major deserts (Chapter 7). Some plants are associated with the Sonoran Desert, whereas others are affiliated with either the Mojave Desert or the Great Basin. In part, this mix is evidence that the river has acted as a corridor along which plants have dispersed, upstream as well as downstream, thereby extending their distributions to include sites within the giant chasm.

A subspecies of prairie rattlesnake is endemic to the Grand Canyon, where it is widely distributed within the confines of the canyon walls. Curiously, the color of this variant is decidedly vermilion to salmon pink, perhaps reflecting a degree of color matching with the similarly colored sandstones found in much of its habitat. Although these colorful snakes occur in many places within the

Grand Canyon, most are encountered in desert scrub communities. The "pink" rattlesnake is not aggressive, but tests of its venom with laboratory animals suggest that it is among the more toxic of all rattlesnakes.

Burros, highly popularized in western lore as faithful companions of lonely prospectors, rendered considerable damage to the Grand Canyon. Once escaped or released, those in the Grand Canyon overgrazed vegetation deep within the chasm. Burros eliminated some species of shrubs and cropped others to their roots. Soil erosion was also noticeably greater where burros grazed, hard-packed trails criss-crossed the terrain, and seeps and springs were disturbed. Park officials thereafter steadily thinned the herd, eventually removing more than 2800 burros (most were shot) but public furor ended this kind of control in 1969; in 1972, the Wild Horse and Burro Act made killing these animals a felony.

The extent of the damage to the Grand Canyon (which continued) was measured by comparing two sites facing each other across the Colorado River. One had a burro population; the other did not. The study measured the composition and density of vegetation and rodent communities. Without burros, 28 species of vascular plants covered 80% of the area, whereas burros reduced the composition to 19 species and cover to just 20%. More species of small mammals were also recorded at the burro-free site, and their density was almost four times greater than where burros occurred. Moreover, the overgrazed range favored an excessively large population of a single species, the cactus mouse, a situation that often develops in abused communities (i.e., disturbance reduces biodiversity). Such data clearly indicate the detrimental ecological conditions produced by feral burros in the Grand Canyon.

Fortunately, both the story and the dilemma end happily. The troublesome burros were corralled under the auspices of a private organization and distributed to citizens for pets. While expensive, the "Adopt a Burro" program eventually eliminated the problem and the burro-damaged range in the Grand Canyon is slowly returning to its original condition.

Tree squirrels living in the ponderosa pine forests on opposite sides of the Grand Canyon have distinct tassels on their ear tips – a feature not found on other tree squirrels in North America. However, those on the North Rim generally have solid white tails and were originally identified as Kaibab squirrels. On the South Rim, however, the tails of tassel-eared squirrels typically have dark upper surfaces, with white appearing only on the underside. These differences led to the designation of Abert's squirrel for all tassel-eared squirrels except for those on the North Rim. Although originally considered as two species, the striking white-tailed population on the North Rim is now considered a subspecies of Abert's squirrel.

The presence of the two forms fits well with the concept of genetic isolation, one developing on either side of the Grand Canyon. The canyon itself did not split a once contiguous squirrel population into separate areas; the chasm existed eons before the squirrels evolved. Nonetheless, when squirrels eventually appeared, nuclear populations somehow reached the pine forests on opposite sides of the canyon and thereafter evolved the distinctive features we see today. Cooler climatic conditions during the Pleistocene perhaps created conditions suitable – at least temporarily – for dispersal around or across the Grand Canyon (i.e., pine forests once existed where there are now deserts). For the subspecies once recognized as the Kaibab squirrel, the Grand Canyon now represents a barrier virtually without equal as an agent of reproductive isolation.

Caves

Thousands of caves lie hidden in many areas of North America. Many occur in the Interior Plateau from Indiana to Alabama and in the Appalachian and Ozark mountains, but caves are also found in Texas and other western states. Caves are also common in Mexico, although they are relatively rare in Canada. Any estimate of cave numbers, however, is conservative because only those with entrances accessible to humans are enumerated. Countless caves with passages far too small to accommodate humans provide suitable habitat for many kinds of organisms.

Caves can form in response to various geological processes, but many result from the dissolution of limestone producing a geological formation known as **karst**. Rainwater acquires some carbon dioxide from the atmosphere and as it passes through decaying organic matter in the soil, thereby forming increasingly stronger carbonic acid. Cave formation also may occur when underground oil deposits release hydrogen sulfide gas, which seeps upward and forms sulfuric acid in contact with water. Although weak, both carbonic acid and sulfuric acid slowly dissolve limestone primarily along cracks that eventually erode into channels, which then widen into passages. Connected systems of passages (caverns) may include chambers of immense size in labyrinths of great length. Because of their immensity and stunning formations, including icicle-like stalactites, a few caves are included in the US National Park System (e.g., Carlsbad Caverns and Mammoth Cave).

Environmental conditions in caves are significantly unlike those in communities above ground. In general, temperatures and humidity in deep caves remain essentially unchanged throughout both a 24-hour period and a calendar year. However, factors such as the number of entrances and passage size affect air flow, which in turn influence the relative stability in cave environments. Because light is totally absent from the deep recesses of caves, cave environments lack day and night and rarely

Figure 12.2 Immense numbers of Mexican free-tailed bats roost on the ceiling at a cave in Uvalde County, Texas. The cave floor beneath this and other bat roosts accumulate huge quantities of guano, which commonly provides the energy necessary to establish food chains in the lightless interior of caves A lethal fungal disease called "white nose syndrome," which produces a white fuzz on a bat's nose, ears, and wings (inset), spreads rapidly when introduced into cave environments and could eventually eliminate entire colonies of bats. Reproduced by permission of Brian R. Chapman and (inset) Marvin Moriarty of the US Fish and Wildlife Service.

reflect the seasons typical of a biological year. Indeed, the static nature of cave environments only exists elsewhere in the abyssal depths of the ocean.

Because green plants are absent, energy in the form of organic matter must be imported into cave ecosystems. Bats sometimes fulfill this function (Fig. 12 2). After foraging nocturnally outside the cave, many thousands (up to millions) of individuals cluster in relatively small areas within a cave where their droppings accumulate beneath roosts. The deep layers of this material – guano – form the first link in the food chains of some caves in temperate regions. The guano deposits are supplemented by a rain of urine, parasites, and the carcasses of dead bats. Cave floors beneath bat roosts therefore become sites where energy from the outside world begins its journey into the structure of cave communities.

Food chains in caves are short and consist mainly of generalists, species that take advantage of as many foods as possible from the limited supply in caves. Beetles and crickets are among the arthropods that consume guano on the cave floor, and these in turn become prey for predators, some of which are other species of insects. In some caves, guano falling into permanent pools nourishes flatworms, blind crayfish, and other aquatic organisms. Mites and other arachnids are among the relatively few species that forage directly on guano and the steady supply of bat carcasses, and their populations may be quite large. The rich deposits of guano also provide nutrients for fungi and bacteria, which add more links in cave food chains.

Other organisms may transport energy into caves, although few of these match the importance of bats.

Cave crickets are a significant exception, however, and regularly acquire energy from external sources before returning to the darkness of the cave interior. Because their populations may be quite large, cave crickets and their droppings provide important sources of organic matter in the food webs of many caves. Droppings of porcupine, raccoon, and rodents also support arthropod communities inside some caves.

Underground streams also carry energy in the form of organic debris into cave ecosystems, especially during periodic floods. Some cave communities gain sources of energy from leaves and other organic materials falling into sink holes. Cave floors beneath collapsed sink holes may support patches of green vegetation where sunlight shines downward.

Communities inside caves develop in zones, which are defined by temperature and light penetration. A twilight zone occurs just inside entrances, where temperature and light regimes change daily and seasonally. A few species of green plants may survive in the twilight zone, especially ferns and mosses that do not require much light. Some herbivores complement detritivores and scavengers in the twilight zone. As a result, the twilight zone includes the greatest diversity of species. A middle zone lies in complete darkness, but temperature varies somewhat in keeping with seasonal changes in the outside world. Finally, in the deep interior, a zone of relatively constant temperature accompanies the inky blackness. The temperature reflects the annual average for the region (i.e., this zone is warm in tropical caves and cool in temperate caves). Relative humidity also is unchanging in the innermost zone.

(a)   (b)

Figure 12.3 Cave faunas include various species of troglobites, such as (a) the Texas blind salamander and (b) cave-adapted Mexican tetra. Each exhibits the classic features of lost pigmentation and blindness. Note the tetra's fully eyed counterparts that live in streams aboveground in the same vicinity. Reproduced by permission of (a) Joe N. Fries and (b) Richard Borowsky.

Biospeleology is the branch of science dealing expressly with the lives of cave organisms, including their evolutionary history (e.g., comparisons with related species living at the surface). Moreover, the features of some cave organisms seemingly represent cases of evolution working backwards – the loss of eyes and other structures – which makes biospeleology all the more interesting. Most animals never venture into caves, but the few that do are grouped using a classification system based on the degree to which each species has adapted to cave life.

The first of these are **trogloxenes** (literally, "cave guests"), that spend much but not all of their lives in caves. Trogloxenes typically use caves for reproduction, roosting, or hibernation but regularly visit the surface, usually for feeding. Bats and cave crickets are good examples of this group. Among birds in North America, only cave swallows depend on caves but they build nests only in the twilight zone near entrances of caves.

Troglophiles (literally, "cave lovers") can successfully complete their life cycles in caves or, with equal success, on the surface, where they select cool, dark, and damp habitats similar to cave environments. This group includes representative species of salamanders, crayfish, spiders, and insects.

Finally, animals known as **troglobites** (literally "cave livers") spend their entire lives exclusively in the blackest recesses of caves (Fig. 12.3). Troglobites have evolved from ancestors that once were troglophiles and, because their adaptations to cave life are so extreme, most troglobitic species are unable to survive elsewhere. Troglobites typically lack pigmentation; they require no camouflage or protection from the sun, and skin color for courtship rituals are unnecessary in perpetual darkness. Likewise,

troglobites are often blind and, with the loss of functional eyes, these animals evolved other types of highly developed sense organs. Elaborate sense organs on the lateral lines of one kind of cave fish, for example, detect minute changes in water movements, and the fish thereby locate their prey and avoid swimming into rocks and other obstacles. Worldwide, fishes representing nine families are adapted to aquatic systems in caves. At least one species of Mexican cave fish – white and blind – is commonly included in home aquarium collections.

The Texas blind salamander, a troglobite, was discovered in 1895 when some were thrust to the surface in artisan well water tapped for a fish hatchery in Texas. These unusual salamanders, while pinkish-white in coloration, are so translucent that their internal organs are often visible. They have spoon-shaped snouts, thin legs like toothpicks, and sensory pits on their heads, which apparently are adaptations for locating food. Because the Texas blind salamander is found only in cave waters at San Marcos, Texas the species represents a classic example of an endemic organism. Fortunately, a refuge offers some protection for the limited habitat in which this species evolved, although heavy pumping threatens the aquifer supplying water to the cave system. Several other species of troglobitic salamanders have been discovered elsewhere in North America (e.g., caves in Georgia).

Some general ecological features emerge from studies of troglobites. To survive in their food-poor environments, when compared to their above-ground counterparts troglobites: (a) grow slowly because of low metabolic rates; (b) may often be smaller; (c) experience relatively long life spans; and d) produce fewer, but larger eggs (e.g., fishes and insects). Their populations grow slowly and they often exist at low densities, conditions that minimize competition within troglobitic communities. Troglobites also lack the rhythmic behavioral patterns typical of surface animals; that is, the constant environment in caves preludes behavior normally associated with alternating periods of day and night, daily behavior know as **circadian rhythm**.

Unfortunately, a devastating disease associated with the fungus *Pseudogymnoascus destructans* has triggered mass mortality in at least seven species of bats overwintering in caves (or mines) in eastern North America. In particular, plummeting numbers of the little brown myotis may soon lead to its regional extirpation, and fears are growing for two endangered species, Indiana and gray myotis. Discovered near Albany, New York in 2006, the fungus had spread to 25 additional states and five provinces in Canada by early 2014. Known as "white-nose syndrome," or WNS, the disease characteristically produces white fuzz on its victims' muzzle and/or ears and wings (Fig. 12.2, inset). The fungal growth apparently awakens bats from their winter torpor, which in turn expends vital reserves of body fat at the coldest time of the year and ends in starvation and death. By the beginning of 2012, WNS had claimed an estimated 5.7–6.7 million bats, posing not only intrinsic concern for the animals *per se* but also ecological consequences. The latter primarily focused on greater insect damage to crops and forest vegetation, perhaps followed by additional use of insecticides.

Only a small portion of the caves in North America have been fully explored and studied by biospeleologists. The recent (2010) discovery of a previously undescribed spider in an Oregon cave suggests that much remains to be learned about the subterranean environment. The large spider, representing a new genus and family, could hardly be overlooked. When extended, the legs with their hook-like claws on the last segment reach up to 7.6 cm (3 inches) in length. The spider hangs from the cave ceiling by a self-produced thread and snatches flying insects out of the air as they pass underneath.

Arctic ice cap

The Arctic ice cap, a physical realm of frozen and liquid water, consists of the permanently frozen parts of the Arctic Ocean, the seasonally frozen areas along its perimeter, and associated waters. Life exists in, on, and under this icy expanse, but its diversity is low and food chains are relatively simple and short. Fundamentally different from most of the ice in Antarctica where a large landmass (a continent) underlies the southern pole, the Arctic ice cap is a frozen ocean floating on a liquid ocean. Pack ice forms when floes of sea ice ram together to produce an expansive and continuous unit

that may enlarge a previously existing area of ice (e.g., outer edges of shelf ice).

Sea ice forms from ocean water and, because of its salt content, sea ice is denser than frozen freshwater (i.e., it tends to sink instead of float). Sea ice may remain permanent all year round, but expands in winter and melts at its edges in summer. However, melting sea ice does not contribute to rising sea levels, given that it already displaces its own mass in the seawater in which it floats; in contrast, melting ice from other sources, particularly glaciers, does affect water levels.

Albedo is a phenomenon of importance, especially regarding growing concerns about changes in global climate. Because light-colored surfaces absorb less heat than darker backgrounds, ice reflects a large percentage of the incoming solar energy back into the atmosphere; snow-covered ice is even more reflective. With a warming globe, more of the ice cap melts each year which increases the surface area of ice-free – and darker – seawater, which subsequently warms and melts even more ice. This "albedo effect," produces a powerful feedback mechanism that steadily reduces ice-covered areas at an ever-increasing rate.

Polynyas are unfrozen openings that persist in sea ice, usually in the same location for extended periods of time; the term is Russian for "natural ice hole." Some form when wind or currents regularly push ice away from certain locations, but larger polynyas arise from upwellings that maintain open water and carry nutrients upward in the water column. This, plus the penetration of sunlight for photosynthesis and improved circulation of oxygen downward in the water column, creates highly productive areas filled with abundant sources of food. Polynyas therefore attract seabirds that often nest within flying distance of these resources, and a single polynya in the Bering Sea overwinters virtually the entire population of spectacled eiders. Bowhead whales also remain in or near large polynyas, an association that once improved the hunting success of whalers.

Communities of algae and other phytoplankton live within and on the underside of sea ice, forming a basic part of an ice-associated or **sympagic** food web. Some are numerous enough to stain the ice brown, which decreases albedo in the immediate area. Melting results, creating a honeycomb of ice channels. Layers of algae up to 2 cm (<1 inch) thick attached to the underside of sea ice may intercept light passing through the ice, absorbing solar energy otherwise available for primary production in the underlying water column. Ice algae, however, contributes to primary production and supports food webs in which copepods and other zooplanktkton represent primary consumers.

Arctic cod occur farther north than any other species of marine fishes, and form immense schools that number in many millions. Although related to the larger and

Figure 12.4 The function of narwhal tusks is not completely understood, but the elongated tooth in males may play a role in establishing social dominance. In a behavior known as "tusking," shown here, males sometimes rub their tusks above the ocean surface. Reproduced by permission of Glenn Williams and the National Institute of Standards and Technology.

commercially valuable Atlantic cod, Arctic cod are of little economic value. Instead, they are of critical ecological importance to the food chains for many species in the Arctic ecosystem and they may account for as much as 75% of the energy transfer between the trophic levels represented by plankton and higher vertebrates. Sea birds such as the common murre depend on Arctic cod as do narwhals and, to a lesser extent, Atlantic cod and Atlantic salmon. Arctic cod spawn under sea ice, where the rough under-surface provides these fish with cover.

Greenland sharks, the northernmost species of the world's sharks, occur in the Arctic waters of the North Atlantic and Arctic Oceans where they have been tracked under ice. They are huge (6.4 m or 24 feet) but swim slowly and, while predators of a wide range of fishes and squid, they also scavenge on whatever might be available. Females of a parasitic copepod often attach to the corneas of Greenland sharks, significantly clouding their eyesight. This impairment seems minimal however, as Greenland sharks chiefly rely on smell to locate their food. The host-parasite relationship perhaps represents an example of **mutualism** in this case, in which the copepods (which may be bioluminescent) attract prey into striking distance for their slow-moving hosts, but this relationship remains unproven.

Narwhals (the so-called "unicorns of the sea") are endemic to Arctic seas and spend much of the year under pack ice, surfacing for air in lengthy fissures known as leads. The males of these medium-sized whales sport a hollow, spiral tusk 1.5–2.7 m (5–9 feet) in length that develops from the left canine; some females may have much smaller tusks. Rarely, some individuals produce two tusks. Narwhal tusks apparently represent a secondary

sexual characteristic (i.e., used for attracting mates). Additionally, groups of males at times project their tusks above the surface, rubbing them against each other in what is known as tusking (Fig. 12.4); this behavior suggests the tusks play a role in establishing and maintaining dominance in social hierarchies. Otherwise, no evidence indicates that the tusks are used for fighting, defense, or securing food, although some recent work proposes that the tusks are uniquely innervated as well as remarkably flexible. Narwhal tusks have been the subject of mythology for centuries, among them the belief that poisoned drinks would be rendered harmless if served in cups made from the tusks. Narwhals seem particularly vulnerable to the effects of climate change and may serve as useful yardsticks for marking faunal responses to warming conditions in arctic regions.

No animal serves better as a symbol of the Arctic than polar bears, and few others rely more heavily on ice. Polar bears prey on seals, particularly ringed seals, which the bears can secure only on a platform of sea ice; in water the seals can easily outmaneuver even the most agile bears. Female polar bears spend the winter months in dens where their cubs are born. When the families emerge in late spring, the semi-starved females hunt for seals not only for their own nourishment, but for sustaining adequate production of enriched milk. In recent years, however, the Arctic ice cap has grown steadily smaller and, with less ice, the hunting grounds for polar bears likewise diminished. The bears must swim more often and farther between floes, expending much-needed energy in the process. In all, the relationship between diminished sea ice and the welfare of polar bears includes increased drowning deaths, fewer and smaller cubs, poorer body condition, and lower survival rates for cubs as well as older age classes. If the warming trend continues to eliminate sea ice in the years ahead, the current distribution of polar bears will be correspondingly reduced, leaving an uncertain future for a far smaller population literally clinging to whatever then remains of the Arctic ice cap.

Global climate change poses a serious threat to virtually all types of Arctic biota, especially those dependent on sea ice. Marine mammals, hooded seals, narwhals, and polar bears appear to be highly sensitive to climate change, which seems closely tied to the production of greenhouse gasses (see *Sea-level rise*, this chapter). Further, as sea ice diminishes, other anthropogenic influences will emerge, namely the opening of new shipping routes (such as the fabled "Northwest Passage") and access to untapped resources such as oil and natural gas. The environmental protection policies of the Arctic Council, a group of eight countries including Canada and the United States, may provide important safeguards regulating these and other human activities in what was once a vast and frozen natural sanctuary.

Niagara Escarpment

A distinctive cliff formation known as the Niagara Escarpment forms the exposed edge of an ancient seabed, the Michigan Basin, which formed more than 400 million years ago when a tropical sea receded and exposed the underlying sediments. The cliffs, which rise vertically for 25 m (82 feet), rim the northern edges of lakes Michigan and Huron (excluding Georgian Bay) and extend into New York State along the southwestern shore of Lake Ontario. The escarpment runs for about 1200 km (750 miles), most of it in Ontario, but the exposed cliff face and its unique ecosystem occupies a total area of just 700 ha (1700 acres). Niagara Falls is a prominent and certainly well-known site along the escarpment which, for much of its length, lies in close proximity to the industrial heartland of North America. Indeed, human activities place heavy burdens on the ecological integrity of the Niagara Escarpment, which various agencies and legislation manage and protect. It was designated as a World Biosphere Reserve by UNESCO (The United Nations Educational, Scientific, and Cultural Organization) in 1990, and is one of 631 such sites with extraordinary ecological and cultural value in 119 nations throughout the world.

A surface layer of hard dolomitic limestone, which overlays softer rock and keeps erosion from rounding the edge of the escarpment, maintains the cliff face. The escarpment contains excellent exposures of 450-million-year-old fossils representing the Ordovician and Silurian eras. Despite the geological prominence of this formation, however, ecologists have only recently studied the cliff communities of the Niagara Escarpment (Fig. 12.5a). The ecologists soon learned (usually while suspended by ropes) of the rich plant life occupying the cracks, ledges, and other microhabitats on the cliff face. Also among the discoveries were some of the oldest trees in North America.

Typical regional communities, usually beech-maple, occur on the plateau atop the Niagara Escarpment. About 20 m (33 feet) from the edge, however, conditions

(a)

(b)

Figure 12.5 (a) Research initiated on the Niagara Escarpment revealed a unique environment, long overlooked as a site of ecological interest. (b) Among other discoveries, rock-climbing ecologists found centuries-old trees growing on the cliff face. Reproduced by permission of (a) Jarmo Jalava and (b) Douglas W. Larson.

begin to change. The soils steadily thin and, although the species composition remains unchanged, the forest canopy is noticeably shorter about 5 m (16 feet) from the edge. A human or animal trail is a common feature about 2 m (6.5 feet) from the cliff edge where a community of cedar, ferns, and shrubs replaces the forest vegetation. Environmental conditions become quite harsh near the cliff edge, where: (a) organic soils give way to bare rock; (b) temperatures fluctuate widely; (c) water availability is unpredictable; and (d) snow cover is absent in the winter, conditions collectively resembling the Tundra Biome instead of forest. Plants (e.g., polypody fern) which are singularly tolerant to extreme dessication grow in the austere zone at the cliff edge.

Eastern white cedar and other species of cliff-face trees are visible over the edge. These grow erect, but their trunks are twisted and stunted where they enter the rocky face of the escarpment. On occasion, when rocks fall from beneath a trunk, a tree hangs inverted on the cliff face, anchored by enough roots to keep it alive for decades or even centuries longer.

Rocky fields known as talus slopes lie at the base of the escarpment. Soil develops and accumulates between the rocks, maintained by the steady fall of organic debris from the cliff face. Conditions are somewhat less severe in the talus slope, and berry bushes and other vegetation grow among the rocks and fallen logs. Finally, at the foot of the talus slope, the habitat again resembles the forest on the plateau atop the escarpment.

The cliff face is not a hospitable site, and the dominant species – eastern white cedar –grows better elsewhere within its range. Nonetheless, the age structure of the stunted trees indicates the cliff-forest has persisted for many hundreds of years (Fig. 12.5b). The oldest trees seldom exceed heights of 3 m (10 feet) and trunk diameters of 30 cm (12 inches). Some cedars, based on tree-ring counts, exceed 100 years of age; a few are 1600 years old. The cliff-forest therefore represents one of the oldest forests and least-disturbed communities in eastern North America. Because of its unique location, the cliff-forest has survived the adversities of European settlement in North America and, except for the rarity of rock slides, persists virtually free from all natural disturbances (e.g., fire).

Although stunted, trees in the cliff-forest are not deprived of either nutrients or water. Given the lack of soil and ground water on the cliff surface, the trees instead gain both water and nutrients such as phosphorus from large colonies of mycorrhizal fungi. Equally amazing, the crystalline structure of the limestone rocks in the escarpment is enriched by communities of nitrogen-fixing algae, and these apparently fertilize the cliff vegetation with significant amounts of nitrogenous compounds.

Interestingly, the twisted trunks of the old cedars are deformed in ways resembling those affecting the ancient bristlecone pines described in Chapter 9 (i.e., growth continues on one side of the trunk, producing a highly asymmetrical stem). In the case of cliff-dwelling cedars, the odd-shaped trunks apparently result from the configuration of their root systems. Several roots anchor each tree to cracks and crevices in the cliff surface, but each of the roots connects to just one part of the trunk. When one root dies, the tree continues growing but nourishment thereafter reaches only part of the trunk and produces an asymmetrical stem.

The cliff fauna does not include any truly endemic species, but the escarpment nonetheless provides habitat not otherwise available in the region. Some animals (e.g., the small species of rattlesnake, the eastern massasauga) are therefore associated with limestone formations and are characteristic of the escarpment fauna. Various hawks and, especially, turkey vultures, nest, roost, and perch on the cliff faces. Because it traverses the ranges of two closely related species, the Niagara Escarpment has become a corridor in which Jefferson and blue-spotted salamanders regularly hybridize. The escarpment contains Canada's richest flora of ferns – at least 50 species, some of which occur in few other locations – as well as its greatest concentration and variety of orchids. In sum, the Niagara Escarpment supports an ecological community whose uniqueness has only been recently discovered.

As a landform, the Niagara Escarpment: (a) adds abiotic richness to the surrounding landscape (e.g., rocky fissures and unique microclimates); (b) provides a corridor of forest habitat through a deforested region; and (c) serves as a refugium (e.g., cover from predators). Moreover, the ancient cedars represent valuable tools for reconstructing climatic conditions in eastern North America for past centuries. Because much of the cliff forest grows near heavily urbanized locations where other forest vegetation was eliminated long ago, the old trees will remain useful to determine the importance of human activities on future climatic events.

The "Father of Waters"

From modest headwaters at Lake Itasca, Minnesota the second-largest river in North America – the Mississippi – meanders sinuously through the middle of the continent on its 4070 km (2530 miles) journey to the Gulf of Mexico. Along the way, major tributaries including the Missouri (the longest river on the continent), Ohio, Arkansas, Red, White, and Tennessee rivers and innumerable lesser streams, swell its flow. The Mississippi River and its tributaries drain 4.8 million km^2 (1.8 million square miles), which is the largest drainage basin in North America and the second largest in the world. The drainage includes 31 states and two Canadian provinces and provides more than 18 million people with drinking water, recreational opportunities, outlets for discharging industrial and municipal wastes, and shipping avenues.

The upper drainage basin of the Mississippi River formed from glacial pressures and meltwater during the Pleistocene Epoch. Massive ice sheets, some up to 3048 m (10,000 feet) thick, gouged out valleys as they pushed southward. When the glaciers melted, huge glacial lakes formed behind terminal moraines from which torrents of water eventually deposited deep layers of fertile soil as they carved the landscape, creating the immense river we know today.

In 1541, the Spanish explorer Hernando de Soto (1496–1542) became the first European to discover the Mississippi River. More than 500 years earlier however, Native Americans occupied a huge city on the river's banks. The ancient city, known as Cahokia, was located across the Mississippi from present-day St Louis, Missouri. At its zenith, 40,000 citizens occupied Cahokia and built at least 120 large earthen mounds over an area of about 15.5 km^2 (6 square miles). The city's occupants farmed the rich soils of the Mississippi Valley and transported trade goods on the river and its tributaries. Strangely, the city declined rapidly and was abandoned more than a century before Europeans arrived in North America.

Cahokia's decline did not diminish the importance of the Mississippi to Native Americans. Indeed, the Objibwe Tribe of the Algonquin Nation provided the earliest-recorded name – *Misi-ziipi* (Great River) – which French explorers transformed into the name used today. The Mississippi River has always provided a vital link for transportation and commerce, including the Confederate Army's shipments of munitions during the Civil War. In July 1863, Union victories at the last two Confederate strongholds on the river permanently severed the Confederate supply route. Upon learning that the Union Army controlled the Mississippi, President Abraham Lincoln wrote, "The Father of Waters again goes unvexed to the sea." Lincoln's nickname for the Mississippi, derived from a Native American reference, was displaced when the popular song "Old Man River" highlighted the 1927 Broadway musical, "Showboat."

A more appropriate moniker, "The Big Muddy," acknowledges the huge volume of sediments the river carries and deposits along its path. However, attempts to maintain shipping lanes and control flooding have reduced the sediment load carried by the river from 400 million metric tons (440 million tons) prior to 1900 to about 145 metric tons (160 million tons) today. In spring, melting snow and heavy rainfall swell the Mississippi and its northern tributaries, inundating vast areas of bottomlands bordering its banks. When the Mississippi overflows, the main course of the river often changes (a process known as **avulsion**) as the water finds steeper routes to the Gulf. Across a broad, flat plain stretching from Cape Girardeau, Missouri to Natchez, Mississippi the river rambles in broad loops bypassing oxbow lakes and shallow marshes that mark previous channels.

In Tennessee, Reelfoot Lake developed in a depression produced by earthquakes in 1811–1812, the largest of which may have registered 8.0 on the Richter Scale in use today. One of the shocks notably reversed the river's course for 10–12 hours. About 61 km^2 (15,000 acres) of land subsided and quickly filled with water from the Mississippi River. Thousands of upland trees drowned in the flooded area, and even bald cypress failed where the water level increased by 3 m (10 feet) or more. However, at sites where the increases reached about 2 m (6.5 feet), the flooded stands of cypress survived and responded to the deeper water by producing new buttresses. The new structures presumably allowed the trees to exploit the dissolved oxygen in the upper strata of the heightened water column. They were not revealed until drought lowered the lake level more than a century later, and were subsequently dubbed "hanging buttresses" because of their high-and-dry position above the diminished water level (Fig. 12.6). Today, the shallow lake includes a variety of aquatic and semi-aquatic communities with oxbows, cypress sloughs, and floodplain forests along its borders. Because of the richness of the lake and adjacent areas, Reelfoot Lake remained high on the list of field trips for classes conducted by the famed ecologist Victor Shelford (Infobox 1.1).

Freshwater mussels occur throughout much of the world but more species live in North America than elsewhere, with the Mississippi River and its tributaries alone serving as home for three times as many species (281 species and 16 subspecies) as the Amazon. The Mississippi's drainage also forms one of North America's important centers of fish distribution. The river basin hosts about 260 species of freshwater fishes in 13 families, about 43% of all freshwater species in North America, the highest diversity of freshwater fishes for any region at a comparable latitude. The Mississippi River system is still inhabited by so-called ancient species such as the long-nosed and short-nosed gar, bowfin, American paddlefish, and sturgeon that lived there well before the Ice Age. These ancient fishes survived periods of glaciation by moving south as the waters became too cold, subsequently recolonizing the northern reaches of the river during periods of glacial retreat.

The Mississippi drainage served as an "incubator" for fish speciation. As glaciations ebbed and flowed, ice-isolated tributaries later reconnected when the ice masses retreated. In addition, low sea levels during glacial periods permitted streams to join before reaching salt water, thus dispersing fish fauna in the lower reaches of the basin. When the ice sheets thawed, sea levels rose, the Mississippi resumed its former configuration, and some fish stocks were isolated and evolved into new forms. Of seven families endemic to North America (e.g., sunfish family), all occur in the Mississippi and its tributaries.

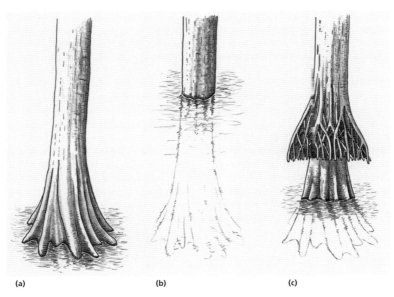

(a)   (b)   (c)

Figure 12.6 A comparison of normal structure at (a) the base of a bald cypress and (c) a "hanging buttress" that developed on some cypress trees at Reelfoot Lake, Tennessee. (b) A series of earthquakes in 1811–1812 formed the lake, which flooded a cypress swamp; the deeper water initiated the growth of new buttresses higher on the trunks of the flooded trees. These were exposed years later (1930) when a prolonged drought lowered the lake's water level. Illustration by Brian R. Chapman based on Kurz and Demaree (1934).

Figure 12.7 By swimming slowly with their large mouths wide open, American paddlefish capture tiny organisms in rake-like structures on their gills. The namesake flattened rostrum is used to orient movements and locate concentrations of planktonic prey. Reproduced by permission of the US Fish and Wildlife Service.

Arguably, the most unique inhabitant of the river is the American paddlefish, a large shark-like fish with a cartilaginous skeleton featuring an elongate, paddle-like snout (Fig. 12.7). As one of only two species of paddlefish in the world (the Chinese paddlefish of the Yangtze River is now considered extinct) the American paddlefish initially occurred only in the Mississippi River and its tributaries, but it has since been introduced elsewhere. Paddlefish feed by swimming leisurely near the surface where, with their large mouths wide open, they sieve tiny aquatic organisms in closely set gill rakers in their throats. The flattened rostrum (paddle) is covered with electroreceptors used to orient their movements and to detect prey. A **potamadromous** species, American paddlefish migrate within the river system to specific habitats (recently inundated gravel bars) where they spawn. After hatching, larval paddlefish are swept downstream to deep pools where they mature. During an average life span of 20–30 years, paddlefish may attain a length of 2.2 m (7 feet) and a weight of 100 kg (220 pounds). Populations of American paddlefish have declined in recent decades as a result of overfishing for their meat and roe. Channelization, impoundments, and other alterations to the flow of the Mississippi River have diminished the habitat for these unusual fish, further contributing to their falling numbers.

Fertile plains, bottomland forests, and wetlands bordering the river provide habitats for a diverse flora and fauna. More than 325 species of North American birds use these habitats during spring and fall migrations, and about 145 species of amphibians and reptiles can be found along the river's banks. Although the riparian habitats sustain many species, the width and current of the Mississippi River present a considerable biogeographic barrier to their east–west dispersal across the river. Repeated glacial melting during the Pleistocene widened the Mississippi and eventually separated populations of species such as striped skunk, spotted skunk, raccoon, northern short-tailed shrew, and five-lined skink. Restricted from further contact between the **conspecifics** on either side of the barrier, distinct genetic lineages have developed and continue to develop in the isolated populations. Populations of the carnivorous pale pitcher plant on opposite sides of the Mississippi River, though cryptic (i.e., they are morphologically indistinct and occupy similar habitats), differ enough genetically to warrant classification as separate species.

The biogeographic barrier generated by the river extends well beyond its terrestrial boundaries. Discharge rates of between 7000 and 20,000 m^3 s^{-1} (200–700 thousand cubic feet per second) contribute about 90% of the fresh water that flows into the Gulf of Mexico. However, fresh water from the Mississippi does not immediately mix with the Gulf's salt water; instead, it flows atop the heavier salt water across the Gulf to the Straits of Florida. The freshwater lens extending far into the Gulf of Mexico forms an ecological barrier preventing larval scleractinian corals (*Madracis decactis* and *Tubastraea coccinea*) produced on Florida's reefs from colonizing oil platforms in the western Gulf.

The watershed of the Mississippi River drains vast areas of agricultural production (e.g., the "corn belt"). Agricultural runoff eventually reaches the Gulf, where concentrations of nutrients such as phosphorous and nitrogen produce algal blooms, deplete dissolved oxygen, and lead to the deaths of organisms in a large "dead zone" off the coast of Louisiana and Texas (see a discussion of dead zones in *Chesapeake Bay*, Chapter 11). This hypoxic zone encompassed an area of 17,521 km^2 (6765 square miles) in 2011, the largest in North America. During periods of flooding, the dead zone enlarges but shrinks when droughts occur in the Mississippi watershed.

Dredging for navigational channels and construction of numerous levees, locks, and dams have greatly changed the characteristics of the river. To control floods, part of the Mississippi's flow is diverted to the Atchafalaya River. Should the Mississippi eventually shift its main channel to the Atchafalaya (i.e., a natural river course

alteration by avulsion), extensive changes will affect plant and animal communities downstream and the numerous port facilities in southern Louisiana. Dams and flood-control structures also reduce the sediment load required to build – and replenish – wetland habitats in the river's delta. As a consequence, large areas of marshlands in the delta are disappearing with little chance of recovering (*Marshes of the Gulf Coast*, Chapter 11) while backwaters, side-channels and pools behind dams trap and fill with sediments. These changes in sedimentation eliminate critical spawning habitats for species such as American paddlefish.

In ecological terms, the Father of Waters no longer flows "unvexed" but the river nonetheless remains a critical environmental and navigational resource. Still, ample evidence indicates that the ecological health of the Mississippi River is gradually declining, and decisions followed by actions are necessary if the river is to sustain its importance as a natural resource.

The Everglades

One of the world's largest freshwater wetlands lies between Lake Okeechobee and the southern tip of Florida, where the "River of Grass" merges with saltwater marshes and mangrove thickets. The Everglades originally covered about 10,500 km^2 (4054 square miles), much of it characterized by vast swales of sawgrass dotted with elevated sites known as tree islands or bayheads. The Everglades represents the largest subtropical region in North America north of Mexico. However, the ecology of the region has been shaped by the joint forces of subtropical temperatures, fire, and a hydrological cycle punctuated by alternate wet and dry periods. From the time of their "discovery" by Spanish explorers, these wetlands remained shrouded with superstition and fear; the characterization "mysterious Everglades" still lingers.

Soils in the Everglades consist of rich peat and muck overlying a limestone floor. These highly organic soils are about 3.5 m (12 feet) thick near Lake Okeechobee, but become much thinner southward where the underlying limestone is sometimes exposed. Beginning early in the last century, large areas of the Everglades were drained for sugarcane, vegetables, and other agricultural crops. Canals diverted water draining from Lake Okeechobee, thereby curtailing much of the southward flow into the Everglades. Eventually, the flow of water was reduced to about 20% of the amount received in 1900. The Everglades steadily diminished as its original land and water patterns changed. However, thanks largely to the perseverance of Marjory Stoneman Douglas (1890–1998), a large part of the remaining area (but only about 25% of the original area) was declared a national park in 1947.

Sawgrass – a sedge and not a true grass – dominates the Everglades landscape and attains its greatest density

(a)

(b)

Figure 12.8 (a) 'Gator holes are among the more important ecological features in the Everglades. These ponds, excavated and maintained by alligators, become refugia for fishes and waterbirds during dry periods. (b) Distinctive cypress domes characterize some areas of the Everglades. Dwarfed cypress trees are also scattered in the sawgrass marsh in the foreground. Reproduced by permission of Paul N. Gray of the National Audubon Society.

at sites covered by surface water for most of the year. This characteristic plant of the Everglades is named for the sharply toothed margins on its long, grass-like leaves. Although sawgrass produces seeds, it usually reproduces vegetatively from the lateral spread of its vigorous root system. Sawgrass is tough and, once established, it seldom provides forage for herbivores. Sawgrass communities burn regularly forming pure stands, which some ecologists regard as fire climaxes. After burning, it can regain its original density within a year or two after burning to ground level. When unburned, however, sawgrass is invaded by a large number of other species, including woody plants.

Communities known as wet prairies develop where the **hydroperiod** is relatively short. These occupy sites known as flats, which are characterized by vegetation of low stature. Of the various types, beakrush flats are especially representative of the wet prairies once typical of pre-drainage Everglades vegetation. Wet prairies are important feeding areas for wading birds.

Alligators contribute an important ecological feature to the Everglades. During the dry season, alligators excavate soil and vegetation forming ponds known as "'gator holes" (Fig. 12.8a). These are up to 10 m (33 feet) across and 1 m (3.3 feet) deeper than the surrounding marshes. Most are situated above depressions in the limestone bedrock underlying most wetland soils in the Everglades, but they are not common where peat is particularly deep. Some 'gator holes, perhaps centuries old, have been maintained by generations of alligators. The debris from these excavations is pushed to the edge of the pond, forming a rim where trees, shrubs, and herbaceous plants produce additional habitat and diversity.

Willows are among the pioneer plants on these rims of disturbed substrate and, after spreading, develop into a thick hedge. In time, cypress and other trees (all with floating seeds) take over the periphery of 'gator holes. Herbaceous plants known as flags grow in fresh water inside the rim surrounding 'gator holes, but they cannot survive the saltwater tides sometimes flooding the marshes outside the rim. The stalks of these plants reach heights of 3 m (10 feet), and their large banana-like leaves turn brown just as the winter dry season begins. In the past, poachers located 'gator holes at this time because the tall growths of flag are so conspicuous against the greener background of other marsh vegetation.

'Gator holes are therefore relatively permanent ponds where fresh water persists during the winter dry season in the Everglades. As many as 23 species of fishes crowd into these refugia when the water recedes each winter, as do immense numbers of aquatic invertebrates. These concentrations provide a base for food chains extending throughout the Everglades ecosystem. When the winter drought ends, organisms in the 'gator holes disperse and repopulate the surrounding marshes.

Alligators are now ranked among the animals using tools, in this case for hunting waterbirds. Egrets and herons at times nest in trees at ponds where alligators concentrate. Recent observations suggest that alligators balance sticks on their snouts to lure adult birds in search of nesting materials, a fatal attraction when the birds attempt to gather the bait. This behavior occurs seasonally during periods when the birds are building their nests, and is limited to alligators living near active rookeries; the origin of the behavior is unclear however (i.e., is it learned and transmitted culturally at specific locations or has it evolved as a genetically based instinct?). Mugger crocodiles in India exhibit the same behavior, which indicates that the phenomenon occurs widely among crocodilians.

Wood storks, among other wading birds, have developed a particularly close link with 'gator holes. The breeding cycle of these large wading birds coincides with the normal dry season each winter, precisely when food is concentrated in 'gator holes. Because wood storks detect their prey with their bills, they capitalize on the dense fish populations to meet the demands of their nestlings. In exceptionally wet years, however, wood storks delay or cease breeding altogether because fish populations in the Everglades are dispersed and cannot be captured efficiently.

Two birds have diets closely linked with apple snails, whose abundance depends on the quantity and quality of water in the Everglades. The first of these, wading birds known as limpkins, use their long, robust, and slightly decurved bills to hammer on the operculum, a horny lid enclosing the snails when they withdraw into their shells. A curve at the tip of the lower bill snips a muscle attaching the snail's soft body to the interior of the shell and the meat is extracted. After capture in marsh vegetation, the snails are carried to a favorite location on firm ground where they are opened. There, discarded shells form a small midden. Limpkins also prey on other small organisms, but the availability of apple snails appears to be the most important habitat requirement for limpkins; indeed, their geographic distribution is nearly identical to the range of apple snails (i.e., Florida, the Caribbean, and Central and South America).

Snail kites (birds in the hawk family) represent the second species dependent on apple snails. These sleek raptors are widely distributed in Central and South America, but in North America the species is restricted to the Everglades and a few other wetlands in Florida (where it is also known as the Everglades kite). The restricted range of snail kites in Florida, along with their near-total dependence on apple snails, led to their classification as an endangered species. The hawk-like, hooked bills of these birds are slender enough to grasp the edge of the operculum, which is then torn free and removed. This done, the birds probe inside the shell where, like limpkins, they cut the muscle attachment and extract the meat. Unlike the wading behavior of limpkins, however, snail kites seek apple snails in open water either by still-hunting from perches or while flying overhead. After securing a snail in their talons, a snail kite flies to a perch to extract the meat from its victim. Piles of empty shells accumulate beneath these feeding perches, where as many as 2000 shells have been counted in middens in South America.

While largely sharing a common diet, the feeding behavior of these birds represents an example of **resource partitioning**. Limpkins probe for apple snails in vegetation, whereas snail kites scoop snails from surface water in open areas; in short, they utilize the same food but do not compete because they hunt in separate habitats.

Native trees in the Everglades include pond apples, which can develop massive trunks and large, buttressed roots, yet seldom exceed 10.5 m (35 feet) in height. Bald cypresses are somewhat more common in the Everglades ecosystem and, where site conditions are favorable, these flood-tolerant trees grow in stands known as cypress domes (Fig. 12.8b). Cypress are fire resistant but, during severe droughts in the Everglades, fires burn into the peat and kill trees by killing their roots, thereby restricting large stands of cypress to perennially water-logged sites.

Circular sites known as tree islands rise above the surrounding wetlands. These are seldom larger than 2 ha (5 acres), but a linear form – tree island strands or bayheads – may cover much larger areas. Curiously, these strands are frequently oriented along a generally north–south axis and are somewhat teardrop in shape, with their blunt ends directed northward (i.e., toward the flow of water) and their southern ends tapering to a point. Red bay, wax myrtle, and holly are principal trees, but hackberry, strangler fig, and sabal palmetto grow on sites with higher elevations. As well as adding diversity to the vast area of marshes, tree islands represent key refuge and fawning sites for white-tailed deer. Because of their elevations, tree islands provide essential habitat for deer during periods of high water, but deer utilize tree islands all year round as foraging areas and for escape and resting cover. Deer would probably not exist in the Everglades without the presence of tree islands.

Abandoned alligator nests (thick piles of uprooted plants) become platforms on which ferns and various woody plants gain footholds and eventually form larger islands. Sometimes known as "alligator hummocks," the piles of densely packed plant litter resist fire and therefore endure for many years. Many serve as nesting sites for turtles, which otherwise may lack suitable nesting habitat in the surrounding area (turtles also lay their eggs in active alligator nests). When fires sweep through the Everglades, the islands of vegetation growing on alligator hummocks usually remain unburned. Some alligator hummocks persisting deep inside mangrove thickets may be 500 years old.

Other plants are not native, and the rapid spread of these threatens the integrity of the Everglades ecosystem (Infobox 12.2). Chief among these pests is melaleuca, an Australian tree released into the Everglades in 1906 as a means of draining the land for agriculture. Stands of melaleuca transpire more water than marsh vegetation, tolerate fire, crowd out native vegetation and reduce habitat for wildlife. The rapidly growing trees can transform a sawgrass marsh into a closed forest in just 25 years. Melaleuca once occupied about 202,500 ha (500,000 acres) in the Everglades ecosystem, but a program of Integrated Pest Management (IPM) subsequently reduced the infestation to about 110,000 ha (271,000 acres). IPM employs mechanical, chemical, and biological controls in ways that attack noxious

Infobox 12.2 Those darn weeds: Invasive plants in North America

The Columbian Exchange describes the large-scale movements of organisms into and out of North America in the centuries following 1492, some by accident, others by design. As transportation improved from galleon to steamship to jumbo jet, opportunities accelerated for the introduction of foreign species that range from insects harmful to farm and forest to zebra mussels and nutria. The list of exotics includes even more species of invasive plants, of which a few are mentioned elsewhere (e.g., cheatgrass in the *Great Basin Desert*, Chapter 7); what follows here is but a small sample of some others.

Kudzu ("the plant that ate the South") is familiar to residents of the southeastern United States, having been brought to the country from Japan in 1876 for exhibition at the Centennial Exposition in Philadelphia. Touted as a miracle plant suitable for livestock fodder, the woody vine with large, grape-like leaves was also praised as an ornamental for shading porches and arbors. In the 1930s, the Soil Conservation Service recommended kudzu as a cover plant for stabilizing eroding soil; it was also believed to replenish nitrogen in soils worn-out by decades of cotton and tobacco farming. Because kudzu thrives in heat, humidity, and sunshine – and is drought-resistant – it spread rapidly in the South. Today large tracts of land and abandoned buildings are encased in a verdant cloak of kudzu to the exclusion of all other vegetation. Its expansion led to the story that kudzu "covers old cars, barns, and slow-moving children." Practical means for controlling kudzu remain elusive, and the plant continues expanding unchecked except where the bite of winter limits its survival.

North America supports several species of invasive aquatic plants from other parts of the world, including water chestnut (from Asia) and hydrilla (from Africa), but here we mention water hyacinth, a particularly troublesome species from South America, and purple loosestrife. Both species chokes out other, more desirable vegetation that is replaced with monotypic stands lacking value as food or cover for waterfowl or other wildlife associated with freshwater habitat. Water hyacinth is a free-floating plant featuring a rosette of wide leaves on spongy, bulbous stalks that act as buoys, a mass of feathery roots dangle into the water, and a spike of showy purple flowers that make it an attractive ornamental for both natural and artificial ponds. Water hyacinth expands with unmatched rapidity; its explosive growth can double a population in about two weeks into mats whose mass may reach 376 tons per ha (152.2 tons per acre). Such dense growth limits water movements and boat traffic, while also blocking sunlight and depleting oxygen in the water column; the latter may lead to fish kills. Mosquitos thrive in the jumble of plants. Purple loosestrife, supported by woody roots, reaches heights of 2 m (6.5 feet) and produces as many as 50 stems whose canopy dominates other wetland vegetation. A single plant may produce 2 million or more seeds each year, with the density of seedlings numbering in the thousands per square meter. Herbicides control water hyacinth, but these are costly and may have their own unwanted consequences. Biological controls hold the most promise for effectively controlling purple loosestrife. After testing, two host-specific leaf-eating beetles from Europe were subsequently released in regions infested with loosestrife; these proved effective but, like other biological controls, will not completely eliminate the presence of this noxious plant.

For all of his brilliance, Benjamin Franklin (1706–1790) made at least one big mistake: he is credited with introducing Chinese tallow into North America, presumably because of its usefulness for making soap and candles. Tallow trees now occur from North Carolina to Texas where they pose a particular threat to the remnants of coastal prairie. Once established, a grove of Chinese tallow can form a closed canopy within ten years, much to the exclusion of other plants. Besides reducing species richness, stands of Chinese tallow also alter the function of the environments they invade, primarily by changing nutrient cycling.

vegetation at all stages of development (e.g., seedlings, saplings, and mature trees). For example, the introduction of herbivorous insects, among them the melaleuca leaf weevil, enhances the injurious effects of fire, herbicide, or cutting treatments. Mature melaleuca trees dying from herbicides react to the stress by releasing immense numbers of seeds; these rapidly germinate and renew the infestation with thick carpets of seedlings. However, insects thereafter attack and devastate the new growth, with the end result that IPM has effectively removed melaleuca from a large area in the Everglades. Brazilian pepper is another exotic pest in southern Florida, where it was introduced as an ornamental late in the 19th century. The seeds of this tree are mildly hallucinogenic and, on occasion, intoxicate birds to the point they cannot fly. Like melaleuca, Brazilian pepper turns

marshes into closed forests, but it also invaded mangrove thickets after Hurricane Andrew struck Florida in 1992.

Predator–prey relationships and perhaps other ecological interactions in the Everglades have changed following the introduction of exotic snakes (likely by pet owners who no longer wanted their animals), particularly Burmese pythons (Fig. 12.9). Since 2000, the abundance of these powerful constrictors has increased dramatically and may continue expanding geographically until halted by the species' intolerance to cold. Large individuals reach 5.5 m (18 feet) in length, meaning that prey as large as white-tailed deer, alligators, and the Florida panther (an endangered species now almost exclusively limited to Everglades National Park) are vulnerable. Where pythons are abundant in the Everglades, surveys indicated reductions of about 99% in the occurrence of

Figure 12.9 The subtropical environment of the Everglades fosters successful invasions of exotic plants and animals, of which Burmese pythons pose the most recent threat to native communities. Where abundant, these large and powerful constrictors reduce populations of mammals – some as large as deer – as well as prey on birds and other vertebrates. Reproduced by permission of the National Park Service.

both raccoons and opossums, 94% for deer, and almost 88% for bobcats; rabbits completely disappeared. The sharp declines of these once-abundant mammals coincided with the increase in pythons. Likewise, the rich avifauna in the Everglades is also affected by pythons whose stomach contents include a large variety of birds. All told, the introduction of an exotic predator has produced, at least for now, a significant new force into the faunal dynamics of the Everglades ecosystem.

Other environmental problems plague the Everglades. These include heavy nutrient loads from fertilizer runoff. The nutrient-rich runoff enhances the growth of plants such as cattail, which replaces sawgrass and other vegetation more representative of the Everglades communities. This is an example of **eutrophication**, the process of aquatic systems acquiring high concentrations of nutrients and the subsequent, sometimes explosive, increase in plant life. Pesticides also drain into the Everglades along with the fertilizer runoff. However, no problem befalling the Everglades matches the importance of water diversion, which has claimed about half of the original wetland area.

Fossil *Lagerstätten*: Windows into North America's ecological past

The fossil record typically portrays only incomplete glimpses into the past. Countless millions of plants and animals died over geological time, but only a small number of those met their end under conditions that favored preservation. And of these, even fewer have been discovered. Nonetheless, some circumstances, although rare, allow a relatively full picture of the past to emerge. Such sources are known as Fossil *Lagerstätten*, a term adopted from the German mining industry that indicated an unusually rich lode or seam of ore (i.e., *lager*, a repository or storage area, and *stätten*, the plural form of *stätte*, meaning location).

Fossil *Lagerstätten* are subdivided into two categories, the first being Concentration *Lagerstätten*, which are deposits containing large numbers of fossilized material (e.g., bone beds). In contrast, Conservation *Lagerstätten* are those rarer sites where conditions allowed preservation of the soft tissues of plants and animals, although these fossils typically occur in fewer numbers (e.g., those imbedded in amber or permafrost).

Two prominent Fossil *Lagerstätten*, representing distinctive points (a) at the beginning of the Cambrian Period over 500 million years ago when complex life first appeared (Burgess Shale), and (b) a period of the relatively recent Holocene Epoch from around 10,000 years ago (La Brea tar pits), offer bookends for the history of life in what is today North America.

Burgess Shale

A trove of fossils from the Cambrian Era lies on the side of a mountain in the Canadian Rockies in eastern British Columbia. The fossils are unique because they represent both hard and soft body parts of a fauna that flourished 520 million years ago following a burst of speciation known as the Cambrian Explosion. Near the end of the Precambrian Era, about 50 million years earlier, life for the most part consisted of single-celled organisms (e.g., bacteria), but at the beginning of the Cambrian the ancient seas "suddenly" harbored some of the strangest

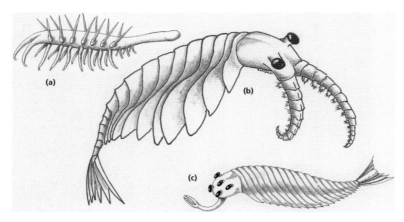

Figure 12.10 In eastern British Columbia, Precambrian deposits known as Burgess Shale preserved the soft body parts of dozens of strange creatures (all now extinct invertebrates), three of which are shown here. (a) *Hallucigenia*, a 3.5 cm long (1.4 inch) marine velvet worm, was likely a scavenger; (b) *Anomalocaris*, a predator that, reaching more than 1 m (3 feet) in length, may have been an arthropod; and (c) although the systematic affinities of *Opabinia* remain unknown, some authorities speculate that it was a small seafloor predator 7 cm (2.8 inches) in length. Illustrations by Brian R. Chapman based on various interpretations of the fossils.

multicellular animals ever to appear on Earth. Moreover, the anatomy of these animals reflects a community replete with predators and prey, burrowers and wanderers, and an otherwise wide range of filled niches. The fossils of the Burgess Shale therefore indicate the presence of a remarkable diversity of body plans and the fabric of a well-developed community structure early in the history of life on our planet.

This *Lagerstätte*, discovered in 1909, lies at the base of what was once a reef of calcareous algae. The community was suddenly entombed when some force, probably a mudslide, suddenly carried the animals downward into deeper water where conditions were hostile. The haphazard positions of the fossils in the shale indicate that the community likely did not live at the site where it was preserved. In all, two conditions prevailed that enabled preservation of the soft tissues: (a) rapid burial in a cloud of fine sediments; in (b) an environment deficient in oxygen.

The fauna of the Burgess Shale includes many species unlike anything known before or since, which has prompted arguments about their classification even at the highest levels of organization (e.g., phylum). One of the strangest animals (described as a "weird wonder") was assigned to the new genus *Hallucigenia* in recognition of its strange anatomy (Fig. 12.10a). Two rows of fleshy limbs provided locomotion and pairs of rigid, stilt-like spines formed protection as the caterpillar-like animal searched for decaying food. The animal was eventually recognized as a type of marine velvet worm. Classification of another creature *Anomalocaris* remains less clear. This animal, a predator reaching 1 m (3 feet) in length and the largest of the Burgess fauna, seemingly represents a now-extinct phylum, although some authorities argue it may

be an arthropod. In any case, this "strange crab" had two anterior appendages adapted for snatching prey, a body fringed with flap-like structures, and a fan-shaped tail (Fig. 12.10b). A large eye extended on a stalk from either side of its head. *Opabinia* likewise defies ready association with any modern phylum and, like some others in the Burgess fauna, has been relegated to an aptly named category, "Problematica" The head of this creature bore five eyes on stalks and a long, flexible proboscis (not unlike the hose of a vacuum cleaner) tipped with spines apparently designed to grasp prey (Fig. 12.10c). Gills extended from the dorsal surfaces of all but the first of 15 body segments, and 3 pairs of flaps formed its tail.

Although the Burgess Shale gained fame for its invertebrate fauna, the Cambrian deposits recently yielded a fossil fish whose stony imprint includes evidence of a notochord, the precursor of a backbone. The discovery pushes the origins of vertebrate evolution far back into the depths of time.

Ecologically, the Burgess fauna typifies a benthic community associated with a muddy seafloor, including a full range of trophic abilities: filter feeders; deposit feeders; scavengers; and predators. The uniqueness of the fossils enables us to appreciate the vast differences that distinguish between the past and present life on Earth. Indeed, with far fewer species, the life entombed in the shale shows greater variety in body plans than occurs in far more recent Fossil *Lagerstätten*. Without it, it would be difficult to appreciate the great array of evolutionary experimentation that preceded life today. Still, science knows little of what triggered the Cambrian Explosion, when nearly all modern phyla appeared, nor why it ended so abruptly.

La Brea tar pits

For eons rich deposits of organic matter settled on a seafloor that today underlies a coastal area in California including the Los Angeles basin, a deep structural depression formed during the Paleogene–Neogene (formally known as the Tertiary Period) by **plate tectonics**. Opposing movements between the Pacific and North American plates produced a ridge that separated the basin from the open ocean, thus forming a stagnant embayment that trapped organic sediments that were later buried by sands and other deposits. The seafloor eventually emerged and formed a coastal plain where streams from the adjacent mountains deposited additional overburden (e.g., alluvial gravel). Meanwhile, the buried deposits of Tertiary organic matter slowly converted into petroleum.

The oil seeped upward through vents and fissures in the overburden and eventually oxidized, thereby losing most of its volatile components. The result was asphalt, which formed sheets and shallow pools (the image of deep ponds of asphalt is a popular misconception). At times, reduced gas pressure underground allowed the asphalt to recede back into the vents, carrying with it any biological materials entrapped in the sticky ooze.

Land animals wandered into the asphalt, particularly when the surface was hidden by leaves and other debris; even mammals as large as mammoths succumbed to entrapment. Ducks and other aquatic birds landed on what they mistakenly perceived as ponds, especially after rains left a glaze of water on the asphalt (sadly, water birds today are similarly lured into collection ponds constructed near oil wells). The trap was most effective in summer when the asphalt softened and, once caught, mired animals seldom escaped. After death and decay, their bones settled and solidified in a matrix of asphalt and alluvial deposits. These vertical, cone-like accumulations preserved bones almost in their original state; soft body parts are not represented, but hairs and feathers are sometimes preserved. The asphalt at times preserved the iridescent colors of insect wings and the tissues of plants (e.g., leaves and cones).

Excavation of the fossil-laden matrix revealed a skewed ratio of predators to prey. Whereas ecosystems typically consist of a few predators atop a pyramid of more-abundant prey, this relationship is reversed at La Brea: 90% of the fossil mammals are those of large cats, wolves and other predators; eagles, hawks, and other carnivorous birds represent about 70% of the avian fossils. This imbalance suggests that a single herbivore, when entrapped, attracted numerous predators and scavengers that likewise became stuck in the asphalt. All told, hundreds of thousands of animals died at La Brea, which accordingly represents a Concentration *Lagerstätte*.

The La Brea fauna included many species of mammals now extinct, including short-faced bears, saber-toothed cats, dire wolves, camels, horses, ground sloths, both mastodons and mammoths, among others, but also species still extant in North America, for example, coyote, puma, bobcat, and jaguar. The unique conditions at La Brea accumulated more fossil birds than anywhere else in the world. The fossils represent many extant species of birds, often those associated with wetlands (e.g., herons, plovers, grebes, and waterfowl) along with condors, turkeys, songbirds, and more than 20 species of eagles, hawks, and falcons. Of these, the golden eagle is the most common. An extinct bird, the huge raptor-like teratorn, stood 0.75 m (2.5 feet) tall and sailed on wings spreading 3.5 m (10.5 feet), making it one of the largest birds known to fly. Modern species of amphibians, reptiles, and fishes also occur in asphalt, as do a number of invertebrates including snails, bivalves, insects, and spiders. Some of these are freshwater species (e.g., rainbow trout), indicating that ponds or streams were sometimes present when asphalt flowed to the surface.

The La Brea fossils indicate the biota that flourished on the western edge of North America at the end of the Wisconsin Glaciation (10,000–40,000 BP), the coldest and last of four major glacial advances in the Pleistocene Epoch in North America. Notably, the mammalian fauna at La Brea includes a large number of what is known as the Pleistocene megafauna, that ism those whose body mass exceeded 44 kg (100 pounds). Virtually all of these species were soon to disappear from North America, either by extinction (e.g., short-faced bears and dire wolves) or emigration (e.g., horses and camels) to other continents, where they continued evolving into their modern forms.

Seeds, leaves, and especially pollen grains preserved in the La Brea asphalt suggest an environment well suited to large herbivores and their predators. Winters were little different than those of today, but the summers were cooler and marked with about twice the current rainfall. These conditions produced four recognizable plant communities: chaparral that formed on the mountain slopes on the eastern edge of the plains as represented by scrub oak and chamise; willows, alder, and others associated with riparian vegetation; a community of redwood, dogwood, and bay that developed in the deepest, most protected canyons; and a plain of seasonal grasses interspersed with sage and other shrubs.

This amenable climate and rich biota at La Brea surely favored human settlement, but the fossils include the remains of only one person, a 9000-year-old skull and partial skeleton of a young woman. Her death remains a mystery. La Brea Woman may have stumbled into the asphalt or perhaps became mired while attempting to retrieve a struggling duck for the evening meal, but another theory proposes murder. Her skull was discovered in several pieces, suggesting a blow to the head with her body dumped into the asphalt to cover the crime. Another idea proposes a ritual burial, although one would expect similar evidence of other funerals to support that notion. About 100 human artifacts – shell jewelry and hide scrapers, among others – have also been recovered in deposits less than 10,000 years in age.

The Florida Keys

A sweeping arc of islands known as the Florida Keys extends 210 km (130.5 miles) from Soldier Key at the Bay of Biscayne southwest into the Gulf of Mexico to Key West. The Keys were a contiguous extension of the Florida peninsula until about 4000 years ago, when rising sea levels isolated the present chain of island areas from the mainland.

In geophysical terms, the Keys lie just north of the Tropic of Cancer (i.e., the northern edge of the true tropics), but because of their climate, geology, and vegetation, they are generally regarded as "tropical." The high point in the archipelago rises to only 5.5 m (18 feet) above sea level, and most of the remaining area lies below 2 m (6.5 feet).

Geographers separate the Keys into three groups: Upper, Middle, and Lower. Key Largo limestone, primarily consisting of materials from fossil coral reefs, runs the entire length of the Keys; the coral formation is topped by **oolitic limestone** beginning at Big Pine Key and continuing southward down the Lower Keys, however. The difference is important as oolitic limestone is relatively impervious to water; surface pools of fresh water therefore form after heavy rains and serve the needs of Key deer and other wildlife. Conversely, Key Largo limestone is porous and rarely allows surface water to accumulate, thereby precluding readily available drinking water for animals that might otherwise prevail on the Upper and Middle Keys. Endemic taxa are especially prominent in the biota of the Lower Keys, which are separated from the Middle Keys by the Moser Channel, a barrier of 11 km (6.8 miles) and the largest gap of water in the archipelago.

The Middle Keys, which are narrower than the other islands, represent a funnel ("bottleneck") where migratory birds – particularly hawks and other raptors – concentrate during their autumn migration. This area is ideal for monitoring and banding birds as they continue *en route* to the West Indies. Six species of raptors (of 16 observed) are common, including the once-endangered peregrine falcon.

Interestingly, the surface soils in the Keys developed largely from parent materials originating as wind-blown dust from Africa, especially from the Bodele Depression in northern Chad. Nutrients in these soils, particularly phosphorus and iron, offset the natural deficiencies in subtropical and tropical soils. They foster the growth of vegetation not only on the Keys but also on other islands in the Caribbean Sea and the rainforests of Brazil, and indeed nourish phytoplankton in the Atlantic Ocean. The long-range transport of African dust did not escape

the notice of Charles Darwin, who collected samples of the fine red soil while sailing aboard the HMS *Beagle* in 1832. He remarked that the amounts of dust "…blown year after year over so immense an area [shows how] a widely extended deposit may be in the process of formation." Today scientists recognize that this **aeolian** process started at the end of the last ice age and transports about 1.5 *billion* tons of African dust each year across the Atlantic Ocean.

Red, black, and white mangroves, along with buttonwood, typically represent the native vegetation at sea level. Hardwood hammocks of species such as gumbo limbo, West Indies mahogany, and Jamaican dogwood – all representative of the West Indian flora in the Keys – develop at higher elevations, as do pinelands of slash pine and saw palmettos. The pineland communities occur on sites known as pine rocklands, a reference to their location on oolitic limestone substrate covered with a shallow layer of surface soils. Patches of thatch palm and silver palm are prominent in the understory, which includes five endemic herbaceous species. Lightning-initiated fires (essential for the continued existence of the endemic flora) once sufficed to maintain the community against hardwood dominance, but controlled burns at 10–20 year intervals are now required to approximate the natural fire regime.

The Key tree cactus is among the more interesting endemic species in the flora of the Lower Florida Keys, although some may also occur in Cuba. Their erect ribbed and spiny stems and branches reach tree-like heights of up to 10 m (33 feet). Their bell-shaped, white flowers open at night and issue a scent similar to garlic. Key tree cacti develop in isolated patches. Whereas seed dispersal by birds is necessary for establishing distant populations, new plants typically develop by vegetative reproduction (e.g., rooting of fallen branches) which fosters the clumped distribution of the species. The species usually grows in tropical hardwood hammocks in the early stages of succession where storm-related flooding is rare, but also occurs in thorn scrub at somewhat lower elevations. Commercial and residential development poses serious threats to the species, which is now listed as endangered. A 3.6-hectare (9-acre) preserve currently offers protection from further development and serves as a site for conservation activities (e.g., propagation research).

Two very different plant communities – hardwood forests and those of red mangroves – weigh heavily in the ecology of the white-crowned pigeon, a species whose northern range extends through the Keys into the southern tip of the Florida peninsula. Unrestricted hunting elsewhere (primarily Caribbean Islands) poses a serious threat to the species but in Florida, where the birds are not legally hunted, habitat loss is the foremost concern. Mangrove forests, many forming offshore islands, provide the pigeons' nesting habitat and are generally well protected on refuges in the Keys. In contrast,

Figure 12.11 Key deer, the smallest subspecies of white-tailed deer, are restricted to the Florida Keys where they wander into urbanized areas. Feeding the deer, especially from a car, reduces their fear of vehicles, the main cause of Key deer mortality. Some are individually marked with collars for study. Reproduced by permission of Roel Lopez.

inland hardwood forests represent foraging areas where, for example, hanging clusters of fruit on poisonwood trees offer lipid-rich foods. The birds are obligate **frugi-vores** and, in turn, disseminate seeds for new growth. Unfortunately, commercial and residential development on privately owned lands in the Keys has claimed large areas of hardwood forest and, while they are unusually swift fliers, the pigeons can no longer reach the remaining hardwood sites on their daily foraging flights. As well as protection of nesting habitat, similar protection of their feeding habitat is equally essential to ensure the future welfare of white-crowned pigeons in North America.

A small subspecies known as the Lower Keys marsh rabbit is endemic to a few of the larger islands in the Lower Keys (e.g., Boca Chica, Sugarloaf, and Big Pine). These rabbits prefer higher elevations of salt- or freshwater marshes with vegetation such as sawgrass, Gulf cordgrass, and seashore dropseed; sea ox-eye is a favorite food. Although any animal with a limited distribution on a handful of islands remains at risk, Lower Keys marsh rabbits face threats from introduced predators in the form of boa constrictors and pythons, although stray housecats pose the greatest immediate threat. Experimental reintroductions of these rabbits proved successful and, if coupled with habitat improvements and land acquisition, may have broader usefulness as a conservation tool in the Keys' fragmented landscapes.

Key deer, the smallest white-tailed deer subspecies (of 28), are also isolated on 20–25 islands in the Lower Keys, with more than half of the population occurring on Big Pine Key, the largest component of the Key Deer National

Wildlife Refuge. Males are about the size of a large dog; females are slightly smaller (Fig. 12.11). During wet periods, Key deer swim to islands with temporary pools of freshwater, but in dry months return to islands with permanent sources of drinking water. Mangroves, black-bead, Indian mulberry, and the berries of thatch palm are important foods, but the diet of Key deer includes a large range of other plants.

Key deer were listed as endangered in 1967 after their population diminished precariously to a low of about 25 in 1955. It subsequently rebounded to more than 700 by the year 2000 and may be close to outstripping its food supply in some areas. They live in close association with humans and frequent yards and roadways. Unfortunately, US Route 1 bisects Big Pine Key, and deer–car collisions once accounted for more than 50% of all deer mortality. Construction of various structures along Route 1, including underpasses designed as safe crossings for deer, subsequently decreased the risk of deer–vehicle collisions.

The northern end of Big Pine Key offers more optimal habitat for Key deer, whereas the habitat at the southern end of the island is more developed and fragmented with roads and fences. Accordingly, dispersal from the growing deer population at the north end supplements the decreasing, non-viable population in poorer habitat to the south. In fact, without dispersal from the north, the southern population has a 97% probability of dropping below 25 individuals within 20 years and thereby reaching the threshold of quasi-extinction. This situation represents an example where **source–sink dynamics** prevail (i.e., an ongoing, one-way flow in the population from abundance to scarcity). In terms of conservation, protection of the remaining high-quality "source" habitat from further development becomes a paramount goal for the future welfare of Key deer.

Sea-level rises represent a viable threat to the biota of the Florida Keys and low-lying coastal islands elsewhere. In particular, intrusion of saltwater into the porous limestone substrate contaminates the drinking water for Key deer and other wildlife. Moreover, because evaporation from an ever-saltier water table increases the salinity of surface soils, the vitality or even the existence of many plants is jeopardized. Indeed, endemic species such as the Key tree cactus already show signs of distress associated with hypersalinity.

In addition to the deer refuge, birds and nesting sea turtles are protected on Key West National Wildlife Refuge (NWR), one of the first refuges declared by President Theodore Roosevelt in 1908. Great White Heron NWR, an extensive area of mangroves, also protects nesting areas for loggerhead and green sea turtles as well as numerous birds, including an isolated population of its namesake white heron (a **morph** of the great blue heron that occurs widely elsewhere in temperate North America). Crocodile Lake NWR on Key Largo includes a diverse tropical hardwood forest of which about 80% of the species are West Indian in origin, but the American crocodile is the focal point. The species differs in significant ways from the American alligators, including its narrow snout, preference for saline water, and its marked intolerance to cold.

The Great Lakes

Massive ice sheets covered the northern half of North America during the Pleistocene Epoch and left permanent marks on the continent when they retreated. Of the many Ice Age glacial legacies, none is as significant as five immense freshwater lakes – Lakes Superior, Michigan, Huron, Erie, and Ontario – known collectively as the Great Lakes. For those who have only experienced the lakes as map features, the enormity of the Great Lakes is difficult to comprehend. For example, nearly 20% of all fresh water on Earth is contained in the combined volume of the five lakes; the pooled volume is about 22,700 m³ (5500 cubic miles) and the total surface area is almost 244,000 km² (94,000 square miles). Because of their immensity, the array of lakes is sometimes described as the "North Coast" of the United States.

Natural watercourses have always connected the five lakes. After entering Lake Superior from one of more than 200 tributaries, water flows by gravity eastward to Lake Huron through the St Mary's River. The Straits of Mackinac form such a wide connection between Lake Michigan and Lake Huron that some consider these as a single body of water. The St Clair River, Detroit River,

and Lake St Clair drain water from Lake Huron into Lake Erie, from where the flow continues down the Niagara River and over Niagara Falls into Lake Ontario. The St Lawrence River completes the long journey to the Atlantic Ocean. Man-made alterations, among them the Soo Canals and the St Lawrence Seaway, now influence movements of water in segments of this huge drainage.

Before glaciers buried the region under an ice blanket nearly 3 km (2 miles) thick, the area now occupied by the Great Lakes developed as a series of river valleys likely associated with an ancestor of the St Lawrence River. When the ice sheets melted, rocks, gravels, and sands once entrained in the ice were deposited in **terminal moraines**. The basins formed by glacial scouring and compression filled with water, forming a series of lakes (precursors to the Great Lakes) between the terminal moraines and the retreating ice fronts.

Lake shorelines changed as glaciers retreated northward and existing shoreline features, such as the extensive sand dune systems on the eastern shore of Lake Michigan, formed during this period. Withdrawal of the glaciers exposed ecological frontiers in each of the lake basins. Pioneer species rapidly invaded the varied habitats – sand dunes, marshes rocky shorelines, lakeplain prairies, savannas, forests, and fens – surrounding the lakes. Immigrant flora and fauna formed new biological communities that promoted diversity as evolution produced new taxa. By the time the first European explorers arrived in the 16th century, several species of ciscoes and a subspecies of walleye, the blue pike, had already differentiated in the lakes. Dune thistle, Houghton's goldenrod, and the Lake Huron locust represent species endemic to shoreline habitats.

The Great Lakes exert considerable climatic influence on adjacent regions. The volume of water retained in each lake moderates the temperatures of the surrounding lands. Compared to other locations at the same latitude, lake basins experience cooler summers and warmer winters. Humidity in the region is consistently high, resulting in increased precipitation in some areas. Winter winds typically sweep across the lakes from the northwest, absorb moisture, and deposit heavy snowfalls in what is known as the "lake effect" on southern and eastern shorelines. "Snow belts" on the downwind coasts receive 1.5–3.4 m (5–11 feet) of snow each winter.

Although usually considered as units in a single system, each lake possesses unique physical characteristics. The smallest lake in the system, Lake Erie, is relatively shallow. Prevailing winds generate intense wave action and marked changes in water levels known as **seiches** that alternately flood, or drain, Lake Erie's shorelines. Strong winds often produce huge waves on the larger lakes. Caught in a 1975 winter storm, the enormous ore freighter SS *Edmund Fitzgerald* sank when battered by near hurricane-force winds and 11 m (35 feet) waves on Lake Superior. Songwriter-singer Gordon

Lightfoot commemorated the loss of the ship and its 29-man crew in a ballad still popular today.

Prior to the arrival of European colonists, an estimated population of 60,000 to 117,000 Native Americans lived around the lakes in small tribal bands that lived compatibly with the natural environments. European explorers searching for a water passage to Asia discovered the Great Lakes in the 16th century and were soon followed by progressions of trappers, loggers, farmers, fishermen, soldiers, and settlers. To support a growing human population, commercial fishing began early in the 17th century with harvests selected from the 150 species of fish in the lakes. Overfishing, habitat loss, pollution, and the introductions of exotic aquatic species thereafter contributed to steady erosion of the lake's environmental quality and productivity.

Lake trout, endemic to cold waters of northern North America, were for more than half a century the most important commercial fish in the upper Great Lakes. Old timers who fished Lake Huron commercially recognized at least 16 distinctive forms. Well adapted to deep, clear waters, lake trout thrived in the cold, oxygen-rich waters of Lake Superior, Lake Michigan, and Lake Huron where their low metabolic rate produced growth in low temperatures, slow movements, and survival for long periods without feeding. As they grow, they switch from an opportunistic diet that includes plankton, shrimp-like crustaceans, and tiny fish, to become piscivorous as adults. Ciscoes, chubs, sculpins, and small whitefish are their primary prey.

In 1932 the unwelcome sea lamprey invaded the upper Great Lakes and their tributaries. Sea lampreys resemble eels but differ in several important ways, paramount of which is their lack of jaws. As parasites, sea lampreys have sucker-like mouths equipped with spiral rows of teeth. They attach themselves to other fishes and, after rasping through the skin, derive nourishment from the body fluids of their hosts.

The unwanted immigration of sea lampreys into the upper Great Lakes became possible when the natural barrier of Niagara Falls was skirted by expansion of the Welland Canal. Curiously, the original canal completed in 1829 somehow kept sea lampreys from reaching the immense inland watershed beyond Lake Ontario, but this changed when the canal was rebuilt a century later. The result of the sea lamprey invasion became apparent when production in the lake trout fishery plummeted by more than 90% between 1935 and 1949. As the numbers of the top predator diminished, populations of a small shad-like fish known as the alewife grew unchecked. Like lampreys, alewives also invaded the lakes through the Welland Canal. The alewife population often experiences seasonal die-offs with huge numbers washing up on shorelines where they decay in noxious piles. Overall, the invasion of lampreys induced complex changes – particularly in predator–prey relationships – in

various fish populations that often resulted in an unstable fishery for all species of commercial value. Millions of dollars have since been expended to control sea lampreys with weirs, shocking devices and, most recently, selective larvicides.

The sea lamprey was certainly not the last exotic species to invade the Great Lakes. Today, the lakes are home to approximately 160 non-indigenous species of algae, invertebrates, fish, and plants. More than 30% of these have entered since the 1960s. The zebra mussel (a native of Russia), for example, was first detected in the Great Lakes system in 1988 when, after spreading across Europe for two centuries, they escaped in Lake St Clair between Detroit, Michigan and Windsor, Ontario. Experts believe they traveled across the Atlantic Ocean either as stowaways in bilge water or attached to the anchor and chains of a transoceanic freighter sailing from Europe.

Adult zebra mussels are relatively small, clam-like bivalves named for the striped pattern on their shells. When two years old, female zebra mussels produce 30,000 to 1 million eggs annually, making the species one of the most prolific of all animals. Dense clusters of zebra mussels tightly attach to surfaces with thread-like "strings" issued from the hinged side of their shells. Huge clusters can completely cover most aquatic substrates (e.g., sand, silt, and, importantly, the shells of native species of mussels, which they ultimately smother) and submerged structures such as pilings, rocks, and the undersides of docks and boats. Densely packed zebra mussels often clog water intakes of municipal water plants, electric power plants, and other consumers of water from the Great Lakes. Estimates of the costs associated with controlling the invasive pest at these facilities range up to about $500 million per year.

Although zebra mussels are considered pests, the species does have some redeeming qualities. As filter-feeders, zebra mussels living in polluted or murky aquatic environments can improve water clarity by removing particulate matter, thereby increasing sunlight penetration and, in turn, greater algae production. As a result, zebra mussels are credited with increasing the population of smallmouth bass in Lake Erie and yellow perch in Lake St Clair. Some ducks and other animals feed on zebra mussels, but the losses to predators do not control mussel populations.

Since their first appearance in North America, zebra mussels have expanded their range into most of the major rivers and lake systems in the Eastern Deciduous Forest (Chapter 4). They now occur at the western margin of the biome in Minnesota and Lake Texoma, Oklahoma. In 2010, the first zebra mussels were discovered in California. Their rapid spread was probably facilitated by mussels attached to the boats, motors, and trailers of recreational watercraft, aided by an ability to survive out of water for up to 7 days. The related invasive bivalve, the quagga mussel, was first observed in the Great Lakes in 1989. Like zebra mussels, these filter-feeders biofoul hard-surfaced structures, but they also colonize deeper benthic habitats where they displace the native burrowing amphipod *Diporeia hoyi*.

The St Lawrence Seaway and the Welland Canal are not the only route of entry for exotic invaders. Two species of Asian carp (silver and big-headed carp) escaped from flooded aquaculture facilities and entered the Mississippi River in the early 1990s. The fishes have steadily moved northward and, in some areas, are now the most abundant species. Should they reach the Great Lakes, the carp may compete for food with valuable sport and commercial fishes. In 2010, the carp were detected just 40 miles from the Chicago Ship and Sanitary Canal, which was constructed to connect Lake Michigan to the Mississippi River. Managers are now struggling to prevent these exotics from reaching Lake Michigan and the lakes beyond. In 2009, however, DNA from Asian carp was discovered on the lakeside of an electric barrier designed to keep carp confined to the Mississippi River system.

Pollution in the Great Lakes reached a peak during the 1960s and 1970s. Many agricultural operations, heavy industries, and municipalities drained sewage and effluents directly into the lakes. As a result, dead zones formed in parts of Lake Erie, and some beaches were declared off-limits because of bacterial contamination. In 1969, a fire on the oily surface of the Cuyahoga River (a tributary of Lake Erie) near Cleveland, Ohio shocked citizens of Canada and the United States into joining hands to clean up the lakes. These recovery efforts – aided in part by the passage of the Clean Water Act of 1972 – took decades, but a welcome sign of success appeared in 1999 when clouds of mayflies, the first seen in nearly 40 years, emerged from Lake Erie. Although zebra mussels remain a problem for many reasons, their filter-feeding helped clear the lake of enough pollutants to allow the survival of mayfly larvae. Recent agreements between Canada and the United States including funding for clean-up and management of invasive species offers hope for continued restoration of the Great Lakes.

Habitat highlights

Rivers of ice

During the Pleistocene Epoch, the Northern Hemisphere experienced recurring periods of glaciation. Today, Greenland's ice sheet and alpine glaciers in North America remind us of a time when the Northern Hemisphere was much colder.

Alpine glaciers form where accumulations of snow and ice exceed ablation (i.e., loss from melting). As layers of snow and ice in alpine glaciers undergo repeated freezing and thawing, the upper layers fuse into firn, a

dense form of ice. Additional layers accumulate over several years, compacting firn into an even denser form of glacial ice at depths below 50 m (160 feet). Gravity controls the slow downward flow of alpine glaciers. The top layer of a glacier generally moves more rapidly than the dense bottom layer, and the center moves more rapidly than the sides – i.e., the sides are slowed by friction with valley walls. Because glaciers move as a solid rather than a liquid river, frictional stresses between layers and between the center and the sides produces deep cracks known as crevasses in the glacial surface. An alpine glacier with deep crevasses presents a formidable barrier that restricts animal dispersal and genetic interchange between closely related organisms separated by the frozen stream.

Spiders and snow fleas are among the few species that regularly occupy glacial ice but one organism, the glacial ice worm, cannot exist anywhere else. The only annelids known to complete their entire life cycle in ice, glacial ice worms inhabit glaciers from the coastal ranges of southern Alaska and British Columbia to Mount Hood, Oregon. In the northern Cascades, ice worms reach densities of 2000 or more per square meter but on Alaskan glaciers densities of 100–200 m^{-2} are the norm. Active during evening and night hours, ice worms eat algae and bacteria on glacial and meltwater stream surfaces; they retreat beneath the snow during daylight hours. Glacial ice worms possess a glycine-rich protein that may act as an anti-freeze. For humans, the protein may prove useful for increasing the storage time for transplant organs as well as improving ice cream.

Changes in glacial mass are sensitive indicators of climate change. Alpine glaciers in North America increase in mass and length during cold years, but recently glacial ablation and retreat have accelerated. With long-term increases in temperature, snowfall lessens in the accumulation zone, the top layers of ice thin, and the glacial terminus recedes (Fig. 12.12). Ironically, Alaska's Tyndall Glacier (named after John Tyndall who in 1861 identified carbon dioxide as a greenhouse gas that could increase global temperatures) has retreated 24 km (15 miles) since 1960 and continues to recede at a rate of 500 m (1600 feet) per year. All 47 glaciers in the North Cascades are currently retreating and four have disappeared completely since 1985.

The retreat of North American glaciers threatens more than ice worms. Glacial meltwater flowing downstream sustains plants and animals requiring cold-water habitat (e.g., cutthroat trout and pink salmon, among others). As glacial mass declines, stream flow gradually diminishes accompanied by corresponding reductions downstream in the dependent biota. Moreover, deep layers of glacial ice sequester pollutants (e.g., DDT and mercury, which were banned or controlled decades ago). With melting, these pollutants are released into aquatic and marine food chains where they again may adversely

(a)   (b)

Figure 12.12 Some glaciers in North America continue to advance, but most have receded during the last century. Two views of Grinnell Glacier, Alaska, from the same vantage point demonstrate the extent of melting between (a) 1938 and (b) 2009. Reproduced by permission of (a) T.J. Hileman and (b) Lindsey Bengston of the US Geological Survey.

affect populations of brown pelicans, peregrine falcons, and other egg-laying species occupying upper trophic levels. Likewise, the addition of heavy metals into marine ecosystems may engender new restrictions on fisheries. Whereas the full range and severity of these and other ecological impacts are not completely known, there is little doubt that the continued loss of glaciers – one of Earth's more powerful forces – will diminish the grandeur of nature.

Hot springs and geysers

Hot springs are distributed widely in North America, with many located in volcanic zones where underground water comes into contact with molten rock (i.e., magma). In areas far removed from volcanoes, geothermal heat from the Earth's mantle generates just as many hot springs. By general definition, hot springs issue water above 36.7°C (98°F), but water temperatures above the boiling point are common in those arising in volcanic zones.

A strong head of steam results from underground water boiled by magma. This pressure is released either as a **fumarole**, when steam escapes at a surface opening, or as a geyser, when steam mixed with water erupts as a fountain. About a thousand fumaroles and nearly 500 geysers (about half of the world's geysers) occur in

Yellowstone National Park, Wyoming. The renowned Old Faithful erupts regularly over a 91-minute cycle and thrusts a spout 32–56 m (106–185 feet) into the air (Fig. 12.13a). An even larger geyser, Steamboat Geyser, generates the world's tallest fountain but it does not erupt on a predictable schedule.

Although originally of interest as geological phenomena, hot springs and geysers also harbor unique organisms able to thrive in such harsh environments. Since microorganisms were discovered in the boiling springs at Yellowstone in 1996, searches have revealed **thermophiles** (organisms living at temperatures of 45–122°C or 113–252°F) in hot springs and geysers all over the world. Even extremely hot springs contain certain viruses, bacteria, and archaea; the latter are single-celled microorganisms that lack a cell nucleus and membrane-bound organelles. Some thermophiles also live in hot waters containing high levels of sulfur, calcium carbonate, or acids otherwise fatal to most forms of life.

Thermophilic microorganisms share a common feature: their enzymes and amino acids continue to function at high temperatures. The heat-stable enzymes of two thermophilic species, *Thermus aquaticus* and *Thermococcus litoralis*, provide sources for the polymerase chain reaction essential for DNA fingerprinting. Indeed, the unique

(a) (b)

Figure 12.13 About 500 geysers occur in Yellowstone National Park, Wyoming. (a) Old Faithful is not the largest in the park, but it is the best-known because of its predictable schedule of eruptions and accessibility to park visitors. (b) Half Dome in Yosemite National Park may be the most widely recognized inselberg in North America. The massive granitic rock rises 1444 m (4737 feet) above the floor of Yosemite Valley. Reproduced by permission of (a) Sandra S. Chapman and (b) Brian R. Chapman.

properties of thermophiles may promote several other advancements in biotechnology. Thermophiles may also provide clues on the origins of life, which may have developed in hot environments about three billion years ago; perhaps they were the first organisms on Earth.

Forest in the clouds

Major elements of the flora and fauna representing the Neotropical and Nearctic biogeographical regions converge near the Tropic of Cancer in the Sierra Madre de Oriental of southern Tamaulipas. This area of environmental heterogeneity marks the northernmost tropical montane cloud forest in North America. Beginning near Gómez Farías, a precarious, bone-jolting ride on a narrow road takes visitors to El Cielo, the "forest in the clouds."

Cloud forests, or "fog forests," develop within a relatively narrow altitudinal zone where moisture-laden clouds blowing inland regularly blanket the slopes. In this case, the Sierra Madre de Oriental obstructs humid winds from the Gulf of Mexico and, when this moist air ascends the slopes, rain and fog develop as the air cools. The cloud forest at El Cielo occurs at elevations of 800–1400 m (2625–4590 feet), and receives an average

rainfall of 2.5 m (9.8 feet) annually. Relative humidity remains consistently above 90%, and water droplets from the persistent mist collect on leaves before dripping onto the substrate below. Such "cloud filtering" increases the amount of water available to roots by as much as 60%. Fog also reduces the influence of direct sunlight, which lowers daytime temperatures and reduces evapotranspiration. In all, El Cielo is a very wet place.

Although tropical montane cloud forests are not as species-rich as tropical lowland forests, they nonetheless harbor many endemic species. Cloud forest trees typically feature more massive trunks, denser crowns, and lower heights than trees in adjacent forests at lower altitudes. Tree branches, often gnarled, are encrusted with epiphytes and support drapes of lianas. A diverse epiphytic community, consisting of lichens, mosses, ferns, bromeliads, and orchids, thrives in the unyielding humidity.

The area supports more than 1000 vascular plant species including more than 30 orchid varieties. The Gómez Farías region is widely known as a birding "hotspot" where observers may see more than half of the species in Mexico's rich avifauna. Visitors are occasionally rewarded with sightings of rare or secretive species such as the great curassow, maroon-fronted parrot, and bat falcon. Small, spotted cats known as margays as well as much larger jaguars prowl in the misty forest.

Loggers plundered El Cielo prior to the 1960s, but rough topography and inaccessibility eventually eliminated the exploitation. Because of its unique location and biodiversity, the cloud forest became part of the Man and the Biosphere Program in 1986. El Cielo Biosphere Reserve now protects 144,530 ha (356,442 acres) of cloud forest as well as zones of tropical semi-deciduous forest, pine-oak forest, and xerophytic scrub. Federal and state laws protect the area's flora and fauna but poachers still exploit some species, especially orchids and cacti. Nonetheless, because of its remoteness, the El Cielo cloud forest remains relatively pristine and, for the determined naturalist, worth the arduous journey.

Granite outcrops and inselbergs

Much of the continental crust is composed of granite, the dense rock used to construct kitchen countertops, gravestones, and courthouses. Formed initially as mineral-rich magma (i.e., molten rock), the hot igneous material pushes upward at some locations producing volcanoes or immense intrusions concealed at depths of 1–5 km (0.6–3 miles) beneath the sedimentary crust. As magma slowly cools and crystallizes under great pressure below the surface, the intrusions solidify into granitic **plutons**. Collisions of tectonic plates or other mountain-building processes subsequently push the granite upward, where erosion of the overlying sedimentary veneer reveals either flattened outcrops – the exposed tops of plutons – or massive mountain-like domes known as **inselbergs.**

Granite appears in many areas of North America, but exposed slabs are concentrated in: the Piedmont Region from Virginia to Alabama whereas granitic domes are particularly common in the western states, especially the Cascade Range of California: the south-central area including Texas, Oklahoma and western Arkansas; and the Piedmont Region of the southeast from Virginia to eastern Arkansas. Some large inselbergs are iconic locations visited by thousands of sightseers daily, including Half Dome and El Capitan in Yosemite National Park, California; Enchanted Rock near Fredricksberg, Texas; and Stone Mountain in Atlanta, Georgia (Figure 12.13b).

At first glance, exposed surfaces of granite appear to be devoid of life. Indeed, the environment on barren rock can be extreme; temperatures in excess of 50°C (120°F) are common during the summer months, light intensity is high, moisture is rarely available, and soil layers, where present, are thin and nutrient-poor. These conditions combine to produce a "micro-environmental desert" in which few species can survive. Crustose lichens and hardy mosses, such as grimmia dry rock moss and haircap moss, initially invade granite faces across much of the southeastern United States. The

corrosive action of these colonizers contributes to soil formation, but exfoliation (a weathering process) also causes granite to flake into particles that eventually become components of soil. Uneven weathering forms small depressions that collect aeolian soils or rainwater. Soil islands forming in these shallow depressions develop pioneer plant communities where a few drought-resistant species, such as Small's stonecrop, can invade.

After rains, "blooms" of fairy shrimp and clam shrimp erupt in freshwater pools forming in depressions on the granite surfaces. The sudden appearance of tiny aquatic animals after long periods (sometimes years) of dry conditions seems miraculous, but the process of **cryptobiosis** allows crustacean eggs to remain dormant until conditions favor hatching. Using a somewhat similar approach, nematodes survive dessication in shallow soils with the adaptation of **anhydrobiosis,** a form of cryptobiosis involving completely reversible dehydration of a living organism. As the soil dries, the nematodes coil into a tight spirals and enter an ametabolic state that can last for more than a year. When moisture again becomes available, the worms rehydrate, uncoil, and resume normal activities.

The disjunct, island-like distribution of granite outcrops, together with the forces of evolution over geological time, have produced numerous endemic species adapted to severe habitats. More than 20 species of endemic plants occur on granite outcrops in the southeast, and most are federally protected. Unfortunately, many landowners view outcrops of granite flatrock as relatively worthless land. Quarrying has destroyed many sites, and others have become unsightly dumps. Although granite flats and inselbergs represent only a small fraction of land area, those that remain offer valuable opportunities for studies of biogeography and plant succession.

Palm forest

The first Spanish explorers visiting the Gulf of Mexico coast discovered an extensive palm forest at the mouth of the Rio Grande. The derivation of the river's original name, *Rio de las Palmas* ("Palm River") was formerly attributed to Alonso Alvarez de Pineda, but historians now doubt this origin. Regardless, the name appropriately reflected the dense palm forest that once extended some 130 km (80 miles) inland and encompassed approximately 16,190 ha (40,000 acres). Today, only small remnants of this unique subtropical habitat remain in the lower Rio Grande Valley of Texas.

Sabal palmettos dominate the palm forest; these reach heights of 18 m (26.25 feet) and produce fan-shaped fronds attached to long, spineless petioles (Fig. 12.14). The fronds drupe downward as they mature, often encasing the top 3 m (9 feet) of the trunk with overlapping layers of dead leaves. The overlapping crowns of mature trees in a dense stands limit the amount of sunlight reaching the

Figure 12.14 Only a small remnant remains of the sabal palmetto forest that once extended far inland from the mouth of the Rio Grande. Although the dense tangle of palm fronds limits light penetration, a rich association of understory plants and wildlife thrives in the forest now protected by the National Audubon Society. Reproduced by permission of Brian R. Chapman.

forest floor. Nonetheless, more than 80 species of native plants push through a tangled accumulation of leaf litter to form an understory community.

Many tropical species reach their distributional limits in the palm forest. White-lipped frogs and black-spotted newts feed on some of the 900 species of beetles in the forest's understory. Speckled racers and black-striped snakes hide in the ground litter to avoid visiting predators such as ocelot, jaguarundi, or white-nosed coati. More than 380 species of birds have been observed in the forest, which serves as a seasonal rest stop for migratory species. Chachalacas ("Mexican pheasants") cackle loudly at sunrise, awakening green jays, Altamira orioles, and olive sparrows. Green kingfishers and groove-billed anis perch along both the river and a **resaca** bordering the forest. Ferruginous pygmy owls signal the close of day, as southern yellow bats and Mexican free-tailed bats emerge from roosts concealed in dead palm fronds to forage on nocturnal insects.

To protect the largest remaining cluster of intact sabal palm forest, the National Audubon Society purchased a 225-ha (527-acre) tract in 1972. The Audubon sanctuary surrounds a 13-ha (32-acre) stand of undisturbed palm forest with a buffer zone consisting of a cresent-shaped resaca and thorn-scrub brushlands. The thorny flora is characteristic of the lower Rio Grande Valley and consists of dense entanglements of retama, huisache, black mimosa, granjeno and Texas ebony. Funding from a private foundation will preserve the palm forest and surrounding habitats for years to come.

Mineral licks

For trailblazer and frontiersman Daniel Boone (1734–1820), Kentucky's treasure of natural resources offered settlers a bounty of rich soil, clear streams, grasslands, forests, and wildlife. He might well have added mineral licks, popularly known as "salt licks," to his list of attractions. Boone hunted bison by following their age-old trails that lead to mineral licks, sites he knew well as had generations of Native Americans before him.

Mineral licks occur in many regions of North America, where naturally occurring deposits of mineral-bearing soils often associated with springs or faults that create seeps, allow elk, moose, bighorn sheep, muskox, and deer and other ruminants to satisfy their "salt drive." Rodents, raccoons, black and brown (grizzly) bears, bees, butterflies and, more rarely, birds (e.g., band-tailed pigeons and goldfinches) are among the other animals reported at mineral licks. Somewhat uniquely, mountain sheep and mountain goats in the Canadian Rockies typically visit licks known as "white earth slopes," which are dry sites of salt-bearing soils on exposed hillsides. In the Black Hills, white-tailed deer paw deep depressions to expose and eat mineral-rich soil, a process known as **geophagy**, which is chewed in the same manner as forage.

Salt drive develops because sodium, while needed for many body functions in animals, is not a component of most forage plants. Salt drive is particularly evident in spring when ruminants begin foraging on fresh vegetation, but is often reduced or absent in winter. This relationship apparently stems from a greater intake of potassium concentrated in succulent vegetation that flushes sodium from mammalian systems; this leads to a temporary sodium imbalance that is satisfied by supplements of sodium, magnesium, and calcium. Interestingly, the vast bison herds roaming the western plains did not visit mineral licks, whereas their eastern counterparts followed well-worn trails to licks throughout the Appalachian Plateau. Western bison offset their sodium deficiencies by ingesting mineral-rich soils adhering to the short grasses in their diet.

Salt licks may serve as sites for other relationships, including social interactions when two or more animals concurrently visit a salt-rich location. Animals likely develop an acquired habit of visiting salt licks even when their salt drive may be reduced or non-existent. The strong attraction of ungulates to salt licks has a dark side, as when diseases and parasites spread among the visitors. Predators, notably those that ambush their prey (e.g., mountain lions), increase their odds of finding prey by hunting at salt licks, as demonstrated by the greater number of mountain goat carcasses occurring near salt licks than elsewhere in Jasper National Park. Regrettably, salt applied to de-ice roads increases collisions between vehicles and moose, especially in spring and early summer when salt drive is greatest and the salt has drained off into roadside pools.

Big Bone Lick in Kentucky has a history all its own. Its namesake is the cache of huge bones from mastodons, ground sloths, and other species of the North American megafauna that sought minerals but died, mired in the waterlogged soil at the site (see *La Brea tar pits* for parallel circumstances). The fossils were known to Native Americans and later to explorers and pioneers, but it remained for William Clark (1770–1838) and Thomas Jefferson (1743–1826) to establish the site as the "cradle of American paleontology." Clark collected fossils for Jefferson and, once in Jefferson's hands, the fossils attracted attention from the era's scientists, some of whom thereafter initiated their own studies of fossilized materials. Clark's visits to Big Bone Lick, Jefferson's intellectual curiosity, and the salt drive of a bygone fauna established paleontology as a science in America.

Bogs and their carnivorous plants

Bogs, also known as herb bogs, grass-sedge bogs, and moist pine barrens among other designations, occur widely in North America from the southeastern coastal plains of Florida to the boreal forests of Minnesota and Canada. Such a widespread distribution indicates profound differences in their flora, but many include one or more species of carnivorous plants, a unique type of vegetation defined by the following criteria: (a) capture insects, spiders, and other types of small animals using specialized structures; (b) absorb nutrients from their prey; and (c) utilize the metabolites they digest for their own growth and development. These plants gain nutrients, primarily nitrogen, from their carnivorous diets, but little energy, which they produce by photosynthesis in the same fashion as other green plants. Remarkably, several unrelated lineages of plants have independently developed their carnivorous characteristics, thereby illustrating an example of **convergent evolution.** A world-wide list includes more than 500 species of carnivorous plants.

About 45 species of carnivorous plants occur in North America, with the greatest diversity (29 species) recorded along the Gulf Coast where the flora of a single bog included a record-setting 13 species. Similar bogs occur along the Atlantic coastal plain (Fig. 12.15a). Sandy soils at these sites are typically deficient in nitrogen and sometimes phosphorus; virtually all are acidic. Small, shallow pools develop in rainy seasons and, in places, provide habitat for the endangered Pine Barrens tree frog. Gulf Coast bogs, like many elsewhere, seldom flood but maintain a watertable at or near or the surface. Water movement is limited, so little organic matter leaves the bog and only small quantities of nutrients flow into these systems. Nutrient cycling thus remains dependent on organic matter released by fires. In the past bogs burned about 2–3 times each decade, but more recently fire suppression has allowed invasions of woody vegetation, resulting in the loss of many bogs along the Gulf Coast. Drainage also contributes to these losses and, together with fire suppression, has reduced the remaining area to just 3% of its former size. Conversely, bogs in Canada remain close to their pre-settlement status.

Some carnivorous plants also grow in somewhat drier environments, but all share the common features of nutrient-impoverished soils and open sites with readily available sunlight. For example, fire-adapted communities of longleaf pine develop and fires that remove the shade-producing understory may offer habitat suitable for Venus fly traps (Fig. 4.8). Because of their heavy investment in unique structures, carnivorous plants lack any advantage in nutrient-rich habitats and instead have adapted to sites where competition is minimal from more aggressive vegetation. Some ecologists offer the analogy "carnivorous plants are to nutrients as cacti are to water."

Based on their respective leaf modifications, carnivorous plants employ one or more passive or active mechanisms to secure their prey: (a) pitfall (pitcher plants); (b) flypaper (sundews); (c) snap trap (Venus fly trap); (d) bladder trap (bladderwort); and (e) lobster pot, a relatively uncommon structure of one-way entries employed by a few uncommon species. Of these, the aquatic bladderworts suck their prey into a vacuum created inside small bladders that open when triggered by the touch of a victim swimming too close to the plant. Most naturalists are more familiar with the terrestrial species, notably the tubular pitcher plants and the much publicized "jaw-snapping" Venus fly trap, which is the only species of its kind. Interestingly, none of the mechanisms directly kill their prey; instead, they die by drowning or some other form of oxygen depletion.

Some genera of pitcher plants secrete enzyme-rich fluids that enrich the pools filling their pitfall traps, but others contain only rainwater (Fig. 12.15b). The pools formed in pitcher plants provide unique microhabitats, known as a **phytotelmata** (singular phytotelma). Animals living in phytotelmata are known as **inquilines**, which include protozoans, fungi, rotifers, crustaceans, arachnids, insects, and, in some locations, amphibians. The larvae of three species of flies, all detritivores but each in different families, coexist in the phytotelmata of purple pitcher plants despite their similar mode of feeding. They rely on a strategy known as **resource partitioning**, which in this

(a)   (b)

Figure 12.15 (a) Bogs on the Atlantic coastal plain typically include a variety of pitcher plants and sundews confined to low, damp sandy soils. (b) Pitcher plants attract and trap insects to obtain nitrogen and phosphorus from their digested bodies, but the larvae of some species, such as this flesh fly, develop within a pitcher plant's phytotelma. Reproduced by permission of (a) Elizabeth D. Bolen and (b) Brian R. Chapman.

case is spatial in nature. The pitcher plant midge feeds on drowned prey accumulating at the bottom of the phytotelmata, whereas the pitcher plant scarcophagid (a type of flesh fly) remains at the top of the water column where it feeds on floating, newly drowned prey. In between, the pitcher plant mosquito filters bacteria and protozoans from the middle of the water column. The spatial separation of these insects minimizes competition for the available food resources within the confines of the microhabitat where they develop as larvae. These organisms are obviously not affected by either the plants' enzymes or the acidic water in the phytotelmata.

A variety of animals fall into pitcher plants although ants are among the most common, reaching nearly 75% of the prey captured in purple pitcher plants. Contrary to expectations, the efficiency of capture is extremely low at <1% overall and just 0.37% for ants. The low efficiency may represent a means of precluding prey from developing avoidance behavior, which would deny the plant any success at all. Additionally, some spiders spin their webs across the trap's opening, thereby intercepting for their own use prey that otherwise would be available to the plant.

In addition to their distinctive plant communities pitcher plant bogs support unusually diversified assemblies of insects, even in those peat bogs in the cool, northern latitudes where diversity is diminished in comparison with communities in the warmer areas of North America (Chapter 1). These include obligate herbivores of carnivorous plants, which feed on nothing but the tissues of these

interesting plants. Bumblebees commonly pollinate pitcher plants, but this is not an obligate relationship; nonetheless, bumblebees may be the only pollinator to provide this vital function in extensive stands of pitcher plants. Regrettably, disturbances may fragment pitcher plant communities and the resulting patchwork may enhance hybridization when the bumblebees pollinate more than one species; the hybrids are fertile.

No truer words were ever penned about North America's vegetation when the pioneer ecologist B.W. Wells (1884–1978) observed that, in carnivorous plants, "nature has carried out the most amazingly different modification of leaf parts, to arrive at the same end – the capture and ingestion of insects."

Readings and references

The Grand Canyon

Brown, B.T. and L.E. Stevens. 1992. Winter abundance, age structure, and distribution of bald eagles along the Colorado River, Arizona. Southwestern Naturalist 37: 404–408.

Bureau of Reclamation. 1995. Operation of Glen Canyon Dam, Colorado River Storage Project, Arizona. Final Environmental Impact Statement. US Department of Interior, Washington, DC. (Describes ecological conditions associated with water released into the Grand Canyon.)

Carothers, S.W. and B.T. Brown. 1991. *The Colorado River Through Grand Canyon, Natural History and Human Change*. University of Arizona Press, Tucson, AZ.

Carothers, S.W., M.E. Stitt, and R.R. Johnson. 1976. Feral asses on public lands: an analysis of biotic impact, legal considerations, and management alternatives. Transactions of the North American Wildlife and Natural Resources Conference 41: 396–406. (Provides comparative data for communities in the Grand Canyon with and without burros.)

Clover, E.U. and L. Jotter. 1944. Floristic studies in the Grand Canyon of the Colorado and tributaries. American Midland Naturalist 32: 591–642.

Davis, R. and D.E. Brown. 1989. Role of post-Pleistocene dispersal in determining the modern distribution of Abert's squirrel. Great Basin Naturalist 49: 425–434.

Flowers, R.M., B.P. Wernicke, and M.A. Farley. 2008. Unroofing, incision, and uplift history of the southwestern Colorado Plateau from apatite (U–Th)/He thermochromometry. Geological Society of America Bulletin 120: 571–587. (Suggests that the Grand Canyon formation was about 17 million years ago.)

Hoffmeister, D.F. and V.E. Diersing. 1978. Review of tassel-eared squirrels of the subgenus Otosciurus. Journal of Mammalogy 59: 402–413.

Johnson, R.R. and S.W. Carothers. 1987. External threats: the dilemma of resource management on the Colorado River in Grand Canyon National Park, USA Environmental Management 11: 99–107.

Kaeding, L.R. and M.A. Zimmerman. 1983. Life history and ecology of the humpback chub in the Little Colorado and Colorado rivers of the Grand Canyon. Transactions of the American Fishery Society 112:577-594.

Lucchitta, I. 1988. Canyon maker: a geological history of the Colorado River. Plateau 59(2): 2–32.

Marsh, P.C. 1985. Effects of incubation temperature on survival of embryos of native Colorado River fishes. Southwestern Naturalist 30: 129–140.

Mells, T.S., J.F. Hamill, G.E. Bennett, L.G. Coggins, et al. (eds) 2010. Proceedings of the Colorado River Basin Science and Resource Management Symposium, November 18–20, 2008, Scottsdale, Arizona. US Geological Survey Scientific Investigations Report 2101-5135. (Describes effects of past and future releases of water from Glen Canyon Dam.)

Powell, J.W. 1875. Exploration of the Colorado River of the West and its Tributaries, Explored in 1869, 1870, 1871, and 1872. Originally published by the Smithsonian Institution; reprinted in 1961 by Dover Publications, New York, NY.

Ramey, C.A. and D.J. Nash. 1976. Coat polymorphism of Abert's squirrel, Sciurus aberti, in Colorado. Southwestern Naturalist 21: 209–217.

Unitt, P. 1987. Empidonax traillii extimus: an endangered subspecies. Western Birds 18: 137–162.

Young, R.A., D. M. Miller, and D.C. Ochsner. 1980. The Grand Canyon rattlesnake (Crotalus viridis abyssus): comparison of venom protein profiles with other viridis subspecies. Comparative Biochemistry and Physiology 66B: 601–604.

Caves

Barr, T.C., Jr. 1968. Cave ecology and the evolution of troglobites. Evolutionary Biology 2: 35–102.

Barr, T.C., Jr and J.R. Holsinger. 1985. Speciation in cave faunas. Annual Review of Ecology and Systematics 16: 313–337.

Boyles, J.G., P.M. Cryan, G.F. McCracken, and T.H. Kunz. 2011. Economic importance of bats in agriculture. Science 332: 41–42. (Estimates that the loss of insectivorous bats in North America may cost the agriculture industry at least $3.7 billion per year.)

Clark, B.K., B.S. Clark, and D.M. Leslie, Jr. 1994. Use of caves by eastern wood rats (Neotoma floridana) in relation to bat populations, internal cave characteristics, and surface habitats. American Midland Naturalist 131: 359–364.

Culver, D.C. 1970. Analysis of simple cave communities. I. Caves as islands. Evolution 24: 463–474.

Culver, D.C. 1982. Cave Life: Evolution and Ecology. Harvard University Press, Cambridge, MA.

Frick, W.F., J.F. Pollock, A.C. Hicks, et al. 2010. An emerging disease causes regional population collapse of a common North American bat species. Science 329: 679–682.

Griswold, C., T. Audisio, and J. Ledford. 2012. An extraordinary new family of spiders from caves in the Pacific Northwest (Araneae, Trogloraptoridae, new family). ZooKeys 215: 77–102.

Holsinger, J.R. 1988. Troglobites: the evolution of cave-dwelling organisms. American Scientist 76: 147–153.

Mitchell, R.W. 1969. A comparison of temperate and tropical cave communities. Southwestern Naturalist 14: 73–88.

Mitchell, R.W. 1971. Food and feeding habits of the troglobitic carabid beetle Rhadine subterranea. International Journal of Speleology 3: 249–270. (Notes an exception to the rule specialists seldom exist in food chains of cave communities.)

Mitchell, R.W., W.H. Russell, and W.R. Elliot. 1977. Mexican eyeless characin fishes, genus Astyanax: environment, distribution, and evolution. Special Publication 12, The Museum, Texas Tech University, Lubbock, TX.

Palmer, A.N. 1991. Origin and morphology of limestone caves. Geological Society of America Bulletin 103: 1–21.

Poulson, T.L. 1963. Cave adaptation in amblyopsid fishes. American Midland Naturalist 70: 257–290.

Poulson, T.L. and W.B. White. 1969. The cave environment. Science 165: 971–981.

Tuttle, M.D., and D.E. Stevenson. 1978. Variation in the cave environment and its biological implications. In: National Cave Management Symposium Proceedings (R. Zuber, J. Chester, S. Gilbert, and D. Rhodes, eds). Adobe Press, Albuquerque, NM, pp. 108–121.

Arctic ice cap

Borucinska, J.D., G.W. Benz, and H.E. Whiteley. 1998. Ocular lesions associated with attachment of the parasitic copepod Ommatokoita elongata (Grant) to corneas of Greenland sharks Somniosus microcephalus (Bloch & Schneider). Journal of Fish Diseases 21: 415–422.

Dunbar, M.J. 1981. Physical causes and biological significance of polynyas and other open water in sea ice. Canadian Wildlife Service Occasional Paper 45: 29–43.

Heide-Jørgensen, M.P., R. Dietz, K.L. Laidre, et al. 2003. The migratory behavior of narwhals (Monodon monoceros). Canadian Journal of Zoology 81: 1298–1305.

Laidre, K.L. 2003. Diving behavior of narwhals (Monodon monoceros) at two coastal localities in the Canadian High Arctic. Canadian Journal of Zoology 80: 624–635.

Laidre, K.L., I. Stirling, L.F. Lowery, et al. 2008. Quantifying the sensitivity of Arctic marine mammals to climate-induced habitat change. Ecological Applications 18: S97–S125.

MacNeil, M.A., B.C. McMeans, N.E. Hussey, et al. 2012. Biology of the Greenland shark Somniosus microcephalus. Journal of Fish Biology 80: 991–1018.

Maykut, G.A. and T.C. Greenfell. 1975. The special distribution of light beneath first-year sea ice in the Arctic Ocean. Limnology and Oceanography 20: 554–563.

Nweeia, M.T., F.C. Eichmiller, P.V. Hauschka, et al. 2014. Sensory ability in the narwhal tooth organ system. Anatomical Record 297: 599–617.

Pagano, A.M., G.M. Durner, S.C. Amstrup, et al. 2012. Long-distance swimming by polar bears (*Ursus maritimus*) of the southern Beaufort Sea during years of extensive open water. Canadian Journal of Zoology 90: 663–676.

Peterson, M.R., W.W. Larned, and D.C. Douglas. 1999. At-sea distribution of spectacled eiders: a 120-year-old mystery solved. Auk 116: 1009–1120. (A polynya in the Bering Sea overwinters a single flock of 333,000 eiders.)

Peterson, M.R., D.C. Douglas, H.M. Wilson, and S.E. McCloskey. 2012. Effects of sea ice on winter fidelity of Pacific common eiders (*Somateria mollissima v-nigrum*) Auk 129: 399–408.

Regehr, E.V., C.M. Hunter, H. Caswell, et al. 2010. Survival and breeding of polar bears in the southern Beaufort Sea in relation to sea ice. Journal of Animal Ecology 79: 117–127.

Sale, R. 2008. *The Arctic, The Complete Story*. Francis Lincoln Ltd., London.

Skomal, G.B. and G.W. Benz. 2004. Ultrasonic tracking of Greenland sharks, *Somniosus microcephalus*, under Arctic ice. Marine Biology 145: 489–498.

Stirling, I. 1997. The importance of polynyas, ice edges, and leads to marine mammals and birds. Journal of Marine Systems 10: 9–21.

Stirling, I. and A.E. Derocher. 2012. Effects of climate warming on polar bears: a review of the evidence. Global Change Biology 18: 2694–2706.

Watanabe, Y.Y., C. Lydersen, A.T. Fisk, and K.M. Kovacs. 2012. The slowest fish: swim speed and tail-beat frequency of Greenland sharks. Journal of Experimental Marine Biology and Ecology 426–427: 5–11.

Niagara Escarpment

Bartlett, R.M., U. Matthes-Sears, and D.W. Larson. 1990. Organization of the Niagara Escarpment cliff community. II. Characterization of the physical environment. Canadian Journal of Botany 68: 1931–1941.

Gerrath, J.F., J. A. Gerrath, and D.W. Larson. 1995. A preliminary account of endolithic algae of limestone cliffs of the Niagara escarpment. Canadian Journal of Botany 73: 788–793.

Kelly, P.E. and D.W. Larson. 1997. Effects of rock climbing on populations of presettlement white cedar (*Thuja occidentalis*) on cliffs of the Niagara Escarpment, Canada. Conservation Biology 11: 1125–1132.

Kelly, P.E. and D.W. Larson. 2007. *The Last Stand: A Journey Through the Ancient Cliff-Face Forest of the Niagara Escarpment*. National Heritage Books, Toronto, ONT.

Kelly, P.E., E.R. Cook, and D.W. Larson. 1994. A 1397-year tree ring chronology of *Thuja occidentalis* from cliff faces of the Niagara Escarpment, southern Ontario, Canada. Canadian Journal of Forest Research 24: 1049–1057.

Larson, D.W. and P.E. Kelly. 1991. The extent of old-growth *Thuja occidentalis* on cliffs of the Niagara Escarpment. Canadian Journal of Botany 69: 1628–1636.

Larson, D.W., S.H. Spring, U. Matthes-Sears, and R.M. Bartlett. 1989. Organization of the Niagara Escarpment cliff community. Canadian Journal of Botany 67: 2731–2742.

Larson, D.W., U. Matthews, and P.E. Kelly. 2000. *Cliff Ecology: Pattern and Process in Cliff Ecosystems*. Cambridge University Press, New York, NY.

Matthes-Sears, U. and D.W. Larson. 1995. Rooting characteristics of trees in rock: a study of *Thuja occidentalis* on cliff faces. International Journal of Plant Science 156: 679–686.

Tovell, W.M. 1992. *Guide to the Geology of the Niagara Escarpment*. Niagara Escarpment Commission, Georgetown, ONT.

The "Father of Waters"

Ambrose, S.E. and D. Brinkley. 2002. *The Mississippi and the Making of a Nation From Louisiana Purchase to Today*. National Geographic Society, Washington, DC.

Ballard, M.B. 2004. *Vicksburg: The Campaign that Opened the Mississippi River*. University of North Carolina Press, Chapel Hill, NC.

Barton, H.D. and S.M. Wisely. 2012. Phylogeography of striped skunks (*Mephitis mephitis*) in North America: Pleistocene dispersal and contemporary population structure. Journal of Mammalogy 93: 38–51. (Includes descriptions of other species isolated geographically by the Mississippi River.)

Carstens, B.C. and J.D. Sadler. 2013. The carnivorous plant described as *Sarracenia alata* contains two cryptic species. Biological Journal of the Linnean Society 109(4): 737–746.

Cullingham, C.I., C.J. Kyle, B.A. Pond, and B.N. White. 2008. Genetic structure of raccoons in eastern North America based on mtDNA: implications for subspecies designation and rabies disease dynamics. Canadian Journal of Zoology 86: 947–958.

Grinnell, J. 1926. Geography and evolution in the pocket gopher. University of California Chronicle 28: 247–262.

Iseminger, W. 2010 *Cahokia Mounds: America's First City*. The History Press, Charleston, SC. (Includes speculation about Cahokia's increase in population and demise.)

Kurz, H. and D. Demaree. 1934. Cypress buttresses and knees in relation to water and air. Ecology 15: 36–41.

Meade, R.H. and J.A. Moody. 2010. Causes for the decline of suspended-sediment discharge in the Mississippi River system, 1940–2007. Hydrology Processes 24: 35–49.

Morris, C. 2012. *The Big Muddy: An Environmental History of the Mississippi and its Peoples from Hernando de Soto to Hurricane Katrina*. Oxford University Press, New York, NY.

Nelson, W.A. 1924. Reelfoot Lake – an earthquake lake. National Geographic 45: 94–114

Robison, H.W. 1986. Zoogeographic implications of the Mississippi River Basin. In: *The Zoogeography of North American Freshwater Fishes* (Hocutt, C.H., and E.O.Wiley, eds). John Wiley and Sons, New York, NY, pp. 267–285.

Rosen, R.A. and D.C Hales. 1981. Feeding of paddlefish, *Polyodon spathula*. Copiea 1981: 441–455.

Sammarco, P.W., D.A. Brazeau, and J. Sinclair. 2012. Genetic connectivity in scleractinian corals across the northern Gulf of Mexico: oil/gas platforms and relationship to the Flower Garden Banks. PLoS ONE 7(4): e30144.

Stahle, D.W. and M.K. Cleaveland. 1992. Reconstruction and analysis of spring rainfall over the southeastern U.S. for the past 1000 years. Bulletin of the American Meterological Society 73: 1947–1961. (See for development of "hanging buttresses.")

Stokstrad, E. 2013. Crafty mollusks use mimicry and muscle. National Wildlife 51(6): 16, 18. (See for diversity of freshwater mussels.)

Theiling, C.H. 1995. Habitat rehabilitation on the Upper Mississippi River. Regulated Rivers Research and Management 11: 227–238.

Williams, J.D., M.J. Warren, Jr, K.S. Cummings, J.L. Harris, and R.J. Neves. 1993. Conservation status of freshwater mussels of the United States and Canada. Fisheries 18: 6–22.

Wilkens, L., M. Hoffman, and W. Wojtenek 2002. The electric sense of the paddlefish: a passive system for the detection and capture of zooplankton prey. Journal of Physiology 96: 363–377.

The Everglades

Alexander, T.R. 1971. Sawgrass biology related to the future of the Everglades. Soil and Crop Science Proceedings 31: 72–74.

Bryan, D. 1996. Limpkin. In: *Rare and Endangered Biota of Florida*. Vol. V. Birds (J.A. Rodgers, Jr, H.W. Kale II, and H.T. Smith, eds). University Press of Florida, Gainesville, FL, pp. 485–496.

Center, T.D., R.F. Doren, R.H. Hofstetter, R.L. Myers, and L.D. Whiteaker. (eds) 1991. *Proceedings of the Symposium on Exotic Pest Plants*. National Park Service, Washington, DC. (Contains information concerning melaleuca, Brazilian pepper, and other troublesome plants in the Everglades.)

Center, T.D., M.F. Purcell, P.D. Pratt, et al. 2011. Biological control of *Melaleuca quinquenervia*: an Everglades invader. BioControl 57: 151–165. (Assesses the role of insects as part of the IPM program.)

Craighead, F.C., Sr. 1968. The role of the alligator in shaping plant communities and maintaining wildlife in the southern Everglades. Florida Naturalist 41: 2–7, 69–74, 94.

Davis, S.M. and J.C. Ogden. (eds) 1994. *Everglades. The Ecosystem and its Restoration*. St. Lucie Press, Delray Beach, FL.

Dinets, V., J.C. Brueggen, and J.D. Brueggen. 2013. Crocodilians use tools for hunting. Ethology, Ecology & Evolution 27(1), doi: 10.1080/03949370.2013.858276.

Dorcas, M.E, J.D. Willson, R.N. Reed, et al. 2012. Severe mammal declines coincide with proliferation of invasive Burmese pythons in Everglades National Park. Proceedings of the National Academy of Sciences 109: 2418–2422.

Douglas, M.S. 1988. *The Everglades: River of Grass*. Revised edition. Pineapple Press, Sarasota FL.

Dove, C.J., R.W. Snow, M.R. Rochford, and F.J. Mazotti. 2011. Birds consumed by the invasive Burmese python (*Python molurus bivittatus*) in Everglades National Park, Florida, USA. Wilson Journal of Ornithology 123: 126–131.

Gunderson, L.H. 1995. Everglades, human transformations of a dynamic ecosystem. In: *Encyclopedia of Environmental Biology* (W.A. Nierenbach, ed.). Elsevier, Vol. 1, pp. 705–715.

Kahl, M.P. 1964. Food ecology of the wood stork (*Mycteria americana*) in Florida. Ecological Monographs 34: 97–117.

Lodge, T.E. 1994. *The Everglades Handbook. Understanding the Ecosystem*. St. Lucie Press, Delray Beach, FL. (An overview for readers of every background.)

Loveless, C.M. 1959. A study of the vegetation in the Florida Everglades. Ecology 40: 1–9.

Loveless, C.M. 1959. *The Everglades Deer Herd, Life History and Management*. Technical Bulletin No. 6, Florida Game and Fresh Water Fish Commission, Tallahassee, FL.

Perez, L. 2012. *Snake in the Grass: An Everglades Invasion*. Pineapple Press, Sarasota FL.

Silvers, C.S., P.D. Pratt, A.P. Ferriter, and T.D. Venter. 2007. T.A.M.E. Melaleuca: a regional approach for suppressing one of Florida's worst weeds. Journal of Aquatic Plant Management 45: 1–8.

Snyder, H.F. and H.A. Snyder. 1969. A comparative study of predation by limpkins, Everglade kites, and boat-tailed grackles. Living Bird 8: 177–223. (Cites record of 2000 snail shells in a single midden.)

Voous, K.H. and T. van Dijk. 1973. How do snail kites extract snails from their shells? Ardea 61: 179–185.

Fossil Lagerstätten

Conway, M.S. 1998. *The Crucible of Creation: The Burgess Shale and the Rise of Animals*. Oxford University Press, New York.

Flickinger, E.L. and C.M. Bunck. 1987. Number of oil-killed birds and fate of bird carcasses at crude oil pits in Texas. Southwestern Naturalist 32: 377–381.

Friscia, A.R., B. van Valkenburgh, L. Spencer, and J. Harris. 2008. Chronology and spatial distribution of large mammal bones in Pit 91, Rancho La Brea. Palaios 23: 35–42.

Gould, S.J. 1989. *Wonderful Life: The Burgess Shale and the Nature of History*. W. W. Norton & Company, New York.

Harris, J.M. and G.T. Jefferson. (eds) 1985. *Rancho La Brea: Treasures of the Tar Pits*. Science Series 31. Natural History Museum of Los Angeles County.

Morris, S.C. and J.B. Caron. 2014. A primitive fish from the Cambrian of North America. Nature 512: 419–422.

Selden, P.A. and J.R. Nudds. 2004. *Evolution of Fossil Ecosystems*. University of Chicago Press, Chicago, Illinois. (Deals with Fossil *Lagerstätten* in North America.)

Stock, C. 1992. *Rancho La Brea, A Record of Pleistocene Life in California*. 7th edition, revised by J. M. Harris. Science Series No. 37, Natural History Museum of Los Angeles County (The first edition appeared in 1930.)

The Florida Keys

Bancroft, G.T. and R. Bowman. 2001. White-crowned pigeon (*Columbia leucocephala*). In: *The Birds of North America*, No. 595. (A. Poole and F. Gill, eds). The Birds of North America, Inc., Philadelphia, PA.

Braden, A.W., R.R. Lopez, C.W. Roberts, et al. 2008. Florida Key deer *Odocoileus virginianus clavium* underpass use and movements along a highway corridor. Wildlife Biology 14: 155–163.

Carlson, P.C., G.W. Tanner, J.M. Wood, and S.R. Humphrey. 1993. Fire in Key deer habitat improves browse, prevents succession, and preserves endemic herbs. Journal of Wildlife Management 57: 914–928.

Darwin, C. 1846. An account of the fine dust which falls upon vessels in the Atlantic Ocean. Quarterly Journal of the Geological Society of London 2: 26–30.

Faulhaber, C.A., N.D. Perry, N.J. Silvy, et al. 2006. Reintroduction of Lower Keys marsh rabbits. Wildlife Society Bulletin 34: 1198–1202.

Goodman, J., J. Maschinski, P. Hughes, et al. 2012. Differential response to soil salinity in endangered Key tree cactus: implications for survival in a changing climate. PLoS ONE 7(3): e32528.

Harveson, P.M., R.R. Lopez, N.J. Silvy, and P.A. Frank. 2004. Source-sink dynamics of Florida Key deer on Big Pine Key, Florida. Journal of Wildlife Management 68: 909–915.

Hoffman, W. and H. Darrow. 1992. Migration of diurnal raptors from the Florida Keys into the West Indies. Hawk Migration Studies 17: 1–14.

Hoffmeister, J.E. and H.G. Multer. 1968. Geology and origin of the Florida Keys. Geological Society of America Bulletin 79: 1487–1502.

Lazell, J.D., Jr. 1984. A new marsh rabbit (*Sylvilagus palustris*) from Florida's Lower Keys. Journal of Mammalogy 65: 26–33.

Lazell, J.D., Jr. 1989. *Wildlife of the Florida Keys: A Natural History*. Island Press, Washington, DC.

Lidz, B.H. and E.A. Shinn. 1991. Paleoshorelines, reefs and a rising seas: south Florida. Journal of Coastal Research 7: 203–229.

Lima, A.N. and R.M. Adams. 1996. The distribution and abundance of *Pilosocereus robinii* (Lemaire) Byles and Rowley in the Florida Keys. Bradleya 14: 57–62.

Lopez, R.R., M.E.P. Vieira, N.J. Silvy, et al. 2003. Survival, mortality, and life expectancy of Florida Key deer. Journal of Wildlife Management 67: 34–45.

Lopez, R.R., N.J. Silvy, R. F. Labisky, and P. A. Frank. 2003. Hurricane impacts on Key deer in the Florida Keys. Journal of Wildlife Management 67: 280–288.

Lopez, R.R., N.J. Silvy, B.L. Pierce, et al. 2004. Population density of the endangered Florida Key deer. Journal of Wildlife Management 68: 570–575.

Lott, C.A. 2006. A new raptor migration monitoring site in the Florida Keys: counts from 1999–2004. Journal of Raptor Research 40: 200–209.

Muhs, D.R., J.R. Budahn, J.M. Prospero, and S.N. Carey. 2007. Geochemical evidence for African dust inputs to soils of western Atlantic island: Barbados, the Bahamas, and Florida. Journal of Geophysical Research 112, F002009.

Owen, C.B., R.R. Lopez, C.W. Roberts, et al. 2008. Florida Key deer *Odocoileus virginianus clavium* underpass use and movements along a highway corridor. Wildlife Biology 14: 155–163.

Parker, I.D., A.W. Braden, R.R. Lopez, et al. 2008. Effects of US 1 Project on Florida Key deer mortality. Journal of Wildlife Management 72: 354–359.

Pulliam, H.R. 1988. Sources, sinks, and population regulation. American Naturalist 132: 652–661.

Ross, M.S., J.J. O'Brien, and L.J. Flynn. 1992. Ecological site classification of Florida Keys terrestrial habitats. Biotropica 24: 488–502.

US Fish and Wildlife Service. 1998. Multi-species recovery plan for the threatened and endangered species of south Florida. In: Volume 1, *The Species*. Vero Beach, FL, pp. 4-111–4-118. (Source for information about the Key tree cactus.)

The Great Lakes

Beamish, F.W.H. 1980. Biology of the North American anadromous sea lamprey. Canadian Journal of Fisheries and Aquatic Sciences 37: 1924–1943.

Bence, J.R., R.A. Bergstedt, G.C. Christie, P.A. Cochran, et al. 2003. Sea lamprey (*Petromyzon marinus*) parasite–host interactions in the Great Lakes. Journal of Great Lakes Research 29: 253–282.

Brown, E.H. 1968. Population characteristics and physical condition of alewives, *Alosa pseudoharengus*, in a massive dieoff in Lake Michigan, 1967. Great Lakes Fisheries Commission Technical Report 13, Great Lakes Fisheries Commission, Ann Arbor, MI.

Butts, L. and B. Krushelnicki. 1988. The Great Lakes: an environmental atlas and resource book. United States Environmental Protection Agency, Chicago, IL, and Environment Canada, Toronto, ONT.

Grady, W. and E. Damstra. 2011. *The Great Lakes: The Natural History of a Changing Region*. Greystone Books, Vancouver, BC.

Kunkel, K., N. Westcott, and D. Kristovich. 2002. Effects of climate change on heavy lake-effect snowstorms near Lake Erie. Journal of Great Lakes Research 28: 521–526.

Larson, G. and R. Schaetzl. 2001. Origin and evolution of the Great Lakes. Journal of Great Lakes Research 27: 518–546.

Lightfoot, G. 1976. "The Wreck of the Edmund Fitzgerald." Reprise Records, Warner Music Group, New York, NY. (Song was released originally on Lightfoot's "Summertime Dream" album.)

Limburg, K.E., V.A. Luzadis, M. Ramsey, et al. 2010. The good, the bad, and the algae: perceiving ecosystem services and disservices generated by zebra and quagga mussels. Journal of Great Lakes Research 36: 86–92.

Mazak, E.J., H.J. MacIsaac, M.R. Servos, and R. Hesslein. 1997. Influence of feeding habits on organochlorine contaminant accumulation in waterfowl on the Great Lakes. Ecological Applications 7: 1133–1143.

Miller, R.R., J.D. Williams, and J.E. Williams. 1989. Extinction of North American fishes during the past century. Fisheries 14: 22–48.

Mills, E., J.H. Leach, J.T. Carlton, and C.L. Secor. 1994. Exotic species and the integrity of the Great Lakes: lessons from the past. BioScience 44: 666–676.

Rasmussen, J.L., H.A. Regier, R.E. Sparks, and W.W. Taylor. 2011. Dividing the waters: the case for hydrologic separation of the North American Great Lakes and Mississippi River basins. Journal of Great Lakes Research 37: 588–592.

Shaetzl, R.J. and S.A. Isard. 2002. The Great Lakes. In: *The Physical Geography of North America* (R.A. Orme, ed.). Oxford University Press, Inc. New York, NY, pp. 307–334. (Describes effects of the Great Lakes on the climate of surrounding areas.)

Spring, B. 2002. *The Dynamic Great Lakes*. Independence Books, Baltimore, MD.

Habitat highlights

Rivers of ice

Arendt, A.A., K.A. Echelmeyer, W.D. Harrison, et al. 2002. Rapid wastage of Alaska glaciers and their contribution to rising sea level. Science 297: 382–386.

Bogdal, C., P. Schmid, M. Zennegg, et al. 2009. Blast from the past: melting glaciers as a relevant source for persistent organic pollutants. Environmental Science and Technology 43: 8173–8177.

Dyurgerov, M.B. and M.F. Meier. 2000. Twentieth century climate change: evidence from small glaciers. Proceedings of the National Academy of Science 97: 1406–1411.

Fagan, B.M. 2001. *The Little Ice Age: How Climate Made History, 1300 to 1850*. Basic Books, New York, NY.

Hambrey, M. and J. Alean. 2004. *Glaciers*, second ed. Cambridge University Press, Cambridge, UK. (Primary source for explanation of glacial formation and movements.)

Hartzell, P.L., J.V. Nghiem, K.J. Richio, and D.H. Shain. 2005. Distribution and phylogeny of glacier ice worms (*Mesenchytraeus solifugus* and *Mesenchytraeus solifugus rainierensis*). Canadian Journal of Zoology 33: 1206–1213.

Paul, F., A. Käb, M. Maisch, et al. 2004. Rapid disintegration of Alpine glaciers observed with satellite data. Geophysical Research Letters 31(L1402): 1–4.

Shain, D.H., M.R. Carter., K.P. Murray, et al. 2000. Morphologic characterization of the ice worm, *Mesenchytraeus solifugus*. Journal of Morphology 246: 192–197.

Hot springs and geysers

Madigan, M.T., J.M. Martinko, D. Stahl, and D.P. Clark. 2010. *Brock Biology of Microorganisms*, 13th edition. Benjamin Cummings, Boston, MA.

Pentecost, A., B. Jones, and R.W. Renault. 2003. What is a hot spring? Canadian Journal of Earth Science 40: 1443–1446.

Plummer, C., D. Carlson, and L. Hammersley. 2009. *Physical Geology*. McGraw-Hill, New York, NY.

Robb, F., G. Antranikian, D. Grogan, and A. Driessen. 2007. *Thermophiles: Biology and Technology at High Temperatures*. CRC Press, Boca Raton, FL.

Turner, J. 2008. *Travels in Greater Yellowstone*. St. Martin's Press, New York, NY.

Forest in the clouds

Alcántara, O., I. Luna, and A. Velázquez. 2002. Altitudinal distribution patterns of Mexican cloud forests based upon preferential characteristic genera. Plant Biology 161: 167–171.

Carvajal-Villereal, S., A. Caso, P. Downey, A. Moreno, M.E. Tewes, and L.I. Grassman, Jr. 2011. Spatial patterns of the margay (*Leopardus weidii*; Felidae, Carnivora) at "El Cielo" Biosphere Reserve, Tamaulipas, Mexico. Mammalia 76: 237–244.

Castro-Arellano, I. and T. Lacher, Jr. 2005. A new record and altitudinal extensions for El Cielo Biosphere Reserve mammals, Tamaulipas, Mexico. Revista Mexicana de Mastozoologia 9: 150–154.

Hamilton, L.S., J.O. Juvik, and F.N. Scatena. (eds) 1995. *Tropical Montane Cloud Forests*. Springer-Verlag, Inc., New York, NY.

Hernández-Xolocotzi, E., H.A. Crum, W.B. Fox, Jr, and A.J. Sharp. 1951. A unique vegetation area in Tamaulipas. Bulletin of the Torrey Botanical Club 78: 458–463.

Hietz, P. and U. Hietz-Seifert. 1995. Structure and ecology of epiphyte communities of a cloud forest in central Veracruz, Mexico. Journal of Vegetation Science 6: 719–728.

Leopold, A.S. 1950. Vegetation zones of Mexico. Ecology 31: 507–513.

Puig, H., R. Bracho, and V.J. Sosa. 1983. Composicíon florística y estructure del bosque mesófilo en Gómez Farías, Tamaulipas, México. Bióteca 8: 339–359.

Sutton, G.M. and O.S. Pettingill, Jr. 1942. Birds of the Gomez Farias Region, southwestern Tamaulipas. Auk 59: 1–34.

Vogelmann, H.W. 1973. Fog precipitation in the cloud forests of eastern Mexico. BioScience 23: 96–100. (Describes "cloud filtering.")

Webster, F. and M.S. Webster. 2002. *The Road to El Cielo*. University of Texas Press, Austin, TX.

Granite outcrops and inselbergs

Anderson, R.C., J.S. Fralish, and J.M. Baskin. (eds) 2007. *Savannas, Barrens, and Rock Outcrop Plant Communities of North America*. Cambridge University Press, Cambridge, UK.

Baskin, J.S. and C.C. Baskin. 1988. Endemism in rock outcrop plant communities of unglaciated eastern United States: an evaluation of the roles of the edaphic, genetic, and light factors. Journal of Biogeography 15: 829–840.

Burbanck, M.P. and D.L. Phillips. 1983. Evidence of plant succession on granite outcrops of the Georgia Piedmont. American Midland Naturalist 109: 94–104.

Burbanck, M.P. and R.B. Platt. 1964. Granite outcrop communities of the Piedmont Plateau in Georgia. Ecology 45: 292–305.

Burke, A. 2003. Inselbergs in a changing world – global trends. Diversity and Distributions 9: 375–383.

Jocque, M., B. Vanschoenwinkel, and L. Brendonck. 2010. Freshwater rock pools: a review of habitat characteristics, faunal diversity, and conservation value. Freshwater Biology 55: 1587–1602.

McVaugh, R. 1943. The vegetation of the granitic flat-rocks of the southeastern United States. Ecological Monographs 13: 119–166.

Wessels, T. 2002. *The Granite Landscape: A Natural History of America's Mountain Domes, from Acadia to Yosemite*. Countryman Press, Woodstock, VT.

White, W.A. 1945. Origin of granite domes in the southeastern Piedmont. Journal of Geology 53: 276–282.

Wyatt, R. 1981. Ant-pollination of the granite outcrop endemic *Diamorpha smallii* (Crassulaceae). American Journal of Botany 68: 1212–1217.

Wyatt, R. and J.R. Allison. 2000. Flora and vegetation of granite outcrops in the southeastern United States. Ecological Studies 146: 409–433.

Palm forest

Brush, T. 2005. *Nesting Birds of a Tropical Frontier: The Lower Rio Grande Valley of Texas*. Texas A&M University Press, College Station, TX.

Chapman, S.S. and B.R. Chapman. 1990. Bats from the coastal region of southern Texas. Texas Journal of Science 42: 13–22.

Chipman, D.E. 1995. Alonso Alvarez de Pineda and the Río de las Palmas: scholars and the mislocation of a river. Southwestern Historical Quarterly 98: 369–385.

Clover, E.U. 1937. Vegetational survey of the lower Rio Grande Valley, Texas. Madroño 4: 41–66; 77–100. (Lists 81 plant species in the palm forest.)

Diamond, D.D., D.H. Riskind, and S.L. Orzell. 1987. A framework for plant community classification and conservation in Texas. Texas Journal of Science 39: 203–221.

Dixon, J.R. 2013. *Amphibians and Reptiles of Texas: With Keys, Taxonomic Synopsis, Bibliography, and Distribution Maps*, third edition. Texas A&M University Press, College Station, TX.

Everitt, J.H., F.W. Judd, D.E. Escobar, et al. 1996. Using remote sensing and spatial information technologies to map sabal palm in the lower Rio Grande Valley of Texas. Southwestern Naturalist 41: 218–226.

Lockett, L. 1995. Historical evidence of the native presence of *Sabal mexicana* (Palmae) north of the lower Rio Grande Valley. Sida 16: 717–719.

Lonard, R.I. and F.W. Judd. 2002. Riparian vegetation of the lower Rio Grande. Southwestern Naturalist 47: 420–432.

Schmidly, D.J. 2004. *The Mammals of Texas*, revised edition. University of Texas Press, Austin, TX.

Mineral licks

Cowan, I. McT. and V.C. Brink. 1949. Natural game licks in the Rocky Mountain National Parks of Canada. Journal of Mammalogy 30: 379–387. (Cites increased predation of mountain goats at salt licks.)

Fraser, D. and E.R. Thomas. 1982. Moose–vehicle accidents in Ontario: relation to highway salt. Wildlife Society Bulletin 10: 261–265.

Hedeen, S. 2008. *Big Bone Lick: The Cradle of American Paleontology.* University of Kentucky Press, Lexington, KY. (Describes the collaboration between Clark and Jefferson.)

Holl, S.A. and V.C. Bleich. 1987. Mineral lick use by mountain sheep in San Gabriel Mountains, California. Journal of Wildlife Management 51: 383–385. (Supports the belief that lick use is related to moisture content of spring forage.)

Jones, R.L. and H.C. Hanson 1985. *Mineral Licks, Geophagy, and Biogeochemistry of North American Ungulates.* Iowa State University Press, Ames, IA.

Kennedy, J.F., J.A. Jenks, R.L. Jones, and K.J. Jenkins. 1995. Characteristics of mineral licks used by white-tailed deer (*Odocoileus virginianus*). American Midland Naturalist 134: 324–331. (Describes sites in the Black Hills.)

Morgan, R. 2007. *Boone: A Biography.* Algonquin Press, Chapel Hill, NC.

Sanders, T.A. and R.L. Jarvis. 2000. Do band-tailed pigeons seek a calcium supplement at mineral sites? Condor 102: 855–863.

Poole, K.G., K.D. Bachmann, and I.E. Teske. 2010. Mineral licks use by GPS radio-collared mountain goats in southeastern British Columbia. Western North American Naturalist 70: 208–217.

Bogs and their carnivorous plants
Adlassnig, W., M. Peroutka, and T. Lendl. 2011. Traps of carnivorous pitcher plants as a habitat: composition of the fluid, biodiversity, and mutualistic activities. Annals of Botany 107: 181–194.

Atwater, D.Z., J.L. Butler, and A.M. Ellison. 2006. Spatial distribution and impacts of moth herbivory on northern pitcher plants. Northeastern Naturalist 13: 43–56.

Brewer, J.S. 2002. Why don't carnivorous pitcher plants compete with non-carnivorous plants for nutrients? Ecology 84: 451–462.

Cresswell, J.E. 1991. Capture rates and composition of insect prey of the pitcher plant *Sarracenia purpurea.* American Midland Naturalist 125: 1–9. (Notes spider webs spanning the pitfall traps.)

Dahlem, G.A. and R.F.C. Naczi. 2006. Flesh flies (Diptera: Sarcophagidae) associated with North American pitcher plants (Sarraceniaceae), with descriptions of three new species. Annals of the Entomological Society of America 99: 218–240.

Ellison, A.M. and N.J. Gotelli. 2009. Energetics and the evolution of carnivorous plants: Darwin's "most wonderful plants in the world." Journal of Experimental Botany 60: 19–42.

Folkerts, D.R. 1999. Pitcher plants in wetlands of the southern United States: arthropod associates. In: *Invertebrates in Freshwater Wetlands of North America: Ecology and Management* (D.P. Batzer, R.B. Rader, and S.A. Wissinger, eds). John Wiley & Sons, New York, NY, pp. 247–275.

Folkerts, G.W. 1982. The Gulf Coast pitcher plant bogs. American Scientist 70: 260–267.

Giberson, D. and M.L. Hardwick. 1999. Pitcher plants (*Sarracenia purpurea*) in eastern Canadian peatlands, ecology and conservation of invertebrate inquilines. In: *Invertebrates in Freshwater Wetlands of North America: Ecology and Management* (D.P. Batzer, R.B. Rader, and S.A. Wissinger, eds). John Wiley & Sons, New York, NY, pp. 401–422. (Includes the spatial relations and coexistence of three dipteran detritivores.)

Spitzer, K. and H.V. Danks. 2006. Insect biodiversity of boreal peat bogs. Annual Review of Entomology 51: 137–161.

Stephens, J.D. and C.R. Folkerts. 2012. Life history aspects of *Exyra semicrocea* (pitcher plant moth) (Lepidoptera: Noctuidae). Southeastern Naturalist 11: 111–126. (An obligate herbivore of purple pitcher plants.)

Wells, B.W. 1932. *The Natural Gardens of North Carolina.* University of North Carolina Press, Chapel Hill, NC.

Infobox 12.1. Eugene P. Odum (1913–2002), founder of ecosystem science
Barrett, G.W. 2003. Resolution of respect. Eugene P. Odum, pioneer of ecosystem science, 1913–2002. Bulletin of the Ecological Society of America 84: 11–12.

Ewel, J.J. 2003. Resolution of respect. Howard Thomas Odum, 1924–2002. Bulletin of the Ecological Society of America 84: 13–15. (The brother of Eugene and, in his own right, a major figure in ecology.)

Odum, E.P. 1968. Energy flow in ecosystems: a historical review. American Zoologist 8: 11–18.

Odum, E.P. 1969. The strategy of ecosystem development. Science 164: 262–270.

Odum, E. P. 1953. *Fundamentals of Ecology.* W.B. Saunders, Philadelphia, PA. (First of five editions of a landmark text.)

Odum, H.T. and E.P. Odum. 1955. Trophic structure and productivity of a windward coral reef community on Eniwetok Atoll. Ecological Monographs 25: 291–320.

Odum, E.P. and G.W. Barrett. 2004. *Fundamentals of Ecology.* Fifth edition. Thompson Brooks/Cole, Pacific Grove, CA.

Infobox 12.2. Those darn weeds: Invasive plants in North America
Bell, M. 1966. Some notes and reflections upon a letter from Benjamin Franklin to Noble Wimberly Jones, October 7, 1772. Ashantilly Press, Darien, GA. (Letter that accompanied a shipment of Chinese tallow seeds.)

Blossey, B., L.C. Skinner, and J. Taylor. 2001. Impact and management of purple loosestrife (*Lythrum salicaria*) in North America. Biodiversity and Conservation 10: 1787–1807.

Bruce, K.A., G.N. Cameron, and P.A. Harcombe. 1995. Initiation of a new woodland type on the Texas coastal prairie by the Chinese tallow tree (*Sapium sebiferum* (L.) Roxb.). Bulletin of the Torrey Botanical Club 122: 215–225.

Bruce, K.A., G.N. Cameron, P.A. Harcombe, and G. Jubinsky. 1957. Introduction impact on native habitats, and management of a woody invader, the Chinese tallow tree (*Sapium sebiferum* (L.) Roxb. Natural Areas Journal 17: 255–260.

Crosby, A.W. 1972. *The Columbian Exchange: Biological and Cultural Consequences of 1492.* Greenwood Press, Westport, CN.

Groth, A.T., L. Lovett-Doust, and J. Lovett-Doust. 1996. Population density and module demography in *Trapa natans* (Trapaceae), an annual, clonal aquatic macrophyte. American Journal of Botany 83: 1406–1415.

Johnson, F.A. and R. Montalbano, III. 1987. Considering waterfowl habitat in hydrilla control policies. Wildlife Society Bulletin 15: 466–469.

Malecki, R.A., B. Blossey, S.D. Hight, et al. 1993. Biological control of purple loosestrife. BioScience 43: 680–686.

McCormick, C. 2005. *Chinese Tallow Management Plan for Florida.* Florida Exotic Pest Plant Council, Center for Aquatic and

Invasive Plants, University of Florida, Gainesville, FL. (A major source of information.)

Mullin, B.G. 1998. The biology and management of purple loosestrife (*Lythrum salicaria*). Weed Technology 12: 397–401.

Penfound, W.T., and T.T. Earle. 1948. The biology of water hyacinth. Ecological Monographs 18: 447–472.

Renne, I.J., and S.A. Gauthreaux, Jr. 2000. Seed dispersal of the Chinese tallow tree (*Sapium sebiferum* (L. Roxb.) by birds in coastal South Carolina. American Midland Naturalist 144: 202–215.

Thompson, D.Q., R.L. Stuckey, and E.B. Thompson. 1987. *Spread, Impact, and Control of Purple Loosestrife (Lythrum salicaria) in North America.* US Fish and Wildlife Service, Washington, DC.

Toft, J.D., C.A. Simenstad, J.R. Cordell, and L F. Grimaldo. 2003. The effects of introduced water hyacinth on habitat structure, invertebrate assemblages, and fish diets. Estuaries 26: 746–758.

Wolverton, B.C. and R C. McDonald. 1979. Water hyacinth (*Eichhornia crassipes*) productivity and harvesting studies. Economic Botany 33: 1–10.

Appendix

This Appendix provides the list of scientific names of organisms mentioned in the text. With some exceptions, each entry contains an abbreviation enclosed in parentheses to indicate the biome (or chapter) where the common name of the organism appeared in the text. Some species, of course, occur in more than one ecological association. Scientific names cited in older sources but no longer in use also are included, and some accounts are accompanied by informative notations. Extinct species are noted by an asterisk (*) following the scientific name.

BF	Boreal Forest	I	Introduction
CE	Coastal Environments	MF	Montane Forest
CPJ	Chaparral and Pinyon-Juniper Woodlands	RG	Regional Grasslands
D	Deserts	T	Tundra
EDF	Eastern Deciduous Forest	TRF	Temperate Rain Forest
G	Grasslands: Plains and Prairies	SE	Special Environments

Bacteria, algae, fungi, and lichens

Brown algae, class Phaecophyceae (CE). Large, multicellular seaweeds.

Brown spot needle blight, *Mycosphaerella dearnessii* (EDF). Formerly *Scirrhia acicola* or *Lecanosticta acicola*.

Brown tide, *Aureoumbra lagunesis* (CE).

Chestnut blight, *Cryphonectria parasitica* (EDF). Formerly *Endothia parasitica*.

Cyanobacterium, phylum Cyanobacteria (BF, CE). Plural, Cyanobacteria: photosynthetic bacteria (or blue-green algae) causing turbidity and anoxia in some oligotrophic lakes; also form the base of food chains in marine wind-tidal flats.

Diatoms, class Bacillariophyceae, phylum Heterokontophyla (CE). A group of more than 200 genera of phytoplanktonic algae characterized by silicon in their cell walls.

Dogwood anthracnose, *Discula destructiva* (EDF).

Green algae (CE). A group of eukaryotic, unicellular or multicellular algae capable of photosynthesis; some have flagella and are capable of locomotion.

Lichen (T, BF). Many species exist, including *Caloplaca elegans*, a bright orange species associated with tundra sites enriched by bird droppings. Lichens occur worldwide in virtually all terrestrial environments. Part algae or cyanobacteria, part fungi, these composite organisms are classic examples of **mutualism**, a common type of symbiosis. See also Reindeer moss.

Mycobacterium, *Mycobacterium* spp. (CE). Ten species of *Mycobacterium* have been identified as causing the disease known as mycobacteriosis in striped bass.

Phytoplankton (CE). Collective name for all one-celled, floating photosynthetic marine organisms.

Red heart rot, *Phellinus pini* (EDF). Formerly *Formes pini*.

Red tide, *Karenia brevis* (CE).

Reindeer moss, *Cladonia* spp. (T, BF). A predominant genus of lichen; see Lichen.

Slime mold, *Labyrinthula* spp. (CE). Species in this group are famously difficult to determine and often are not formally specified; *L. zosterae* is sometimes listed as the pathogen for eelgrass-wasting disease.

White heart rot, *Phellinus igniarius* (AP). Formerly *Formes igniarius*.

White pine blister rust, *Cronartium ribicola* (EDF, MF).

Plants

Agave, *Agave* spp. (D). Any of 194 species of desert plants of which *Agave americana* is the most common in the southwestern United States and northeastern Mexico. See also Blue agave.

Alder, *Alnus* spp. (T, BF, TRF).

Alfalfa, *Medicago sativa* (MF).

Alkali sacaton, *Sporobolus airoides* (RG).

Alligator weed, *Alternantheria philoxeroides* (CE).

Ecology of North America, Second Edition. Brian R. Chapman and Eric G. Bolen.
© 2015 John Wiley & Sons, Ltd. Published 2015 by John Wiley & Sons, Ltd.

American basswood, *Tilia americana* (EDF).

American beach grass, *Ammophila breviqulata* (CE). Identified as American marram grass in some accounts.

American beech, *Fagus grandifolia* (EDF).

American chestnut, *Castanea dentata* (EDF).

American shoalgrass, *Halodule beaudettei* (CE). Identified as *H. wrightii* in older accounts.

American sweetgum. See Sweetgum.

Arborvitae. See Northern white cedar.

Arctic lupine, *Lupinus arcticus* (T).

Arctic willow, *Salix arctica* (T).

Arrowhead, *Sagittaria lancifolia* (SE). Also known as duck potato or bulltongue arrowhead.

Ashe juniper, *Juniperus ashei* (RG).

Atlantic white cedar, *Chamaecyparis thyoides* (EDF).

Bald cypress, *Taxodium distichum* (EDF, SE). Written as baldcypress in some sources.

Balsam fir, *Abies balsamea* (BF).

Balsam poplar, *Populus balsamifera* (G).

Balsam root, *Balsamorthiza sagittata* (RG).

Barb goatgrass, *Aegilops triuncialis* (RG).

Beach cordgrass, see Smooth cordgrass.

Beach strawberry, *Fragaria chiloensis* (CE).

Beach tea, *Croton punctatus* (CE).

Beaked hazelnut, *Corylus rostrata* (BF).

Beakrush, *Rhynchospora tracyi* (SE).

Bear oak, *Quercus ilicifolia* (EDF). Also known as scrub oak.

Bearded wheatgrass, *Agropyron subsecundum* (G).

Beech. See American beech.

Berlandier ash, *Fraxinus berlandieriana* (RG).

Big bluestem, *Adropogon gerardii* (G). Sometimes called "turkey feet."

Bigleaf maple, *Acer macrophyllum* (TRF).

Bitter panicum, *Panicum amarum* (CE). Also known as bitter panicgrass.

Blackbead, *Pithecellobium keyense* (SE).

Black grama, *Bouteloua eriopoda* (D, RG).

Black grass, *Juncus gerardi* (CE). A rush, family Juncaceae, not a true grass.

Black gum, *Nyssa sylvatica* (EDF).

Black hickory, *Carya texana* (G).

Black mangrove, *Avicennia germinans* (CE).

Black mimosa, *Mimosa nigra* (SE). Also known as catclaw and catclaw acacia.

Black needle rush, *Juncus roemerianus* (CE).

Black oak, *Quercus velutina* (EDF).

Black spruce, *Picea mariana* (T, BF).

Blackjack oak, *Quercus marilandica* (EDF).

Bladderwort, *Utricularia* spp. (SE). A genus of 215 or more species; *Utricularia vulgaris* occurs widely in freshwater wetlands of North America.

Blowout grass, *Redfieldia fluexuosa* (G).

Blue agave, *Agave tequilana* (D).

Blueberry, *Vaccinium* spp. (T, EDF).

Bluebunch wheatgrass, *Agropyron spicatum* (RG, D).

Blue grama, *Bouteloua gracilis* (G, RG).

Blue gum, *Eucalyptus globulus* (MF). Introduced into California from Australia.

Blue oak, *Quercus douglasii* (RG).

Blue spruce, *Picea pungens* (MF).

Boojum, *Fouquieria columnaris* (D). Some place the boojum in genus *Idria*.

Brazil, *Condalia hookeri* (RG). Sometimes called blackbrush.

Brazilian pepper, *Schinus terebinthifolius* (SE).

Bristlecone pine, *Pinus aristata* (MF). Many taxonomists recognize two species of bristlecone pines: *P. longaeva* in the Great Basin and *P. aristata* in the Rocky Mountains.

Broom crowberry, *Corema conradii* (EDF).

Buckthorn, *Rhamnus cathartica* (EDF).

Buffalo grass, *Buchloe dactyloides* (G, RG).

Buffalo pea, *Astragulus crassicarpus* (G). Also known as groundplum milkvetch.

Bufflegrass, *Cenchrus cilaris* (D). An exotic species. Formerly *Pennisetum ciliare*.

Bush muhly, *Muhlenbergia porteri* (RG).

Buttercup, *Ranunculus* spp. (T). Many species, widely distributed in several biomes.

Buttonwood, *Conocarpus erectus* (SE).

California scrub oak, *Quercus berberidifolia* (CPJ). Closely related species *Q. dumosa*, once known by this common name, is now Nuttall's scrub oak.

Camas, *Camassia quamash* (RG).

Candelilla, *Euphorbia antisephilitica* (D). Also known as as candelaria, wax plant, or wax weed.

Carolina hemlock, *Tsuga caroliniana* (EDF).

Cattail, *Typha latifolia* (CE).

Ceanothus, *Ceanothus* spp. (CPJ). Hoaryleaf ceanothus *C. crassifolius* is a representative of coastal chaparral, whereas desert ceanothus *C. greggi* is a counterpart in the interior.

Cedar elm, *Ulmus crassifolia* (RG).

Century plant, see Agave.

Chamise, *Adenostoma fasciculatum* (CPJ).

Cheatgrass, *Bromus tectorum* (RG, D). An exotic species, sometimes known as downy broomgrass.

Chestnut oak, *Quercus montana* (EDF).

Chinese chestnut, *Castanea mollissima* (EDF).

Chinese tallow, *Sapium sebiferum* (SE). Occasionally cited as *Triadica sebifera*.

Cinchona, *Cinchona* spp. (TRF). About 38 species of *Cinchona* occur in tropical forests of South America.

Cinquefoil, *Potentilla* spp. (T). Low-growing herbs; some taxonomists place cinquefoils in genus *Dasiphora*.

Clubmosses, *Lycopodium* spp. or *Lycodiella* spp. (EDF, CE). Not true mosses, but more closely allied with ferns.

Coahuila scrub oak, *Quercus intricata* (CPJ).

Common cane, *Phragmites australis americanus* (CE). Some botanists recognize four species. Some authorities prefer *P. communis* for this species.

Common reed, see common cane.

Common ryegrass, *Lolium multiflorum* (RG).

Compass plant, *Silphium lancinatum* (G).

Coontail, *Ceratophyllum demersum* (CE).

Cordgrass, *Spartina* spp. (CE). Any of 14 species of coastal grasses that often form dense colonies in salt marshes.

Cottongrass, *Eriophorum vaginatum* (T). Not a grass but a sedge, family Cyperaceae.

Creosote bush, *Larrea tridentata* (D).

Curly grass fern, *Schizaea pusilla* (EDF).

Curly mesquite grass, *Hilaria belangeri* (RG).

Currant, *Ribes* spp. (EDF).

Cypress. See Bald cypress.

Dandelion, *Taraxacum* spp. (T).

Desert ironwood, *Olneya tesota* (D).

Desert willow, *Chilopsis linearis* (D).

Digger pine, *Pinus sabiniana* (CPJ).

Dogwood. See Flowering dogwood.

Douglas fir, *Pseudotsuga menziesii* (MF, TRF). Because this species is not a true fir, the common name is sometimes hyphenated.

Duck potato, *Sagittaria* spp. (CE). *S. lancifolia* is typical of fresh marsh on the Gulf Coast.

Dune thistle, *Cirsium pitcheri* (SE).

Dwarf mistletoe, *Arceuthobium cyanocarpum* (MF).

Dwarf palmetto, *Sabal minor* (SE).

Eastern hemlock, *Tsuga canadensis* (EDF).

Eastern redbud, *Cercis canadensis* (SE).

Eelgrass, *Zostera marina* (CE). Other species include *Z. asiatica* and the introduced *Z. japonica*.

Elephant tree, *Bursera microphylla* (D).

Englemann spruce, *Picea engelmannii* (MF).

European beach grass, *Ammophila arenaria* (CE).

Fescue, *Festuca* spp. (G). Taxonomists recently moved some of the 300 species of *Festuca* to the related genus *Lolium* (ryegrasses).

Fetterbush, *Lyonia lucida* (EDF).

Filaree, *Erodium* spp. (RG). Exotic forbs, including broadleaf filaree *E. botrys*.

Fireweed, *Chamerion angustifolium* (MF, TRF). Listed as *Epilobium angustifolium* in some references.

Flag, *Thalia geniculata* (SE). Also known as fireflag or alligator flag.

Flowering dogwood, *Cornus florida* (EDF, SE).

Foxtail pine, *Pinus balfouriana* (MF).

Fraser fir, *Abies fraseri* (BF).

Giant kelp, *Macrosystis pyrifera* (CE).

Giant sequoia, *Sequoia giganteum* (MF). Sometimes known as bigtree.

Glasswort, *Salicornia* spp. (CE). Several species, all oligate halophytes.

Globemallow, *Spaeralcea* spp. (D). Desert globemallow *S. ambigua* occurs in the Sonoran and Great Basin deserts.

Gooseberry, *Ribes* spp. (EDF).

Goose grass. *Puccinellia phryanodes* (T).

Grand fir, *Abies grandis* (MF, TRF).

Granjeno, *Celtis pallida* (RG, SE).

Grape, *Vitis* spp. (EDF). About 60 species of vines including the muscadine, *V. rotundifolia*.

Greasewood, *Sarcobatus vermiculatus* (D).

Greenbriar, *Smilax* spp. (EDF).

Grimmia dry rock moss, *Grimmia laevigata* (SE).

Guayule, *Parthenium argentatum* (D).

Gulf cordgrass, *Spartina spartinae* (G).

Gumbo limbo, *Bursera simaruba* (SE).

Gyp grama, *Bouteloua briseta* (D).

Haircap moss, *Polytrichium commune* (SE).

Hairy grama, *Bouteloua hirsuta* (G, RG).

Hairy Orcutt grass, *Orcuttia pilosa* (RG).

Halogeton, *Halogeton glomeratus* (D).

Hazelnut, *Corylus* spp. (EDF). American hazel *C. americana* is common in southeastern North America.

Heath, family Ericaceae (T).

Holly, *Ilex* spp. (EDF, SE). Inkberry *I. glabra* and gallberry *I. coriacea* are common in evergreen forests.

Honey mesquite, *Prosophus glandulosa* (RG).

Hornbeam, *Carpinus caroliana* (EDF).

Houghton's goldenrod, *Solidago houghtoni* (SE).

Huckleberry, *Vaccinium* spp. (TRF). Red huckleberry *V. parvifolium* occurs in the Olympic Peninsula; *Gaylussacia* spp. occur in southeastern North America.

Huisache, *Acacia smellii* (RG, SE).

Hydrilla, *Hydrilla verticillata* (SE).

Idaho fescue, *Festuca idahoensis* (D).

Indiangrass, *Sorghastrum nutans* (G).

Indian mulberry, *Morinda royoc* (SE).

Indian ricegrass, *Oryzipsis hymenoides* (D).

Ironwood, *Carpinus caroliniana* (SE). Also known as American hornbeam.

Jamaican dogwood, *Piscidia piscipula* (SE).

Jefferey pine, *Pinus jeffreyi* (MF).

Jojoba, *Simmondsia chinensis* (D).

Joshua tree, *Yucca brevifolia* (D).

Juniper, *Juniperus* spp. (CPJ, PJW). *J. deppeana* occurs on desert mountain slopes in the southwestern United States.

Kelp, Order Laminariales (CE). A group of large, thallus-forming brown algae. Giant kelp *Macrocystis pyrifera* and bull kelp *Nereocystis luetkeana* are common species of the kelp forests along the Pacific Coast of California.

Key tree cactus, *Pilosocereus robinii* (SE). A taxon variously designated at the generic level as *Cereus*, *Cephalocereus*, and *Pilocerceus*.

Kudzu, *Pueraria* spp. (EDF, SE). An introduced vine represented by *P. lobata* and *P. montana*.

Lance-leaved psoralea, *Psoralea lanceolata* (G).

Larch, *Larix laricina* (T). Also known as tamarack.

Laurel, *Kalmia* spp. See Mountain laurel.

Laurel oak, *Quercus laurifolia* (SE).

Lead plant, *Amorpha canescens* (G).

Leatherleaf, *Chamaedaphne angustifolia* (T, EDF).

Lechuguilla, *Agave lechuguilla* (D).

Limber pine, *Pinus flexilis* (MF).

Little bluestem, *Schizachyrium scoparium* (G, RG). Formerly *Andropogon scoparium*.

Live oak, *Quercus virginiana* (EDF, SE).

Loblolly bay, *Gordonia lasianthus* (EDF, SE).

Loblolly pine, *Pinus taeda* (EDF).

Lodgepole pine, *Pinus contorta* (MF).

Longleaf pine, *Pinus palustris* (EDF). Also known as southern yellow pine.

Lupine, *Lupinus lepidus* (TRF). Also known as Pacific lupine.

Maidencane, *Panicum hemitomon* (CE, SE).

Manateegrass, *Syringodium filiforme* (CE).

Mangrove, medium-sized trees growing in saline environments (CE). Red mangrove, black mangrove, and white mangrove occur in North America.

Manzanita, *Acrostaphylos* spp. (C). *A. glauca* is one of about a dozen species of manzanita.

Marsh elder, *Iva frutescens* (CE).

Marsh hay, *Spartina patens* (CE). Also known as salt-meadow cordgrass.

Melaleuca, *Melaleuca quinquenervia* (SE).

Mesquite, *Propsophis* spp. (RG, D). See Honey mesquite.

Mexican pinyon, *Pinus cembroides* (CPJ).

Milfoil, *Myriophyllum* spp. (SE). Genus with 45 species of submerged aquatic plants.

Monterey pine, *Pinus radiata* (MF).

Moss, Division Bryophyta (EDF). Tiny ground-hugging plants forming small carpets on acidic soils.

Moss-campion, *Silene acaulis* (T).

Mountain hemlock, *Tsuga mertensiana* (MF).

Mountain laurel, *Kalmia latifolia* (BF, EDF). *K. angustifolia* is another common species.

Mountain mahogany, *Cerocarpus montanus* (MF, G, C).

Mountain maple, *Acer spicatum* (SE).

Mountain muhly, *Muhlenbergia montana* (G).

Mountain oat grass, *Danthonia compressa* (BF).

Muhly grass, *Muhlenbergia reverchonii* (RG).

Napa thistle, *Centaurea melitensis* (RG).

Needlegrasses. *Stipa* spp. (G). Needle-and-thread, *S. compacta*, is a well-known example; also *S. spartec*.

Nodding needlegrass, *Stipa cerna* (RG).

Northern purple pitcher plant, *Sarracenia purpurea purpurea* (SE).

Northern red oak, *Quercus rubra* (EDF).

Northern white cedar, *Thuja occidentalis* (EDF, BF, SE). Also known as arborvitae, or "tree of life."

Norway spruce, *Picea abies* (BF, MF).

Ocotillo, *Fouquieria splendens* (D).

Olney three-square, *Scirpus olneyi* (CE).

One-seed juniper, *Juniperus monosperma* (CPJ). This and other western junipers are widely known as cedars.

Oyamel fir, *Abies religiosa* (MF).

Pacific silver fir, *Abies amabilis* (TRF).

Pacific yew, *Taxus brevifolia* (TRF).

Paddle-bladed seagrasses, *Halophila decipens*, *H. engelmanni*, and *H. johnsonii* (CE).

Pale pitcher plant, *Sarracenia alata* (SE). Recent evidence suggests two morphologically identical, but genetically distinct, species occur on opposites side of the Mississippi River.

Palo verde, *Parkinsonia* spp. (D). Formerly listed in genus *Cercidium*. See also Retama.

Pan American balsamscale, *Elyonurus tripsacoides* (G).

Parry pine, *Pinus quadrifolia* (CPJ).

Pearly everlasting, *Anaphalis margaritacea* (TRF).

Pin oak, *Quercus palustris* (EDF).

Pinyon pine, *Pinus edulis* (CPJ).

Pitcher plant, *Sarracenia* spp. (EDF).

Pitch pine, *Pinus rigida* (EDF).

Poison ivy, *Toxicodendron radicans* (SE).

Poisonwood tree, *Metopium toxiferum* (SE).

Polypody fern, *Polypodium vulgare* (SE).

Ponderosa pine, *Pinus ponderosa* (MF).

Pond pine, *Pinus serotina* (EDF).

Pondweeds, *Potamogeton* spp. (CE). Potamogetonaceae is a diverse family of aquatic plant.

Port Orford cedar, *Chamaecyparis lawsoniana* (TRF).

Post oak, *Quercus stellata* (G).

Prairie dropseed, *Sporobolus heterolepis* (G).

Prickly pear, *Opuntia* spp. (RG, D). A common genus of cacti.

Purple borage, *Coldenia hispidissima* (D).

Purple loosestrife, *Lythrium salicaria* (SE).

Purple needlegrass, *Stipa pulchra* (G, RG).

Purple prairie clover, *Dalea purpurea* (G).

Pyxie flowering-moss, *Pyxidanthera barbulata* (EDF). Not a true moss.

Quaking aspen, *Populus tremuloides* (G, MF).

Rabbitbrush, *Chrysothamnus* spp. (D). Dwarf blue rabbitbush *C. nauseosus* is a common representative of this genus.

Railroad vine, *Ipomoea pes-caprae* (CE).

Red bay, *Persea borbonia* (EDF, SE).

Red cedar, *Juniperus virginiana* (SE).

Red fescue, *Festuca rubra* (CE).

Red fir, *Abies magnifica* (MF).

Red hickory, *Carya ovalis* (EDF).

Red mangrove, *Phizophora mangle* (CE).

Red maple, *Acer rubrum* (EDF).

Red pine, *Pinus resinosa* (EDF).

Red root, *Ceanothus ovatus* (G).

Red shanks, *Adenostoma sparsifolium* (CPJ).

Red spruce, *Picea rubens* (BF).

Redwood, *Sequoia sempervirens* (MF).

Retama, *Parkinsonia aculeata* (RG, SE). Sometimes classified in genus *Genista*; also listed as Mexican palo verde *Retama sphaerocarpa* in some references.

Rhododendron, *Rhododendron* spp. (T, BF). Lapland rosebay, *R. lapponicum*, is a circumpolar species in Arctic tundra; *R. catawbiense* and *R. arborescens* characterize heath balds in the southern Appalachian Mountains.

Rocky Mountain juniper, *Juniperus scopulorum* (CPJ).

Rose, *Rosa* spp. (RG). Many wild species, of which Wood's rose *R. woodsii* is representative.

Rough fescue, *Festuca scabrella* (G).

Russian thistle, *Salsola tragus* (D). Also known as "tumbleweed" or "wind witch." Botanists disagree on the scientific name, but there is some consensus on *S. tragus* for the inland form and *S. kali* for the coastal form.

Sabal palmetto, *Sabal palmetto* (SE).

Sacaton, *Sporobolus wrightii* (RG).

Sagebrush, *Artemisia* spp. (D). Big sagebrush *A. tridentata* is a widespread dominant in the Great Basin; others include sandsage *A. filifolia* and black sage *A. nova*.

Sago pondweed, *Potamogeton pectinatus* (CE).

Saguaro, *Carnegiea gigantea* (D).

Saltbrush, *Altriplex nuttallii* (D).

Salt grass, *Distichlis spicata* (CE).

Saltcedar, *Tamarix ramosissima* (D, SE).

Saltwort, *Batis maritima* (CE). Also known as turtle weed.

Sandberg bluegrass, *Poa secunda* (RG).

Sand bluestem, *Andropogon hallii* (G).

Sand reedgrass, *Calamovilfa longifolia* (G).

Saw palmetto, *Serenoa repens* (SE).

Sawgrass, *Cladium jamaicense* (SE). Not a true grass, but a sedge; also known as Everglades river grass.

Saxifrage, *Saxifrage* spp. (T). A large genus; purple mountain saxifrage *S. oppositifolia* occurs in Tundra.

Scarlet oak, *Quercus coccinea* (EDF).

Seacoast bluestem, *Schizachyrium scoparium*, var. *littoralis* (G). Formerly *Andropogon scoparium.*

Sea-lavender, *Limonium carolinianum* (CE).

Sea lettuce, *Ulva lactuca* (CE). A large algae.

Sea oats, *Uniola paniculata* (G, CE).

Sea ox-eye, *Borrichia frutescens* (CE).

Seashore dropseed, *Sporobolus virginicus* (SE).

Sedges, *Carex* spp., including Hoppner's sedge, *C. subspathacea* (T). A large group herbaceous plants, some of which resemble grasses.

Shadscale, *Altriplex confertifolia* (D).

Shoalgrass, *Halodule beaudettei* (CE). Formerly *H. wrightii.*

Shoreline sea purslane, *Sesuvium portulacastrum* (CE).

Shortleaf pine, *Pinus echinata* (EDF).

Shrub honeysuckle, *Diervilla sessifolia* (BF).

Shrub live oak, *Quercus turbinella* (CPJ).

Sideoats grama, *Bouteloua curtipendula* (G).

Silver beachweed, *Ambrosia chamissonis* (CE). Also known as beach bur.

Silver palm, *Coccothrinax argentata* (SE).

Silverling, *Baccharis halmifolia* (CE).

Silvery lupine, *Lupinus argenteus* (MF).

Sitka spruce, *Picea sitchensis* (TRF).

Slash pine, *Pinus elliottii* (EDF, SE). Formerly *P. caribaea*. A variety *P. e.* var. *densa* occurs in the Florida Keys.

Slender oat, *Avena barbata* (G).

Slough grass, *Spartina pectinata* (G).

Small's stonecrop, *Diamorpha smallii* (SE).

Smoke tree, *Psorothamnus spinosa* (D). Formerly *Dalea spinosa.*

Smooth cordgrass, *Spartina alterniflora* (CE). Also known as saltmarsh cordgrass or beach cordgrass.

Snowberry, *Symphoricarpos* spp. (G, RG). Western snowberry *S. occidentalis* is representative.

Soapweed, *Yucca elata* (D). Also known as soaptree.

Soft chess, *Bromus mollis* (RG).

Sotol, *Dasylirion leiophyllum* (RG).

Spanish moss, *Tillandsia usneoides* (EDF). Not a moss, but an epiphytic flowering plant in the pineapple family, Bromeliaceae.

Spike rush, *Eleocharis* spp. (CE).

Spruce, genus *Picea* (CE). Any of 35 species of large coniferous evergreen trees.

Squaw seed, *Senecio integerrimus* (RG).

Squirreltail, *Sitanion hystalk* (D).

Strangler fig, *Ficus aurea* (SE).

Subalpine fir, *Abies lasiocarpa* (MF). Corkbark fir *A. l. arizonica* is a subspecies endemic to the Pinaleño Mountains of Arizona.

Sugar hackberry, *Celtis laevigata* (RG).

Sugar maple, *Acer seccharum* (EDF).

Sundew, *Drosera* spp. (SE). Any of about 194 species, including *D. rotundifolia.*

Surfgrasses, *Phyllospadix scouleri*, *P. serrulatus*, and *P. torreyi.* (CE). The group is also known as rockgrasses.

Sweetbay, *Magnolia virginiana* (EDF).

Sweetgum, *Liquidambar styraciflua* (EDF, SE). Sometimes called "star oak" by ambitious realtors; known as "American sweetgum" in Mexico.

Switchgrass, *Panicum virgatum* (G).

Tarbush, *Flourensia cernua* (D).

Texas ebony, *Ebenopsis ebano* (RG, SE). Formerly *Pithecellobium flexcaule.*

Texas grama, *Bouteleua rigidiseta* (RG).

Thatch palm, *Thrinax morrisii* (SE).

Three-awn grass, *Aristida* spp. (G, RG). Wright's threeawn *A. wrightii* is representative; see also Wiregrass.

Titi, *Cyrilla racemiflora* (EDF).

Tobosa grass, *Hilaria mutica* (D).

Trembling aspen, *Populus tremuloides* (CE). Also known as quaking aspen or poplar.

Trillium, any of several species in the genus *Trillium* (EDF). Also known as "wakerobin."

Turkey oak, *Quercus laevis* (EDF).

Turtlegrass, *Thalassie testudinum* (CE).

Vine maple, *Acer circinatum* (TRF).

Violet, *Viola* spp. (EDF). Many species, including those in the understory of deciduous forests.

Water chestnut, *Trapa natans* (SE).

Water hyacinth, *Eichornia crassipes* (SE).

Water hyssop, *Bacopa monnieri* (CE).

Water milfoil, *Myriophyllum spicatum* (CE).

Water oak, *Quercus nigra* (EDF).

Wax myrtle, *Myrica cerifera* (CE, EDF, SE).

West Indies mahogany, *Swietenia mahagoni* (SE).

Western hemlock, *T. uga heterophylla* (MF).

Western redcedar, *Thuja plicata* (TRF).

Wheatgrass, *Agropyron* spp. (G, RG).

Whitebark pine, *Pinus albicaulis* (MF).

White bursage, *Ambrosia dumosa* (D). Also known as burro weed.

White fir, *Abies concolor* (MF).

White mangrove, *Laguncularia racemosa* (CE).

White oak, *Quercus alba* (EDF).

White pine, *Pinus strobus* (EDF).

White spruce, *Picea glauca* (T, BF, MF).

White water lily, *Nymphaea odorata* (CE).

Wigeongrass, *Ruppia maritima* (CE). Not a grass, but a flowering, herbaceous monocot.

Wild celery, *Vallisneria americana* (CE).

Wild indigo, *Baptisia lactea* (G).

Wild oat, *Avena fatua* (RG).

Willow, *Salix* spp. (SE). Coastal plain willow, *S. caroliniana*, pioneers the rims of alligator holes in the Everglades.

Willow oak, *Quercus phellos* (EDF).

Wiregrass, *Aristida stricta* (EDF).

Witch hobble, *Viburnum lantanoides* (EDF).

Yaupon, *Ilex vomitoria* (CE).

Yellow birch, *Betula alleghaniensis* (BF). Formerly *B. lutea*.

Yellow poplar, *Liriodendron tulipifera* (EDF). Also known as tuliptree or tulip poplar.

Yellow sand verbena, *Abronia latifolia* (CE).

Insects and other invertebrates

Acorn barnacle, *Balanus glandula* (CE).

Amphipods (CE). A large group of small, shrimp-like crustaceans represented by Grammaridae, Ampeliscidae, and other families in the order Amphipoda.

Ants, family Formicidae (T). A family of social insects with about 12,500 known species.

Aphids, family Aphidae (TRF). Popularly known as plant lice.

Apple snail, *Pomacea paludosa* (SE).

Arctic bumblebee, *Bombus polaris* and *B. hyperboreus* (T). Four species of bumblebees occur in the Arctic. A kleptoparasitic species *B. hyperboreus* usurps the nest of *B. polaris*.

Arctic woollybear moth, *Gynaephora groenlandica* (T). This species should not be confused with the orange and black woollybear caterpillar of the Isabella tiger moth (*Pyrrharctia isabella*) whose distribution is limited to more temperate regions in North America.

Army cutworm moth, *Euxoa auxilaris* (MF).

Atlantic surf clam, *Spisula solida* (CE).

Balsam wooly adelgid, *Adelges picea* (BF).

Banana slug, *Ariolimax columbianus* (TRF).

Barnacle, infraclass Cirripedia, phylum Arthropoda (CE).

Bay scallop, *Argopecten irradians* (CE). One of several scallops.

Bivalve mollusks, class Bivalvia, phylum Mollusca (CE). Clams, oysters scallops, mussels, and any of the more than 15,000 species of mollusks with a shell divided into right and left valves connected by a hinge.

Black fire beetle. *Melanophilia acuminata* (CPJ).

Black flies, family Simulidae (T, BF).

Blue crab, *Callinectes sapidus* (CE).

Blue mussel, *Mytilus edulis* (CE).

Brine shrimp, *Artemia fransiscana* (D).

Bumblebees, *Bombus* spp. (T, SE). A genus of about 300 species, of which about 46 occur in North America north of Mexico; See Arctic bumblebee.

Butterflies, order Lepidoptera (T, MF). An order of winged insects, including moths, with over 15,000 species.

Cicada, see Periodic cicada or Desert cicada.

Clam shrimp, order Cyclestherida, phylum Arthropoda (CE). Tiny shrimp-like crustaceans with shells resembling bivalve mollusks.

Copepods, subclass Copepoda, phylum Arthropoda (CE). A large group (10 orders and 13,000 or more species) of tiny shrimp-like crustaceans, including *C. glacialis*, important in Arctic food webs.

Crab, order Decapoda, phylum Arthropoda (CE). Decapod crustaceans having a shell, claws, and a tail coiled underneath the body.

Crustacean, subphylum Crustacea (CE). Large group of arthropods including crabs, lobsters, crayfish, shrimp, barnacles, and woodlice.

Delta green ground beetle, *Elaphrus viridus* (RG).

Desert cicada, *Diceroprocta apache* (D).

Earthworms, order Oligochaeta, phylum Annelida (EDF). Includes twelve families; of some 183 taxa in the USA and Canada, 60 are invasive species; *Lumbricus rubellus* and "night crawlers" *L. terrestris* are representative of the latter.

Eastern oyster, see Oyster.

Eelgrass limpet, *Lottia alveus* (CE).

Fairy shrimp, order Anostraca, phylum Arthropoda (CE). Small shrimp-like crustaceans (class Branchiopoda).

Fire ant, *Solenopis invicta* (EDF).

Fire beetle, see Black fire beetle.

Flesh fly, family Sarcophagidae, order Diptera. (SE). Any of 108 genera and more than 2500 species.

Forest tent caterpillar, *Malacosoma disstria* (BF).

Ghost crab, *Ocypode quadrata* (CE).

Glacial ice worm, *Mesenchytraeus solifugus* (SE).

Green sea urchin, *Strongylocentrotus droebachiensis* and *Lytechinus variegates* (CE). *S. droebachiensis* is the urchin of kelp forests, but also occurs on the northern Atlantic coast; *L. variegates* is found in seagrass beds of the southern Atlantic and Gulf coasts.

Gypsy moth, *Lymantria dispar* (EDF).

Harvester ants, *Pogonomyrmex* spp. (D).

Hemlock wooly adelgid, *Adelges tsugae* (BF).

Hermit crab, order Decapoda, phylum Arthropoda (CE). Crabs with long, curved abdomens attached for protection in abandoned shells of sea snails.

Horseshoe crab, *Limulus polyphemus* (CE). This species, of which four are known, occurs on the Atlantic Coast of North America.

Ice worm, see Glacial ice worm.

Isopod, order Isopoda, phylum Arthropoda (CE). Most members of this large group are flattened dorsoventrally and the body is slightly arched; many armor-like plates cover the back.

Jack pine budworm, *Choristoneura pinus* (BF).

Ladybird beetles, family Coccinellidae (TRF). About 150 species of these distinctive insects, also known as ladybugs, occur in the United States. Most species are highly beneficial.

Lake Huron locust, *Trimerotropis huroniana* (SE).

Leaf-eating beetles, *Galerucella calamariensis* and *G. pusilla* (SE).

Lettered olive, *Oliva sayana* (CE).

Limpet (CE). A common name applied to snails with a conical shell and clinging tightly to a hard substrate; true limpets are often found on rocky seashores.

Mayfly (SE). Any of three species: *Ephemera simulans*, *Hexagenia rigida*, and *H. limbata*.

Melaleuca leaf weevil, *Oxyops vitiosa* (SE)

Mole crab, *Emerita* spp. (CE). Small decapod crustaceans; three species occur on North American seashores: *E. talpoidea* (Atlantic), *E. analoga* (Pacific), and *E. benedicti* (Gulf of Mexico).

Monarch butterfly, *Danaus plexippus* (MF).

Moon snail, see Shark eye.

Mosquitoes, family Culculidae (T).

Mountain pine beetle, *Dendroctonus ponderossae* (MF).

Mussel, see Bivalve mollusks.

Nose bot, *Cephenomyia trompe* (T).

Olympia oyster, *Ostrea* (*Ostreola*) *conchaphila* (CE). Some sources use *Ostreola* as the generic name.

Osedax (CE). A genus of marine worm discovered in 2002 feeding on the bones of a gray whale carcass.

Oyster, *Crassostrea virginica* (CE). Also known as Atlantic oyster, Virginia oyster, or American oyster. See also Pacific oyster and Olympia oyster.

Oyster pea crab, *Zaops ostreum* (CE).

Parasitic copepod, *Ommatokoita elongata* (SE).

Periodic cicada, *Magicicada* spp. (EDF). Any of seven species, three of which maintain 17-year cycles and four with 13-year cycles.

Periwinkle, *Littorina* spp. (CE). Small marine snails.

Pitcher plant midge, *Metriocnemus knabi* (SE).

Pitcher plant mosquito, *Wyeomyia smithii* (SE).

Pitcher plant moth, *Exyra semicrocea* (SE).

Pitcher plant sarcophagid, *Fletcherimyia fletcheri* (SE). Formerly *Blaesoxipha fletcheri*; see also Flesh fly.

Polychaete, class Polychaeta, phylum Annelida (CE). Segmented marine worms, often brightly colored, with well-developed heads and pairs of fleshy protrusions on each body segment aft of the head.

Poplar borer, *Saperda calcerata* (G).

Quagga mussel, *Dreissena bugensis* (SE). Once considered *D. rostiformis bugensis*.

Queen conch, *Strombus gigas* (CE).

Sand dollar, order Clypeasteroidea, phylum Echinodermata (CE). Flattened, burrowing organisms with an endoskeleton covered with skin of small spines that propel the animal just beneath the seafloor; named for their likeness to tarnished silver dollars.

Sea anemone, order Actiniaria, phylum Cnidaria (CE). A cylindrical, predatory animal anchored to the substrate at one end by a basal disc and with a mouth surrounded by tentacles at the other end.

Sea cucumbers, class Holothuroidea, phylum Echinodermata (CE). Elongated marine organisms resembling a cucumber in shape; a leathery skin covers their endoskeleton.

Sea louse, family Caligidae, phylum Arythropoda (CE). Marine crustaceans that are ectoparasites on fish.

Sea slug (CE). A name commonly used for saltwater snails that either lack a shell or possess an internal shell; the name is also applied to various other marine organisms lacking a shell.

Sea snail (CE). Common name for any snail that inhabits brackish or salt water.

Sea stars, class Asteroidea, phylum Echinodermata (CE). Marine predators having a central disc from which radiate five or more arms; moves using numerous tube feet on the underside of the arms; also known as starfish. Species with elongated, slender arms are known as brittle stars.

Sea urchin, *Strongylocentrotus polyacanthus* (CE).

Serpulids, family Serpulidae, phylum Annelida (CE). Tube-building worms whose tubes sometimes coalesce to build reefs.

Shark eye, *Neverita duplicata* and *N. delessertiana* (CE). Also known as moon snails.

Shellfish (CE). A common name for any aquatic or marine invertebrate having an exoskeleton and used for food, including snails, crabs, lobsters, shrimp, and starfish.

Shrimp (CE). A common name applied to a variety of decapod crustaceans, but most commonly to organisms with long antennae, ten pairs of slender legs, and a muscular abdomen capable of both forward and backward movements.

Snail, class Gastropoda, phylum Mollusca (CE). An animal possessing a coiled shell into which it can withdraw its entire body.

Snow flea, *Hypogastrura nivicola* (SE). Not a flea, but a springtail in order Collembola.

Spiders, order Araneae, phylum Arthropoda (TRF).

Sponge, phylum Porifera (CE). Multicellular organisms having only two layers of cells and a body with many pores for circulating water; body shape varies with habitat conditions.

Spruce budworm, *Choristoneura fumifera* (BF).

Surf clam, see Atlantic surf clam.

Tube worm, class Polychaeta, phylum Annelida (CE). Worm-like animals that anchor their posterior end and secrete a mineral tube around their body into which they can withdraw.

Vernal pool tadpole shrimp, *Lepidurus packardi* (RG).

Warble fly, *Oedamagena tarandi* (T).

Water boatman, *Trichocorixa verticalis* (D).

Whelk, *Buscyon* spp. (CE). Mollusks also known as conch.

White pine weevil, *Pissodes strobi* (EDF).

Zebra mussel, *Dreissena polymorpha* (SE).

Fishes

Alewife, *Alosa pseudoharengus* (SE).

American eel, *Anguilla rostrata* (CE).

American paddlefish, *Polyodon spathula* (SE).

American shad, *Alosa sapidissima* (CE).

Arctic cod, *Boreogadus saida* (SE).

Atlantic cod, *Gadus morhus* (SE).

Atlantic salmon, *Salmo salar* (SE).

California grunion, *Leuresthes tenuis* (CE)

Channel catfish, *Ictalurus punctatus* (SE).

Chinese paddlefish, *Psephurus gladius* (SE).

Chinook salmon, *Oncorhynchus tshawytscha* (TRF).

Codfish, See Arctic cod.

Common carp, *Cyprinus carpio* (SE).

Big-headed carp, *Hypophthalmichthys nobilus* (SE).

Blenny, suborder Blennioidei (CE). Any of 833 species in 6 families; small, elongate fish common on ocean floors, rocky seashores, and coral reefs

Bluefish, *Pomatomus salatrix* (CE).

Bluegill, *Lepomis macrochirus* (SE).

Blue pike, *Sander vitreus glaucus* (SE).

Bowfin, *Amia calva* (SE).

Brook trout, *Salvelinus fontinalis* (SE).

Bucktooth parrotfish, *Sparisoma radians* (CE).

Ciscoe, *Coregonus* spp. (SE).

Cutthroat trout, *Onycorhynchus clarkii* (SE).

Devils Hole pupfish, *Cyprinodon diabolis* (D).

Eel, see American eel.

Greenland shark, *Somniosus microcephalus* (SE).

Hagfish (CE). Jawless, eel-like scavengers feeding on carcasses lying on the seafloor; several genera recognized, including *Myxine* and *Eptatretus*.

Hairlip sucker, *Lagochila lacera* (G).

Humpback chub, *Gila cypha* (SE).

Lake trout, *Salvelinus namaycush* (SE).

Least chub, *Lotichthys phlegethontis* (D).

Long-nosed gar, *Lepisosteus osseus* (SE). Also known as needlenosed gar or billyfish.

Mackerel, *Scomberomorus* spp. (CE).

Mountain sucker, *Panosteus platyrhynchus* (D).

Ozark cavefish, *Amblyopsis rosae* (SE).

Pacific salmon (TRF). Any of five species of salmon, all in genus *Oncorhynchus*, that occur on the Pacific Coast of North America.

Parrotfish (CE). Any of about 90 species of herbaceous marine fishes in the family Scaridaewhose teeth are adapted to rasp algae from the surfaces and recesses of coral.

Pickerel, *Esox americanus* (CE).

Pink salmon, *Onyorhynchus gorbuscha* (SE).

Rainbow trout, *Oncorhynchus mykiss* (SE). Formerly *Salmo gairdneri*.

Red drum, *Sciaenops ocellatus* (CE). Also known as redfish.

Salmon shark, *Lamna ditropis* (TRF).

Sea lamprey, *Petromyzon marinus* (EDF).

Shark (CE). Any of some 470 species in eight orders, including those associated with reefs (e.g., silvertip shark *Carcharhinus albimarginatus* and gray reef shark (*C. amblyrhynchos*).

Short-nosed gar, *Lepisosteus platostomus* (SE).

Silver carp, *Hypophthalmichthys molitrix* (SE).

Siscowet, *Salvelinus namaycush siscowet* (SE). A genetically distinct form of lake trout, most common in Lake Superior.

Smallmouth bass, *Micropterus dolomieu* (SE).

Snapper (CE). Any of about 100 species of predaceous marine fishes in the family Lutjanidae, including the red snapper, *Lutjanus campechanus*.

Sockeye salmon, *Oncorhynchus nerka* (TRF).

Splake (SE). A reproductively sterile hybrid between lake trout and brook trout.

Spring cavefish, *Chologaster agassizi* (SE).

Striped bass, *Morone saxatilis* (CE).

Sturgeon (SE). Three closely-related species (shovelnose sturgeon *Scaphirynchus platorhynchos*, pallid sturgeon *S. albus*, and Alabama sturgeon *S. suttkusi*) occupy the Mississippi River and its tributaries.

Suckermouth minnow, *Phenacobius mirabilis* (G).

Sunfish (SE). Any of numerous species, along with two species of freshwater bass, in the family Centrarchidae.

Swampfish, *Chologaster cornuta* (SE).

Utah chub, *Gila atraria* (D).

Utah sucker, *Catostomus ardens* (D).

Walleye, *Sander vitreus* (SE). *Stizostedion vitreum* in some sources.

Yellow perch, *Percia flavescens* (SE).

Amphibians

Black-spotted newt, *Notopthalamus meridionalis* (SE).

Blue-spotted salamander, *Ambystoma laterale* (SE).

California tiger salamander, *Ambystoma californiense* (RG). Once considered *A. tigrinum californiense*.

California giant salamander, *Dicamptodon ensatus* (TRF).

Coastal giant salamander, *Dicamptodon tenebrosus* (TRF). Formerly *D. ensatus*. One of three species of *Dicamptodon* occuring in the mountains of western North America.

Coastal tailed frog, *Ascaphus truei* (TRF).

Gopher frog, *Lithobates capito* (EDF). Formerly *Rana capito*.

Grotto salamander, *Eurycea spelaea* (SE). Formerly *Typhlotriton spelaeus*.

Jefferson salamander, *Ambystoma jeffersonianum* (SE).

Jordan's salamander, *Plethodon jordani* (BF). Red-cheeked salamander in many references; most subspecies once attributed to *P. jordani* are now recognized as distinct species using genetic signatures.

Mexican burrowing frog, *Rhynophrinus dorsalis* (RG). Mexican burrowing toad in some references.

Mole salamander, *Ambystoma talpoideum* (EDF).

Olympic torrent salamander, *Rhyacotriton olympicus* (TRF).

Oregon slender salamander, *Batrachoseps wrightorum* (TRF).

Ornate chorus frog, *Pseudacris ornata* (EDF).

Pine Barrens tree frog, *Hyla andersoni* (SE).

Red-backed salamander, *Plethodon cinereus* (EDF).

Red-legged salamander, *Plethodon shermani* (BF).

Red-spotted newt, *Notophthalmus viridescens* (SE).

Southern Appalachian salamander, *Plethodon teyahalee* (BF).

Southern chorus frog, *Pseudacris nigrita* (EDF).

Southern gray-cheeked salamander, *Plethodon metcalfi* (BF).

Southern leopard frog, *Lithobates sphenocephalus* (EDF). Formerly *Rana utricularia*.

Spadefoot toad, *Scaphiopus* spp. (D). Three species in North America, of which Couch's spadefoot toad *S. couchii* is representative; four species of western spadefoots are placed in the genus *Spea*.

Tailed frog, *Ascaphus* spp. (TRF). Two species are recognized: the coastal tailed frog *A. truei* from the Pacific coastal ranges, and the Rocky Mountain tailed frog *A. montanus*.

Texas blind salamander, *Eurycea rathbuni* (SE). Formerly *Typhlomolge rathbuni*.

Tiger salamander, *Ambystoma tigrinum* (G).

Wandering salamander, *Aneides vagrans* (MF).

White-lipped frog, *Leptodactylus labialis* (SE).

Wood frog, *Lithobates sylvaticus* (T). Formerly *Rana sylvatica*.

Reptiles

Agassiz's desert tortoise, *Gopherus agassizii* (D).

American alligator, *Alligator mississippiensis* (SE).

American crocodile, *Crocodylus acutus* (SE).

Black-striped snake, *Coniophanes imperialis* (SE).

Blue spiny lizard, *Sceloporus cyanogenys* (RG). Listed as *S. serrifer cyanogenys* in some older references.

Burmese python, *Python molurus bivittatus* (SE).

Caiman, subfamily Caimaninae (D). Any of several species found in Central and South America.

Common Chuckwalla, *Sauromalus ater* (D). Formerly *S. obesus*.

Desert grassland whiptail lizard, *Aspidoscelis uniparens* (RG). Formerly *Cnemidophorus uniparens*.

Desert iguana, *Dipsosaurus dorsalis* (D).

Diamond-backed terrapin, *Malaclemys terrapin* (CE).

Eastern indigo snake, *Drymarchon couperi* (EDF). Formerly *D. corais couperi*.

Eastern massasauga, *Sisturus catenatus catenatus* (SE).

Five-lined skink, *Plestiodon fasciatus* (SE). Formerly *Eumeces fasciatus*.

Gila monster, *Heloderma suspectum* (D).

Glass snakes, *Ophisaurus* spp. (CE). Not snakes, but legless lizards with smooth, shiny scales.

Gopher tortoise, *Gopherus polyphemus* (EDF).

Grand Canyon rattlesnake, *Crotalus oreganus abyssus* (SE). Formerly *C. viridis abyssus*.

Green turtle, *Chelonia mydas* (CE). Also known as green sea turtle.

Little striped whiptail lizard, *Aspidoscelis inornata* (RG). Formerly *Cnemidophorus inornatus*.

Loggerhead, *Caretta caretta* (SE). Also known as loggerhead sea turtle.

Mojave fringe-toed lizard, *Uma scoparia* (D).

Morafka's desert tortoise, *Gopherus morafkai* (D).

Mugger crocodile, *Crocodylus palustris* (SE).

New Mexico whiptail lizard, *Aspidoscelis neomexicana* (RG). Formerly *Cnemidophorus neomexicanus*.

Plains garter snake, *Thamnophis radix* (G).

Sidewinder, *Crotalus cerastes* (D). *C. c. cerastes* occurs in the Mohave Desert; *C. c. cercobombus* occurs in the Sonoran Desert.

Spectacled racer, *Drymobius margaritiferus* (SE).

Texas horned lizard, *Phrynosoma cornutum* (D). Often called a horned "toad."

Tiger whiptail lizard, *Aspidoscelis tigris* (RG). Formerly *Cnemidophorus tigris*.

Birds

Altamira oriole, *Icterius gularis* (RG, SE).

American black duck, *Anas rubripes* (CE).

American goldfinch, *Spinus tristis* (SE).

American woodcock, *Scolopax minor* (BF). Formerly *Philohela minor*.

Arctic tern, *Sterna paradisaea* (T).

Atlantic brant, *Branta bernicla hrota* (T, CE). This and its Pacific counterpart, black brant, now classified as subspecies, were once regarded as a separate species.

Atlantic puffin, *Fratercula arctica* (CE).

Attwater's prairie chicken, *Tympanuchus cupido attwateri* (G).

Bald eagle, *Haliaeetus leucocephalus* (TRF, SE).

Band-tailed pigeon, *Columbia fasciata* (SE).

Barred owl, *Strix varia* (TRF).

Bat falcon, *Falco rufigularis* (SE).

Brant brant, *Branta bernicla nigricans* (CE).

Blackburnian warbler, *Setophaga fusca* (BF). Formerly *Dendrocia fusca*.

Black duck, see American black duck.

Black skimmer, *Rhynchops niger* (CE).

Blue jay, *Cyanocitta cristata* (EDF).

Brown-headed cowbird, *Molothrus ater* (G, EDF).

Brown pelican, *Pelecanus occidentalis* (CE).

Burrowing owl, *Athene cunicularia* (G). *A. c. hypugaea* is closely associated with prairie dog and ground squirrel burrows in western North America; *A. c. floridana* occurs in southern Florida and is not affiliated with rodent habitations. Placed in the monotypic genus *Speotyto* by some.

Cackling goose, *Branta hutchinsii* (I).

Cactus wren, *Campylorhynchus brunneicapellus* (D).

California condor, *Gymnogyps californicus* (CPJ, RG).

California thrasher, *Toxostoma redivivum* (CPJ).

Canada goose, *Branta canadensis* (I, T, CE).

Canvasback, *Aythya valisineria* (D, CE).

Cave swallow, *Petrochelidon fulva* (SE).

Chachalaca, see Plain chachalaca.

Clark's nutcracker, *Nucifraga columbiana* (MF).

Common eider, *Somateria mollissima* (T).

Common murre, *Uria aalge* (SE).

Common tern, *Sterna hirundo* (CE).

Cormorant (CE). Fish-eating birds in order Pelecaniformes.

Dark-eyed junco, *Junco hyemalis* (MF). White-winged juncos, once identified as *J. aikeni*, are now a regional variant included in this species.

Darwin's finches, subfamily Geospizinae (Fringillidae). A "singular group" of birds on the Galapagos Islands and often credited as a major stimulus for Darwin's concepts.

Dipper, *Cinclus mexicanus* (MF). Also known as water ouzel.

Dowitcher, *Limnodromus* spp. (CE). Short-billed dowitcher *L. griseus* is more likely to be seen in coastal habitats than the long-billed dowitcher *L. scolopaceus*.

Ducks, family Anatidae (SE). E.g., northern pintail.

Dunlin, *Calidris alpine* (CE).

Eastern meadowlark, *Sturnella magna* (G).

Egret, family Ardeidae (CE). Long-necked, long-legged wading birds; egrets differ from herons only in name.

Elf owl, *Micrathene whitneyi* (D).

Eskimo curlew, *Numenius borealis* (T).

Everglade kite. See Snail kite.

Ferruginous hawk, *Buteo regalis* (G).

Ferruginous pygmy owl, *Glaucidium brasilianum* (SE).

Gadwall, *Anas strepera* (CE).

Gambel's quail, *Lophortyx gambelii* (D).

Giant Canada goose, *Branta canadensis maxima* (I, T).

Gila woodpecker, *Centurus uropygialis* (D).

Gilded flicker, *Colaptes chrysoides* (D).

Golden-cheeked warbler, *Setophaga chrysoparia* (CPJ). Formerly *Dendroica chrysoparia*.

Golden-crowned kinglet, *Regulus satrapa* (TRF).

Golden eagle, *Aqila chrysaetos* (T).

Gray hawk, *Buteo nitidus* (RG).

Great blue heron, *Ardea herodias* (SE).

Great curassow, *Crax rubra* (SE).

Great white heron, *Ardea herodias occidentalis* (SE). Once considered as separate species.

Greater prairie chicken, *Tympanuchus cupido* (G).

Grebes, family Podicipedidae (SE). E.g., pied-billed grebe, *Podilymbus podiceps*.

Green jay, *Cyanocorax yncas* (SE).

Green kingfisher, *Chloroceryle americana* (SE).

Gull, family Laridae, subfamily Larinae (TRF, CE). A large group of robust birds with webbed feet; most are scavengers associated with aquatic or marine habitats; related to terns. The glaucous-winged gull, *Larus glaucescens*, is representative.

Groove-billed ani, *Crotophaga sulcirostris* (SE).

Gyrfalcon, *Falco rusticolus* (T).

Heath hen, *Tympanuchus cupido cupido** (EDF).

Hermit thrush, *Catharus guttatus* (EDF). Formerly *Hylocichla guttata*.

Heron, see Egret.

Hudsonian godwit, *Limosa haemastica* (T).

Indigo bunting, *Passerina cyanea* (EDF).

Killdeer, *Charadrius vociferous* (G).

King eider, *Somateria spectabilis* (T).

Kirtland's Warbler, *Setophaga kirtlandii* (EDF). Formerly *Dendroica kirtlandii*.

Lark sparrow, *Chondestes grammacus* (G).

Least tern, *Sterna antillarum* (G, CE).

Lesser Canada goose, *Branta canadensis parvipes* (I).

Lesser golden plover, *Pluvialis dominica* (T). Also known as the American golden plover.

Lesser prairie chicken, *Tympanuchus pallidicinctus* (G).

Lesser sandhill crane, *Grus canadensis canadensis* (G).

Lesser scaup, *Aythya affinis* (CE).

Lesser snow goose, *Chen caerulescens hyperborea* (T, CE). Formerly assigned to genus *Anser*.

Limpkin, *Aramus guarauna* (SE).

Long-tailed duck, *Clangula hyemalis* (T). Also known as Oldsquaw.

Mallard, *Anas platyrhynchos* (EDF, CE).

Marbled murrelet, *Brachyramphus marmoratus* (TRF).

Maroon-fronted parrot, *Rhynchopsitta terrisi* (SE).

Masked bobwhite, *Colinus virginianus ridgway* (RG).

Merlin, *Falco columbarius* (T). Once known as pigeon hawk.

Mountain quail, *Oreotyx pictus* (CPJ).

Mountain plover, *Charadrius montanus* (G). *Eupoda montana* in some references.

Mourning dove, *Zenaida macroura* (RG).

Mute swan, *Cygnus olor* (CE).

Northern flicker, *Colaptes auratus* (G).

Northern bobwhite, *Colinus virginianus* (RG).

Northern goshawk, *Accipiter gentilis* (BF).

Northern pintail, *Anas acuta* (CE). Often identified simply as "pintail."

Northern spotted owl, See Spotted owl.

Ovenbird, *Seiurus aurocapillus* (EDF).

Olive sparrow, *Arremonops rufivigatus* (SE). Also known as green finch.

Passenger pigeon, *Ectopistes migratorius** (EDF).

Peregrine falcon, *Falco peregrinus* (CPJ). Formerly known as the duck hawk.

Pinyon jay, *Gymnorhinus cyanocephala* (CPJ).

Piping plover, *Charadrius melodus* (CE, G).

Plain chachalaca, *Ortalis vetula* (SE, RG).

Pomarine jaeger, *Stercoraius pomarinus* (T).

Ptarmigan. *Lagopus* spp. (T). White-tailed ptarmigan *L. leucurus* is an alpine species, whereas willow ptarmigan *L. lagopus* and rock ptarmigan *L. mutus* occur exclusively in Arctic Tundra.

Rails, Rallidae (SE) E.g., king rail, *Rallus elegans*.

Raven, *Corvus corax* (T, BF, TRF).

Red-breasted nuthatch, *Sitta canadensis* (EDF).

Red-cockaded woodpecker, *Picoides borealis* (EDF).

Red crossbill, *Loxia curvirosta* (BF).

Redhead, *Aythya americana* (CE).

Red-headed woodpecker, *Melanerpes erythrocephalus* (EDF).

Red knot, *Calidris canutus* (CE).

Ringed kingfisher, *Ceryle torquata* (RG).

Robin, *Turdus migratorius* (T).

Roseate tern, *Sterna dougallii* (CE).

Ross goose, *Anser rossi* (T). Some place this species in the genus *Chen*.

Rough-legged hawk, *Buteo lagopus* (T).

Ruddy turnstone, *Arenaria interpres* (CE, T).

Ruffed grouse, *Bonasa umbellus* (BF).

Rufous hummingbird, *Selasporus rufus* (TRF).

Sage hen, *Centrocercus urophasianus* (G, D).

Saltmarsh sparrow, *Ammodramus caudacutus* (CE).

Sanderling, *Calidris alba* (CE).

Scarlet tanager, *Piranga olivacea* (EDF).

Scrub jay, *Aphelocoma coerulescens* (CPJ).

Semipalmated plover, *Charadrius semipalmatus* (T).

Semipalmated sandpiper, *Calidris pusilla* (CE).

Sharptailed grouse, *Tympanuchus phasianellus* (G).

Snail kite, *Rostrhamus sociabilis* (SE). Also known as the Everglades kite.

Snow goose, see lesser snow goose.

Snowy owl, *Nyctea scandiaca* (T).

Southwestern willow flycatcher, *Empidonax traillii extimus* (SE).

Spectacled eider, *Somteria fischeri* (SE). Sometimes placed in monotypic genus *Lampronetta*.

Spotted owl, *Strix occidentalis* (TRF). The northern subspecies, *S. o. caurina*, is the focus of controversy in old-growth forests.

Spruce grouse, *Canachites canadensis* (BF).

Starling, *Sturnus vulgaris*.

Teratorn, *Teratornis merriami** (SE).

Tern, family Laridae, subfamily Sterninae (CE). A large group of slender, mostly fish-eating birds with long narrow wings, forked tails, and pointed bills; closely related to gulls.

Thick-billed murre. *Úria lómvia* (SE).

Trumpeter swan, *Cygnus buccunator* (T). Some place this species in the genus *Olor*.

Tundra swan, *Cygnus columbianus* (T, CE). Formerly known as the whistling swan.

Turkey vulture, *Cathartes aura* (SE).

Varied thrush, *Ixoreus naevius* (TRF).

Vaux's swift, *Chaetura vauxi* (TRF).

Waterfowl, family Anatidae (CE). The family of ducks, geese and swans

Western kingbird, *Tyrannus verticalis* (G).

Western meadowlark, *Sturnella neglecta* (G).

Whimbrel, *Numenius phaeopus* (T).

White-crowned pigeon, *Patagioenas lecucocephala* (SE). Earlier literature places the species in the genus *Columbia*.

White ibis, *Eudocimus albus* (SE).

White pelican, *Pelecanus erythrorhynchos* (CE).

White-winged crossbill, *Loxia leucoptera* (BF).

White-winged dove, *Zenaida asiatica* (RG, D).

White-winged junco, See Dark-eyed junco.

Whooping crane, *Grus americana* (G).

Wild turkey, *Meleagris gallopavo* (EDF). Several races, including *M. g. silvestris*, widespread in eastern North America north of Florida.

Willet, *Catotrophoris semipalmatus* (CE).

Willow ptarmigan, *Lagopus lagopus* (T).

Wilson's plover, *Charadrius wilsonia* (CE).

Winter wren, *Troglodytes troglodytes* (TRF).

Wood duck, *Aix sponsa* (EDF).

Wood stork, *Mycteria americana* (SE).

Wood thrush, *Hylocichla mustelina* (EDF).

Wood warblers, family Parulidae (EDF). Numerous species, especially in eastern North America; many winter in Central and South America.

Worm-eating warbler, *Helmitheros vermivorus* (EDF).

Wrentit, *Chamaea fasciata* (CPJ). Not a true wren.

Yellow warbler, *Setophaga petechia* (T). Formerly *Dendroica petechia*.

Mammals

Abert's squirrel, *Sciurus aberti* (SE). See Kaibab squirrel.

American lion, *Panthera atrox** (SE).

American mastodon, *Mammut americanum** (SE).

American pika, *Ochotona princeps* (MF). Also known as cony or rock rabbit. A disjunct population in Alaska and adjacent Canada is a separate species, the collared pika, *O. collaris*.

Apache pocket mouse, *Perognathus flavescens apache* (D). Formerly assigned full species status, *P. apache*.

Arctic ground squirrel, *Urocitellus parryii* (T). The generic name was recently changed from *Spermophilus* (and was previously *Citellus*).

Arctic hare, *Lepus arcticus* (T).

Armadillo, *Dasypus novemcinctus* (D). Also known as the nine-banded armadillo.

Badger, *Taxidea taxus* (G).

Banner-tailed kangaroo rat, *Dipodomys spectabilis* (RG).

Beaver, *Castor canadensis* (BF, G).

Beluga, *Delphinapterus leucas* (I).

Bison, *Bison bison* (G). Incorrectly known as "buffalo."

Black bear, *Ursus americanus* (EDF, MF).

Blackbuck, *Antelope cervicapra* (RG).

Black-footed ferret, *Mustela nigripes* (G).

Black-tailed jackrabbit, *Lepus californicus* (G, MF).

Blue whale, *Sibbaldus musculus* (CE).

Bobcat, *Lynx rufus* (G, SE). Listed as *Felis rufus* in older sources.

Bowhead whale, *Balaena mysticetus* (SE .

Brown (grizzly) bear, *Ursus arctos* (TRF . *U. a middendorfi* is the brown or Kodiak bear of the Alaskan coast.

Brown lemming, *Lemmus sibiricus* (T). Formerly *L. trimucronatus*.

Brush rabbit, *Sylvilagus bachmani* (CPJ).

Cactus mouse, *Peromyscus eremicus* (SE).

California mouse, *Peromyscus californicus* (CP .

Capybara, *Hydrochoerus hydrochaeris* (D).

Caribou, *Rangifer tarandus* (T, BF). The subspecies barren-ground caribou *R. t. groenlandicus* occurs in Tundra, whereas woodland caribou *R. t. caribou* live in Boreal Forest. Other subspecies include the Peary caribou *R. t. pearyi* of the High Arctic.

Chipmunks, *Neotamias* spp. Taxonomists place all 21 species of western chipmunks in this genus, including: the lodgepole chipmunk of the Sierra Nevada; the alpine chipmunk *N. speciosus*; yellow-pine chipmunk *N. alpinus*; *N. amoenus*; and the least (or sagebrush) chipmunk, *N. minimus*. Only one species, *Tamias striatus*, occurs in eastern North America. Formerly, all chipmunks were placed in the genus *Eutamias*. See Eastern chipmunk.

Collared peccary, *Tayassu tajacu* (D, RG . Also known as the javelina.

Coyote, *Canis latrans* (G, MF).

Dire wolf, *Canis dirus** (SE).

Douglas squirrel, *Tamiasciurus douglasi* (TRF). Also known as the chickaree.

Dusky-footed woodrat, *Neotoma fuscipes* (CPJ .

Eastern chipmunk, *Tamias striatus* (MF, SE . Formerly *Eutamias striatus*.

Eastern cottontail, *Sylvilagus floridanus* (D).

Eastern gray squirrel, *Sciurus carolinensis* (EDF).

Eastern woodrat, *Neotoma floridana* (SE)

Florida manatee, *Trichechus manatus latirostris* (CE). Also known as sea cow.

Florida panther, *Puma concolor coryi* (SE)

Fox squirrel, *Sciurus niger* (EDF).

Gray myotis, *Myotis grisescens* (SE). Also known as the gray bat.

Gray-footed chipmunk, *Tamias canipes* (D).

Gray fox, *Urocyon cinereoargenteus* (EDF).

Gray squirrel. See Eastern gray squirrel.

Gray whale, *Eschrichtius robustus* (CE).

Gray wolf, *Canis lupus* (T, BF). Several subspecies, including a white form known as the tundra wolf.

Grizzly bear, *Ursus arctos* (TRF, MF). Formerly *U. horribilis*. "Barren-ground grizzly" is a popular ecological designation, but is not a recognized subspecies. See also Brown bear.

Ground squirrels (T). All ground squirrels were once placed in genus *Spermophilus* or (earlier) *Citellus*, but recent genetic studies revealed that ground squirrels should be separated into 8 genera. See also Arctic ground squirrel.

Hares. See Jackrabbits.

Harlan's ground sloth, *Glossotherium harlani** (SE).

Hooded seal, *Cystophora cristata* (SE).

Humpback whale, *Megaptera novaeangliae* (CE).

Imperial mammoth, *Mammuthus imperator** (SE).

Indiana myotis, *Myotis sodalis* (SE). Also known as the Indiana bat.

Island fox, *Urocyon littoralis* (RG).

Jackrabbits and hares, *Lepus* spp. (G, D). See also Black-tailed jackrabbit and Snowshoe hare.

Jaguar, *Panthera onca* (D, SE). Although once separated into several subspecies, including *P.o. arizonensis*, taxonomists now recognize only a single form throughout the species' extensive distribution.

Jaguarundi, *Puma jagauoroundi* (RG, SE). Until recently, *Hepailurus* or *Felis* was the generic name.

Javelina, see Pecarries.

Kaibab squirrel, *Sciurus aberti* (SE). Formerly *S. kaibabensis*, but now united with Abert's squirrel into a single species.

Kangaroo rats, *Dipodomys* spp. (D).

Key deer, *Odocoileus virginianus clavium* (SE).

Killer whale, *Orcinus orca* (CE). Sometimes identified simply as "Orca."

Kit fox, *Vulpes velox* (D). Formerly *V. macrotis*, but now united with the swift fox into a single species.

Labrador collared lemming, *Dicrostonyx hudsonicus* (T).

Least weasel, *Mustela nivalis* (T, BF). Formerly *M. rixosa*.

Leopard. *Panthera pardus* (D, SE).

Little brown myotis, *Myotis lucifugus* (SE). Also known as the little brown bat.

Long-nosed bat, *Leptonycteris nivialis* (D).

Lower Keys marsh rabbit, *Sylvilagus palustris hefneri* (SE).

Lynx, *Lynx canadensis* (T, BF).

Manatee, see Florida manatee.

Margay, *Leopardus weidii* (SE).

Marmots, *Marmota* spp. (T). Hoary marmot *M. caligata* occupies alpine areas in western Canada northward into Alaska, whereas the yellow-bellied marmot *M. flaviventris* occurs in similar habitat in the western United States.

Marsh rabbit, *Sylvilagus palustris* (CE).

Mexican free-tailed bat, *Tadarida brasiliensis* (CV, SE).

Mink, *Mustela vison* (TRF).

Moose, *Alces alces.* (BF).

Mount Graham red squirrel. See Red squirrel.

Mountain goat, *Oreamnos americanus* (T).

Mountain sheep, *Ovis canadensis* (MF). Also known as bighorn sheep.

Mule deer, *Odocoileus hemonus* (MF, CPJ, TRF). *O. h. sitkensis* is the Sitka black-tailed deer of old-growth forests in the Pacific Northwest.

Muskox, *Ovibos moschatus* (T).

Muskrat. *Ondatra zibethicus* (CE).

Narwhal, *Monodon monoceros* (SE).

North Atlantic right whale, *Eubalaena gracialis* (CE).

Northern flying squirrel, *Glaucomys sabrinus* (TRF).

Northern short-tailed shrew, *Blarina brevicauda* (SE).

Nutria, *Myocastor coypus* (CE). Also known as coypu.

Ocelot, *Leopardus pardalis* (RG, SE).

Oldfield mouse, *Peromyscus polionotus* (CE). Coastal inhabitants are known as the beach mouse.

Opossum, *Didelphis virginiana* (SE).

Ord's kangaroo rat, *Dipodomys ordii* (D).

Peary Land collared lemming, *Dicrostonyx groenlandicus* (T). Also known as the Greenland collared lemming.

Pecarries, family Tayassuidae, the New World Pigs (D). The family includes the collared peccary (*Tayassu tajacu*) in the southwestern United States, and Central and South America.

Pika. See American pika.

Pocket gopher, family Geomyidae (G, RG, D). Heavily-built **fossorial** rodents with fur-lined cheek pouches used to transport food.

Polar bear, *Ursus maritimus* (T). Formerly *Thalarctos maritimus*.

Porcupine, *Erethizon dorsatum* (D, SE).

Pronghorn, *Antilocapra americana* (G).

Pygmy rabbit, *Brachylagus idahoensis* (D).

Raccoon, *Procyon lotor* (MF, SE).

Red squirrel, *Tamiasciurus hudsonicus* (BF, MF). *T. h. grahamensis*, occurs only in the Pinaleño (Graham) Mountains of Arizona.

Red tree vole, *Arborinus longicaudus* (TRF). Formerly *Phenacomys longicaudus*.

Richardson ground squirrel *Urocitellus richardsonii* (G). Formerly placed in genus *Spermophilus* or *Citellus*.

Ringed seal, *Pusa hispida* (SE).

Rock pocket mouse, *Chaetodipus intermedius* (D). Formerly *Perognathus intermedius*.

Roosevelt elk, *Cervus elaphus roosevelti* (TRF).

Saber-toothed cat, *Smilodon californicus** (SE).

Sea cow. See Manatee.

Sea otter, *Enhydra lutris* (CE).

Seals (CE). Marine mammals with front and hind limbs modified into flippers.

Short-faced bear, *Arctodus simus** (SE).

Shrews, family Scricidae (EDF). Small insectivorous mammals that also occur in Arctic, desert, and other regions of North America.

Snowshoe hare, *Lepus americanus* (BF, G). Also known as the snowshoe rabbit or varying hare.

Southern yellow bat, *Lasiurus ega* (RG, SE).

Sperm whale, *Physeter macrocephalus* (CE).

Spotted skunk, *Spilogale putorius* (SE).

Striped skunk, *Mephitis mephitis* (SE).

Tassel-eared squirrel, *Sciurus aberti* (SE). A taxon uniting Abert's and Kaibab squirrels, despite clear differences in their pelage.

Townsend's chipmunk, *Neotamias townsendii* (TRF). Formerly *Tamias townsendii*.

Weasels (D). Any of three species of *Mustela*, excluding mink (e.g. longtail weasel *M. frenata*).

White-nosed coati, *Nasua narica* (SE).

White-tailed deer, *Odocoileus virginianus* (EDF, CPJ).

White-throated woodrat, *Neotoma albigula* (D, RG).

Wolf. See Gray wolf.

Glossary

A horizon: The surface layer of mineral soil typically characterized by large accumulations of organic materials and the site of much biological activity; iron and aluminum oxides in this horizon are leached to lower layers. See **Soil horizon**; **Humus**.

Abiotic level: The non-living foundation of an ecosystem; includes mineral soil, air, and water. See **Ecosystem**.

Abcission layer: The cell zone where leaf stalks (petioles) detach themselves from branches at the end of the growing season.

Adaptation: An evolutionary process through which a species or population acquires traits allowing each successive generation to improve its fitness (i.e., better reproductive success and greater survival).

Adiabatic: An event occurring without heat entering or leaving a system, as sometimes happens when air moves in the atmosphere.

Aeolian: Term indicating transport by wind, typically applied in reference to loess and other soils relocated from another location (often at considerable distances) by wind action.

Aerobic: Existence or activity that takes place in the presence of oxygen. Aerobic bacteria require oxygen to exist. For comparison, see **Anaerobic**.

Agrostology: The study of grasses, especially their taxonomy and classification.

Albedo: The ratio between the amount of incoming solar energy and the amount of that energy reflected back into the atmosphere, which is often expressed as a percentage. An albedo of 1.0 (or 100%) represents perfect reflective power, whereas 0 indicates no reflective power at all (i.e., all solar energy is absorbed). Typical values for ice are 0 5–0.7, fresh snow 0.8–0.9, and water <0.1.

Albinism: The lack of natural pigmentation, often a genetic trait and typical of some kinds of cave-dwelling organisms. For comparison, see **Melanism**.

Alcid: Any bird in the family Alcidae, which includes several species of murres, guillemots, auklets, and puffins, among others; all are associated with marine environments.

Allelopathy: The adverse effect of one plant on another, resulting from the production of toxic chemicals. Chaparral communities include some species of allelopathic plants.

Allen's Rule: An ecological generalization describing a trend among endotherms to have longer appendages (e.g., legs and ears) in warmer than in colder climates. This is most apparent in species having wide north–south geographic ranges, and is

an adaptation for releasing body heat. Posited in 1877 by American ornithologist Joseph Asaph Allen (1838–1921). For comparison, see **Bergmann's Rule**. See also **Endotherm**.

Allopatric: Refers to two or more species or populations having distributional ranges that do not overlap. The desert cottontail and the New England cottontail have allopatric ranges. For comparison, see **Sympatric**.

Allopatric speciation: Evolution that takes place in separate geographic locations, typically because an intervening physical factor prevents genetic interchange that eventually leads to new taxa on either side of the barrier. Rivers, mountains, and glaciers are common types of physical barriers. See **Speciation**; **Taxon**.

Altitudinal migration: Regular vertical movements triggered by seasonal changes in the environment. Elk migrate up (spring and summer) and down (autumn and winter) the Rocky Mountains each year. For comparison, see **Latitudinal migration**.

Amphitropical disjunction: An unusual type of distribution in which the geographical range of an organism occurs on either side of the tropics, but not in a tropical region.

Anadromous: A life cycle pattern in which fish spend most of their lives in the sea and, as adults, ascend freshwater streams and rivers (usually to the same rivers where they hatched) to spawn. Pacific salmon and American shad are examples of anadromous fish. For comparison, see **Catadromous**.

Anaerobic: Existence or activity that takes place in the absence of oxygen. Some bacteria exist without oxygen and therefore persist under anaerobic conditions. For comparison, see **Aerobic**.

Anhydrobiosis: A temporary suspension of metabolic activity because of extreme dehydration. For comparison, see **Cryptobiosis**.

Anthropogenic: Generated or influenced by humans. Anthropogenic fires result from careless campers, but natural fires result from lightning.

Apex predator: Any predator at the top of a food chain which has no predators (except for humans). Mountain lions, wolves, and most sharks are examples.

Arctic: Generally applied, in ecological terms, to environments north of the tree line (i.e., Tundra), and often divided into High Arctic and Low Arctic. In geographical terms, the region north of the Arctic Circle (i.e., latitudes above 65°N). See **Sub-Arctic**; **High Arctic**; **Low Arctic**.

Avulsion: Literally, tearing away. Ecologically, the creation of a new river channel when a previous channel is abandoned.

Ecology of North America, Second Edition. Brian R. Chapman and Eric G. Bolen.
© 2015 John Wiley & Sons, Ltd. Published 2015 by John Wiley & Sons, Ltd.

Backshore: The area of a shoreline between the high tide line and the ridge of large dunes.

Basin and Range Province: A large geographic region in western North America characterized by mountains rising from a broad plain (i.e., extensive, relatively flat basins, not valleys, intervene between the mountains).

Benthic: Term associated with the bottom of a lake, ocean, or other aquatic environment. Clams are benthic organisms, and the peat and muck in a salt marsh is a benthic environment.

Bergmann's Rule: An ecological generalization stating that populations of endotherms inhabiting cooler climates tend to have larger body sizes and smaller surface-area-to-volume ratios than populations of the same species living in warmer areas; an adaptation for conserving body warmth. Posited in 1847 by German physiologist Karl Georg Lucas Christian Bergmann (1814–1865). For comparison, see **Allen's Rule**. See also **Endotherm**.

Biodiversity: Popular term for biological diversity, which is measured by: (a) species richness (= number of species per location); and (b) evenness (= abundance of each species per location).

Biogeography: The study of past and present distributions of plants and animals. Addresses large-scale (e.g., continental drift) and small-scale (e.g., seed dispersal) factors. See **Phytogeography**; **Zoogeography**.

Biological control: A pest-control method directed toward some "weakness" in the biology of the target species. Includes use of host-specific diseases, parasites, predators, or some aspect of breeding biology (e.g., pheromone-baited traps) to reduce the pest population.

Biomagnification: The accumulation of persistent materials as they pass upward through the food chain. Particularly important in the case of toxic substances such as pesticides, which may be applied at apparently "safe" rates, but become harmful to vertebrates eating quantities of contaminated foods (e.g., falcons eating songbirds that eat insects).

Biomass: The mass (weight) of organisms per unit of area, typically determined to evaluate the extent of a trophic level or primary productivity. Best measured on a dry-weight basis. See **Trophic level**; **Primary productivity**.

Biome: The largest ecological unit, which includes all areas of the globe with similar characteristics (e.g., Grasslands and Deserts each represent a biome).

Biospeleology: The study of cave life.

Biota: The combined plant and animal life occurring in a defined area (e.g., the biota of Newfoundland). Flora + Fauna = Biota.

Biotic community: See **Community**.

Bolide: A extraterrestrial object (either an icy comet or stony asteroid) striking Earth and leaving evidence of its impact, usually a crater.

Bolson: A flat-floored, semi-arid desert depression or valley surrounded by higher terrain such as hills or mountains. Waters draining from higher elevations collect in the basin and evaporate leaving shallow, salty playa lakes or salt-encrusted pans.

Browse: Woody vegetation (e.g., twigs and stems) that serves as food for herbivores, usually in winter. Aspen and willows provide browse for moose. Continued heavy cropping creates a browse line, an indicator of reduced carrying capacity. See **Carrying capacity**.

Burl: Enlarged woody growth that forms at the root crown in some species of chaparral; buds in these structures sprout after fires. Burls are harvested in some parts of the world as material for manufacturing pipes, bowls, and furniture.

Cecum (plural ceca): Dead-end pouches, often in pairs, extending from the intestinal tracts of animals with fibrous diets high in cellulose; these contain microbial biota specifically adapted to digest fibrous foods. The human appendix is likely a rudimentary, non-functional cecum.

Canopy: Uppermost layer in any vegetative community where the highest layer forms a distinctive habitat, but most often used to describe the cover provided by the tallest trees.

Carnivore: Any meat-eating animals, including diverse species such as bass, hawks, and mountain lions. Also denotes animals included in the mammalian order Carnivora, whether strictly meat eaters or not (e.g., bears). For comparison, see **Herbivore**.

Carrying capacity: The ability of habitat to sustain a population at maximum numbers, measured by density in relation to the availability of food or other resources and denoted K by population ecologists. During droughts, carrying capacity may decrease from three deer to one deer per 10 ha (25 acres).

Caryopsis: Technically a fruit with a fused seed coat but, for our purposes, a grass "seed" or grain.

Catadromous: A pattern of behavior in which fish spend most of their adult life in fresh waters and migrate to the sea for to spawning. Freshwater eels of the genus *Anguilla* – both the American species *A. rostrata* and the European species *A. anguilla* – are examples of catadromous fishes. Remarkably, both species spawn in the Sargasso Sea, after which the larval forms separate and travel to rivers in each continent. For comparison, see **Anadromous**.

Chaparral: A climax of dense brush with leathery, evergreen leaves that develops in a climate of hot dry summers and cool moist winters; replaces itself after fires.

Circadian rhythm Predictable activity patterns based on a 24-hour cycle of sleep followed by activity.

Circumpolar: A pattern of distribution in which the distributional range of an organism encircles the polar axis. Snowy owls and polar bears are found in northern Eurasia as well as in North America.

Climax: A community in dynamic equilibrium, normally regarded as the "end point" of succession. Generally expressed in terms of vegetation, but in ecological terms also includes animals. See **Subclimax**; **Succession**.

Coadaptation: The changes in behavioral, physical, or biological characteristics in one organism in response to changes in characteristics of an interacting organism. The process often results in mutualism between two species. Synonym: Coevolution.

Colonnade: A row of trees established at the top of a nurse log. See **Nurse log**.

Commensalism: A type of symbiosis in which organisms of one species benefit from their association with another species, but the latter is neither helped nor injured by the relationship. Pilot fish (ramoras) harmlessly attach themselves to sharks as a means of finding scraps of food, but the sharks are not affected by these "freeloaders." See **Symbiosis**

Community: The living part of an ecosystem. See **Ecosystem**.

Competitive exclusion: The concept that two or more species cannot coexist in the same habitat when these species depend exactly on the same resources for their survival. One species excludes the other from the habitat.

Composite: Any plant in the family Asteraceae, formerly known as Compositae. This is the largest family of plants and includes dandelions, sunflowers, cockleburs, asters, coneflowers, thistles, sagebrush, and daisies.

Conifer: Term applied to cone-bearing trees, which are popularly called "evergreens" because all except larch shed and replace their needle-like leaves gradually throughout the year. Examples include pines, spruces, and firs. For comparison, see **Deciduous**.

Conservation biology: the study and management of small populations, many of which may be threatened with extinction. The discipline often addresses genetic issues and is supported by a journal of the same name

Conspecific: Of the same species. Can apply to two or more individuals or populations that are of the same species. Cogeners are organisms within the same genus and may apply to two or more species in the genus.

Continental shelf: The undersea extension of a continent or a continental plate. The shelf usually terminates at a point where an abrupt slope descends to much deeper water.

Convergent evolution: Development of similar form and function (i.e., physical or behavioral features) in species that lack close taxonomic affinities but live under the same environmental conditions (e.g., aridity). Many unrelated plants have evolved thorns as a means of protecting their tissues from grazers. Likewise, willow ptarmigan and Arctic foxes each have dark summer "coats" that turn white in winter.

Coppice dune: Small accumulations of sand that are miniature versions of a large dune; usually found on the backshore of a beach behind driftwood, debris, or a cluster of plants.

Coriolis effect: An effect of the rotation of the Earth that caused moving objects at the surface of the plane to be deflected in a counterclockwise motion in the Northern Hemisphere or clockwise in the Southern Hemisphere; also known as Coriolis force.

Cosmopolitan: A distributional pattern that includes several continents.

Critical habitat: Designated areas that contain features essential for the conservation of threatened and endangered species that may also require special management or protection; includes areas not currently occupied by the species but will be needed for its recovery. When approved by the Secretary of the Interior, this means federal agencies cannot fund or authorize activities (e.g., mining) that might adversely modify the area in ways detrimental to the target species.

Cryoplanation: The shaping of a landscape by ice. This is often associated with glacial movements, but can also be the result of freeze–thaw cycles.

Cryptic coloration: Patterns of color that allow an organism to blend undetected into the substrate or background.

Cryptobiosis: A temporary living state in which metabolic activity is suspended or undetectable.

Cryptogam: A plant or plant-like organism that reproduces by spores instead of seeds. Common examples include ferns, mosses, algae, and fungi.

Cultch: Any solid substrate, such as shells, deposited to form attachment sites for oyster spat, a juvenile stage of oyster development.

Dead zone: Any area of marine or fresh water in which fish and other organisms cannot exist, usually because levels of dissolved oxygen are limited.

Deciduous: Term applied to trees or shrubs that shed their leaves at the end of the annual growing season. Most have flattened leaves. Examples include maples, elms, most oaks, and, notably, larch, a needle-bearing species in the pine family. For comparison, see **Conifer**.

Decomposer: A trophic level that consists of fungi and, particularly, bacteria; scavengers are often included.

Decomposition: The process by which the bodies of organisms or organic substances break down into simpler forms of matter. Degradation may involve living organisms or physical and chemical processes.

Decreasers: Plants that decrease in abundance as grazing intensity increases. Typically, these are highly palatable species in a grassland community. For comparison, see **Increasers**.

Decumbent: Botanical term for plants whose stems or trunks lie parallel to the ground even though some branches may be upright.

Detritivores: Organisms that eat organic residues. See **Detritus**.

Detritus: Organic residues, typically those decomposing parts of dead plants and animals that sink to the floor of an aquatic or terrestrial community. See **Detritivores**.

Diapause: A period of dormancy and arrested development, typically employed by some invertebrates to endure periods of harsh environmental conditions (e.g., drought or cold).

Disclimax: A community in which some type of disturbance prevents succession from continuing toward the typical climax. Heavy grazing or repeated fires are common causes of disclimax communities.

Discontinuous distribution: A pattern in which the geographical range of an organism occurs in two or more widely separated areas, but not in between.

Disjunct: A marked separation between two natural populations. See **Discontinuous distribution**.

Drupe: A fleshy fruit containing a hard pit surrounding a seed (e.g., a peach or cherry).

Ecesis: A measure of effective dispersal and survival, an attribute of successful colonists such as dandelions, coconuts, and mangroves.

Ecological amplitude: A measure of the niche in which a species can survive; a greater range of habitats is available to species with wide amplitudes. White-tailed deer thrive in the hardwood-conifer communities of southern Canada southward to the Florida Keys and into the shrub environments of

Texas and beyond (wide amplitude), whereas black-footed ferrets occur only in active prairie dog towns (narrow amplitude). Species with narrow ecological amplitudes often face the threat of extinction.

Ecological equivalents: Different species that play similar roles in separate communities, such as hooved grazers on grasslands in Africa and North America. Saguaro cacti and trees also represent ecological equivalents in their structure for woodpeckers in the Sonoran Desert and forests, respectively.

Ecosystem: The functional unit for considering the flow of energy and materials through the living and non-living components in a defined area at a given time. See **Trophic level**.

Ecotone: The transition zone between two communities. Sharply delineated transitions are known as "knife-edge" ecotones, but most ecotones are gradual.

Ectotherm: An organism whose body temperature varies, in large measure, with ambient temperatures; in popular terms, a "cold-blooded" animal. Most invertebrates, fishes, amphibians, and reptiles are ectotherms. The geographical or ecological distribution of ectotherms is often limited by extremes in ambient temperatures. Poikilotherm is a synonym. For comparison, see **Endoderm**.

Edaphic: Adjective referring to soil, especially features such as texture, drainage, and fertility. Edaphic conditions are of direct importance to the development of vegetation, but also influence the activities of burrowing animals.

Effective population size: A concept reflecting the genetic diversity within a population. In small, inbred populations, this number usually represents far fewer individuals than actually in the population (e.g., an inbred population of 20 animals may collectively reflect the genetic diversity of just 5 individuals).

Emery's Rule: An ecological generalization stating that social parasites tend to be closely related to, and closely resemble, the host species (i.e., the species they parasitize). Posited in 1909 by Italian entomologist Carlo Emery (1848–1925). See **Kleptoparasitism**.

Endemic: Term that indicates a limited distribution of a taxon. Islands or isolated bodies of water often contain endemic species. On a larger scale, Australia is famous for its endemic fauna of kangaroos and other marsupials.

Endophylls: The microorganisms that live inside the needles of trees in an old-growth forest. For comparison, see **Epiphylls**. See also "**Scuzz**."

Endotherm: An organism whose body temperature is independent of ambient temperature; in popular terms, a "warm-blooded" animal. Birds and mammals are endotherms and therefore have representatives living in a wide range of temperature extremes (e.g., Arctic to tropical). Homeotherm and Homoiotherm are synonyms. For comparison, see **Ectotherm**.

Ephemeral: In botany, a description for short-lived plants appearing irregularly (not seasonally), usually from seeds germinating after heavy rains. Also applied to ponds and other habitats that do not persist all year round (e.g., seasonal sites such as vernal pools, described in Chapter 6).

Epiphylls: The microorganisms that live on the surface of leaves, particularly the needles of trees in old-growth forest. For comparison, see **Endophylls**. See also "**Scuzz**."

Epiphyte: Term for a plant that derives its nutrients directly from the atmosphere without rooting or parasitizing vegetation on which it drapes; sometimes called an "air plant." Spanish moss is a well-known example.

Epizootic: An outbreak of a disease that affects large numbers of animals in a short time. Equivalent to an epidemic in human populations.

Epizootiology: The "how" and "why" of wildlife diseases; in short, the study of disease ecology (e.g., why do some diseases occur only at certain times of the year and how do diseases spread through animal populations?). The equivalent of epidemiology in studies of human populations.

Estivation: A prolonged state of dormancy (similar to hibernation) during times of heat or drought that is characterized by lowered metabolic and breathing rates. The word is sometimes spelled aestivation.

Estuary: The lower reaches of a river where it meets the intertidal zone of an ocean, featuring an upper layer of freshwater extending over a heavier, lower layer of salt water. Typically areas of high productivity.

Eurythermal: Tolerance for a broad range of temperatures. Conversely, a stenothermic organism survives only in a narrow range of temperatures. The prefix *eury-* (wide) is also applied to other environmental factors (e.g., euryhaline, tolerance for a wide range of soil or water salinities). For comparison, see **Stenothermal**.

Eustatic: Global change in sea level, as opposed to local change in sea level.

Eutrophication: The process by which nutrients, especially phosphates and nitrates, accumulate in lakes and other aquatic systems, thereby accelerating the growth of algae and other plant life. It is a natural, slow-aging mechanism, but human activities often rush the process (e.g., fertilizer runoff and, at one time, wastewater-borne phosphates from detergents).

Evolution: The process by which the inherited characteristics of a biological population change over successive generations, typically in response to the forces of natural selection.

Exclosure: A study site surrounded by a fence, which is designed to exclude one or more kinds of animals from entering; usually erected for the purpose of determining the effects of herbivores on vegetation (i.e., a comparison between vegetation inside and outside of the exclosure).

Exotic: Any non-native species, usually introduced intentionally or accidently by humans and often becoming pests (e.g., gypsy moths, fire ants, and nutria).

Extinct: No longer exists anywhere (e.g., passenger pigeons). Often misused for species no longer present locally or regionally but still survive in other locations (e.g., wolves). For comparison, see **Extirpated**.

Extinction: The process of becoming or state of being extinct. See **Extinct**.

Extinction vortex: A series of events that leads to the extinction of small populations. For example, a population weakened by starvation and then infected by a serious disease has fallen into an extinction vortex. Imbalanced age or sex distributions, as well as inbreeding, are sometimes components of this phenomenon.

Extirpated: Description for a taxon that is no longer present at a location but still occurs elsewhere. Wolves have been extirpated from much of their former range in North America. For comparison, see **Extinct**.

Facultative: Term for a species whose existence is usually, but not completely, tied to a set of conditions or habitat. For comparison, see **Obligate**. See also **Mutualism**.

Facultative mutualism: See **Mutualism**.

First-order stream: Ecologists, geographers, and others classify streams in a hierarchal system based on the convergence of tributaries. First-order streams are the smallest and lack any tributaries. When two first-order streams merge, a second-order stream is formed; the junction of two second-order streams marks the beginning of a third-order stream, and so on. Tenth to twelfth-order streams include the Amazon, the Mississippi, and other of the world's largest rivers.

Floristics: The field of botany concerned with the distribution and composition of vegetation, commonly on a regional scale and/or under special conditions (e.g., the floristics of glacially disturbed soils in New England).

Food chain: The pathway of matter and energy (= food) flow through the trophic levels of an ecosystem, typically plant to herbivore to carnivore. Because of alternate routes, most ecosystems have many food chains which together are a food web. See **Ecosystem**; **Trophic level**.

Food web: See **Food chain**.

Forb: A broad-leaved herbaceous plant (i.e., not a woody plant and not a grass). Dandelions, goldenrods, and milkweeds are examples. See **Herbaceous**.

Fossorial: Describes organisms that live totally or mostly underground. Examples include earthworms, moles, pocket gophers, and prairie dogs.

Founder effect: Describes the process by which a population reflects the genetic diversity of its founders. Commonly applied to populations established by a few individuals and therefore possessing a limited gene pool. Such populations, even when later reaching large numbers, still have no more genetic diversity than present in the few original founders.

Frequency of occurrence: A statistic in food habits research based on the number of samples (e.g., stomachs, crops, or droppings) that contain a certain food, expressed as a percentage of all samples examined. Provides a measure of how common a food is in the diet but not how much is consumed. If a sample of 20 raccoon stomachs included 15 with remains of crayfish, the frequency of occurrence for crayfish is 75%.

Frugivore: A fruit-eating animal.

Frugivorous: Term describing the diet of a frugivore. See **Frugivore**.

Fundamental niche: The greatest breadth of a niche as defined by an organism's ability to tolerate a wide range of environmental conditions (e.g., wet and dry, hot and cold, etc.).

A species may be able to live in several habitats (its functional niche) but does not because it is prevented from doing so by other species. See **Niche**; **Realized niche**; **Competitive exclusion**.

Fumarole: An opening in the Earth's crust from which steam and other gases escape on a regular or continuing basis. Fumaroles often are associated with areas of volcanic activity.

Geophagy: Ingestion of earthy substances such as soil, mud, clay, or chalk to obtain salt or other minerals.

Gigantothermy: Term describing the ability of certain large ectothermic animals (e.g., leatherback sea turtle and some large snakes) to maintain a constant, relatively high body temperature because their great size results in a high surface-to-volume ratio; sometimes known as ectothermic homeothermy. See **Ectotherm**; **Homeothermy**.

Glacial refugium: A site within a glaciated region that escapes being covered by ice, thereby providing a refuge where the biota might remain intact and continue speciation. When the glacier retreats, new species or the remaining population of older species may spread outward into the habitat previously covered with ice. Plural: refugia. See **Speciation**.

Glacial till: The unsorted gravel and rock debris deposited by a melting glacier.

Graminoid: Term for grass-like plants (e.g., some sedges resemble grasses, but represent separate taxa). Graminoid vegetation characteristically occurs in meadows.

Granivore: A grain-eating organism. Most sparrows and finches have granivorous diets.

Gravid: A term describing an animal that is carrying eggs or young at an early stage of development.

Guild: A group of organisms, not of the same species, that share a common resource. Birds gleaning insects from leaves in a deciduous forest belong to the same guild.

Gyre: Ocean-wide circulation pattern of rotating currents caused by the Coriolis effect.

Halophyte: A plant tolerant of a saline environment.

Herbaceous: Term for the soft tissues of plants (e.g., leaves and soft stems). Forbs and most grasses are herbaceous. Trees and shrubs, although they have herbaceous foliage, are regarded as woody vegetation. See **Forb**.

Herbivore: An animal that eats plants (e.g., grasshopper, rabbit, and bison). See **Trophic level**.

Hibernaculum: A den or other protected area, such as a cave or tree hollow, where an organism spends the winter in hibernation. Plural: hibernacula.

High Arctic: In ecological terms, the region of polar deserts and perpetual ice and snow. Includes Ellesmere and most other islands in the Canadian Archipelago and Greenland, but little of mainland North America. Some consider 75°N latitude as the southern edge of High Arctic.

High-energy beach: A sandy seashore facing the open ocean and pounded by heavy surf, resulting in alternating periods of expansion and contraction. For contrast, see **Low-energy beach**.

Holarctic: A pattern of distribution for organisms found in both North America and Eurasia. See **Nearctic**; **Palaearctic**.

Homing behavior: The return of a migratory organism to a breeding site where it has had previous experience. Adult Pacific salmon return from the ocean to breed in the same freshwater stream where they hatched years before. Some birds return year after year to the same nesting site.

Horizon: A layer or zone within a soil profile, typically varying by color as well as by distinctive physical and/or chemical features. In mature (= well-developed) soil, the profile consists of an A (upper), B, and C (lower) horizons. Immature soils are usually azonal.

Humus: Organic matter, typically a component of soil derived from the decomposition of plant and animal tissues.

Hurricane: A massive cyclonic storm with a cloud-free "eye" in the center of circulation and sustained winds above $127 \, \text{km} \, \text{hr}^{-1}$ (74 mph); also known as a tropical cyclone or cyclone.

Hydric: Term that refers to wet conditions. Hydric soils are inundated for at least a part of the year. For comparison, see **Xeric**; **Mesic**.

Hydroperiod: The length of time a wetland is actually inundated each year. In tidal wetlands, the hydroperiod is based on the extent of inundation during a 24-hour cycle.

Hyporheic flow: The mix of subsurface stream water with ground water, both beneath and lateral to the stream bed, and an active ecotone where nutrients may be exchanged (e.g., those from salmon carcasses). It may continue in droughts when the visible flow of a stream disappears, especially in those with gravel beds, and support organisms (e.g., amphibians, mollusks, and crustaceans) finding refuge under deep-seated rocks or ledges.

Hypoxia: The aquatic condition in which dissolved oxygen falls below the level necessary to sustain life; generally defined as levels below 2–3 parts per million.

Inbreeding depression: An undesirable result of repeated matings within small, isolated populations of related individuals that lessens genetic diversity and produces abnormalities, often those affecting reproduction. See **Fitness**.

Increasers: Plants that increase as grazing intensity increases; typically unpalatable species in a grassland community. For comparison, see **Decreasers**.

Individual distance: The space around an individual in which the presence of another individual of the same species causes discomfort or intolerance. Individual distance varies by species. The regular spacing among birds perched on wires is a result of this behavior.

Inquilines: Species living in water enclosed in plants (e.g., pitcher plants). See **Phytotelmata**.

Inselberg: An isolated mountain rising above the surrounding terrain and having such difference characteristics to the adjacent region that it acts an island. Some references substitute the term monadnock.

Integrated pest management (IPM): Employment of several methods and strategies for pest control. For insects, IPM often employs a biological control and light doses of insecticides. See **Biological control**; **Pheromone**.

Intertidal zone: Area of a shoreline exposed during low tides and submerged during high tides; a harsh environment.

Introgression: The incorporation of genes from one species into the gene pool of another when hybrids steadily backcross with the remaining parental stock of genetically "pure" individuals, with the result that the latter lose their genetic identity as a species. Commonly known as "genetic swamping."

Invaders: Plants that invade severely disturbed sites; brush typically invades overgrazed grasslands. Invaders often include exotic species, many of which are noxious.

Invasive species: Any plant or animal species that becomes abundant in the absence of natural controls to levels that disrupt natural environments that they invade.

Island biogeography: The concept that larger islands support more species than smaller islands; the distance to the nearest mainland, which is the source of colonists, also influences the number of species on the island. Islands may be real (e.g., surrounded by water) or functional, such as parks in cities, ponds in prairies, or mountain peaks.

Karst: A landscape or geological formation resulting from the dissolution of rock layers by mildly acidic water. The long-term action of acids on strata composed of soluble rock such as limestone, gypsum, or dolomite often produces sinkholes, caves, and underground aquatic systems.

Keystone species: A species whose presence and activities profoundly influence the structure of the community in which it lives. Unfortunately, this role is sometimes not discovered until the keystone species is reduced or eliminated. The sea otter is a keystone species in the kelp forests of the Pacific Coast.

Kleptoparasite: Literally means "stealing parasite." A form of social parasitism in which one species exploits the food or nests of another species to the detriment of the latter; in some birds and bumblebees, the parasite lays its eggs in the nest of the host and allows the host to provision the young.

Krummholz: Literally means "crooked wood." The gnarled, twisted, and diminutive form of trees growing in severe environments at the timber line. Whitebark pine, subalpine fir, and other krummholz species develop normally into trees of full stature where conditions are not harsh (i.e., below tree line).

Latitudinal migration: Regular north–south movements of animals triggered by seasonal changes in the environment. Many birds migrate north (spring and summer) and south (autumn and winter) across North America in response to changing photoperiods. See **Photoperiod**.

Law of the minimum: An ecological generalization stating that, of all the resources needed by a plant or animal (e.g., water, nutrients, cover), the resource available in the least amount exerts the greatest control on growth and survival of an organism and its distribution.

Law of tolerance: An ecological generalization stating that the success and distribution of an organism is determined by the organism's ability to tolerate the maximum and minimum extremes of all the factors needed for the organism's survival. Most freshwater fishes are restricted to habitats with relatively low salt concentrations because these species are unable to survive in salt water.

Layering: A type of asexual reproduction in which the lower branches of a plant touch the ground and take root, thereby establishing new and independent individuals.

Legume: Any member of the pea family Fabaceae (formerly Leguminosae) which typically bear seeds in pods. Familiar species include clover, lupine, locust, vetch, and commercial plants such as beans, peas, and peanuts.

Lek: A site where males gather to display visual and/or acoustic behaviors for the purpose of attracting females for breeding. Also known as courtship arenas. Lek behavior is best known in birds (e.g., the "booming grounds" in prairie chickens) but also occurs in other vertebrates and some insects.

Lichen: Symbiotic organisms that are part fungi, part algae. Widespread in distribution. they are common in the Arctic flora where they persist in extreme cold and aridity. Lichens typically develop into one of three forms: foliose (leaf-like), crustose (like paint splashes), or fruticose (shrub-like).

Limnology: A branch of ecology dealing with the chemical, physical, and biological properties of water in lakes, rivers, and other freshwater systems.

Loess: Deep, wind-blown (Aeolian) deposits of silty soils, accompanied by clay and fine sand; highly productive and valued for agriculture. Dried beds of glacial water are typical sources for these dusty materials, which winds may carry for long distances before they settle.

Longshore current: An ocean current that flows parallel and close to the shore.

Low Arctic: In ecological terms, the region immediately north of the tree line (i.e., Tundra) but south of polar deserts and perpetual ice and snow, which are considered High Arctic. See **High Arctic**.

Low-energy beach: A sandy seashore protected from the direct effects of large waves and experiencing little or no erosion; typically found in a sheltered locations such as bays, estuaries or reefs (see **High-energy beach** for contrast).

Macrofauna: Small invertebrates that live in marine or freshwater sediments; by definition, macrofauna are organisms that are retained on a sieve with mesh size 0.5 mm. See **Meiofauna**.

Magma: Molten rock still below ground. Magma becomes lava when it reaches the surface.

Mariculture: Specialized type of aquaculture involving the cultivation of marine organisms for food. Also known as ocean farming.

Mast: Collective term for woody fruits such as acorns, walnuts, and hazelnuts.

Meiofauna: Small invertebrate animals that live in marine or freshwater sediments; by definition, meiofauna can pass unharmed through a 0.5–1.0 mm mesh sieve, but are retained by a 30–45 μm mesh. See **Macrofauna**.

Melanism: A genetically determined condition in which black pigmentation predominates in the pelage, plumage, or skin of organisms more typically characterized by lighter-colored features. For comparison, see **Albinism**.

Mesic: A relative term used to describe the "middle" between wetter and drier conditions. Climax communities are generally mesic in comparison with pioneer communities.

Mesohaline: Term used to describe water with salinity levels between 11 parts per thousand (ppt) and 18 ppt.

Mesopredator: Any medium-size predator in aquatic or terrestrial food chains. Examples include foxes, pickerel, and mackerels. Mesopredators typically increase, often with harmful consequences, when the large apex predators (e.g., wolves or sharks) of a community are removed. See **Trophic cascade**.

Microclimate: Highly localized regime of temperature, moisture, and other conditions. In the High Arctic, some plants survive only on the lee side of rocks where the microclimate provides greater soil moisture.

Microhabitat: A restricted site where the immediate conditions provide favorable habitat for certain organisms (e.g., the lee side of a rock provides a microhabitat for mosses).

Micronutrients: Nutrients needed for normal growth and development but required only in small amounts; usually minerals (e.g., zinc and copper).

Midden: An ancient deposit of domestic refuse that may include shells, bones, broken pottery, and other discarded materials. Middens indicate past sites of human occupation and offer clues about diet and social development.

Monotypic taxon: A taxon that includes only one species. Ironwood occurs in a monotypic genus (*Olynea*); pronghorn are placed in a monotypic family (Antilocapridae).

Morph: A variant, usually in color, from the more typical appearance of the species. Examples include the blue phase of the normally white Arctic fox, the white phase of the great blue heron, and the blue phase of the snow goose. Screech owls also occur in either a gray or red phase.

Muskeg: A marsh or bog typical of Arctic and Boreal Forest. Muskeg areas usually contain sphagnum moss and decaying plants above wet acidic soils.

Mutualism: A type of symbiotic relationship in which two species benefit from a close relationship with each other. In some cases, the relationship is mandatory for the survival of each species (obligate), whereas in others, the relationship is desirable but not required (facultative).

Natural selection: The process of evolution; individual organisms with greater fitness survive and reproduce, whereas organisms with characteristics that are less conducive for survival do not pass on their genes.

Nearctic: A pattern of distribution for organisms found in only in North America. For comparison, see **Palearctic**.

Neotropical: Refers to the distribution of organisms found in Central and South America. Also describes migrant birds that breed in North America but overwinter in Central or South America.

Niche: The functional role of a species within its multidimensional environment (i.e., time, space, cover, food, and water). Bison and jackrabbits occupy grazing niches in grassland communities. Some ecologists also include a physical component in the definition, thereby regarding the niche as an organism's "job and address."

Nitrogen fixation: A process by which an organism removes nitrogen from the air and adds it to the soil. Nitrogen-fixing bacteria are associated with the root systems of certain plants, especially legumes. See **Legume**.

Nurse log: A log on which tree seedlings begin a new generation; common in Temperate Rain Forest. Trees established on nurse logs mature in rows known as colonnades. See **Colonnade**.

Nurse tree: A tree providing shelter for another plant. During their early years, saguaro cacti require the microhabitat of nurse trees. See **Microhabitat**.

O horizon: When present, the uppermost layer in a soil profile that lies above the A horizon, made up primarily of organic matter in the form of decomposing leaves, branches, animal wastes, and dead animals. See **Soil profile**; **A horizon**.

Obligate: Term for a species whose existence requires a specific set of conditions or habitat. Some plants grow only in mists near waterfalls. Similarly, some parasites survive in association with only one kind of host—no other host species will suffice. For comparison, see **Facultative**. See also **Mutualism**.

Oligohaline: Term applied to the salinity level of water that is essential considered "fresh," that is, water having salinity content between 0.5 parts per thousand (ppt) and 10 ppt.

Oligotrophic: Term used to describe a body of water low in nutrients and low in productivity. Most oligotrophic lakes are clear.

Ombrotrophic: Describes sites where precipitation provides the nutrient base; as a result, such locations are typically nutrient-poor. Sphagnum bogs are a typical example.

Oolitic limestone: Sedimentary rock typically formed in shallow seas during interglacial periods and named for its composition of tiny egg-shaped grains of calcium carbonate, which were rolled by ocean currents prior to their consolidation.

Palearctic: A pattern of distribution for organisms found in northern Eurasia. For comparison, see **Nearctic**.

Paludification: Process of bog (muskeg) expansion resulting from slowly rising water table, typically caused by the accumulation of sphagnum peat.

Parthenogenesis: A form of asexual reproduction in which embryos develop from unfertilized eggs.

Pass: A storm-cut channel bisecting a barrier island or, in some cases, forming an island from a peninsula. Over time, passes often fill in (sometimes quickly) but other later open at the same site. Man-made cuts are also known as passes; these are dredged for boating access and/or to improve water exchange between an inner lagoon and the outlying sea.

Patterned ground: Distinctive geological features, commonly a network of polygons, often marking the surface of Arctic landscapes.

Peat: An accumulation of partially decomposed vegetation or other organic matter in unique damp areas called muskegs, mires, quagmires, or bogs. Sphagnum moss is a prevalent component of peat, especially in northern bogs.

Periphyton: The complex association of detritus and microorganisms, typically cyanobacteria, algae, fungi, and diatoms, forming a felt-like coating on surfaces of plants and objects submerged in marine or aquatic environments. Host plants (e.g., wigeongrass) only provide a site for attachment, not nutrition. Periphyton is an important link in aquatic food chains.

Permafrost: Commonly used term for perennially frozen ground, typically associated with Arctic Tundra.

Pheromone: Chemical sex attractant naturally emitted by an insect or other organism; sometimes used as lures for pests because the chemicals are species-specific. See **Biological control**.

Philopatry: Literally, "love of home;" the behavioral tendency to return to a natal area for a part of the life cycle.

Photoperiod: The amount of daylight during a 24-hour period. In North America, photoperiods are longer in summer and shorter in winter. Changing photoperiods trigger biological responses such as migration and flowering in some organisms.

Photosynthesis: The process whereby green plants and other organisms containing chlorophyll pigments convert light energy from the sun, carbon dioxide, water, and minerals into chemical energy. Cyanobacteria and ceratin sea slugs are also capable of photosynthesis. The process releases oxygen as a byproduct, and the chemical energy it produces is used for an organism's own metabolic activities or stored in its tissues.

Phreatophyte: A plant whose roots penetrate deeply enough to reach the underlying water table; such plants often out-compete other vegetation where water is limited.

Physiognomy: Term for structure, usually applied to vegetation. Although the species of grasses may vary by location, for example, prairies have a common physiognomy whether in Asia, Africa, or North America.

Physiological ecology: A branch of ecology concerned with the physiological adaptations of organisms to their environment (e.g., how plants and animals cope with heat and limited water in deserts).

Phytogeography: A branch of biogeography that deals with the past and present distribution of plants. See **Biogeography**.

Phytotelmata: Bodies of water held and enclosed in terrestrial plants, including those located in cavities in trees and stumps, the base of bamboo and similar grasses, leaf axils, and, notably, pitcher plants. These support distinctive communities, some species of which are unique to these habitats. Singular: phytotelma. See **Inquilines**.

Pioneer: The first stage in the successional development of vegetation. In many areas, a pioneer community on bare soil typically consists of annual weeds; lichens may pioneer rock surfaces. See **Lichens**; **Succession**.

Plate tectonics: The concept, now widely accepted, that the Earth's surface consists of plates, some of which (e.g., the continents or parts thereof) move atop other plates, thereby changing the position and sizes of continents and oceans over geological time. Today's arrangement of North America and other landforms was therefore vastly different in the distant past.

Pluton: A massive body of solid rock protruding from deep layers of the Earth's crust to the surface or near the surface.

Podzol: Former term for soils that develop when iron and aluminum are leached from the upper horizon (i.e., podzolization); distinguished by a gray-colored zone lying just below a thin crust of sandy soil. Typical of the Boreal Forest. See **Podzolization**; **Spodosol**.

Podzolization: A process of soil formation in which acids from decomposition leaches aluminum, iron, silicas, and clays from upper layers leaving a grayish-colored layer.

Polyhaline: Term describing water salinities varying between 19 parts per thousand (ppt) and 35 ppt, the salinity level of many ocean waters.

Polymorphic: Literally, "many forms." Describes populations in which individuals of the same species differ in appearance, usually coloration (e.g., fox squirrels and snow geese). The trait is permanent (i.e., screech owls with gray plumage remain gray-colored and those with rusty plumage remain rust-colored). Each form is commonly known as a "morph." See **Morph**.

Polynya: A naturally occurring area of open water surrounded by sea ice, often formed by upwellings of nutrient-rich waters. Among others, North Water Polynya in northern Baffin Bay, Canada represents an important feeding area for Arctic wildlife.

Potamadromous: Migration by fish from one freshwater region or habitat to another without ever leaving freshwater. American paddlefish *Polyodon spathula* migrate from deep, flowing channels within the Mississippi River to shallow gravel beds to spawn. A similar term, oceanodromous, denotes fish migration within saltwater systems. For comparison, see **Catadromous**; **Anadromous**.

Predator satiation: An adaptive strategy by some prey species that lessens the risk of predation on an individual existing in a population of huge numbers. In essence, the prey is so numerous that predators can eat their fill without limiting the breeding success (and survival) of their prey species. Examples include cicadas, but large schools of small fish also reflect this strategy.

Primary consumers: The trophic level represented by plant-eating animals (e.g., grasshoppers and rabbits). See **Trophic level**; **Herbivore**.

Primary producers: The first trophic level in an ecosystem, represented by green plants. Often shortened simply to "producer." See **Ecosystem**; **Trophic level**.

Primary productivity: The conversion of gaseous carbon (from CO_2) into plant tissue via photosynthesis; in essence, the formation of living matter (plant tissues) from non-living materials. Typically measured by the biomass of new growth per unit of area. See **Biomass**.

Primary succession: Succession on previously unvegetated locations such as recently deglaciated land, dunes, or the surface of a new volcano (i.e., sites with no soil development). For comparison, see **Secondary succession**. See **Succession**.

Productivity: See **Primary productivity**.

Rain shadow: The dry region on the lee side of high mountain ranges. Rain and snow fall when the air is cooled at higher elevations on the windward side of the range, leaving little moisture for precipitation on the opposite side. The Great Basin lies in the rain shadow of the Sierra Nevada.

Raptor: Any bird of prey, typically a hawk or owl, but the term has only a functional meaning (e.g., an avian predator) and no taxonomic significance.

Realized niche: The relatively narrow range of environmental conditions in which an organism actually exists, despite being capable of living within a broader functional niche. Competition from other species keeps these organisms from expanding their niche. See **Niche**; **Functional niche**; **Competitive exclusion**.

Recruitment: The increase in populations resulting from successful reproduction, in some cases supplemented by immigration. Usually applied to the annual growth of animal populations.

Redirected behavior: A response, although appropriate to the mood, directed toward an inappropriate object. Humans, when angry, might kick a wall instead of kicking the object of their wrath.

Refugia: Geographical areas where pockets of plants and animals survive large-scale environmental changes that otherwise eliminate entire biotas. Commonly used in reference to locations escaping the regional or continental effects of glaciers. Singular: refugium.

Relict: A remnant population of a once larger distribution; may be applied to a species or, sometimes, a community.

Resaca: Synonymous with ox-bow lake, a (usually crescent-shaped) lake formed when a river changed course and deposited sediments restricting water flow from the river except during floods. The term is commonly used in the southwestern United States and Mexico.

Resilience theory: Explains the persistence of natural systems in response to changes resulting from natural causes; the capacity to absorb disturbance and reorganize in ways that retain the same functions, structure, and feedbacks as before. Proposed in 1973 by ecologist H. C. Holling and expanded in later work.

Resource partitioning: The result of evolutionary changes in behavior or other specialization that lessens direct competition between species or members of a population for the same resource. Various grazing animals forage on the same grasses, but select different stages of grass growth or different plant parts (e.g., new shoots versus mature leaves).

Rhizome: Specialized form of stem running outward from a "mother" plant just under the soil surface, from which new plants sprout along its length. A common structure in grass sod. Stolons are similar, but run above the ground.

Riparian: Term for streamside habitats or conditions. Some animals are closely associated with riparian vegetation (e.g., southwestern willow flycatchers nest exclusively along riparian corridors in Arizona and do not nest outside of this narrow zone).

Rip current: Seaward movement of water perpendicular to a beach caused by rapid backwash of wave brought to a beach by waves; also known as "undertow."

Rotation: The interval of time between commercial cuttings of timber. Forest practices leading to shorter rotations yield greater financial returns. However, short rotations alter the composition and structure of forests, typically reducing biodiversity.

Savanna: An open, usually grassy habitat studded with widely spaced trees; natural vegetation sometimes described as "park-like" in appearance. *Savannah* is an alternate spelling.

Scarified: Process in which the outer covering of a seed is breached ("scarred"), thereby allowing water to enter and initiate germination. The seeds of some plants are scarified by passing through the digestive tract of animals.

Sclerophyll: Any plant with tough, leathery leaves, but usually evergreen; often covered with a waxy cuticle. Sclerophyllic plants are resistant to water loss; some also tolerate salt aerosols. Examples include holly, live oak, and brush in chaparral communities.

"Scuzz": A nickname for the complex community of algae, bacteria, and other microorganisms living on (and in) needles of trees in old-growth forests. See **Endophylls; Epiphylls.**

Senescent death: Literally "dying of old age," used to characterize the death of Pacific salmon after spawning. The fish do not feed on their upriver journey and instead expend their body reserves for energy, eventually exhausting these and dying less of "old age" than of "burning out."

Secondary consumers: The third trophic level in an ecosystem, represented by predators that consume plant-eating animals (e.g., a fox preying on a rabbit). See **Ecosystem; Herbivore; Trophic level.**

Secondary succession: Succession at a site previously occupied by vegetation; typically occurs on abandoned farmland or in burned or cutover forests (i.e., sites where soil is present and remains essentially unchanged). For comparison, see **Primary succession.** See also **Succession.**

Seed bank: Residual supply of viable seeds which persist in the soil while awaiting conditions favorable for germination. The seed bank at a given location often includes several species, each with its own ecological tolerance (e.g., some species favoring wet habitats, others favoring drier conditions).

Seiche: Alternate flooding and draining of coastal areas caused by wind-generated wave action.

Selection pressure: An environmental force affecting the survival of individuals and their production of viable offspring based on fitness, the combination of genes that influences survival and successful reproduction. Such a force continually shapes the physical, behavioral, and/or physiological features in each succeeding generation. Individuals with favorable genes are "selected for," whereas those with unfavorable genes are "selected against." White pelage in winter is the evolutionary response of some mammals to the selection pressure of predation.

Serotinous: Adjective describing cones that remain closed with intact seeds until an environmental stimulus (usually fire) triggers their release. Serotinous cones typically remain attached for many years or until fire occurs.

Serpulid reef: Complex reef-like structures formed by aggregations of serpulids, worms that build tubes composed of calcium carbonate. Such reefs can be composed of living organisms or fossilized remains.

Sexual dimorphism: Morphological differences between males and females other than those normally associated with primary sexual characteristics. White-tailed deer are sexually dimorphic because the males grow antlers and the females do not.

Shade-intolerant: Term for trees and other vegetation whose seedlings cannot survive in shade. Seedlings of shade-intolerant species cannot survive in the understory of mature plants of their own kind and do not produce another generation. In succession, shade-intolerant vegetation is replaced by a climax community of shade-tolerant species.

Slack: A shallow trough, or depression, between adjacent dunes where higher humidity and moisture conditions promote a distinctive community of plants and animals. Sometimes called a swale.

Slot canyon: A narrow canyon that is significantly deeper than wide. Formed by the rush of water through limestone or sandstone, many slot canyons are less than 2 m (6 feet) wide and more than 20 m (66 feet) deep.

Snag: An upright trunk of a dead or dying tree; key habitat for various kinds of forest animals. Sometimes known as "stubs."

Soil order: Term used to describe the various types of soil based on combinations of particle sizes and chemical composition. Sand, gravel, and clay are examples of soil orders.

Soil profile: The cross-section of soil layers from the uppermost layer (O horizon) to the bedrock (parent material).

Solifluction: Movement of saturated soils as a result of regular freezing and thawing. On Arctic slopes, temporarily warmed soil slumps downhill before freezing again, forming lobes known as solifluction terraces. This produces unstable vegetation and azonal soils.

Source–sink dynamics: An ecological model used to assess how high- versus low-quality habitat affects the growth or decline of animal populations. An ecological trap is an associated situation, such as when a hayfield (otherwise high-quality habitat) attracts ground-nesting grassland birds whose nests are later destroyed when the field is mowed.

Speciation: The formation of a new species; commonly results from the isolation of populations that subsequently develop their own genetic composition and characteristics. See **Allopatric speciation.**

Species richness: The term describing the number of species found in a given habitat, region, or volume.

Spodosol: A soil that develops after water percolation leaches organic compounds, iron, and aluminum from the upper horizon in a process known as podzolization; distinguished by a gray-colored zone lying just below a thin crust of sandy soil. Typical of the Boreal Forest.

Staging areas: A stopover location for migratory animals, usually birds, where large numbers accumulate before continuing onward.

Stenothermal: A stenothermic organism survives within a limited range of temperatures. *Steno-* indicates a narrow tolerance to some environmental condition (recall the fine-lined penmanship of a stenographer). For comparison, see **Eurythermal.**

Stereotypic behavior: Predictable, unvarying behavioral responses released by a specific stimulus. Red spots on the bills of some gulls stimulate nestlings to open their mouths for food.

Stochastic: Describes events that occur on the basis of probability (random chance). Forest fires, hurricanes, and floods are stochastic events.

Stolon: See **Rhizome**.

Sub-Arctic: In ecological terms, somewhat loosely applied to the region immediately south of the tree line known as Forest Tundra.

Subclimax: A community maintained at a stage of development below climax. Without fire, subclimax communities of longleaf pine would reach a climax of oak-hickory. See **Climax**; **Succession**.

Succession: The concept of vegetational development following largely predictable stages from pioneer to climax communities. See **Climax**; **Pioneer**; **Subclimax**.

Surf zone: The region close to a shore where waves break.

Supratidal zone: Area on rocky seashores above the high tide line that is regularly splashed by waves, but not submerged by ocean waters. Also known as supra-littoral zone or splash zone.

Swash zone: The intertidal zone on a beach where waves wash in and out.

Symbiosis: A general term for any of several types of interactions between two organisms of different species. Parasitism is one type of symbiosis, but another type favors both species. See **Mutualism**; **Commensalism**.

Sympagic: Descriptor for habitat characterized by ice or its associated biota. Sympagic species include algae, fungi, flatworms ("ice worms"), and crustaceans (e.g., certain amphipods) occurring in alpine as well as Arctic environments. Small channels within ice provide habitat for many of these species; others live either on the upper or lower surfaces of ice.

Sympatric: Refers to the overlapping distributions of two or more species or populations. The geographic ranges of black bears and grizzly bears are sympatric in much of western North America. Sympatric organisms generally occupy ecologically separate areas (i.e., different habitats) within their shared range. For comparison, see **Allopatric**.

Talus slopes: Rock-strewn areas on mountainsides (i.e., rock slides), often beneath cliffs. The rocks (sometimes called "scree") are smaller on the upper reaches of the slope and increase in size toward the bottom. Because they are unstable, the sites generally lack soil and plant development.

Taxon: Any unit or group within the hierarchy of classification, whether a species, genus, family, order, class, or phylum. Plural: taxa.

Tephra: A collective term for the ash, cinders, and pumice that erupt from a volcano.

Terminal moraine: An accumulation of rocks and other debris marking the farthest extent of glacial advance.

Territorial behavior: Vocal and/or visual displays by males used to defend their territories; a means of repelling males while concurrently attracting females. See **Territory**.

Territory: A defended area that is typically maintained with displays whose meaning is of special significance to others of the same species. See **Territorial behavior**.

Tertiary consumers: The uppermost trophic level in a typical ecosystem and consisting of top predators (e.g., mountain lions and polar bears). See **Ecosystem**; **Trophic level**; **Carnivore**.

Thermokarst: A soil-slumping process associated with the thawing of frozen ground, including permafrost; sometimes the result of natural forces but also caused by human activities. Also defines pitted landscapes affected by this process. In some cases, thermokarst lakes form in the thawed depressions.

Thermophile: Literally, "heat loving." An organism that exists in habitats having extremely high temperatures. Many thermophiles are bacteria or archaea.

Tree line: A rather uncertain demarcation, variously defined, separating Tundra from Boreal Forest (see Chapter 2). "Timber line" is a common synonym, but is sometimes reserved for the termination of forested habitats on mountaintops.

Troglobite: Animal that can live only in caves (i.e., obligate cave dwellers). Troglobitic species usually lack pigmentation and are blind (e.g., Ozark cavefish). A similar term, troglodyte, refers to humans who in earlier times reputedly lived in caves and is the basis for the generic name of some wrens (e.g., house wren, *Troglodytes aedon*), even though they do not live in caves.

Troglophiles: Animals that can complete their lives equally well in caves or above ground. Species with these adaptations often represent evolutionary "stepping stones" to becoming troglobites.

Trogloxenes: Animals that spend considerable time in caves, often while roosting or hibernating, but must also visit the surface, particularly to feed (e.g., many bats).

Trophic cascade: Collapse of a food chain and community structure precipitated by the loss or significant reduction of a principal species known as an apex predator (i.e., top-down effects on lower trophic levels). See **Apex predator**; **Trophic level**.

Trophic level: That part of a food chain or web represented by organisms sharing a common type of food. Green plants, herbivores, and carnivores each represent separate trophic levels. Trophic levels provide a basis for assessing the flow of energy and materials through an ecosystem. See **Carnivore**; **Ecosystem**; **Decomposer**; **Herbivore**; **Primary consumer**; **Primary producer**; **Secondary consumer**; **Tertiary consumer**.

Tropical cyclone: See **Hurricane**.

Tropical cyclogenesis: The term describing the development of storms that eventually strengthen to hurricane force.

Tropical depression: A low-pressure area that generates thunderstorms in an area within 10–20° north or south latitudes, i.e., the "tropics;" some tropical depressions enlarge to eventually become hurricanes.

Tropical storm: A storm generated initially in a tropical region that has developed a cyclonic circulation of sustained winds between 63 and 127 km hr^{-1} (39–74 mph).

Tsunami: A water wave or series of waves caused by abrupt displacement of a large amount of water; earthquakes are the most common tsunami generator but volcanic eruptions, landslides, meteor impacts, and other sudden displacements of water may cause a tsunami.

Ungulate: Term for herbivores that have specialized stomachs, hooves, and either horns or antlers (e.g., bison, elk, deer, and pronghorn). Not a taxonomic designation.

Weathering: The chemical, physical, and biological processes that act singly or in combination to produce soil by reducing underlying parent material (e.g., "bedrock") to smaller particles.

Wrack: Uprooted vegetation and dead marine animals (e.g., seagrasses and jellyfish deposited linearly at high tide, forming a wrack line running parallel to the shoreline. Storms often produce a wrack line well beyond the high-tide mark.

Xeric: Term that refers to dry conditions. Deserts represent a highly xeric biome. For comparison, see **Hydric**; **Mesic**.

Zoogeography: A branch of biogeography that deals with the past and present distribution of animals. See **Biogeography**.

Index

Note: Page numbers followed by "f" indicate a figure reference. Page numbers followed by "t" indicate a table reference. Page numbers followed by "b" indicate inclusion in an Infobox.

Ecology of North America, Second Edition. Brian R. Chapman and Eric G. Bolen.
© 2015 John Wiley & Sons, Ltd. Published 2015 by John Wiley & Sons, Ltd.